NATURE AND HUMAN SOCIETY

The Quest for a Sustainable World

Peter H. Raven, Editor
Tania Williams, Associate Editor

Proceedings of the 1997 Forum on Biodiversity

Board on Biology
National Research Council

NATIONAL ACADEMY PRESS
Washington, D.C.

National Academy Press   2101 Constitution Avenue, NW   Washington, DC 20418

NOTICE: The project that is the subject of this report was approved by the Governing Board of the National Research Council, whose members are drawn from the councils of the National Academy of Sciences, the National Academy of Engineering, and the Institute of Medicine. The members of the committee responsible for this report were chosen for their special competences and with regard for appropriate balance.

This study was supported between the National Academy of Sciences and Monsanto Company; The John D. and Catherine T. MacArthur Foundation through grants 97-50855 and 97-48904; The Winslow Foundation; National Science Foundation through grant DEB-9729452; The David and Lucile Packard Foundation through grant 97-8124; Homeland Foundation through grant 3-97-085; Liz Claiborne and Art Ortenberg Foundation; V. Kann Rasmussen Foundation; The World Conservation Union; Trillium Corporation; The Jenifer Altman Foundation through grant 231. Any opinions, findings, conclusions, or recommendations expressed in this publication are those of the author(s) and do not necessarily reflect the views of the organizations or agencies that provided support for the project.

This material is not an official report of the Board on Biology or the National Research Council and the opinion reports are solely those of the individual forum participants. The papers presented in this volume are based upon presentations made at the October 27–30, 1997 meeting.

Library of Congress Cataloging-in-Publication Data

Forum on Biodiversity (1997 : National Academy of Sciences)
  Nature and human society : the quest for a sustainable world :
proceedings of the 1997 Forum on Biodiversity / Board on Biology,
National Research Council.
      p. cm.
Includes bibliographical references and index.
  ISBN 0-309-06555-0 (hardcover)
  1. Biological diversity--Congresses. 2. Nature—Effect of human
beings on—Congresses. 3. Human ecology—Congresses. 4. Sustainable
development—Congresses. I. National Research Council (U.S.). Board on
Biology. II. Title.
  QH541.15.B56 F685 1997
  333.95'11—dc21
                                    99-50565

Cover: Art by Bert Dodson.

Nature and Human Society: The Quest for a Sustainable World is available from the National Academy Press, 2101 Constitution Avenue, N.W., Box 285, Washington, DC 20055; 1-800-624-6242 or 202-334-3313 (in the Washington metropolitan area). Internet: www.nap.edu

Printed in the United States of America

THE NATIONAL ACADEMIES

National Academy of Sciences
National Academy of Engineering
Institute of Medicine
National Research Council

The **National Academy of Sciences** is a private, nonprofit, self-perpetuating society of distinguished scholars engaged in scientific and engineering research, dedicated to the furtherance of science and technology and to their use for the general welfare. Upon the authority of the charter granted to it by the Congress in 1863, the Academy has a mandate that requires it to advise the federal government on scientific and technical matters. Dr. Bruce M. Alberts is president of the National Academy of Sciences.

The **National Academy of Engineering** was established in 1964, under the charter of the National Academy of Sciences, as a parallel organization of outstanding engineers. It is autonomous in its administration and in the selection of its members, sharing with the National Academy of Sciences the responsibility for advising the federal government. The National Academy of Engineering also sponsors engineering programs aimed at meeting national needs, encourages education and research, and recognizes the superior achievements of engineers. Dr. William A. Wulf is president of the National Academy of Engineering.

The **Institute of Medicine** was established in 1970 by the National Academy of Sciences to secure the services of eminent members of appropriate professions in the examination of policy matters pertaining to the health of the public. The Institute acts under the responsibility given to the National Academy of Sciences by its congressional charter to be an adviser to the federal government and, upon its own initiative, to identify issues of medical care, research, and education. Dr. Kenneth I. Shine is president of the Institute of Medicine.

The **National Research Council** was organized by the National Academy of Sciences in 1916 to associate the broad community of science and technology with the Academy's purposes of furthering knowledge and advising the federal government. Functioning in accordance with general policies determined by the Academy, the Council has become the principal operating agency of both the National Academy of Sciences and the National Academy of Engineering in providing services to the government, the public, and the scientific and engineering communities. The Council is administered jointly by both Academies and the Institute of Medicine. Dr. Bruce M. Alberts and Dr. William A. Wulf are chairman and vice chairman, respectively, of the National Research Council.

FORUM ON BIODIVERSITY COMMITTEE

Peter H. Raven (*Chair*), Missouri Botanical Garden, St. Louis, MO
Michael J. Bean, Wildlife Program, Environmental Defense Fund, Washington, DC
Colin W. Clark, Mathematics Department, University of British Columbia, Vancouver, British Columbia, Canada
Rita R. Colwell, University of Maryland Biotechnology Institute, University of Maryland System, College Park, MD
Joel L. Cracraft, American Museum of Natural History, Department of Ornithology, New York, NY
Frank W. Davis, Department of Geography, University of California, Santa Barbara, CA
Prosser Gifford, Director of Scholarly Programs, Library of Congress, Washington, DC
Gary S. Hartshorn, Organization for Tropical Studies, Duke University, Durham, NC
Olga F. Linares, Smithsonian Tropical Research Institute, Miami, FL
Thomas E. Lovejoy, Counselor for Biodiversity and Environmental Affairs, Smithsonian Institution, Washington, DC
Jane Lubchenco, Department of Zoology, Oregon State University, Corvallis, OR
Dan Martin, World Environment and Resources Program, John D. & Catherine T. MacArthur Foundation, Chicago, IL
Nalini Nadkarni, The Evergreen State College, Olympia, WA
Michael H. Robinson, National Zoo, Smithsonian Institution, Washington, DC
Daniel Simberloff, Department of Biological Sciences, Florida State University, Tallahassee, FL
David B. Wake, Museum of Vertebrate Zoology, University of California, Berkeley, CA
Edward O. Wilson, Museum of Comparative Zoology, Harvard University, Cambridge, MA
Joy B. Zedler, Pacific Estuarine Research Laboratory, San Diego State University, San Diego, CA

Advisor

Stuart Pimm, Department of Zoology and Graduate Program in Ecology, The University of Tennessee, Knoxville, TN

Convener Liaison

Lynne Corn, Environment and Natural Resources Division, Congressional Research Service, Washington, DC
Victoria Dompka, American Association for the Advancement of Science, Washington, DC
Don E. Wilson, Neotropical Biodiversity Program, National Museum of Natural History, Smithsonian Institution, Washington, DC

Staff

Paul Gilman, Project Co-Director
Donna M. Gerardi, Project Co-Director
Kathleen A. Beil, Administrative Assistant
Norman Grossblatt, Editor
Erika Shugart, Research Aide
Susan S. Vaupel, Editor
Tania Williams, Program Officer

PREFACE

THE 1986 NATIONAL FORUM on BioDiversity carried the urgent warning that the habitats and environments necessary to foster biodiversity were rapidly being altered. The Second National Forum on Biodiversity was held in Washington, DC, on October 27-30, 1997, under the auspices of the National Academy of Sciences (NAS), the Smithsonian Institution, the Library of Congress, and the American Association for the Advancement of Science (AAAS). It conveyed the positive message that we had learned and were making efforts to conserve biodiversity—that it does not have to be a win-lose situation. It highlighted a number of outstanding efforts to conserve biodiversity in ways that are amenable to all parties involved.

The second forum was envisaged to celebrate how much we had achieved since the 1986 forum. We hoped to target the general public as the audience, using dynamic means to catch their interest. It was to be a dialogue, using, for instance, a town meeting, live chat rooms on the Web, and a live-action camera in the Amazon rain forest canopy. The speeches would be peppered throughout to convey our progress and the direction we needed to head in. Although we could not secure the funds necessary to support such a venture, we believe that that format should be used for a third forum. It will be valuable to assemble top scientists to discuss where we are and where we should go. We were impressed and pleased by how easily we secured eminent speakers; many of them had to rearrange their schedules to speak but did so eagerly because of the importance of the topic.

We were confounded by the difficulty of presenting all the desired topics at the 3-day forum in such a way that there would be enough time to cover them fully and to allow question and answer sessions with the audience. To fit more topics in, we held several brown-bag luncheon discussions each day; these discussions received favorable comments because they allowed adequate give and take in an intimate atmosphere. When we were putting this volume together, we took the opportunity to address some of the lesser-known groups of organisms that had not been well covered, such as protists, mites, and fungi. We also held a number of events to increase outreach to the public and Congress: several speakers were sent to Capitol Hill to brief congressional members and staff, others participated in radio news events, and all participated in a lunch with the press.

The body of the program, including lectures and brown-bag sessions, was held at NAS. An opening evening lecture was held at the Baird Auditorium of the National Museum of Natural History. The Library of Congress hosted a special dinner and exhibit for the speakers. And the premier screening of the National Geographic film, *Don't Say Goodbye*, and an accompanying exhibit of the photographic work of Susan Middleton and David Liittschwager were held at AAAS. Over 750 people registered for the 3-day forum, and all the events were well attended.

Numerous people were involved in organizing the forum. The National Research Council empaneled a committee to serve as science advisers. That panel enlisted the help of David Wilcove, George Woodwell, and Walt Reid to finalize the program. Staff of the convening organizations did the brunt of the planning: Tania Williams of the National Research Council directed the staff efforts with the invaluable assistance of Donna Gerardi, Erika Shugart, and Kathleen Beil, also of the National Research Council; Prosser Gifford of the Library of Congress; Lynne Corn of the Congressional Research Service; Don Wilson of the Smithsonian Institution; and Dick Gertzinger, Victoria Dompka, and Lars Bromley of AAAS. Ruth O'Brien of the National Research Council organized the complicated arrangements that led to a smoothly conducted meeting; she was assisted by Stacey Burkhardt of the National Research Council. Authors were sent completed and edited manuscripts in late 1998 so that they could update the references. Hence in this volume, there are many references to work published after the forum was held.

We wish to thank the Mansanto Company, The John D. and Catherine T. MacArthur Foundation, The Winslow Foundation, National Science Foundation, The David and Lucile Packard Foundation, Homeland Foundation, Liz Claiborne and Art Ortenberg Foundation, V. Kann Rasmussen Foundation, The World Conservation Union, Trillium Corporation, and The Jenifer Altman Foundation for their support of this effort.

Tania Williams served as managing editor for this volume, Norman Grossblatt was senior manuscript editor, and Karen Phillips edited several of the manuscripts.

The beautiful art created for the forum, which serves as the cover of this volume, was the work of Bert Dodson.

Peter H. Raven
Chair

CONTENTS

PART 5

THREATS TO SUSTAINABILITY

PART 6

INFRASTRUCTURE FOR SUSTAINING BIODIVERSITY—SCIENCE

PART 7

INFRASTRUCTURE FOR SUSTAINING BIODIVERSITY—SOCIETY

INTRODUCTION

PETER H. RAVEN

Missouri Botanical Garden, P.O. Box 299, Saint Louis, MO 63166-0299

HUMAN EXISTENCE depends inextricably on other life forms. All humans need Earth's flora, fauna, and microorganisms for sustenance, materials, energy, and even the air they breathe. We all have the capacity to learn from and enjoy life on Earth through its diverse beauty, complexity, and invention. Some humans—particularly scientists—dedicate themselves to exploring the secrets of the incredible array of biota on our small, blue planet. Through the deeper understanding that their work provides, all humans can directly or indirectly derive benefits. But how much of Earth's biotic complexity do we, as citizens or scientists, understand, and what more do we most need to find out to ensure that Earth's biota can continue to provide for us and for generations to come?

The extent and variability of life on Earth is referred to as "biodiversity." Scientists in many disciplines have engaged in extensive exploration of biodiversity. Many exciting advances in understanding have occurred in the last decade, since the National Forum on BioDiversity was held in Washington, DC, in 1986, under the auspices of the National Academy of Sciences and the Smithsonian Institution. The advances have taken place because scientists have identified new theoretical frameworks, developed new technologies to observe life in the field, and analyzed new data while discovering tens of thousands of new kinds of organisms. Thus, our collective knowledge is growing rapidly. But many scientific advances are still needed, and much current information is not widely known beyond the community of scientists who study biodiversity.

The National Academy of Sciences, the Smithsonian Institution, the Library of Congress, and the American Association for the Advancement of Science rec-

1

ognize that advancing the biodiversity sciences and improving public understanding of it require effective communication between the public and the scientific community. To that end, those four organizations convened the Second National Forum on Biodiversity—Nature and Human Society: The Quest for a Sustainable Future—on October 27–30, 1997. The 3-day conference was held at the National Academy of Sciences in Washington, DC. The papers presented in this volume are based upon presentations made at the conference. The material in this book is not an official report of the Board on Biology or the National Research Council and any opinions expressed are solely those of the individual forum participants.

The second forum provided a venue for the world's leading experts in the biodiversity sciences—ranging from agronomy to zoology—to discuss their understanding and future scientific directions. Through the World Wide Web, it engaged the experts in a dialogue with the American public about biodiversity, especially its relevance to humans, our understanding of it, and the challenges that lie ahead. The forum had three goals, as follows:

• **Review state-of-the-art science that helps us to understand Earth's biological diversity**. The forum accomplished that goal by engaging scientists who work in fields that focus on different aspects of the extent and variability of life on Earth. The activities of the forum provided opportunities for scientists to share new information with each other and the public, to confirm some theories and refute others, to discuss emerging fields that need new information, and to develop strategies to learn more about biodiversity and the proper management of it.

• **Engage scientists and nonscientists in a discussion of what science is, how it works, and the issues that scientists should address, including issues of practical importance to the public.** That goal was accomplished by holding brown-bag lunch sessions on each day of the forum where the speakers were available to discuss general questions posed to them by forum attendees.

• **Make the information discussed at the forum accessible to the general public in an understandable way.** This proceeding's volume accomplishes that goal. It is derived from the research literature and forum activities, and it explains biodiversity to the general public in lay terms.

Given those goals, what was perhaps most striking about the ideas presented at the forum was the discovery of the convergence that had occurred over the preceding decade between the concept of biodiversity, which used to be taken loosely to mean a roster of species, and the concept of "sustainable development." It is now widely understood that biodiversity is what makes our planetary home what it is and makes our life here possible; in turn, it is biodiversity that we must use to build our sustainable future. The living systems of Earth are powered by perhaps 350,000 of the estimated 7 million or more species that share the planet with us: the plants, algae, and photosynthetic bacteria that alone have the ability to capture a small portion of the Sun's energy and transform it into chemical bonds, which in turn provide the energy needed for the metabolism of those organisms and indirectly for all others, including humans.

Our planet is 4.5 billion years old, and life existed at least 3.8 billion years ago. The earliest life forms were bacteria, and for at least 3.5 billion years of Earth history, cyanobacteria—photosynthetic bacteria—have been changing the nature of the atmosphere from a reducing one to the oxidizing one we have today. The accumulated bodies of cyanobacteria have likewise over the years been transformed into the oil and natural-gas deposits that humans have been using, with coal, to power their industrial processes for more than 200 years. By about 1.5 billion years ago, the first eukaryotic cells (cells with nuclei) had appeared, in part as a result of processes of serial symbioses that provided the basis for their intracellular complexity. Eukaryotic cells had aggregated to form multicellular organisms by about 700 million years ago, and these multicellular organisms—the ancestors of terrestrial vertebrates, arthropods, fungi, and plants—invaded the land, first becoming terrestrial about 430 million years ago.

On land, with its greater array of distinct habitats, organisms proliferated greatly. Today, some 85% of all living species occur on land, even given the much greater fundamental diversity of marine organisms. As terrestrial organisms evolved in part into larger and more complex forms, forests came into existence, by at least 300 million years ago; the masses of decaying vegetation from the forests, under suitable circumstances, became coal. In the forests and other vegetation types that characterized the world of the Mesozoic Era (65–245 million years ago), biological diversity increased greatly. The Mesozoic Era began with the most extensive extinction event recorded, the great majority of all living species disappearing forever, and ended with the most recent extinction event, that at the end of the Cretaceous Period, the third and final geological period into which the Mesozoic is divided.

About 65 million years ago, it is estimated that two-thirds of all terrestrial organisms disappeared; the character of life changed permanently. Perhaps 500,000 kinds of organisms survived the extinction event at the end of the Cretaceous, and they have gradually given rise to what has conservatively been estimated as 7 million kinds of living eukaryotic organisms today and an unknown number of kinds of prokaryotic ones (cells without nuclei). This great elaboration of life has resulted not only in the elaboration of species and the forms of individual organisms, but also in the development of increasingly complex biological communities, particularly at low latitudes.

We are now participating in the sixth great extinction event; again, an estimated two-thirds of the kinds of terrestrial organisms are threatened with extinction in the near future. The extinction event that closed the Mesozoic Era seems almost certainly to have resulted from the collision of an asteroid Earth somewhere off the end of what is now the Yucatan Peninsula, but humans are the active force driving the wholesale massacre of living things that is taking place today. How has that come to be?

Our genus, *Homo*, evolved from *Australopithecus* in Africa some 2 million years ago but existed at relatively low population densities until quite recently, geologically speaking. At the time when our ancestors were developing crop agriculture, starting about 10,000 years ago, the human population of Earth numbered several million fewer than visit the Smithsonian museums each year—far fewer than the population of the Washington, DC, area—scattered over Eurasia, Africa, Australia, North America, and South America at about the density of aboriginal people in Australia before European contact. With a dependable supply of food that could be stored, however, humans developed increasingly complex societies, went to war with one another, developed technologies to harness power of diverse kinds, formed states, and began to exert pressure on the living world to a degree that had not been experienced earlier.

With the accelerated growth in the human population, there were some 130 million people by the time of Christ, about 500 million in early Renaissance times, and about 1 billion at the start of the 19th century, when an English clergyman named Thomas Malthus was warning of the danger that population growth might outstrip our ability to feed ourselves. Our numbers had grown to 2.5 billion by 1950 and then the growth really underwent great acceleration: over just 50 years, 3.5 billion people have been added to the human population, and we shall enter the 21st century with more than 6 billion people spread throughout the world.

The illusion of abundant, cheap energy that we have created has fueled population growth by increasing the rate and intensity of all of our activities. Consider some of the changes that have occurred during the last 50 years. We have lost about one-fourth of the topsoil and one-fifth of our agricultural land, so that we are feeding 6 billion instead of 2.5 billion people with greatly decreased natural resources. We have altered the character of the atmosphere, adding about one-sixth to the atmospheric carbon dioxide, the primary greenhouse gas, and diminishing the stratospheric ozone layer by about 7%, thus increasing the incidence of skin cancer at middle latitudes in the Northern Hemisphere by about one-fifth. About one-third of the forests that existed in 1950 have been cut without being replaced, and human pressures continue to grow at such a pace—fueled by growing numbers of people, increased affluence (consumption), and the use of inappropriate technologies—that it is increasingly difficult for any ecosystem to regenerate itself.

Worst of all is the increased level of extinction, now hundreds of times above the background rate for the last 65 million years, as reviewed by Stuart Pimm and Thomas Brooks for this volume, and likely to lead to the disappearance of about two-thirds of all kinds of living organisms by the end of the next century—an extinction event that would be, as mentioned above, roughly equivalent to the one that occurred at the end of the Cretaceous Period, but in this case driven by only one species: humans. Violating the principle enunciated by conservationist Aldo Leopold years ago—"The first rule of intelligent tinkering is to save all the cogs and wheels"—we are in the unfortunate situation of trying to build an "age of biology" while wasting the organisms that are of fundamental importance in meeting our objective, whether individually or collectively. There is ample reason to try to reverse the trend, but we seem to have neither the collective wisdom nor the will to attempt to do so in a meaningful way.

One of the most damaging illusions that is being perpetrated in our time is the one that because humans have solved other problems in the past history of Earth, they certainly can solve the ones that we are facing now. Such a view ignores the scale of the problem that we face, when we are consuming or wasting an estimated 45% of all terrestrial net photosynthetic productivity, and using about 55% of the available fresh water. No pressures remotely like those have existed before. To build a world that is sustainable—one in which animals, plants, fungi, microorganisms, and people will be able to continue to exist peacefully, harmoniously, and sustainably over the long run will require every ounce of wisdom, science, common sense, and affection for one another that we can possibly muster. Anyone who denies this conclusion is simply misinformed or uninterested in our common future; since you, dear reader, know better, you are obligated to tell them that they are wrong, to comport yourself, your business, your country, your neighborhood, your church, your National Academy of Sciences, or any other organization in which you are involved, in such a way as to help to make the future a pleasant, abundant, and prosperous reality instead of an increasingly devastated, homogeneous, and exhausted one.

This all became very personal for me when I saw Susan Middleton, that talented photographer, in the National Geographic movie that premiered at the forum, *Don't Say Goodbye*, stepping gingerly on rocks around my manzanita. I call it *my* manzanita, because I discovered that plant in 1952, the only known survivor of *Arctostaphylos hookeri* subsp. *ravenii*, when I was 15 years old and collecting plants in the Presidio of San Francisco, never imagining that it might be the only one or that so many species of organisms soon would be threatened with extinction. Instead of stepping gingerly around the plant as Susan did, I just walked right up to it, cut off a few branches for herbarium specimens, pressed them, and went on my way.

People weren't worried about extinction then, and few worried about human population growth. America was bustling with energy after the conclusion of World War II and after capturing nearly 40% of the world's economic activity, giving our nation an incredible level of prosperity that has never been duplicated in relative terms since and cannot realistically be duplicated in the future. In the United States we have about 4.5% of the world's population and control about one-fourth of the world's economy, but sometimes we seem incapable of recognizing our interdependence with the other nations of the world.

But to move forward with our story—we became increasingly concerned with population in the 1960s, when we had begun to recognize and to worry about our impact on the environment, both local and global. The first Earth Day was held in 1970; President Nixon signed the key environmental legislation, under which our country has operated since, in the early 1970s; and we moved forward, our national population and that of the entire world growing rapidly. As we have seen, we live in a nonsustainable world that we are destroying and homogenizing as we use up its resources; and the characteristics of our world are both unjust and unstable. Of the 6 billion people alive today, 2 billion live in abject poverty, with less than $1 per day in income; some 700 million are malnourished to the extent that their brains do not develop properly and their bodies are wasting away. The proportion of people living in industrialized countries has fallen from one in three in 1950 to one in five today; but the industrialized countries, with only one-fifth of the global population, control 85% of the world's economy, use comparable proportions of its resources, and cause a proportionate amount of pollution and environmental degradation. Put another way, 80% of the people in the world live in developing countries, which have about 80% of the world's biodiversity (and by far the less well-known part of that biodiversity), 15% of the world's wealth, and no more than a tenth of the world's scientists, most of them in a few countries. Most of the world's biodiversity is in countries where there is little scientific basis for dealing with it or using it sustainably for the benefit of their people, so their biological heritage, which is of great common value to all people on Earth, is being lost without any chance of doing anything about it. By paying so little attention to so many poor people, we are discriminating against them, and particularly against the women and children who live among them, and thus denying ourselves the benefit of their creativity in addressing the serious problems that we are confronting together. We cannot afford to do that, whether we recognize the problem or not.

But what has happened since 1986, since we held the National Forum on Biodiversity? In general, I think we have come to recognize with Dan Janzen that the world is indeed a garden, and that humans are, for better or worse, responsible for all of it, dominating every ecosystem, depositing manufactured chemicals onto every square centimeter every minute. We dominate every ecosystem on Earth, as abundantly demonstrated by the papers included in this volume, and we ought to accept the responsibility of managing our planet much better and more sustainably than we are now: we owe such behavior to the future. A majority of all lands are intensively incorporated into human activities of one kind or another, and the proportion and intensity continue to grow rapidly. Parks and other protected areas must be viewed as special parts of human-dominated ecosystems and managed as such; there is no turning back to an earlier world, in which such areas might have been segregated and held apart from human activities.

In the year after First National Forum on BioDiversity, the report of the World Commission on the Environment and Development, the "Brundtland report", was published as *Our Common Future*. That report, more than any other, popularized the concept of sustainable development and began to delineate the issues debated at Rio de Janeiro's Earth Summit in 1992, on which the future of all countries, all companies, all institutions, and all of nature ultimately depend. In putting together the concept of sustainable development with that of biodiversity, we have come to see that to preserve, nurture, rebuild, restore, and refresh the increasingly modified living systems that support us, we have one primary tool and that tool is biodiversity.

The concept that DNA is the carrier of genetic information was first reported in 1944, about 50 years ago. The first transfer of one gene from one kind of organism to an unrelated kind of organism took place in 1973, only about 25 years ago. And it is only in the last few years that we have begun to determine the complete sequences of genetic material of prokaryotic and eukaryotic organisms. Until now, no one could make a reasonable estimate of how much and how the genome of, say, corn differs from that of a human; we are just beginning to learn. If the 21st century ushers in the age of biology, we shall need to understand biodiversity—the incredible diversity of life on Earth—if we are to be able to achieve the kinds of results that we confidently expect and that potentially have such a great bearing on the human prospect. In that context, the impending extinction of perhaps one-fifth of the species of organisms within a quarter-century or two-thirds of the total within the next century is unacceptable, especially considering that we have even cataloged and named only about one-fourth of the total that we estimate exist.

Looking at the planet Earth from somewhere else, observers would find it impossible to believe that we have no common, well-organized international effort devoted to preserving our organisms. They would be incredulous at the thought that the rapidly shrinking 20% of us who live in industrialized countries have not long since joined hands with the people who live in developing countries, and by doing so created a collegial and mutually supporting economic situation within which it would be possible to save as complete as possible a selection of the world's biodiversity. The way we are behaving amounts to sheer madness, and we must find a way to stop it. Can we not find a way to do so?

As David Suzuki points out vividly, our response to the ecological crises we face is not appropriate, given the enormity of those crises; but we are facing thousands of ecological Pearl Harbors today, mostly without even noting them, much less making any effort to avert them. His powerful analogy presenting the state of the world's ecology as though we were all passengers in a huge car going as fast as possible toward a brick wall and just sort of chatting amiably as it speeds along, with most of the people in the world actually locked in the trunk: that's something to think about! At the very core of human existence, or of human prospects for the future, are the kinds of values that Jim Morton discusses in this forum. Why do we find them so hard to embrace and act on?

The situation that I have outlined briefly here, and that is discussed in a number of the papers in this volume, demands the reformulation of both philosophical systems and human actions around the principles of sustainability, with a proper appreciation of biodiversity at the heart of the matter.

Many responses are possible to the crises that are so well laid out in the papers included in this volume. It is clear that our knowledge of biodiversity is incomplete—it is in no way adequate for the challenges that we face in trying to build a sustainable world. Even according to the conservative estimates presented by Bob May, we have named no more than one-fourth of the world's eukaryotic species, and the prokaryotic species are so poorly known that we cannot reasonably provide even an order-of-magnitude estimate of their numbers. We must therefore accelerate and at the same time make more selective our approaches to learning about global biodiversity: there is no hope of completing an inventory during a century in which up to two-thirds of the species are likely to vanish permanently. Completing the inventories of some better-known groups and those of economic importance—such as vertebrate animals, butterflies, ticks, mosquitoes, and plants—seems reasonable; taking appropriate steps to gain an appreciation of the diversity and patterns of geographic distribution of others—especially those of ecological or economic significance, such as fungi, nematodes, mites, and selected groups of insects—also seems reasonable. The simple fact is that if we do not do so, we shall never have an idea of how many species there were or where they existed; we shall certainly be less able save them over the coming years.

It will be necessary for the industrialized countries of the world—20% of the world's population with 80% of the wealth, about 90% of the scientists and engineers, and 20% of the biodiversity—to recognize that the world's biodiversity is both our common heritage and the key to our future sustainability. By understanding it, learning how to use it sustainably, protecting it, and preserving it, we shall be making a priceless gift to future generations and acting responsibly in the face of one of the greatest challenges that ever confronted humanity. To the extent that this forum has contributed to that goal, it should be judged a success and a helpful building block along the way to a sound and sustainable future.

P A R T

1

DEFINING

BIODIVERSITY

BARRIERS TO PERCEPTION:
FROM A WORLD OF INTERCONNECTION
TO FRAGMENTATION

DAVID T. SUZUKI

The Suzuki Foundation,
2211 West 4th Ave., Suite 219, Vancouver, BC V6K 4S2, Canada

MAKING SENSE OF THE COSMOS

THE GREAT molecular biologist and Nobel laureate Francois Jacob has stated that the human brain has a built-in need for order. From earliest times, human beings looked out and recognized cycles, repetitive patterns in nature—day following night, the seasons, tides, lunar cycles, plant succession, animal migration— that conferred the ability to predict their recurrence, and thus people acquired a semblance of understanding of and control over the cosmic forces impinging on their lives. Gifted with an enormous brain, our distant ancestors were inquisitive, experimental, and inventive. Over time, they acquired profound insights into their immediate surroundings that had conferred survival value. No doubt they pondered many of the same cosmic questions that we ask today: How did we get here? Where are we going? What is the meaning of life? As the great French anthropologist Claude Lévi-Strauss wrote:

> I see no reason why mankind should have waited until recent times to produce minds of the caliber of a Plato or an Einstein. Already over two or three hundred thousand years ago, there were probably men of similar capacity (Lévi-Strauss 1968).

From the dawn of human awareness, people accumulated insights and understanding and superstitions that were woven into their mythologies, into the fabric of their culture and identity. Anthropologists call this a "worldview"; in it, nothing exists in isolation from anything else. The rocks, the wind, the stars, the rivers, the forests,and people are all inseparably intertwined. The past, the present, and the future form a seamless flowing continuum. In such a world, human beings often were saddled with enormous responsibility to keep it all going. They had to behave properly, say the right prayers, and follow the proper rituals and ceremonies, or the world could collapse. So the great bounty of the world of which humans partook was laden with responsibility.

FROM INTERCONNECTION TO FRAGMENTATION

When Francis Bacon recognized that knowledge (*scientia*) is power, he began a fundamental shift in how we perceive our surroundings. Science is a radically different way of seeing the world. Instead of trying to understand the whole universe, scientists focus on a part of nature, separate it from its surroundings, control everything impinging on it, measure everything within it, and thereby acquire profound insights into that isolated bit of nature. Ever since Newton described the universe as an immense clockwork mechanism, scientists have been motivated by the notion that by analyzing nature in fragments, we could eventually understand the whole by putting the pieces together as in a giant jigsaw puzzle. Reductionism is at the heart of modern science.

Physicists recognized in this century that reductionism does not work. The universe is not like a giant machine. Quantum mechanics revealed that at the most fundamental level of subatomic particles we could not know their precise location with certainty, only by statistical probability. Furthermore, as Nobel laureate Roger Sperry pointed out, properties emerge from the interactions of parts of nature that cannot be predicted on the basis of the individual properties of the parts (Sperry 1968). However, most of biology and medicine remains predicated on reductionism.

In this century, humankind has undergone massive changes with explosive speed. Harnessing the enormous power of technology, increasing in number exponentially, and accepting a global economy based on endless growth and productivity, we have become a superspecies capable as no other species has ever been of modifying the biophysical features of the planet on a geological scale. In a moment of evolutionary time, great rivers can be diverted or dammed, wetlands drained, ancient forests cleared, and air, water, and soil polluted. As technology and the economy have become the dominant elements of our lives, worldviews have been shattered, and we are no longer able to recognize the exquisite interconnections that mean that every human action has enormous repercussions throughout the biological world. As Thomas Berry says:

> It's all a question of story. We are in trouble just now because we do not have a good story. We are in between stories. The old story, the account of how we fit into it, is no longer effective. Yet we have not learned the new story (Berry 1988).

The challenge we face is to rediscover those connections and recognize that we remain embedded in nature so our every action is laden with consequences and ramifications. The difficulty is that we have barriers that blind us to those interconnections. If we are to pass through the barriers, we first have to recognize them.

BARRIERS TO INTERCONNECTEDNESS

The Move to Cities

If we look at humankind over the vast sweep of evolutionary time, one of the monumental transitions has been the change in this century in how we live. In 1900, only 16 cities had a million or more people. The largest was London, with 6.5 million. Tokyo was seventh, with 1.5 million. More than 95% of humanity lived in rural village communities. We were an agrarian species. Today, over 400 cities have a million or more people. The top 10 all have more than 11 million, and Tokyo is the largest with 26.5 million! Over half of all people now live in large urban settings, and the proportion is increasing all the time (World Almanac 1996).

Designed properly, cities could be ecologically far more benign in energy use, pollution, use of cars, and so on. But in cities, we live in a human-created habitat that is severely diminished in biological diversity. Our surroundings are dominated by one species—us—and the few plants and animals that we decide to share space with or cannot quite eliminate. In such an environment, it becomes easy to think that we are special, that our creativity has enabled us to escape the constraints of our biological nature. It is easy to forget that we remain absolutely dependent on air, water, soil, energy, and biodiversity for our survival and good health.

I have been shocked while making television programs by the number of urban children (and adults) who have little idea of the source of their food. Many do not know that vegetables grow in the "dirt" or that wieners, hamburgers, and drumsticks are the muscles of animals. They do not know where electricity, water, plastic, or glass comes from or where sewage and garbage go. Yet they are all services delivered not by the economy, but by Earth itself.

Science and Technology

As a student in the immediate post-Sputnik years, I was taught and believed that science enables us to push back the curtains of ignorance to unlock the deepest secrets of the universe and thereby to acquire the understanding that is vital to control and manage the world around us. Progress in science during this century has been spectacular; in my field of genetics, it takes my breath away to see techniques used in undergraduate laboratories that I never dreamed would be available in my lifetime. And the technological prowess that accompanies our insights is truly phenomenal. But in our understandable exuberance over our discoveries, we forget how science progresses, and we forget the extent of our ignorance.

When I graduated as a fully accredited geneticist in 1961, I thought I was pretty hot. I knew about DNA and operons and cistrons. But now when I tell students about our 1961 ideas of chromosome structure, gene function, and regulation, they laugh in disbelief. Seen through the perspective of what we know in 1997, our hotshot ideas of 1961 are naive and far off the mark. But students are stunned when I remind them that when they have been professors for 20 years and tell their students what the hottest ideas of 1997 were, those students will also be highly amused. The very nature of science is that we know that most of our current ideas, models, and hypotheses are wrong, in need of major modification, or irrelevant. As we rush to patent and apply ideas and techniques in molecular biology, we remain ignorant about the makeup and extent of biological diversity on the planet. As E.O. Wilson has argued, the 1.6 million species identified may be less than 20% of all species on Earth (Wilson 1992). And identification of a species merely means that a biologist has classified and named a dead specimen; it does not mean that we know anything about how many individuals there are, the distribution of the species, how it interacts with other species, or anything about its basic biology. We are tearing at the intricate web of living things before we have any understanding of its components or how they interact to maintain the planet's productivity. Our basic descriptive research is imperative. Currently, the strength of scientists is description: because we know so little, we make discoveries wherever we look. But for the same reason, we cannot be prescriptive in recommending meaningful action for environmental problems that we encounter.

Rachel Carson's 1962 seminal book *Silent Spring* was a warning that technology, however beneficial, invariably has costs, and because our knowledge is still so limited, our capacity to anticipate or predict all consequences and costs is extremely restricted. When the insecticidal properties of some molecules were discovered, the benefits of killing insect pests were obvious. At that time, geneticists knew enough to predict that resistant mutants would quickly render an insecticide ineffective, and ecologists understood that the use of broad-spectrum insecticides made little ecological sense when fewer than one-thousandth of all insect species are pests to human beings. But no one could have anticipated the biomagnification of insecticides, because scientists discovered the phenomenon only when populations of some birds, such as eagles, decreased drastically. If we cannot anticipate the consequences of powerful new technologies and if our knowledge of the basic biological and physical makeup of Earth is minuscule, can we go on embracing new technologies with the hope that the inevitable problems that they create will be correctable by further technological innovation? I don't see how we can.

The Information Explosion

Today, as we prepare to leap into a new millennium, our leaders wax eloquent and ecstatic about the information superhighway that will take us there. But having worked as both a university professor and a host in television and radio since 1962, I can tell you that the challenge we face today is not a need for access to more information but a way of wading through information overload. The average person today is confronted with "info-glut," and most of what passes as information is junk. On an anecdotal level, I encounter many people who regale me with fantastic ideas—Bermuda triangles, extraterrestrial abductions, or scientific breakthroughs—and when I ask the source of their stories, the answer is often "I read it" or "I saw it on TV." But if people do not make a distinction between information obtained from the *National Enquirer* and information obtained from *Scientific American* or the *New Scientist*, or between Geraldo Rivera and *The Nature of Things* or *Nova*, then information is validated simply on the grounds that it exists.

And the nature of the electronic media is that they create a virtual reality that is *better* than the real thing. After all, you can now experience the kinkiest sex without fear of being caught or catching AIDS; you can lose a gunfight and live to fight again; you can have a horrendous crash in a car race and walk away. When I began my career in television, I had the great conceit to think that through this medium I would create films that would stand out like jewels, entertaining while educating the viewing public. My hope was that with good natural-history films people would grow to love and value the wonderful diversity of other species and complex ecosystems. But I have learned that our programs, too, are a form of virtual reality.

Years ago, I was on a talk show on national television, and the host asked me, "As a scientist, what do you think the world will be like in 100 years?" I responded that if human beings are still around in a century, I would hazard a guess that they would curse us for two things—nuclear power and television. Ignoring the nuclear issue, the host did a double take and stammered, "Why television?" My response was "Bob, you asked me a very tough question. If I had responded 'Gee, Bob, that's a hard one' and then proceeded to think for 10 seconds, you would have cut to commercials within 3 seconds. Because television is not serious, it cannot tolerate dead air." Now in reflecting on that exchange, I have recognized that when we assemble a nature film, we create an artifact: we send a photographer to the Arctic or the Amazon for months to get all kinds of shots-to-end-all-shots. Then in an editing room, we string them all together to produce an illusion that a tropical rain forest or the Arctic is a blur of activity. But the one ingredient that is indispensable to experience the real world is *time*. As telecommunication technology jams more and more information into less and less space, it delivers more jolts per second to an audience now hooked on and demanding more and more adrenaline-charged jolts. And the overriding message within the medium, even for a public-supported medium like the Canadian Broadcasting Corporation, is consume, consume, consume.

As Thomas Veltre of the New York Zoological Association has pointed out, the underlying message in television is diametrically opposed to that of environmentalism (Veltre 1990). Those of us concerned with sustainable futures look at the world on a geological time scale; we try to see the whole picture, and we urge conservation. Information conveyed by the electronic media is conveyed as a series of unrelated bullets conveying little sense of the context and history that give us an understanding of why they matter. We are assaulted by instant and fragmented factoids; and throughout, we are exhorted to buy, buy, buy.

Politics, Politicians, and Bureaucracy

Now that the ideological battle and insane arms race between the Soviet Union and the United States has ended, we revel in the apparent triumph of democracy and the efficiency of the global market. But there are enormous ecological problems that governments on any side of the political spectrum are ill equipped to handle.

To begin with, political action is predicated on the need to obtain tangible results in time for the next election, a timeframe that is too short to deal seriously with many of our most important challenges, such as species extinction and climate change. Thus, for example, in a study initiated by Prime Minister Brian Mulroney in 1988, it was found that Canada could readily achieve a 20% reduction in CO_2 emission in 15 years for a net savings of $150 billion! That apparent good news has never been formally released, and nothing was ever done to implement it. That is because to achieve the CO_2 reduction and save an enormous sum, an initial $74 billion has to be invested. It would be political suicide to announce such an up-front expenditure; besides, the political beneficiary of the savings would be someone else 15 years later.

A further problem that I have found in Canada is that elected politicians come primarily from two professions: business and law. In part, that reflects the fact that few people from labor, farming, homemaking, teaching, and so on can afford to run for office and *lose*. But this skewed representation distorts perceptions of government priorities. It is not an accident that in my country there is excessive concern with economic and jurisdictional issues. In the last session of Parliament, of more than 600 questions asked during Question Period, a mere seven were on the environment, but many concerned Quebec separation, gun control, and athletics. To compound the limited perspectives of government, when 50 members of Parliament were tested for their comprehension of scientific and technological terms and concepts, lawyers and businesspeople scored at the absolute rock bottom of the heap. Yet they will make decisions about the future of old-growth forests, climate change, ozone depletion, toxic pollution, genetic engineering, artificial intelligence, and many other issues requiring an understanding of science and technology. Clearly, the challenge is to make science and technology a fundamental part of every citizen's education.

Perhaps the greatest challenge is that political priorities are defined by a profound species chauvinism that blinds us to larger ecological principles. Once elected to office, politicians are beholden to financial backers, their party, and the electorate, apparently in descending order of importance. But children do not vote. For that matter, future generations do not vote. Yet they are the ones with the most at stake in the decisions now being made by governments. In addition, our governments' priorities are too restricted along species lines to enable them to assess ecological problems adequately. Thus, we create political boundaries that we then deploy every effort to protect. But human borders make little ecological sense to air, water, plants, and animals. Watersheds, mountaintops, ozone layer, valley bottoms, jet streams, wetlands, flyways, ocean currents—these are the real ecological determinants of meaningful boundaries.

Nothing illustrates better the ludicrousness of our political attempts to manage nature than Pacific salmon, which currently inflame American and Canadian political rhetoric. Adults of the five species of salmon know very well where they "belong": in the natal rivers and streams that they left 2–5 years before. But because fishing fleets intercept them at sea, we must establish an International Salmon Commission to set quotas for each nation. As the animals move from Alaska past British Columbia to Washington, Oregon, and California, fishers take them in the open sea as though the fish belong to them. Even when the fish reach their river homes in British Columbia, the federal government decrees that they fall under the Department of Indian Affairs for the aboriginal food fishery and the Department of Fisheries and Oceans for the commercial fishers, while the provincial government claims the highest revenue from sport fishing, which falls under the Department of Tourism. As the salmon move up the rivers, activities administered under the Departments of Urban Affairs, Mining, Agriculture, Forestry, and Science and Technology impinge on their fate. So human categories and priorities transform what is a single biological issue into a multiplicity of bureaucratic turf wars, thereby making it certain that the fish will never be dealt with in a way that will ensure their long-term survival and abundance.

When politicians attempt to bring "all the stakeholders" to the table to hash out a contentious issue—such as clear-cutting old-growth forests, damming a river, or building a new nuclear facility—the most important stakeholders are not present. Where are the children, the unborn generations, the fish, air, trees, water, or topsoil? Our minister of forests does not speak on behalf of the forest, nor the minister of agriculture on behalf of the soil, nor the minister of fisheries on behalf of the fish. Instead, we attempt to shoehorn nature into the demands of human economic, political, and social priorities, often rationalizing our actions by claiming that environmental assessments permit them. In Canada, environmental regulations are often suspended because of the need to stimulate the economy or create jobs.

In our position of dominance, we now assume that the planet is a massive resource that is ours to exploit as we wish. Thus, the 1987 UN Commission on the Environment and Development report *Our Common Future* suggested a goal of protecting 12% of the land in every country. Canada does not come close to that target either federally or provincially, and there has been vehement opposition to attempting to achieve it. It is assumed that human beings—one of perhaps 10–30 million species—have the right to exploit 88% of the land!

The Global Economy

Finally, we are being sold on a kind of global economics that runs counter to what we have learned from biology in the second part of the century. In the early 1960s, geneticists began to apply the tools of molecular biology to look at the products of single genes within a species. To their amazement, they discovered that there was a tremendous amount of genetic polymorphism. Now we understand that genetic diversity is the key to a species's resilience and adaptability as the environment changes. It also appears that species diversity within ecosystems and ecosystem diversity around the world are also critical elements in life's resilience. Humans have added another level of diversity that is important for our species's resilience: culture. Human cultures are profoundly local and have enabled groups of our species to survive and flourish in environments as different as the Arctic, grasslands, mountain ranges, steaming jungles and rain forests, and arid deserts. We even flourish in New York, Tokyo, and London, for Heaven's sake!

We have learned that when we attempt to raise large numbers of organisms of a single species or one genetic strain of animal or plant, that population becomes extremely vulnerable to pests, infection, or environmental change. Monoculture runs counter to the fundamental biological principle of maximal diversity as the key to adaptability, and we have learned that at great cost in agriculture, forestry, and fisheries. In spite of this insight, we continue to ignore the importance of maximizing diversity and thus sacrifice long-term resilience and sustainability for the sake of immediate human needs. And we are drastically reducing diversity, not just in the natural world but in human societies around the world. A single notion of economics and development has been spread throughout the globe as nations ignore the 1933 warning of the father of the International Monetary Fund and World Bank, John Maynard Keynes:

> I sympathize with those who would minimize rather than maximize economic entanglement between nations. Ideas, knowledge, art, hospitality, travel—these are the things which should of their nature be international. But let goods be homespun whenever it is reasonable and conveniently possible; and above all, let finance be primarily national (Keynes 1933).

The economic monoculture that is pursued by every government in the world makes no ecological sense. Most economists externalize the very support systems of life—air, ozone layer, topsoil, water, and biodiversity itself. Small wonder, then, that it is cheaper for a Toronto restaurant-owner to serve lamb imported from New Zealand than mutton purchased from a farm 40 km north.

Even though we live in a finite world, economics is predicated on the notion that it is not only possible but necessary to strive for steady, endless growth. It is suicidal for a single species that is increasing in numbers exponentially and that has already co-opted 40% of the net primary productivity (NPP) of the planet to demand further economic growth that will come from increasing its share of the NPP (Vitousek and others 1986).

The destructive consequences of this mindless fixation on economic growth as society's most important goal are exacerbated by the measurements of economic success. Any transaction of goods and services resulting in an exchange of money registers as an increase in GDP, whether it is the purchase of weapons to counter high crime rates, hospital and funeral costs of homicides and cigarette-smoking, or cleaning up after an oil or chemical spill. In the GDP, whether money is spent to correct social or ecological damage is irrelevant. As shown by the organization Redefining Progress, which uses an economic indicator that subtracts for such costs, the per capita GDP has more than doubled since 1950, but the Genuine Progress Indicator (GPI) rose slowly to a peak about 1970 and has been declining ever since (Cobb and others 1994).

The global economy that Keynes warned about is dominated by speculators and transnational corporations (TNCs) that are no longer tied to local populations or ecosystems. The current attempt by the OECD to gain passage of the Multilateral Agreement on Investments will open each country to the depredations of the TNCs while freeing them of responsibility to provide jobs or income for local communities or environmental protection of local ecosystems. Maximizing profit appears to be sufficient rationale for globalization of markets and economies.

Where once currency represented something tangible, increasingly it stands for itself. Today, we can buy money, sell money, and make more money without adding anything of value to society or the planet. The $1.3 trillion in daily currency speculation is bigger than most government treasuries, as we see when governments attempt without success to stop the fall in the franc and peso. This global currency flows electronically across all borders and grows far more quickly than real things. So now, as companies diversify, they can deplete one sector and then move to the remaining areas of income. The great temperate rain forests of British Columbia add "fiber" at the rate of 2–3% per year. Obviously, by cutting only 2% or 3% of the trees each year, forest companies could remove the equivalent of the entire forest in 35 or 23 years, respectively, and still have the entire forest left. But it makes no economic "sense" to take only 2% or 3% per year if a company can make 8% or 9% on its investment by clear-cutting an entire forest and putting the money in the bank. If the money is invested in forests in other countries, it might be possible to make far more; and when the forests are gone, the money can be put into fish; and when they are gone, the money can go into biotechnology or computers. So the economics drive a company to maximize profit without regard to long-term sustainability.

RECONNECTING OURSELVES BY SETTING THE BOTTOM LINE

Today, governments around the world pursue a "bottom line" that is driven by an economy that is disconnected from the real world and fundamentally destructive of local communities and local ecosystems. Global competitiveness, efficiency, debt, deficit, and profit are buzzwords defining bottom lines. But it is a bottom line that omits the fundamental basic needs of all human societies. To see what our real nonnegotiable needs are, we must first recognize and surmount the barriers to the interconnections between our activities and the rest of the world that nurtures us.

The first level of human need is defined by our biology—as animals, we have fundamental requirements, and failure to meet them adequately results in death or truncated lives. These needs are so important that our bodies have a multitude of safety devices to ensure that they are met. I am speaking, of course, of our need for clean air, clean water, clean soil, and clean energy, all of which are delivered by the planet's collective biodiversity. We need only hold our breath for 1 minute to recognize the life-giving nature of air. Deprived of air for 3 minutes, we are permanently brain-damaged; after 5 minutes, we die. From the moment of our birth to the instant of our death, we need air. We take each breath of air deep into the most intimate moist, warm parts of our body, where we literally fuse with the air at the surfactant layer lining the alveoli of the lungs. And when we exhale, our breath rushes out and into the noses of our neighbors! We inhale atoms that once were parts of trees, birds, worms, and snakes. We inhale atoms that were once breathed in by Joan of Arc and Jesus Christ. Air is not empty space; it is a physical substance, a matrix in which we are embedded and linked to all terrestrial life on Earth.

We can make a similar case for water, which is at least 60% of our body weight. Water inflates us, enters into metabolic reactions, cools us, and delivers atoms and molecules that we need to survive. Through the hydrologic cycle, water cartwheels endlessly around the planet, purified by soil and plants, transpired back into the air by forests. Water is another glue that holds all of life together; we only have to go without a drink for a day to know how important it is.

Every bit of our nutrition that builds and renews our bodies was once alive. As botanist Martha Crouch says, our relationship with food is the most intimate relationship we have with other beings in that we take them into our bodies and incorporate them into our cells and tissues. And all of our food ultimately comes from the soil. It is remarkable then, when our absolute survival and quality of life depend on the quality of air, water, and soil, that we use them freely as dumping grounds for our toxic wastes.

As living beings, we need energy; and all the energy that we use ultimately comes from the sun. The capacity to capture that energy and send it to us in a usable form resides in Earth's great forests and ocean systems. Ultimately, it is the sum total of all of life's forms—Earth's biodiversity—that somehow purifies and renews our real necessities.

We have another level of fundamental needs, for we are social animals. As the young field of ecopsychology emphasizes, we are deeply embedded in the natural world, and it is an illusion to suggest that we are truly independent beings. Whatever we do to our surroundings, we do to ourselves. Numerous studies show that as social animals, we need the early experience of love for the full development of our potential. Studies done in Romania after Ceaucescu's fall indicate that children raised in orphanages and provided with food, clothing, and shelter but never held or cuddled grow up physically and psychically damaged (Johnson and others 1992). The best way to ensure the love that humanizes us is to provide the opportunity for stable family relationships, and that is generally ensured by strong local communities. Employment is a fundamental need, and numerous scientific studies document the medical, physical, and psychological problems that arise from chronic unemployment or unexpected loss of a job (Lin and others 1995). We must be able to ensure justice and security to avoid the problems that can result from their absence. These are the fundamental social needs that must be met for long-term sustainable futures.

Finally, we are spiritual animals that need to be connected to the natural world. E.O. Wilson has called our need to be with other species *biophilia*, an innate requirement (Wilson 1984). As mortal beings, we are sustained by the knowledge that our kind will live on and that nature itself will continue to thrive after our individual deaths.

I suggest that by re-examining the fundamental needs on which a truly sustainable future can be built, we will also rediscover the incredible interconnections that once held people together in their surroundings.

REFERENCES

Berry T. 1988. The dream of the Earth. San Francisco CA: Sierra Club Bk.

Carson R. 1962. Silent spring. Boston MA: Houghton Mifflin.

Cobb C, Halsted T. 1994. The genuine progress indicator: summary of data and methodology. San Francisco CA: Redefining Progress Inst.

Johnson DE, Miller LC, Iverson S, Thomas W, Franchino B, Dole K, Kiernan MT, Georgieff MK, Hostetter MK. 1992. The health of children adopted from Romania. J Amer Med Asso 268:3446–51.

Keynes JM. 1933. National self-sufficiency. In: Moggeridge D (ed). The collected writings of john Maynard Keynes Vol 21. London UK: Cambridge Univ Pr.

Lévi-Strauss C. 1968. The concept of primitiveness in man the hunter. New York NY: Aldine Publ. [Lee RB and de Vore I (eds)].

Lin RL, Shah CP, Svoboda TJ. 1995. The impact of unemployment on health: a review of the evidence. Can Med Asso J 153:529–40.

Sperry R. 1968. Changed concepts of brain and consciousness some value implications, Zygon. J Relig Sci 20.

World Almanac: Funk and Wagnalls Co.

Veltre T. 1990. Speech delivered at the Wildlife Filmmakers' Symposium. Bath UK.

Vitousek PM, Ehrlich AH, Ehrlich PR, Matson PA. 1986. Human appropriation of the products of photosynthesis. Bioscience 36:368–73.

Wilson EO. 1984. Biophilia: the human bond with other species. Cambridge MA: Harvard Univ Pr.

Wilson EO. 1992. The diversity of life. New York NY: WW Norton.

THE CREATION OF BIODIVERSITY

EDWARD O. WILSON

Museum of Comparative Zoology, Harvard University,
26 Oxford Street, Cambridge, MA 02138-2902

THE TERM *biodiversity*, short for *biological diversity*, was introduced by the National Research Council staff at the first National Forum on BioDiversity, held in Washington in September 1986; and it gained rapid global currency after the publication of the forum proceedings in 1988. *Biodiversity* means, in simplest terms, the variety of life found in the Creation; it is the entirety of life on the planet.

Biologists rescue this conception from vacuity by analyzing biodiversity at different levels of organization, from biosphere downward to gene, and integrating the information to address the fundamental questions of its breadth and origin. More recently, with growing alarm, they have widened their focus to include the causes of the accelerating decline of biodiversity in the human-saturated environment. The first process, creation, is the concern of evolutionary theory; the second is the subject of the new discipline of conservation biology. I will now address the first process.

Researchers have found it most useful to stress diversity at just three levels of biological organization, namely, ecosystem, species, and gene. An ecosystem is a local community of species organisms plus their physical environment. Familiar examples are a New England pond, an old-growth forest in Oregon, and a deep-sea thermal vent off the Pacific coast. Although broad *types* of ecosystems, such as old-growth conifer forests and thermal vents, can be roughly defined by properties they have in common, no two particular ecosystems belonging to a given type are ever exactly alike, either in their species composition or in their physical environment. Throughout the world, individual ecosystems are highly endangered or have disappeared. When a forest is cut, others of the same ecosystem type can persist nearby, but its unique properties have vanished forever. Moreover, many ecosystems contain endemic species, native to that place and environment and found nowhere else. A threatened individual ecosystem or aggregate of ecosystems with many endemic species is called a "hot spot." The rain forest of Kauai is a hot spot; and because so many other kinds of ecosystems on Kauai and the surrounding islands contain threatened endemic species, all of Hawaii is justifiably called a hot spot.

Because ecosystems are difficult to classify and even in many cases to delimit geographically, they are seldom used in quantitative studies of biodiversity. The unit of choice is the species; species are relatively easy to describe and have been the focus of more than 2 centuries of research in classification and biogeography. The traditional definition of the species is the one given in the "biological species concept": a population or series of populations of individuals capable of freely interbreeding with one another under natural conditions—in short, a closed gene pool. The occurrence of an occasional hybrid is not enough to combine two species into one under this definition; only free interbreeding can do that. Also, the ready production of hybrids in zoos and botanical gardens—for example, between lions and tigers—does not suffice. Gene flow must occur under natural conditions, which apparently never occurred between lions and tigers where they coexisted in the past.

The biological species concept works very well for most kinds of animals and for some plants, such as the orchids, but it has serious problems. In a large percentage of cases, there is no way to know whether two populations that occupy different geographic ranges would interbreed if somehow they met under natural conditions. A population of birds on Oahu, for example, cannot be judged with certainty against a somewhat different population on Kauai. The usual taxonomic solution to the dilemma under the biological species concept is to classify the two populations as subspecies, or geographic races, of the same species.

Yet another problem with the biological species concept is its irrelevance to the vast assemblage of life forms that do not reproduce sexually—or else do so rarely enough to reduce sexuality to marginal importance in the life cycle. Thus bacteria, which with the asexual Archaea are both the most primitive and the most numerous organisms on Earth, cannot be classified by the biological species concept. A bacterial species is instead defined as a lineage with 30% or more difference from other lineages in DNA base pairs or else subjectively different enough in traits of biochemistry and structure to justify such recognition. As a result, and with insufficient technology to impose even these loose criteria, no one knows to within a factor of less than 100 how many bacterial species exist on the planet.

Understandable dissatisfaction with the biological species concept has encouraged the devising of an alternative definition, that of the phylogenetic species. In this view, the most meaningful species is a distinctive population with a monophyletic lineage—in other words, derived from a single ancestral species. It is of little concern in this view if the populations have indeterminate breeding potential with other populations. As long as the population comprises individuals of the same coherent lineage that are distinguishable to a subjectively agreed-on degree from that of other populations, it can be ranked as a species.

The advantage of the biological species concept is its recognition that closed gene pools are entities that have for the most part been irreversibly launched on an independent course of evolution by mutation and recombination. Given enough time, and without regression by hybridization, species that are at first barely distinguishable are destined to become very distinguishable. The advantage of the phylogenetic species concept is that it reflects rigorously the history of groups of related species without reference to their hypothesized future.

The two cross-cutting criteria, breeding and phylogenetic, can be joined to create a synthetic species concept as follows. A sexual species is a population that is both reproductively isolated and monophyletic. Suppose that a monophyletic sexual population is geographically isolated, so that its reproductive status vis-à-vis similar populations is indecipherable. It can be called a species if it is markedly distinct, or a subspecies if only slightly distinct.

What the new emphasis on molecular markers and phylogenetic analysis comes down to is, I believe, the prospect of increasing the number of formally recognized species through ever finer analyses of the phylogeny of populations, especially such analyses based on DNA sequencing. More subspecies, once they have been found to have substantial differences that are concordant across their ranges, will be raised to species rank. And more sibling species, which are hard to detect with conventional anatomical characters, will be recognized and named. The new emphasis does not, however, in my opinion represent a fundamental shift away from the species concept already used by most practicing taxonomists. As a rule, they have embraced the concepts of both reproductive isolation and monophyly while recognizing as guesswork the assignment of reproductive relationships among closely related but geographically isolated populations.

The current trend of systematic theory is toward a higher degree of objectivity and consensus than existed in the past. A synthetic, truly biological species concept, providing considerable information about each genetically distinguishable population, seems attainable. This aim is of central importance in ecology and conservation biology. How species are delimited and classified determines the number recognized, as well as the number of genera and other higher categories into which they can be defensibly grouped, and hence the magnitude of both local and global biodiversity. It affects the evaluation of the status of individual populations in conservation planning, that is, whether the populations are ranked as species, subspecies, or neither. And finally, the refined species concept conforms more closely to the emerging picture of how biodiversity is created.

Biodiversity is the product of two complementary processes of evolution. The first is vertical evolution of individual populations by changes in chromosome composition and gene frequency. During the process, biodiversity at the level of this hereditary unit grows or declines. But the number of species, the next level up, does not necessarily change as a result. The second evolutionary process, then, is the multiplication of species, often called speciation. In the course of vertical evolution, some species split into two or more daughter species; others do not.

Virtually all biologists closely familiar with the details of vertical evolution give natural selection the dominant role in evolution. In simplest terms, it begins when different forms of the same genes, or alleles, originate by mutations, which are random changes in the long sequences of DNA that compose the genes. In addition to such point-by-point scrambling of the DNA, new mixes of alleles are created by the recombining processes of sexual reproduction. Other forms of mutation occur when entire chromosomes, the carriers of the genes, are duplicated, deleted, broken, fused, or otherwise reconfigured. The mutations, genic and chromosomal, that enhance survival and reproduction of the carrier organisms spread through the population.

The ones that do not enhance fitness fall to very low frequencies or disappear altogether. Chance mutations are the raw material of evolution. Environmental challenge, deciding which mutants and their combinations will survive and reproduce, molds the population from this protean genetic clay.

Although natural selection has the commanding creative role, another force must be mentioned in any account of evolution. By chance alone, substitutions occur through long stretches of time in some of the genes. The continuity of random change is often smooth enough to measure the age of different evolving lines of organisms. But this genetic drift, as it is called, while altering the diversity of genes, adds little to evolution at the level of cells, organisms, and populations. The reason is that the mutants involved in drift must be neutral, or nearly so, in the crucible of natural selection; in other words, they can have little or no effect on the details of higher biological organization on which organisms depend for survival and reproduction.

Driven by natural selection, some species break into daughter species. By the criterion of reproductive isolation, species multiply when populations acquire genetic differences that interfere with mating or the healthy growth of hybrid offspring. These differences are called intrinsic isolating mechanisms. They affect various parts of the life cycle concerned with sexual reproduction, such as differences between populations in times or places of mating, in courtship and mating procedures, and in the developmental physiology of offspring. They can occur singly or in any combination, depending on the biological nature of the species and vagaries of natural selection affecting its evolution.

The classic model of species formation is geographic speciation. Its principal steps, which have been richly documented, are the following. A single population of interbreeding individuals is split into two or more populations by a geographic barrier. Because the barrier is not part of the genomes of the populations, it can consist of almost any feature of the physical environment. It can be the drying of a mesa when the climate enters an arid phase, causing the forest that once covered it to break into fragments sheltered by scattered canyons. It can be the straits that separate two islands of an archipelago. A bird species might only rarely cross this permanent water barrier, but when the event does occur, individuals from one island are able to invade the other island, where the colonists form a population almost entirely isolated from the source population.

As the two populations separated by geographic barriers of whatever nature diverge, they progress from being genetically identical or nearly so to slightly or moderately different, at which point they can be called subspecies—or, meaning the same thing in this context, geographic races. At this stage, the systematist who emphasizes interbreeding capacity says, "The differences are worthy of recognition but not strong enough or involved enough with reproductive traits to call the populations species. If the diagnostic traits were stronger and especially of this nature, I'd call them species." Another systematist, concerned more with phylogenetic criteria, might respond, "All right, but I'll call them species if there are multiple and well-marked diagnostic characters throughout the population, and if the species with which they are compared share an immediate common ancestor and possess their own well-marked and consistent traits. I am not so interested in trying to predict their future."

Even with such clarification, however, the distinctions between subspecies and species are filled with residual ambiguities difficult to explain to impatient students or members of Congress. Here are several:

• A subspecies, or geographic race, can contain genes and traits that are even more distinctive than those of otherwise similar reproductively isolated species, yet not be reproductively isolated or coherent enough to meet the criteria of a phylogenetic species.

• Two species can be separated by numerous genetic differences that nevertheless produce no outward traits easily discerned by investigators. Examples include odors used in communication and internally hidden physiological processes. These "sibling species," even though important elements of biodiversity, are nevertheless consistently undercounted.

• Some species, especially on continents or large islands, are broken into numerous local populations that vary genetically from one another. The temptation exists to recognize many subspecies among the populations, but two outstanding difficulties are often encountered in such cases. First, the geographic limits of each population are often difficult, if not impossible, to define. Second, the traits typically vary discordantly. To take an imaginary but realistic example of discordance, size might decrease from north to south, color from east to west, food preference from northwest to southeast, and so on, indefinitely. The number of geographic character lines that can be drawn and hence the number of subspecies recognized in such discordantly varied species depend on which traits are chosen to follow them. Still, in spite of this difficulty, a large percentage of species comprise local populations that can be easily delimited and whose diagnostic traits are concordant enough to justify subspecific or, by stress on the phylogenetic criterion, specific status.

• To add to the many complications inherent in geographic differentiation, species can also multiply in the absence of geographic barriers. Almost half of living plant species and a smaller number of animal species have arisen by polyploidy, the multiplication of entire chromosome sets. The idea of polyploidy can be quickly grasped as follows. If the number of chromosomes in the egg or sperm of a nonpolyploid organism is N (haploid), then the number in the fertilized egg and ensuing organism is 2N (diploid). In a polyploid, the number in the fertilized egg and ensuing polyploid organism is 3N (triploid), or 4N (tetraploid), and so on. A polyploid with 4N chromosomes can in some cases breed with its 2N ancestor, but the hybrid offspring, which carries 3N chromosomes in each cell, is ordinarily unable to complete the steps of meiosis and hence to produce viable sex cells of its own. As a result, the 2N ancestor and its 4N derivative are distinct species. The splitting of one species into two species in this case occurred across only two generations—a near instant in evolutionary time. A variation of the process can occur when two species create a hybrid that is also a polyploid. With two of each kind of chromosome thus provisioned in each cell of the organism, the chromosomes can pair off with exact equivalents in the first meiotic division, permitting the production of normal sex cells. The polyploid hybrids can as a result breed successfully with one another, but not with their diploid parents; so they are established as a new, reproductively isolated species.

Another form of sympatric speciation, or species multiplication in the absence of geographic barriers, is through host races. The process is hard to detect and harder to prove, but it might be far more important in nature than previously appreciated. It unfolds when a species of say, an insect is specialized to feed on the leaves or fruit of one species of tree, a common situation in nature. It also mates exclusively on this same host plant. A few individuals, either because they are mutants in food preference or because they make an error in plant selection (and then become imprinted on the wrong tree species), move to an alternative host, where they proceed to feed and mate in isolation. As a result, two populations coexist in the same locality. At first, when the differences among them are slight, they are legitimately called host races, or ecological subspecies. But as they diverge genetically, and especially if the host preferences have a hereditary basis, they are classifiable as distinct species.

No one at this time can confidently evaluate the prevalence of sympatric speciation by host races or other highly local splitting of populations. But given that insect species alone number in the millions, many of them specialized as herbivores on plants or as inhabitants of microhabitats, the process might in time prove to be one of the most important in the origin of biodiversity.

As new species originate—sometimes across only two generations, sometimes during a period of hundreds or thousands of generations—other species die. Over large geographic areas and spans of time, the balance of birth and death maintains a roughly equilibrial number of species in major groups, such as birds, ants, conifers, and mosses. The number appears to be a complex correlate of factors summarized by the acronym ESA, not for the Endangered Species Act or the Ecological Society of America, but for Energy, Stability, and Area. In general, the greater the amount of energy available to the ecosystems, the larger the number of species; thus, high levels exist in the energy-rich coral reefs worldwide and the great tropical moist forests of South America, Africa, and Asia. The more environmental stability, as in the tropical forests and bottoms of the oceans, the greater the number. And, finally, the larger the area, the more species that can be sustained within it.

The role of area in particular can be described by the following broad rule: the number of species occurring in physically well-demarcated habitats—such as islands of an archipelago, patches of woodland in a fragmented forest, or clusters of lakes—varies from the sixth to the third root of the area of the habitats. The exact value varies with the kinds of habitat and organisms studied and the part of the world in which they occur. A common central value is the fourth root, which translates to an easily recalled rule of thumb: a 10-fold increase in area results in a doubling of the number of species.

Where the ESA factors combine, an astonishing number of species have typically accumulated. The greatest biodiversity overall in the world appears to occur in the upper Amazon Basin, which is notably high in all the ESA factors. For example, the largest number of butterfly species in the world observed at a single locality is 1,300, recorded within 3,925 hectares of mostly lowland rain forest at Pakitza, Parque Nacional del Manu, Peru, by Robbins and co-workers (1996). By comparison, only 380 species are known from all of western Europe (Higgins and Riley 1970). Similarly, the world record for ants is 365 species, collected within only 8 hectares of lowland rain forest at Cuzco Amazónico, also in Amazonian Peru, by Stefan Cover and John Tobin (personal communication). That diversity can be instructively compared with the 555 species found in all of North America (Bolton 1995).

The assembly of biodiversity at the level of ecosystems encompasses two complementary principles of organic evolution. The first is adaptive radiation, the expansion of multiplying species from individual stocks into niches available to them. The second is convergent evolution, the increasing similarity in anatomy, physiology, or behavior, singly or in combination (but not in the underlying genetic codes), of radiating groups found in different parts of the world.

The Hawaiian archipelago, the most isolated islands on Earth, are appropriately cited as a natural laboratory that displays the two complementary principles with exceptional clarity. Its roughly 8,000 known endemic land and freshwater species (Eldredge and Miller 1995) have been derived from only a few hundred ancestral species that managed to cross the immense barrier of the Pacific Ocean from continents and islands on both sides. Many of the colonists, arriving over a period of several million years, found an array of major niches open that were closed by competitors in other parts of the world. Among the insects that converged dramatically to adaptive types in other places are geometrid moths whose caterpillars abandoned herbivory to become ambush predators of other insects and a dragonfly whose nymphs have left freshwater streams to forage on land. One lineage of ducks, the moa-nalos, now extinct, evolved into large flightless forms with tortoise-like bills. The fullest and best-known radiation among animals is in the Hawaiian honeycreepers of the family Drepanidinae, whose 23 species (living and recently extinct) were derived from a single ancestral fringillid bird species. In anatomy and behavior, they have variously filled the niches of warblers, woodpeckers, finches, nectar-feeding sunbirds, and parrots. The most striking example among plants is in the tarweeds of the sunflower family Asteraceae, whose numerous species vary from low, herbaceous mats to shrubs and trees, to the spectacular silversword of Maui's Haleakala Crater.

It is by countless such radiations and exchanges of species among their own evolutionary headquarters that the ecosystems have assembled. On a grand scale, much of the history of life can be viewed as a succession of adaptive radiations during which major groups displaced previous assemblages or were able to spread into wholly new adaptive zones made possible by the increasing complexity of pre-existing ecosystems. Life has always expanded to fill the space and use the energy offered to it. The glory of this creative process is the biosphere, billions of years old, over which humanity has lately taken command. The tragedy is that we are thoughtlessly tearing it down before we fully understand its origin, how it is sustained, and the essential role that it plays in human welfare.

REFERENCES

Because this essay is a primer of a broad array of topics, it is appropriate to recommend three general texts, of the many available, for a more detailed introduction:

Raven PH, Evert RF, Eichhorn SE. 1999. Biology of plants, 6th ed., New York NY: Worth Publ.
Futuyma DJ. 1997. Evolutionary biology, 3rd ed., Sunderland MA: Sinauer Assoc.
Wilson EO. 1992. The diversity of life. Cambridge MA: Harvard Univ. Pr.

Several specialized citations not covered in these general works are given below.

Bolton B. 1995. A taxonomic and zoogeographical census of the extant ant taxa (Hymenoptera: Formicidae). J Nat Hist 29:1037–56.
Eldredge LG, Miller SE. 1995. How many species are there in Hawaii? Bishop Mus Occas Pap 41:1–18.
Higgins LG, Riley ND. 1970. A field guide to the butterflies of Britain and Europe. Boston MA: Houghton Mifflin.
Robbins RK, Lamas G, Mielke OHH, Harvey DJ, Casagrande M. 1996. Taxonomic composition and ecological structure of the species-rich butterfly community at Pakitza, Parque Nacional del Manu, Perú. In: Wilson DE, Sandoval A (eds). Manu: the biodiversity of Southeastern Peru. Washington DC: Smithsonian Inst Pr. p 217–52.

THE DIMENSIONS OF LIFE ON EARTH

ROBERT M. MAY

Department of Zoology, University of Oxford,
South Parks Road, Oxford OX1 3PS, UK

INTRODUCTION

This paper aims to give estimates of the numbers of living and distinct species of eukaryotes that have been named and recorded. As will be seen, the factual numbers are accurate only to within 10% or more, mainly because we lack a well-documented and synoptic catalog of all named species. Next, I survey estimates of the total numbers of eukaryotic species on Earth today. Here, our ignorance is such that defensible estimates have a range of a factor of over 100—from a few million to 100 million or more. I conclude by asking what fraction of species that have ever lived on Earth are with us today and outlining an approach to an answer that avoids the huge uncertainties in absolute species numbers.

On the one hand, this paper builds on Wilson's (this volume) scene-setting account of the evolutionary and ecological causes and consequences of biological diversity, seeking to quantify the resulting abundance of life forms. On the other hand, the concluding part of the paper prepares the ground for Pimm and Brooks's (this volume) assessment of likely future rates and patterns of extinction.

Throughout the paper, the focus is on species, and eukaryotic species at that.

Why species? As discussed elsewhere (Collar 1997; Groombridge 1992; Heywood 1995; May 1994a; Wilson 1992), biological diversity exists on many levels, from the genetic diversity in local populations of a species or between geographically distinct populations of a species, all the way up to communities or ecosystems. Any level can be predominant, depending on the questions being asked. At the most basic level, genetic diversity in a species is the raw stuff on which evolutionary processes work their wonders. At the opposite extreme, "we do not have to embrace the wilder poetic flights of Gaians to acknowledge that ecosystems can usefully be regarded as supraorganisms for many discussions of the way biological and physical processes entwine to maintain the biosphere as a place where life can flourish" (May 1994a). A different kind of stratification is oriented toward taxonomy, from races and subspecies through genera and families to phyla and kingdoms.

Given the variety of ways of measuring the dimensions of life on Earth, I nevertheless believe that species are usually the best place to begin. For one thing, there is the practical reason that effective conservation action needs public support, and the public identifies more easily with tangible biological species than with abstractions such as gene pools or ecosystems. For another thing, although it is undoubtedly more important to preserve habitats and ecosystems than individual species, the choices that we will increasingly be forced to make are likely ultimately to be species-based (Claridge and others 1997; Wilcove 1994).

Why eukaryotic species? A molecular biologist could justifiably argue that plants, animals and fungi represent only a recently diversified tip of an evolutionary tree whose main flowering is among bacteria and archaea. But what is meant by species among bacteria and the like is vastly different from what is meant among plants and animals (see, for example, Bisby and Coddington 1995; Vane-Wright 1992). For instance, different strains of what is currently classified as a single bacterial species, *Legionella pneumophila*, have nucleotide-sequence homologies (as revealed by DNA hybridization) of less than 50%; this is as large as the characteristic genetic distance between mammals and fishes (Selander 1985). Relatively easy exchange of genetic material among different "species" of such microorganisms means, I think, that basic notions about what constitutes a species are necessarily different between animals and bacteria. That holds even more strongly for viral species, many of which are best regarded as "quasispecies swarms" (Eigen and Schuster 1977; Nowak 1992). Of course, even within well-studied groups of plants and animals, some workers recognize many more species than others, especially when the organisms in question can reproduce asexually; thus, some taxonomists recognize around 200 species of the parthenogenetic British blackberry, others see only around 20, and a "lumping" invertebrate taxonomist might concede only two or three.

Be this as it may, in what follows I restrict attention to numbers of distinct species of living eukaryotic organisms. In academic fashion, I begin by dwelling on a range of problems before turning to group-by-group assessments of known and suspected total numbers.

NUMBERS OF NAMED SPECIES

Patterns of Effort

From Linnaeus's time to our own, it has often been noted that some groups have received much more taxonomic attention than others (see, for example, Hawksworth 1997). One indication of this is the rates at which species are being recorded. Over the span 1978–1987, an average of five new species of birds was described each year, representing an annual average growth rate in the bird species list of 0.05%. For insects, nematodes, and fungi, the corresponding annual averages for newly added species were 7,222, 364, and 1,700, respectively, representing species-list growth rates of 0.76%, 2.4%, and 2.4% (Hammond 1992, table 4.6). From an academic dean's view, the typical bird or mammal species gets about 1.0 scientific paper per year, other vertebrate species get about 0.5 paper per year, and the average invertebrate species is lucky to average 0.1 paper per year and more likely to get 0.01 (May 1988, table 3).

That pattern of attention among groups reflects the distribution of the taxonomic workforce, as summarized in table 1 (condensed from Gaston and May 1992). Taking a very conservative estimate of 3 million invertebrate species as the global total, table 1 shows that the ratio of taxonomists to species is an order of magnitude greater for vertebrates than for plants and two orders of magnitude greater for vertebrates than for invertebrates. This is no way to run a business. It reflects intellectual fashions and bears no relation to the relative importance of taxa either in the sweep of the evolutionary story or in the delivery of ecosystem services.

Reorganizing our priorities rapidly, to learn more about the little things that arguably run a lot of the natural world, will not be easy. Fascination with the furries and featheries goes deep: in the UK, the Royal Society for the Protection of Birds (RSPB) has almost 1 million members; the analogous society for plants (the Botanical Society of the British Isles) has around 10,000; and there is no corresponding society to express affection for nematodes.

Problems with Synonyms

Despite the gross incompleteness of, and biases in, the taxonomic record, a colleague in the physical sciences might reasonably expect that we could at least say how many living species have been named and recorded. It is a simple fact, ascertainable in principle. But the lack of synoptic databases for most groups means that such factual totals are not generally available. Hence the embarrassing situation that "the figures for described species given, even in high profile re-

TABLE 1   Taxonomy of Taxonomists: Rough Estimate of Distribution of Taxonomic Workforce among Broad Taxonomic Groups in Australia, United States, and UK

| | | Animals | | |
	Plants	Vertebrates	Invertebrates	Microorganisms	Fossils
Approximate division of workforce, %	30	25	35	2–3	5
Estimated total number of living species, thousands	300	45	3,000+	?	—

Source: after Galston and May 1992.

ports and ostensibly authoritative works, vary considerably (for examples see Gaston 1991a,b; Hammond 1992, 1995b), and are almost always, and notably with respect to some of the larger invertebrate animal groups, out-of-date, due to delays in cataloguing" (Hammond 1995a).

This is one reason why I cannot provide a crisp and definitive table of recorded species numbers, group by group. There is, however, a more fundamental and nastier problem. The count of recorded species is inflated by synonyms: single species have been independently and differently named and recorded on two or more occasions. Given, for example, that some 40% of all named beetle species are known from only one geographic site, and that no intercollated database exists, the synonymy problem should not surprise us.

For the better-studied groups—such as birds, mammals and many plant families—synonyms have usually been fairly thoroughly resolved. In contrast, among the more poorly known groups, which tend to contain many more species than the better-known ones, synonymy rates can run high. Hammond (1995a) notes that in 1979 some 2,116 beetle species were newly described and 426 named beetle species were recognized as synonymous with others; thus the net gain in known beetle species in 1979 was roughly 80% of the number newly described. Gaston (1991a) surveyed known synonymy rates for the four major insect orders—Coleoptera, Lepidoptera, Hymenoptera, and Diptera—for the period 1986–1989; he found that the rates varied, but averaged about one-third of the number of species newly described over the same period. Another study of particular groups, mainly insects, found typical synonymy rates of around 20% with some groups exceeding 50% (Gaston and Mound 1993). Bland Findlay (Natural Environmental Research Council, Lake Windermere, UK, pers. comm.) and collaborators have focused on six recent taxonomic revisions of six species-rich genera of ciliates and have found that 584 previously recognized species were reduced to 293 when synonyms were removed; this represents a synonymy rate of 50%. The recently published checklist of Nearctic insects (Poole and Gentili 1996) recognizes 95,694 distinct species but acknowledges 152,079 species names, for an overall rate of resolved synonymy of 37%; the rates in individual orders range from 49% for lepidoptera to around 20% for mecoptera, megaloptera, and trichoptera.

Moreover, any such assessment of synonymy rates must be a lower limit; other synonyms are yet to be uncovered or to accumulate in new work. Solow and others (1995) have made a start on estimating the true rate of synonymy. They used the records of thrip (Thysanoptera) species as published each year since 1901. Some 197 workers have named thrip species (and 28 of these have had all their names relegated to synonymy). Of the total of 6,112 thrip names, 1,326 are currently recognized as synonyms, for an observed synonymy rate of 22%. We also know what proportion of the names published each year are known to be synonyms. Not surprisingly, there is a much higher rate among the names assigned in earlier years; it takes time to uncover aliases. Using this information, Solow and others fitted a probability distribution to the time taken to uncover a synonym and then estimated how many more have yet to be revealed. They conclude that the true proportion of synonyms is around 39%, roughly double the observed rate. They also estimate that on the average it takes around 43 years to identify a synonym. Although there can be some technical quibbles about the details of the calculation (May and Nee 1995), it is clearly indicative.

A more serious question concerns the extent to which the thrip data are representative of other groups. Altaba (1996) has noted the great variations in synonymy rates among mollusk taxa in Mediterranean regions: for melanopsids (relatively large freshwater snails), recent work suggests an observed synonymy rate of roughly 40%; for unionoids (freshwater mussels), he estimates a rate of around 93%; but for hydrobiids (minute snails, often living in springs and subterranean waters), he estimates a rate of 5% or less.

Even for mammals, things are not really as simple as suggested above. Using the database for Neotropical mammals that he is compiling, Patterson (1996) notes that three-fourths of the names for all species recognized since 1980 had earlier been regarded as synonyms. Over that period, the number of species resurrected from earlier relegation to synonymy (173) was three times the number newly banished as synonyms (62) or the number newly described (60). These reappraisals derive more from changing emphases in taxonomic research (in particular, the relatively recent shift toward phylogenetic concepts to replace earlier biological species concepts) than from the independent rediscoveries that account for many insect synonyms. But the complexities that they introduce into the listing of numbers of distinct species are nonetheless real.

In summary, even if we could pull together all the catalogs scattered among museums and other institutions around the world, an accurate assessment of the total number of distinct species currently named and recorded would elude us. The synonymy problem varies from group to group, and it tends to be worst for the most species-rich groups. In light of the work of Solow and others, it could be argued that an overall discount factor of something like 20% might be applied to existing species lists (Hammond 1992, 1995a). But other people are entitled to other guesses.

Numbers of Named and Distinct Eukaryotic Species

The list of numbers of described and extant species in table 2 is derived largely from the thorough work of Hammond (1992, 1995a), itself based on wide consultation. Hammond's estimates were around 1.7 million in 1992 and around 1.74 million in 1995; the largest components of the latter assessment are listed in table 2. Hammond (1995a) also estimated that a total of "13,000 or so" new species are described each year, and that this number had been strikingly constant over the preceding decades. Allowing for synonyms, I would place the true rate of addition of new and distinct species at around 10,000 per year (which roughly reconciles Hammond's 1992 and 1995 estimates).

The right-hand column in table 2 gives my own current assessment, modified in the light of discussion at, and immediately arising from, the meeting on which this volume is based. For some of the groups synonymy might not pose a problem, but it undoubtedly does for the species-rich groups that dominate the overall number (particularly insects, but also crustaceans, nematodes, arachnids, and fungi). My estimated total count of distinct living species is 1.5 million, and this number probably contains an uncertainty of about 10% or so.

TABLE 2 Number of Named, Distinct Species of Eukaryotes

| Group | No. Species, thousands | |
	Hammond (1995a)	This Paper
Protozoa	40	40
Algae	40	40
Plants	270	270
Fungi	70	70
Animals	*1,320*	*1,080*
Vertebrates	45	45
Nematodes	25	15
Molluscs	70	70
Arthropods	1,085	855
(Crustaceans)	*(40)*	*(40)*
(Arachnids)	*(75)*	*(75)*
(Insects)	*(950)*	*(720)*
(Others)	*(20)*	*(20)*
Other animals	95	95
TOTAL	1,740	1,500

The estimate of 1.5 million is essentially identical with Wilson's (1988) widely cited figure of 1.4 million (based mainly on expert opinions for various groups), if we update to allow for adding around 10,000 new and distinct species each year over the last decade.

Before commenting on some individual entries in the right-hand column of table 2, it is helpful to draw back and consider the more coarsely grained picture presented in table 3 of metazoan species in different phyla, subdivided by broad habitat (marine, freshwater, symbiotic, and terrestrial). Here we see order-of-magnitude assessments of species numbers, which highlight how any overall estimate of recorded species diversity is dominated by a few groups. Terrestrial arthropod species are roughly ten times more numerous than any other group, and benthic arthropods and annelids, with mollusks and platyhelminths, account for most of the remaining animal species. The table also underlines how diversity measured by species numbers is very different from diversity in terms of basic body plans (reflected at the phylum level). Although more than 85% of all recorded species are terrestrial (Barnes 1989; Briggs 1994), phyla are predominantly aquatic: 32 of 33 are found in the sea (21 are exclusively marine), whereas only 12 are found on land (only one exclusively).

Before presenting some telegraphic comments on table 2, I emphasize that (with a few exceptions for arithmetic clarity) I have given all numbers to only two significant figures. In some cases, the second digit is reasonably secure (for example, the number of distinct plant species currently described is probably 270,000 rather than 280,000 or 260,000), but in other cases—especially the overwhelmingly important insects—even the first digit is unsure. Systematists and conservation biologists have an unfortunate tendency to present estimates that convey a mislead-

TABLE 3   Distribution of Phyla of Metazoans by Habitat

| | No. Species in Habitat[a] | | | | |
| | Marine | | | | |
Phylum	Benthic	Pelagic	Freshwater	Symbiotic	Terrestrial
Acanthocephala	0	0	0	2	0
Annelida	4	1	2	2	3
Arthropoda	4	3	3	2	5
Brachipoda	2	0	0	0	0
Bryozoa	3	0	1	0	0
Chaetognatha	1	1	0	0	0
Chordata	3	3	3	1	3
Cnidaria	3	2	1	1	0
Ctenophora	0	1	0	0	0
Dicyemida	0	0	0	1	0
Echinodermata	3	1	0	0	0
Echiura	2	0	0	0	0
Gastrotricha	2	0	2	0	0
Gnathostomulida	2	0	0	0	0
Hemichorodata	1	0	0	0	0
Kamptozoa	1	0	1	1	0
Kinorhyncha	2	0	0	0	0
Loricifera	1	0	0	0	0
Mollusca	4	2	3	2	4
Nematoda	3	0	3	3	3
Nematomorpha	0	0	0	2	0
Nemertea	2	1	1	1	1
Onychophora	0	0	0	0	1
Orthonectida	0	0	0	1	0
Phoronida	1	0	0	0	0
Placozoa	1	0	0	0	0
Platyhelminthes	3	1	3	4	2
Pogonophora	2	0	0	0	0
Porifera	3	0	1	1	0
Priapula	1	0	0	0	0
Rotifera	1	1	2	1	1
Sipuncula	2	0	0	0	1
Tardigrada	1	0	2	0	1
TOTAL (33)	27	11	14	15	11
ENDEMIC	10	1	0	4	1

[a] 0 denotes absence of phylum from habitat, and 1–5 indicate number of recorded species, to within rough order of magnitude, in phyla that are present: $1 = 1$–100 species; $2 = 10^2$–10^3; $3 = 10^3$–10^4; $4 = 10^4$–10^5; and $5 = 10^5$ and up. Source: after May (1994b).

ing sense of precision; for example, Wilson's actual estimate in 1988 was 1,392,485 named species rather than 1.4 million. This should be avoided.

Table 2 shows that my assessment of 1.5 million species differs from Hammond's (1995a) 1.74 million by virtue of my estimating 0.23 million fewer insect species, and 0.01 million fewer nematode species. Hammond's 950,000 insect species comprise 400,000 beetle species, 150,000 lepidopteran species, 130,000 hymenopteran species, 120,000 dipteran species, and 150,000 other species. Although Hammond gives a good discussion of the problems of synonymy (referred to above), I believe that he does not adequately discount the totals. My suggested 720,000 insect species in table 2 are 300,000 beetles; 300,000 lepidopteran, hymenopteran, and dipteran species combined; and 120,000 other species. This accords roughly with Nielsen's (Australian National Insect Collection, CSIRO, Canberra, Australia, pers. comm.) estimate of around 750,000 insect species and brings the present estimate into accord with Wilson's (1988) earlier one. I have reduced the nematode species total from 25,000 to 15,000 on the basis of discussions and other published estimates.

My other numbers in table 2 agree with Hammond's (1995a) estimates. Most seem reasonably agreed on among the relevant experts. The roughly 80,000 species of Protoctista (protozoans and algae) are mainly in Bacillariophyta (12,000), Foraminifera (10,000), Gamophyta (10,000), Rhodophyta (5,000), Actinopoda (6,000), Ciliophora (8,000), and Sporozoa (5,000). The estimated 270,000 plant species (embryophytes) are mainly in Spermatophyta (240,000), Pteridophytes (10,000), and mosses and liverworts (16,000). The estimate of 70,000 distinct species of mollusks strikes me as having an uncertainty of about 10%. The same is true for the estimate of 75,000 species of arachnids; an estimate of 36,000 distinct spider species is fairly sure, but the very rough estimate of 40,000 distinct mite species might have an uncertainty of 10% or more.

NUMBERS OF SPECIES EXTANT TODAY

The true total of extant species, as distinct from those we have named and recorded, is hugely uncertain. Table 4 shows Hammond's (1995a) excellent summary of the range of estimates of the possible totals in the major groups of eukaryotes and his own "working figures".

My current estimate is presented in the right-hand column of table 4. The most important discrepancies between my best guesses and Hammond's are in my lower numbers for fungi (1 million fewer species) and for insects (4 million fewer). There are other minor differences, but those two account for essentially all the difference between Hammond's estimate of roughly 12 million and mine of roughly 7 million species. Hammond's (1995a, table 3.1.2) estimated total was actually 13.6 million, but this included 1.4 million bacteria and viruses.

Before briefly discussing table 4, I emphasize the great uncertainty in many of its numbers. The overall range of estimates runs from 3 million to more than 100 million species, with a conservative estimate of the likely range being 5–15 million eukaryotic species. Hammond's 12.2 million best guess is remarkably close to Briggs's (1994) independent estimate of 12.3 million, although they differ considerably in detail (Briggs has 10 million insects, 1 million nematodes, but essentially no fungi).

As discussed much more fully elsewhere (May 1988, 1990, 1994a; Hammond 1992, 1995a), there are many ways to estimate species totals. They include subjective expert opinion, extrapolation of trends, assessments of ratios of unknown

TABLE 4   Estimated Total Numbers of Living Species

	No. Species, millions		
	Hammond (1995a)		
Group	High–Low	Working Figure	This Paper
Protozoa	200–60	200	100
Algae	1,000–150	400	300
Plants	500–300	320	320
Fungi	2,700–200	1,500	500
Animals	100,000–3,000	9,800	5,570
Vertebrates	*55–50*	*50*	*50*
Nematodes	*1,000–00*	*400*	*500*
Molluscs	*200–100*	*200*	*120*
Arthropods	*100,000–2,400*	*8,900*	*4,650*
(Crustaceans)	*(200–75)*	*(150)*	*(150)*
(Arachnids)	*(1,000–300)*	*(750)*	*(500)*
(Insects)	*(100,000–2,000)*	*(8,000)*	*(4,000)*
Other vertebrates	*800–200*	*250*	*250*
TOTAL	100,000–3,500	12.2	6.8

Range	:	100–3
Plausible range	:	15–5
Best guess	:	7

to known species in previously unstudied places, and other methods that combine evidence with various degrees of theoretical argument. The remainder of this section outlines some of the salient points of the various approaches, particularly in relation to my choice of lower estimates in table 4.

Insects

As reviewed by May (1994a) and Hammond (1995a), extrapolation of past trends and surveys of expert opinion tend to put insect species totals in the rough range of 5–10 million. Estimates based on detailed keying-out of the fraction of species new to science in previously unexplored regions tend to give lower numbers—around 3 million (for example, Hodgkinson and Casson 1993). Conversely, estimates reached by using a chain of theoretical arguments to scale from numbers of beetle species in the canopies of individual tropical tree species to tropical insect species totals about 30 million (Irwin 1984); reappraisal of such theoretical arguments has, however, suggested totals more like 3 million (May 1988, 1990; Stork 1988).

I have chosen a best guess of 4 million (rather than Hammond's 8 million, or the lower 2 million guess by Nielsen and Mound, this volume) largely on the basis of the new approach developed by Gaston and Hudson (1994). This original method first asks what fractions of the species in particular taxa are found in each of nine biogeographic realms (these nine realms represent a slight extension of the conventional Wallace scheme); the reference taxa range from general categories (such as higher plants, amphibians, birds, and mammals) to very particular ones (such as dragonflies, tiger beetles, and swallowtails). Gaston and Hudson then take a range of estimated total numbers of insect species in the Nearctic and in Australia and scale them up to global totals on these biogeographic bases. For example, given that Nearctic higher plants represent 6.5% of the global total, an estimated total of 200,000 Nearctic insect species would imply around 3 million insect species in total. For their fairly wide range of estimators, Gaston and Hudson arrive at global insect totals in the range of 1–10 million. I favor an assessment of around 150,000–250,000 Nearctic insects (with Australian insect totals less sure), and use of the higher plants as the biogeographic template, which gives 2–4 million insects in total. This estimate tends to accord with those from empirical studies, such as those of Hodgkinson and Casson (1993); hence my choice of 4 million insect species in table 4. It also accords with Erwin's (Smithsonian Institution, Washington, DC, pers. comm.) recent estimate that preliminary keying-out of some of his tropical-canopy beetle collection suggests that around 80% of the species are new; this implies multiplying the insect total in table 2 by 5, which again gives around 4 million.

Fungi

Observing that there are about six to seven fungal species for each indigenous plant species in the United Kingdom, Hawksworth (1991) suggested that the global total of around 270,000 plant species should be scaled up to yield around 1.5 million species of fungi. Given that only some 72,000 fungal species have yet been named, that would imply that 95% remain to be discovered. Put another way, we might expect that in collections from previously unstudied places, only 5% of fungal species would be known, which is very discordant with the facts (May 1991). Seemingly in support of the 1.5 million estimate (Hawksworth and Rossman 1997), Mibey and Hawksworth (1997) cite 43 species new of 61 species of Meliolaceae and 10 new of 14 Asterinaceae studied in Kenya: but if the 71% figure were representative, it would scale from the known 72,000 fungal species to only around 250,000.

I think the inconsistencies here are associated with problems in simply scaling from UK fungus-plant ratios to global totals. As discussed more carefully, and with other examples elsewhere, such scaling up assumes, among many other things, that fungal species and flowering-plant species characteristically have similar geographic ranges and latitudinal distributions (May 1990). I think it more likely that typical fungal species have wider geographic distributions than typical plant species. Witness the study by Rossman and Farr (1997) of four representative groups of fungi, of which the North American species represented 40–50%, 16%, 54%, and 68% of the world total. The corresponding figure for North American flowering-plant species is 6.5%: maybe the North American fungi are vastly better known than those of other parts of the world, but surely not to this extent. Also, the flowering-plant diversity of the United Kingdom is depauperate, still recovering from the last ice age.

Such considerations undercut many other scaling-up exercises. A count of Heliconius butterfly species to Passiflora species in typical Neotropical sites, scaled against the roughly 360 species of Passiflora in South America, would suggest around 500 species of Heliconius. There are in fact only 66. The same butterflies use different Passiflora species in different places. There are many other such cautionary tales (May 1990).

Other Taxa

Some other "high" entries in table 4 also come from scaling-up of one kind or another. Grassle and Maciolek (1992) have suggested 10 million or more marine macrofaunal species (mostly mollusks, crustaceans, and polychaete worms) on the basis of a different kind of extrapolation. As pointed out on ecological (May 1992) and statistical (Solow 1995) grounds, such projections must be treated with considerable caution.

Apart from insects and fungi, my estimates in table 4 differ little from those discussed fully by Hammond (1995a). I have revised protozoa, algae, and mollusks down a bit and nematodes up a bit as a result of input from this forum. Influenced by Platnick (1997), I have revised arachnids down to around two-thirds of Hammond's estimate. These changes, however, have little effect on my best guess of about 7 million species, some 5 million lower than Hammond's (1995a).

SPECIES ALIVE TODAY AS A FRACTION OF
THE HISTORICAL TOTAL

Given the great uncertainties in how many species are alive today, any estimate of the total numbers ever to have lived, or of likely future numbers of extinctions over the coming century, is even more imprecise.

There is, however, an alternative approach that asks about the fraction of species alive today, or about comparative rates of extinction (in terms of probabilities that species in particular groups became extinct recently, or under various assumptions about the future relative to average extinction probabilities over the sweep of the geological record). Such assessments involve dimensionless ratios and thereby factor out the gross uncertainties associated with absolute numbers of species, permitting quite accurate statements to be made.

For an assessment of f, the fraction of all species to have lived since the Cambrian dawn of hard-bodied fossils (some 600 million years ago) that are alive today, we first ask what is the average life span of a species in the fossil record, from origination to extinction. Such life spans vary greatly, both within and among groups. Raup (1978) brought together several studies and then analyzed some 8,500 cohorts of fossil genera to conclude that the average life span of invertebrate species is around 11 million years. A later, and particularly thoughtful, review by Sepkoski (1992) suggests that 5 million might be a better estimate. The top part of table 5 summarizes the studies surveyed by Sepkoski (1992) and some others, giving an overall impression that the average species has a life span of around 5–10 million years, but with much variability (May and others 1995).

TABLE 5   Estimated Life Spans, from Origin to Extinction, of Various Taxa in the Fossil Record

Taxon	Date of Estimate	Average Life Span, millions of years
Part I: references in May and others (1995)		
All invertebrates	(Raup 1978)	11
Marine invertebrates	(Valentine 1970)	5–10
Marine animals	(Raup 1991)	4
Marine animals	(Sepkoski 1992)	5
All fossil groups	(Simpson 1952)	0.5–5
Mammals	(Martin 1993)	1
Cenozoic mammals	(Raup and Stanley 1978)	1–2
Diatoms	(Van Valen 1973)	8
Dinoflagellates	(Van Valen 1973)	13
Planktonic foraminifers	(Van Valen 1973)	7
Cenozoic bivalves	(Raup and Stanley 1978)	10
Echinoderms	(Durham 1970)	6
Silurian graptolites	(Rickards 1977)	2
Part II: information compiled by R. Cocks[a] (pers. comm.)		
Silurian graptolites	(Koren and Rickards 1996)	0.2
Cambrian trilobites	(Davidek and others, in press)	0.4
Brachiopods	(R. Cocks[a], pers. comm.)	0.5
Rodents	(R. Cocks[a], pers. comm.)	0.3–1.0
Perrissodactyls	(R. Cocks[a], pers. comm.)	0.5
Insectivores	(J.J. Hooker[a], pers. comm.)	3
Corals (tertiary to recent)	(Budd and others 1996)	0.2–7.0 (average 4)
	(Buzas and Culver 1984)	14–16
Foraminifers	(J.R. Young[a], pers. comm.)	c. 10
Coccoliths		

[a] Natural History Museum, London, UK

Cocks (Natural History Museum, London, UK, pers. comm.) has recently compiled a somewhat wider range of estimated species life spans, arguing broadly for a shorter average figure than those above. Graptolites in the Lower Palaeozoic seem to evolve particularly quickly: a collection of more than 30 species from the Silurian of Kazakhstan has examples of three successive species within a single graptolite zone, the duration of which is probably 500,000 years; thus, individual species life spans could be as short as 150,000 years. Likewise, Cambrian trilobites in the Acado-Baltic realm show 25 species with an average life span of 500,000 years. Brachiopods also can be short lived, with particular examples (such as *Eocoelia intermedia*) having life spans less than 500,000 years. Turning to vertebrates, small mammals have evolved at such speeds that most rodent species have life spans of less than 1 million years, with even shorter durations (300,000–400,000 years) in times of rapid dispersal. Perrissodactyls also typically have life spans of less than 500,000 years. Insectivore species live longer, averaging maybe

3 million years. A sample of 175 species of tertiary to recent Corals has species life spans ranging from 200,000 years to 7 million years, with an average of about 4 million years. Moving on up to longer life spans, we find an analysis of 131 species of benthic Foraminifera with average life spans of 14–16 million years, although some have shorter spans, around 7 million years. Coccoliths have comparable longevity. Perhaps the longest-lived species that is well documented is a bryozoan that ranges from the early Cretaceous to the present, a span of around 85 million years (PBT Taylor, Natural History Museum, London, UK, pers. comm.). These estimates, and supporting references, are set out in the lower part of table 5.

In short, there is very great variability—over a range of a factor of 100—among species life spans in the documented fossil record. If one is to speak of an average, it might be better to offer a range like 1–10 million years. Forced to produce a more definite guess, Cocks and his colleagues in the Natural History Museum in London produce a figure of 4–5 million years.

If the sweep of the fossil record is around 600 million years and the average life span from origin to extinction of individual species averages around 4–5 million years, then we might conclude that the species living today—or at any other specific instant—represent just under 1% of the total ever to have lived; that is, f is about 0.01.

Such an estimate, however, assumes total species numbers to have been roughly constant over the 600 million years. That, of course, is not so. As has been argued by Sepkoski (1992), and more recently by Rosenzweig (1997), on the grounds of apparent trends, and by others from more recondite analyses (some involving power laws and fractal measurements; for example, Solé and others (1997), in a very broad outline the history of the fossil record is one of roughly linear increase in species numbers. That implies that the number of species living today is roughly twice the average over the fossil record, which suggests that they make up more like 2% of those ever to have lived, or an f of around 0.02. Benton (1995, 1997) has gone further, marshaling evidence in support of an exponential increase in terrestrial species diversity since the end of the Precambrian; I read this work as arguing for an f of 0.03 or higher.

The latter estimate is subject both to the uncertainties in species life spans and to other complications. For instance, given that most living species are terrestrial insects, whose origins were more like 400 million years ago (and whose average life spans might be somewhat longer than the overall average—see May and others 1995), f could be somewhat larger than 0.02.

Whatever the details, today's evolutionary heritage of living species is not a negligible fraction of those ever to have graced the planet. By the same token, only relatively few past species have exited in dramatic mass extinctions (by the above estimate, the "big five" mass extinctions, even if they had each wiped out virtually all extant species, account for only 5%, or at most 10%, of all endings). The sixth wave, on whose breaking tip we stand, is an uncommon evolutionary event, when judged against the geological record.

Pimm and Brooks (this volume) extend earlier work by May and others (1995) and themselves (Pimm and others 1995), applying similar arguments based on comparative species life spans to estimate recent and likely future changes in extinction rates, as seen against the background average of the fossil record.

CODA

Emphasizing the uncertainties, I have estimated that the number of distinct eukaryotic species alive on Earth today lies in the 5–15 million range, with a best guess of around 7 million. Of these, roughly 1.5 million have been recognized. Allowing for the resolution of synonyms, new species are being recorded at around 10,000 each year. At that rate, it will take over 500 years to complete the catalog.

Such a 500-year estimate is, of course, misleading on several grounds. For one thing, recent and likely future extinction rates point toward qualitative reductions in the catalog. Even more important, I believe that advances in automating molecular sequencing, along with more systematic and computerized handling of phylogenetic information, will revolutionize the basic task of taxonomy in ways that we can yet barely imagine. I guess that within 50 years, and possibly much sooner, we will put a small DNA sample from a newly collected specimen into a machine and be told its exact location in a synoptic tree of living species.

The task of inventorying is sometimes mistaken for "stamp collecting" by thoughtless colleagues in the physical sciences. But such information is a prerequisite to the proper formulation of evolutionary and ecological questions, and essential for rational assignment of priorities in conservation biology (Nee and May 1997; Vane-Wright and others 1990). Lacking basic knowledge about the underlying taxonomic facts, we are impeded in our efforts to understand the structure and dynamics of food webs, patterns in the relative abundance of species, or, ultimately, the causes and consequences of biological diversity.

It is interesting to speculate whether the denizens of other inhabited planets— if there are any—share the vagaries of our intellectual history: a fascination with the fate of the universe and the structure of the atom, lagging well behind interest in the living things with which we share our world. A different, but related, question lies in human institutions' difficulties in taking action to address long-term problems at the expense of short-term interests (witness climate change). Such questions do not come readily under Medawar's rubric of science as "the art of the soluble", but they go to the heart of humanity's future, which unwittingly entrains the rest of life on Earth.

REFERENCES

Altaba CR. 1996. Counting species names. Nature 380:488–9.
Barnes RD. 1989. Diversity of organisms: how much do we know? Amer Zool 29:1075–84.
Benton MJ. 1995. Diversification and extinction in the history of life. Science 268:52–8.
Benton MJ. 1997. Models for the diversification of life. Trends Ecol Evol 12:490–5.
Bisby FA, Coddington J. 1995. Biodiversity from a taxonomic and evolutionary perspective. In: Heywood VH (ed). Global biodiversity assessment. p 27–57
Briggs JC. 1994. Species diversity: land and sea compared. Syst Biol 43:130–5.
Budd AF, Johnson KG, Stemann TA. 1996. Plio-Pleistocene turnover and extinctions in the Car-

ibbean reef-coral fauna. In: Jackson JBC, Budd AF, Coates AG (eds). Evolution and environment in Tropical America. Chicago IL: Chicago Univ Pr. p 168–204.

Buzas MA, Culver SJ. 1984. Species duration and evolution: benthic Foraminifera on the Atlantic continental margin of North America. Science 225:829–30.

Claridge MF, Dawah HA, Wilson MR (eds). 1997. Species: the units of biodiversity. London UK: Chapman & Hall.

Collar NJ. 1997. Taxonomy and conservation: chicken and egg. Bull B O C 117:122–36.

Davidek K and others. In press. New uppermost Cambrian U-Pb date from Avalonian Wales and age of the Cambrian-Ordovician boundary. Geolog Magz.

Eigen M, Schuster P. 1977. The hypercycle. Naturewiss. 58:465–526.

Erwin TL. 1982. Tropical forests: their richness in Coleoptera and other arthropod species. Coleopt Bull 36:74–82.

Gaston KJ. 1991. The magnitude of global insect species richness. Cons Biol 5:283–96.

Gaston KJ. 1991. Body size and the probability of description: the beetle fauna of Britain. Ecol Entomol 16:505–8.

Gaston KJ, Hudson E. 1994. Regional patterns of diversity and estimates of global insect species richness. Biod Cons 3:493–500.

Gaston KJ, May RM. 1992. The taxonomy of taxonomists. Nature 356: 281–2.

Gaston KJ, Mound LA. 1993. Taxonomy hypothesis testing and the biodiversity crisis. Proc Roy Soc Br 251:139–42.

Grassle JF, Maciolek NJ. 1992. Deep-sea species richness: regional and local diversity estimates from quantitative bottom samples. Amer Nat 139:313–41.

Groombridge B (ed). 1992. Global biodiversity: status of the Earth's living resources. London: Chapman & Hall.

Hammond PM. 1992. Species inventory. In: Groombridge B (ed). Global biodiversity: status of the Earth's living resources. London: Chapman & Hall. p 17–39.

Hammond PM. 1995a. The current magnitude of biodiversity. In: Heywood VH (ed). Global biodiversity assessment. Cambridge UK: Cambridge Univ Pr. p 113–28.

Hammond PM. 1995b. Described and estimated species numbers: an objective assessment of current knowledge. In: Allsopp D, Colwell RR, Hawksworth DL (eds). Microbial diversity and ecosystem function. Wallingford UK: CAB International. p 29–71

Hawksworth DL. 1991. The fungal dimension biodiversity: magnitude, significance, and conservation. Mycol Res 95:441–456.

Hawksworth DL. 1997. Orphaus in "botanical" diversity. Muelleria 10:111–23.

Hawksworth DL, Rossman AY. 1997. Where are all the undescribed fungi? Phytopathology 87:888–91.

Heywood VH (ed). 1995. Global biodiversity assessment. Cambridge UK: Cambridge Univ Pr.

Hodkinson ID, Casson D. 1991. A lesser predilection for bugs: Hemiptera diversity in tropical rain forests. Biol J Linn Son 43:101–109.

Koren TN, Rickards RB. 1996. Taxonomy and evolution of Llandovery Biserial Graptoloids from the Southern Urals, Western Kazakhstan. Palaeontology 54:5–103.

May RM. 1988. How many species are there on earth? Science 241:1441–9.

May RM. 1990. How many species? Phil Trans Roy Soc Br 330:293–304.

May RM. 1991. A fondness for fungi. Nature 352:475–6.

May RM. 1992. Bottoms up for the oceans. Nature 357:278–9.

May RM. 1994a. Conceptual aspects of the quantification of the extent of biological diversity. Phil Trans Roy Soc Br 345:13–20.

May RM. 1994b. Biological diversity: differences between land and sea. Phil Trans Roy Soc Br 343:105–11.

May RM, Lawton JH, Stork NE. 1995. Assessing extinction rates. In: Lawton JH, May RM (eds). Extinction rates. Oxford UK: Oxford Univ Pr. p 1–24.

May RM, Nee S. 1995. The species alias problem. Nature 378:447–8.

Mibey RK, Hawksworth DL. 1997. Meliolaceae and Asterinaceae of the Shimba Hills, Kenya. Wallingford UK: CAB International.

Nee S, May RM. 1997. Extinction and the loss of evolutionary history. Science 278:692–4.

Nowak MA. 1992. What is a quasispecies? Trends Ecol Evol 7:118–21.

Patterson BD. 1996. The species alias problem. Nature 380:589.

Pimm SL, Russell GJ, Gittleman JL, Brooks TM. 1995. The future of biodiversity. Science 269:347–50.

Platnik NI. 1997. Dimensions of biodiversity: targeting megadiverse groups.

Poole RW, Gentili P (eds). 1996. Nomina Insecta Nearctica: a check list of the insects of North America. Rockville MD:US Entomological Information Service.

Raup DM. 1978. Cohort analysis of generic survivorship. Paleobiology 4:1–15.

Rosenzweig ML. 1997. Tempo and mode of speciation. Science 277:1622–3.

Rossman AY, Farr J. 1997. Towards a virtual reality for plant associated fungi in the United States and Canada. Biod Cons 6:739–51.

Selander RK. 1985. Protein polymorphism and the generic structure of natural populations of bacteria. In: Ohta T, Aoki K (eds). Population genetics and molecular evolution. Berlin: Springer Verlag. p 85–106.

Sepkoski JJ. 1992. Phylogenetic and ecologic patterns in the Phanerozoic history of marine biodiversity. In: Eldredge N (ed). Systematics, ecology, and the biodiversity crisis. New York NY: Columbia Univ Pr. p 77–100.

Solé RV, Manrubia SC, Benton MJ, Bak P. 1997. Self-similarity of extinction statistics in the fossil record. Nature 388:764–67.

Solow AR. 1995. Estimating biodiversity: calculating unseen richness. Oceans 38:9–10.

Solow RR, Mound LA Gaston KJ. 1995. Estimating the rate of synonymy. Syst Biol 44:93–6.

Stork NE. 1988. Insect diversity: facts, fiction and speculation. Biol J Linn Soc 35:321–37.

Vane-Wright RI. 1992. Species concepts. In: Groombridge B (ed). Global biodiversity: status of the Earth's living resources. London: Chapman & Hall. p 13–6.

Vane-Wright RI, Humphries CJ, Williams PH, 1994. What to protect? Systematics and choice. Biol Cons 55:235–54.

Wilcove DS. 1994. Turning conservation goals into tangible results: the case of the spotted owl. In: Edwards PJ, May RM, Webb NR (eds). Large scale ecology and conservation biology. Oxford UK: Blackwell Scientific. p 313–29

Wilson EO. 1988. The current state of biological diversity. In: Wilson EO, Peter FM (eds). Biodiversity. Washington DC: National Acad Pr. p 3–18.

Wilson EO. 1992. The diversity of life. Cambridge UK: Harvard Univ Pr.

THE SIXTH EXTINCTION:
HOW LARGE, WHERE, AND WHEN?

STUART L. PIMM
THOMAS M. BROOKS
Department of Ecology and Evolutionary Biology
University of Tennessee, Knoxville, TN 37996-1610

THE SCIENTIFIC consensus is that if current rates of species extinction continue, the fraction of species lost will be comparable to that of the five major extinction events in Earth's geological past (Leakey and Lewin 1996). Unlike the past episodes—the famous one exterminated the dinosaurs—this sixth extinction is driven by the dominance of one species, humans (Ehrlich and Ehrlich 1981). The powerful ethical (Norton 1988) and economic (Costanza and others 1996) reasons why we should prevent this scenario are well known (Myers 2000). Less clear are the details. How many species will we lose? Will these losses occur across the globe, or are some areas more vulnerable than others? How quickly will species disappear: do we have years, decades, or centuries to mitigate our current actions?

Those are the questions we will address here. They are circumscribed in one obvious way: we count species, in part, because it is easy to do so. How, then, might our answers apply to other levels of biodiversity? The utility of the term *biodiversity* stems from the recognition that there is variety in life between individuals in a given population, between populations of a given species, and between species (Wilson 1992, 2000).

Killing individuals does not necessarily kill a population, exterminating a population does not necessarily eliminate the species, the species its genus, and so on. What happens when we reverse this logic (Raup 1979)? When we exterminate, for example, 10% of all species, we will likely exterminate far more than 10% of all populations. Some species will survive our depredations but with severely pruned populations. As Hughes and others (1997, 2000) point out, many important justifications for protecting biodiversity emerge from populations, not species.

In what follows, whatever statistics we estimate for species must be substantial underestimates of the effects on populations.

Some species have many populations, others few. Similarly, some genera have many species, others few; and so on through the taxonomic hierarchy of families, orders, and classes. This hierarchy is the sometimes imperfect surrogates for evolutionary lineages of increasing depth. Random species kills often fall on genera with many species, so generic diversity will survive. A benevolent species killer might select some of the buntings (*Emberiza*), sandpipers (*Calidris*), and greenbuls (*Phyllastrephus*). This action might remove no genera, but only species (whose loss would be mourned only by us connoisseurs of subtle differences in their shades of brown and green). Humanity, however, can be malevolent. Elsewhere, we show that we have already lost more genera of birds and mammals than one expects to lose by chance on the basis of random species losses (Russell and others 1998). So, perversely, the impacts we estimate for species also underestimate the impacts we might expect on the diversity of higher taxonomic categories.

SPECIES LOSSES PAST AND PRESENT

The vast majority of the species that have ever lived are now extinct. So the question "How many species are going extinct?" has to be rephrased: "How much faster are species going extinct than one would expect?" The contrast is one of rate.

Thirteen studies of the fossil record show that species persist for one to a few million years (May and others 1995). We know the names of about 1.55 million species (May, this volume), so each year we would expect one or at most a few species to expire. Within small subsets of species, we would expect to wait longer to see just one extinction—about a century for the 10,000 species of birds, for example. Calculating extinction rates as 'extinctions per species per year' provides a convenient frame of reference for calculating human impact (Pimm and others 1995). We know the names of only a small fraction of the planet's species (May 2000), and so by design, this measure does not depend on our knowing them all.

The fossil-record estimate of roughly a million-year life span for a species is suspect in two obvious ways. First, most kinds of species are absent from the record while invertebrates with hard shells (mollusks and brachiopods) dominate. Second, rare species are likely to be missed entirely (McKinney and others 1996). So how typical is this estimate of the rare vertebrates that form the core of our subsequent discussion? An important clue comes from the constraint that natural extinction rates cannot greatly exceed natural speciation rates (Pimm and others 1995). If it were otherwise, there would be no species in the group in question for us to observe.

The common model of speciation assumes an interbreeding population, and then a barrier that splits it allowing the daughter populations to diverge evolutionarily. Taxonomists pass judgment on whether this divergence is sufficient to have formed species. Alternatively, the barrier might later dissolve and the two populations, by not interbreeding, unequivocally demonstrate their distinctiveness. The distinctiveness of two populations in the latter alternative (sympatry) informs the taxonomic judgments about the former (allopatry). In this model, barriers make species and geological knowledge allows us to date the barriers (Rosenzweig 1995). On the average, species-making barriers should form half a species's lifetime in the past, for some species are near their births, and others are near their deaths.

In North America, the presence of many pairs of similar bird species in forests on either side of the central prairies suggests the Late Pleistocene glaciation only 10,000 years ago as the species-making barrier. This high speciation rate might have been a fortuitous baby boom in species, with current high extinction rates a natural pruning of evolutionary exuberance. In fact, the suggestion itself is wrong. Klicka and Zink (1997) use molecular data to show that for 35 such species pairs the average divergence time is 2.45 million years. Increasing numbers of similar studies will likely flesh out many other details, but overall they support the million-year life span as a conservative estimate for species in general.

HUMANITY'S IMPACT ON SPECIES' LIFETIMES

The expectation that one should wait a century to observe an extinction among a sample of 10,000 species is rudely rejected by birds. In recent history (the last 2,000 years), the 10,000 bird species have suffered an average of one or a few extinctions each year (Steadman 1997). Humanity has decreased the average species lifetime and consequently increased the extinction rate by a factor of several hundred (Pimm and others 1995). We know birds well, and the details are informative.

Most of the bird extinctions have been in islands in the Pacific (Steadman 1995). The extinctions represented by stuffed skins in museums, collected within the last century or so, are a small fraction of the total. We know of many more species only as bones from archaeological samples. These species persist up to, but not through, the layers indicating the island's human colonization. The archaeological samples are inevitably incomplete. On the basis of what fraction of today's species the samples include, we estimate that they have found only half the extinct species (Pimm and others 1994). In addition, few of the 700 Pacific islands large enough and isolated enough to host unique species have been explored by archaeologists. Once again, we must correct the body count to reflect the incompleteness of the sampling. Statistical corrections from known species and surveyed islands suggest that the Polynesian colonization of the Pacific exterminated at least 1,000 species of birds. Locally, as in Hawai'i, the Polynesians exterminated over 90% of bird faunas (Pimm and others 1994).

Conclusion 1. **Over the last few thousand years, humans have eliminated over 10% of the world's bird species and locally over 90% of them. Double-digit extinction percentages are part of our history, not merely a prediction about our future.**

The obvious question is whether birds are exceptionally wimpy. They are well-known and so provide unusual details, but are they just extinction-prone? The answer is an emphatic no. Those who argue, like Simon (1986), that the only current extinctions among the 1.55 million named species are a few species of birds and mammals each year are simply ignorant of the facts. The data prove that extinctions are much more comprehensive. The examples are extraordinarily diverse, including animals and plants, invertebrate and vertebrate animals, species on islands and those on continents, desert species and rain-forest species, and aquatic and terrestrial species (Pimm and others 1995).

Statisticians know that their craft depends on samples and the inferences made from them. Reliable inferences require that samples be representative. Reading the list of examples in the previous paragraph, it is hard to imagine a more representative selection of samples. (Those who deny the generality of high extinction rates frequently use economic statistics based on samples—sometimes very small samples—of the numbers in question. The uncritical faith in statistics in one field and the denial of their existence in others is incongruous.) The high rates of extinction in so many different groups lead to our second conclusion.

Conclusion 2. Surveys of many groups of plants and animals uncover global rates of extinction at least several hundred times the rate expected on the basis of the geological record. These groups are diverse in their natural histories and evolutionary origins. With high statistical confidence, they are typical of the many groups of plants and animals about which we know too little to document their extinction.

That is an ecologically surprising conclusion. Would it not be more reasonable for some kinds of species, such as birds versus beetles, or some kinds of places, such as forests versus deserts, to display concentrations of extinctions? Certainly, islands are home to many of the groups of species that are endangered. Yet extinction centers are found on continents, too, so there is nothing unique about islands. There *are* some differences between taxa. In North America, The Nature Conservancy has surveyed 18 groups of animals and plants to calculate the fraction that are on the verge of extinction (TNC 1996). Only butterflies are *less* vulnerable than birds. Proportionally, freshwater fishes, amphibians, crayfish, and freshwater mussels have 3–7 times more species at risk.

Despite these differences in places and taxa, we find high extinction rates in almost every group of species and in almost every kind of place. This "ecologically surprising conclusion" suggests that general ecological principles that work across all groups lead to substantial fractions of their constituent species becoming vulnerable to extinction. We suggest that there are three such principles:

• Many species have very small range sizes, relative to the average range size. (In other words, the statistical distribution of range size is highly right-skewed). Among birds, 30% of all terrestrial species have ranges smaller than 50,000 km^2 (Stattersfield and others 1998)— an area half the size of Tennessee—whereas the average size is some 40 times larger.

• Species that have small ranges are typically less abundant within those ranges than are species that have large ranges (Gaston 1994).
• Species with small ranges are often geographically concentrated (ICBP 1992). We call these areas of concentration *hot spots* of endemism (Myers 1988; Reid 1998).

Low numbers make a species vulnerable to disasters. So, too, do small geographic ranges; human impacts destroy habitats locally (Manne and others 1999). Nature has put her eggs—species with small geographic ranges that typically have relatively low densities—into a few baskets, the hot spots. The pattern is general because the ecological principles that generate it are ubiquitous. These features, in turn, might be derived from deeper ecological causes, and indeed, ecologists seek such explanations. Whatever the underlying causes of the patterns, their consequences are obvious.

Conclusion 3. Many species are rare and local and so at particular risk from humanity's impact. Such species are not spread evenly; extinctions will be geographically clumped, like broken eggs in a dropped basket.

The aggregation of range-restricted species is the feature common to all the examples of high extinction rates listed earlier. Fish in East African lakes, freshwater mussels in the Mississippi drainage, mammals in Australia, flowering plants in the Cape Province of South Africa, and just about everything on oceanic islands— are all examples of aggregations of range-restricted species and very high extinction rates. Where there are not aggregations of range-restricted species, extinction rates will be low. There have been few bird extinctions in eastern North America—an example to which we will return.

There could be two classes of exceptions to the common pattern: aggregations of range-restricted species that do not suffer high extinction rates, and extinctions of widely distributed species. Salamanders constitute an example of the first. Some 20% of the world's salamanders are found in the mountains of the eastern United States, but few are threatened. The reason could be simply that the nature of the terrain protected it from logging or that salamanders can survive well in the moist, deciduous second growth typical of the region. Some species aggregations are just lucky. Other amphibians illustrate the second class: species appear to be in decline worldwide (Berger and others 1998).

Such exceptions apart, the concentration of extinctions in hot spots for species with small ranges has two consequences for policies to prevent extinction:

Policy consequence 1. The history of areas that do not have concentrations of range-restricted species (cold spots) does not inform us in any simple way about the likely fate of concentrations (hot spots).

Eastern North America is an illustration. After European settlement in the 1600s, most of the forest was cleared, although not simultaneously. There were few extinctions—only four species of birds, for example. That does not mean that clearing other comparable areas elsewhere will have correspondingly small impacts. For birds, eastern North America is a cold spot: of its 160 forest species, only 25 are found only there (Pimm and Askins 1995). Clearing a roughly equal area of forest in insular Southeast Asia would exterminate nearly 600 species of birds (Brooks and others 1997).

Policy consequence 2. The fraction of species that will go extinct will depend critically on whether we lose or protect aggregations of range-restricted species.

The good news is that vulnerable species are concentrated, so saving them requires relatively little area. The bad news is that many of these areas have rapidly growing human populations and are in less-developed countries that have sparse resources to protect them (Balmford and Long 1995). Combining those statements leads to

Conclusion 4. How large the sixth extinction will be is still a matter of human choice, not of predestination.

WHERE ARE THE HOT SPOTS OF ENDEMISM?

Myers informally identified 18 hot spots (Myers 1988, 1990). More recently, there have been many efforts worldwide to identify these key areas formally. There are now sophisticated algorithms for picking the smallest subset of locations that encompass all or some specified fraction of the species that one must protect (Pressey and others 1993). Some of these provide important exercises in method development (Csuti and others 1997). Others, such as the work of Lombard (1995) in the Cape Province of South Africa, inform practical decisions about where to establish nature reserves in this extraordinarily rich (and threatened) plant hot spot (Pimm and Lawton 1998).

There are several limitations. The most severe is that only a small fraction of the planet's species are named (May 2000), and we have range distributions for only a tiny fraction of them. Stork (1997) found that the great majority of insects are known from only one specimen each, and so only one location. Worse, there are complications even for the species we do know well.

Areas rich in species are typically not those rich in range-restricted species (Prendegast and others 1993; Curnutt and others 1993). Equivalently, areas that have similar numbers of species can differ greatly in their numbers of range-restricted species. The Hawaiian Islands, eastern North America, and Great Britain have broadly similar numbers of forest-living bird species (about 150); the percentages of species restricted to those areas are 100%, 17%, and less than 1% respectively. Nor are areas rich in range-restricted species in one group always rich in another: eastern North America is a hot spot for salamanders but not for birds. Recent work in Uganda suggests that this lack of correspondence might not matter, because key areas for each species group still represent other groups remarkably well (Howard and others 1998). Nevertheless, we have much to learn about the geography of hot spots (Pimm and Lawton 1998).

Policy consequence 3. We cannot protect hot spots if we do not know where they are.

When comprehensive data are available, the algorithms to select areas for protection typically choose samples widely scattered across the study region. The size of the samples is set by the resolution of the range maps and is usually arbitrary. (An exception is Lombard's work [above] where the areas are set by the mosaic of different land uses and ownership.) Obviously, we can apply such methods to an ever-diminishing spatial scale. Two individuals of every species require remarkably little space. Thus, even with comprehensive data on species ranges, we must ask the ecological question: How much space must be set aside to protect species?

The question has a political answer: worldwide, about 5% of the land has been set aside for protection. The allocation of this to small and large areas also has a political answer. In the Americas, from Florida (US) southward through Mexico, Central America, and South America, only 21 national parks are larger than 10,000 km^2—roughly a square of 1 degree of latitude and longitude on each side (Mayer and Pimm 1998).

Policy consequence 4. Even if we know where to protect species, we must determine how much area is necessary.

HOW MUCH AREA FOR HOW MANY SPECIES?

Global extinction is driven by the fate of the hot spots (Myers 1988). As the area of these hot spots shrinks because of habitat loss and fragmentation, how many species do we lose? One way to approach the question is simply to count the numbers of threatened and endangered species. That is the approach taken by the "Red Data Books" (Baillie and Groombridge 1996). But only for a few well-known groups of species is such information available (Pimm and others 1995). Fortunately, we can estimate losses of species by considering the amount of habitat that is being destroyed.

Exhaustive surveys of species in progressively larger areas of continuous habitat show that the larger the area surveyed is, the more species there will be. These surveys make it possible to deduce a mathematical relationship between species and area. Surveys of archipelagoes show the same relationship but with fewer species for an area of given size than in areas of continuous habitat. The derivation of a power function from first principles by Preston (1962) has led to the nearly universal acceptance of a form $S = cA^z$ for this relationship, where S is species number, A is area, and c and z are constants (Rosenzweig 1995). Typical values of z for increasingly large subsets of continuous habitat are about 0.15; values for areas between islands within an archipelago are about 0.25 (Rosenzweig 1995).

We can use this relationship to derive mathematically the species loss after fragmentation of a once-continuous habitat area, A_{total}, initially holding S_{total} species that are found only in this habitat (figure 1A). When we destroy the habitat,

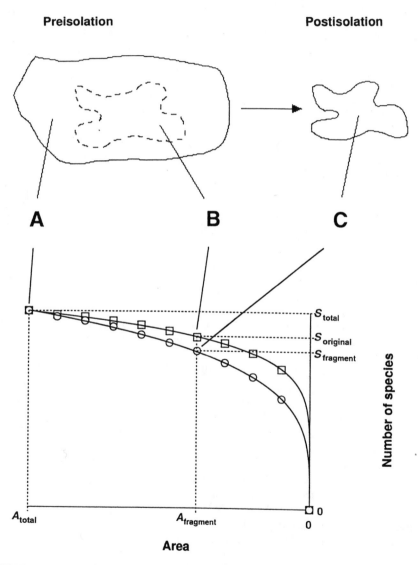

FIGURE 1 Typical species-area relationships. Larger areas (A) have more species than smaller ones (B, C), and areas that have been long isolated—such as islands—have proportionately fewer species (C) than do equal sized areas that are nested within continuous habitat (B).

leaving only an archipelago of fragments, the z value necessary to estimate the number of species that survive, $S_{fragment,}$ in a fragment of area $A_{fragment}$ is the "archipelago value" of 0.25 (figure 1C). In graphical terms, our number of species extinctions is represented by the drop from S_{total} (figure 1A) to $S_{fragment}$ (figure 1C).

We have calibrated this approach for three areas. For eastern North America, a region that has long been deforested and is a cold spot for bird diversity, we find that this recipe exactly predicts the number of bird extinctions (four) that have occurred (Pimm and Askins 1995). For two recently deforested hot spots, insular Southeast Asia (Brooks and others 1997) and the Atlantic forests of South America (Brooks and Balmford 1996), the recipe accurately predicts the numbers of bird species threatened with extinction in the medium term. The recipe is silent, however, about how long the still-surviving but probably doomed species will last. That leads us to our last question.

HOW LONG DOES IT TAKE TO LOSE SPECIES?

There are many ways to answer the question. Extensive modern experience shows that populations numbering in the thousands have risks of extinction observable within human lifetimes. Populations numbering in the tens and hundreds frequently become extinct. Computer and mathematical models provide the theoretical underpinnings of such observations and inform the management of particular species (Pimm 1991).

An entirely different tack comes from looking at the large national parks that are the flagships of their nations' conservation policies. Our experience in advising management about the endangered species in one of these, Everglades National Park in Florida, is that even at this scale, protecting such species requires constant vigilance (Mayer and Pimm 1998). Similar results across similarly large areas have been found elsewhere (Brash 1987; Daniels and others 1990; Diamond 1972; Newmark 1996; Soulé and others 1979; Terborgh 1975).

Between the management of particular endangered species and that of large parks are studies of fragmented habitats. It is on these that we shall concentrate. We can estimate the time that it takes for small patches of natural habitat to lose their species in at least two ways.

The simple way is to find a freshly isolated fragment and then to watch and wait. That is the approach being taken by the exemplary Biological Dynamics of Forest Fragments project in the Brazilian Amazon (Bierregaard and others 1992) and studies of islands isolated by rising waters after the damming of the Lago Gurí, Venezuela (Terborgh and others 1997). The only problem with the approach is that we might not have time to watch and wait. We would like the answers now, not in the future when it is too late to use them.

An alternative approach is to study old fragments of various ages. This approach relies on serendipity; but given the near ubiquity of habitat fragmentation, some fragments, somewhere, will surely provide something close to an ideal experiment. It is also less direct.

Historical collections can provide lists of the total species pool, S_{total}, in the prefragmentation area, A_{total} (figure 1A). Such records rarely distinguish the particular subset that now remains as a fragment, $A_{fragment}$, from the once-continuous habitat that surrounded it. We can estimate the number of species in such prefragmentation subsets, $S_{original}$, using the species-area relationship for the "continuous habitat value" of $z = 0.15$ (figure 1B). Similarly, we can estimate the number that will eventually remain after fragmentation, $S_{fragment}$, using the 'archipelago value' of $z = 0.25$ (figure 1C). Here, we are interested in the species loss from a particular subset, $A_{fragment}$ ("local extinctions" or extirpations), rather than from the entire original area, A_{total} ("global extinctions"). Graphically, the eventual species loss is represented by the drop from $S_{original}$ (figure 1A) to $S_{fragment}$ (figure 1C).

Addressing the issue of "how long" requires more information. We can determine through survey work the number of species surviving now, S_{now}, at any time t after fragmentation. This value should be somewhere between $S_{original}$ and the final number, $S_{fragment}$. From those numbers, we can derive a 'relaxation index' (I), a ratio of the proportion of extinctions yet to occur after time t to the proportion that will eventually occur:

$$I = (S_{now} - S_{fragment})/(S_{original} - S_{fragment}). \qquad (1)$$

Immediately after fragmentation, I will equal 1.0, and it should eventually decline to zero. The final step is to assume a particular form for how I declines with time. As a first approximation, we assume that the decline in species is exponential (Diamond 1972) and therefore that we can characterize it by a fixed time to lose half its species (figure 2). (If the fragment loses 50% of its species in x years, it will lose half of what remains—25% of the total—in another x years, half of what remains (12.5%) of the total in the next x years, and so on.) Thus,

$$I = \exp(-kt), \qquad (2)$$

where k is a decay constant and t is the time since the fragment was isolated. When $I = 0.5$, the fragment has lost half the species that it stands to lose, so t equals the half-life.

Elsewhere, we present data on birds in five rainforest fragments near Kakamega, western Kenya, that we collected over 1996 (Brooks and others 1999). Those data are the results of 8 months of bird surveys through mist-netting, spot counts, and extensive observation; a thorough literature review; an assessment of large quantities of forest-cover data in the form of aerial photographs dating back to 1948, satellite imagery, and anecdotal reports; and a survey of the historical bird specimens in most major museums. For each of the five fragments, we know A_{total}, $A_{fragment}$, S_{total}, S_{now}, and t. From these we can estimate $S_{original}$ and $S_{fragment}$ and then use equation 1 to estimate I.

In figure 3 we plot the proportion of species still expected to be lost, I, against their times since isolation, t. If the declines in species numbers are all exponential with exactly the same half-lives, these points would fall along the same curve.

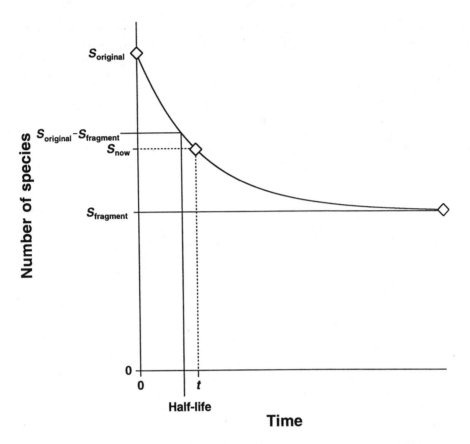

FIGURE 2 Exponential loss of species from fragmented forest. The number of species in an area of once continuous forest ($S_{original}$) declines through the number (S_{now}) at the time (t) when a survey was conducted to the number that will eventually survive ($S_{fragment}$). We can estimate $S_{original}$ and $S_{fragment}$ using the method of figure 1. Because the decay is exponential, we can characterize it by a half-life, the time taken to lose 50% of the species that will eventually be lost.

To a rough approximation, they do so, and their calculated half-times are all broadly similar at between 25 and 75 years—around 50 years.

The long technical details have a short conclusion. Of the species that fragments are going to lose, they lose half in about 50 years. In a century, they will lose 75% of those species.

Conclusion 5. Isolated habitat fragments (certainly fragments of tropical rain forest) will have suffered most of their extinctions by 100 years after isolation.

How do these results compare with other studies? Historical data on forest frag-ments are rare (Laurance and others 1997; Turner 1996), but a few studies do provide dated information on bird communities in fragmented tropical forests (Aleixo and Vielliard 1995; Christiansen and Pitter 1997; Corlett and Turner 1997; Diamond and others 1987; Kattan and others 1994; Robinson 1999; Renjifo 1998). Table 1 summarizes the data from those studies, giving the time between the historical and contemporary surveys (t) and the historical ($S_{historical}$) and con-temporary (S_{now}) numbers of bird species. Assuming a half-life of 50 years, we predict the future equilibrium numbers of species ($S_{fragment}$). Future resurveys of the sites could provide a third point in time along the relaxation curve (figure 2) and therefore test the predictions. Their value now is in suggesting how many more species the sites stand to lose.

How do our results extend globally? We know that over 10% of the world's roughly 10,000 bird species are threatened with extinction, with habitat loss and fragmentation as the main causes (Collar and others 1994). We therefore pre-dict that about 500 of these bird species will go extinct in the next 50 years, pro-ducing an extinction rate of 1,000 extinctions per million species per year. The

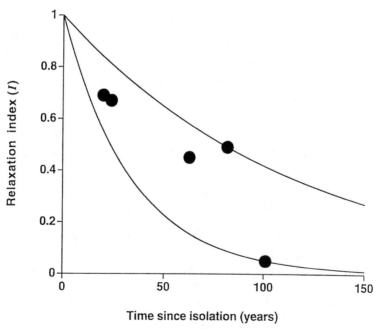

FIGURE 3   How long does it take to lose birds from Kakamega's habitat fragments? We plot a relaxation index (I), which indicates how close a fragment is to suffering so many extinctions that it reaches a new, lower equilibrium of species numbers, against the time (t) since isolation of each fragment. The solid lines indicate exponential decay from the fragments with the shortest (lower line) and longest (upper line) half-lives.

TABLE 1   Published Studies of Changes in Tropical-Forest Bird Communities
After Fragmentation

Fragment	Reference	Size	Date 1	Date 2	t	$S_{historical}$	S_{now}	$S_{fragment}$
Bogor Botanical Garden, Java, Indonesia	Diamond and others 1987	86 ha	−1952	−1985	33	62	42	8*
Sub-Andean region, Colombia	Renjifo 1998	—	~1913	~1998	85	139	~97	78
Santa Genebra, Brazil	Aleixo and Vielliard 1995	251 ha	1977	1993	16	146	134	86*
Lagoa Santa, Brazil	Christiansen and Pitter 1997	285 ha	−1855	1987	132	50	37	35
San Antonio, Colombia	Kattan and others 1994	700 ha	1911	1990	79	128	88	68
Barro Colorado Island, Panama	Robinson 1999	1,500 ha	−1914	−1999	85	121	96	85
Singapore	Corlett and Turner 1997	1,600 ha	~1851	−1991	140	~140	~70	58

NOTE: In each study, the size of the fragments under consideration falls within the same order of magnitude as our Kakamega fragments. Each reports the number of bird species present at the time of a historical survey ($S_{original}$) and the number of species surviving currently (S_{now}), time t after the historical survey. We assume a half-life of 50 years on the basis of our data from Kakamega and therefore a decay constant, k, of 0.014, from equation 2. We then substitute these values into equations 1 and 2 to estimate a future equilibrium number of species ($S_{fragment}$) that will survive in each fragment after complete relaxation. Studies marked * include nonforest species in their counts, so we might underestimate calculated values for the equilibrium numbers of species ($S_{fragment}$).

rates for other groups of species will likely be higher in that they have much greater rates of current endangerment (TNC 1996).

There are sources of uncertainty in these estimates. For example, they assume that habitat destruction will freeze at current levels. Tropical deforestation, in particular, is continuing and accelerating. The worst-case scenario is that we retain only the 5% of the world's tropical forests in protected areas—an event that will happen within 50 years at current rates (Myers 1992). Species-to-area relationships would predict that some 50% of the world's roughly 5,000 forest birds (and millions of other forest animals and plants) would go extinct eventually. Our results above suggest that half the 50% (1,250 species) will be lost before the end of the 21st century, giving an extinction rate of 1,250 extinctions per million species per year.

How do those results compare with other estimates of the magnitude and speed of the extinction? In table 2, we summarize estimates of current global extinction rates produced by seven methods, along with the background rate. The similarity of current rates and their difference by 3–4 orders of magnitude from background rates is striking.

There is a consequence:

Policy Consequence 5. To prevent species extinctions in fragmented habitats we must act immediately, for after a narrow window of only a century, it will be too late.

Finally, we can peer into the dim, more distant future for biodiversity. Prospective extinction rates vary greatly from group to group: 11% of bird species are currently threatened with extinction on the basis of our actions to date (Baillie and Groombridge 1996). Birds appear relatively resistant to extinction. Perhaps one-fourth of all mammal species and even higher proportions of some other groups are now on their way to extinction (Baillie and Groombridge 1996).

If the destruction of natural communities that is now underway throughout the world continues at expected rates, many more species might be similarly doomed to extinction by the end of the next century. The rich tropical forests might contain as many as two-thirds of all the planet's species (Raven 1988). The loss of these forests is rapid and accelerating. We might lose all their species. Suppose that we save 5% of the forests in parks—the average global value for *all* protected areas—and effectively guard them from destructive incursions for the future. Our species-to-area calculations predict that we would eventually lose half the forest species—one-third of all the planet's species. Experience with tropical forests suggests that saving 5% of them will require considerable effort.

We can now give answers to the questions that we posed at the outset. How soon will extinctions occur? Very soon: we can expect to see widespread extinctions in fragmented habitats within 50 years, with the extinction rate about 1,000–10,000 times greater than background rates. Where will extinctions strike hardest? In the hot spots of biodiversity in the tropics. How many species will be lost if current trends continue? Somewhere between one-third and two-thirds of all species—easily making this event as large as the planet's previous five mass extinctions.

TABLE 2 Estimates of Global Extinction Rates, Extinctions per Million Species per Year

Source	Method	Extinction rate (E/MSY)
May and others (1995)	Background rate (13 studies)	0.1-1
This study	Half-life of 50 years for threatened birds	1,000
Mace (1994)	Extinction probabilities from vertebrate Red List categories	1,100-2,200
Smith and others (1993)	Movement of birds and mammals through Red List categories	1,400-2,000
Myers (1979)	Extrapolation of exponentially increasing extinctions	4,000
Myers (1988)	Destruction of ten hotspots by 2,000	7,000
Reid (1992)	Species-area relationship from deforestation rates (6 studies)	1,000-11,000
Ehrlich (1994)	Increasing human energy consumption	10,000

ACKNOWLEDGMENTS

This study was funded by National Geographic Society Research Award 5542–95, a Pew Fellowship in Conservation to SLP, and an American Museum of Natural History Collection Study Grant to TMB. J. Akiwumi, J. Baraza, R. Fox, R. Honea, M. Ibrahim, L. Isavwa, M. Mwangi, K. Orvis, R. Peplies, and J. Robinson helped with forest cover data. In Kakamega, J. Barnes, R. Barnes, L. Bennun, D. Gitau, T. Imboma and his family, C. Jackson, J. Kageche Kihuria, M. Kahindi, S. Karimi, L. Lens, D. Muthui, J. Odanga, D. Onsembe, N. Sagita, J. Tobias, E. Waiyaki, C. Wilder, and the rest of the staff of the Ornithology Department, NMK, and the Kenya Wildlife Service and Forest Department staff were crucial to fieldwork. T., D., and G. Cheeseman, the late G.R. Cunningham-van Someren, M. Flieg, M. Lynch. D. Turner, and D.A. Zimmerman provided further data, as did many museum staff, in particular our hosts P. Sweet, D. Willard, R. Paynter, and P. Angle; and S. Conyne. Thanks to R. May, P. Raven, D. Vázquez, C. Wilder, and an anonymous reviewer for comments on the manuscript.

REFERENCES

Aleixo A, Vielliard JME. 1995. Composição e dinâmica da avifaunda da Mata de Santa Genebra, Campinas, São Paulo, Brasil. Revta bras Zool 12:493–511.

Baillie J, Groombridge B. 1996. 1996 IUCN red list of threatened animals. Gland Switzerland: The IUCN Species Survival Commission.

Balmford A, Long A. 1994. Avian endemism and forest loss. Nature 372:623–4.

Berger L, Speare R, Daszak P, Green DE, Cunningham AA, Goggin CL, Slocombe R, Ragan MA, Hyatt AD, McDonald KR, Hines HB, Lips KR, Marantelli G, Parkes H. 1998. Chytridiomycosis causes amphibian mortality associated with population declines in the rainforests of Australia and Central America. Proc Natl Acad Sci USA 95:9031–6.

Bierregaard RO Jr, Lovejoy TE, Kapos V, Santos AA dos, Hutchings RW. 1992. The biological dynamics of tropical rainforest fragments. Bioscience 42:859–66.

Brash AR. 1987. The history of avian extinction and forest conversion on Puerto Rico. Biol Cons. 39:97–111.

Brooks T, Balmford A. 1996. Atlantic forest extinctions. Nature 380:115.

Brooks TM, Pimm SL, Collar NJ. 1997. Deforestation predicts the number of threatened birds in insular South-East Asia. Cons Biol 11:382–394.

Brooks TM, Pimm SL, Oyugi JO. 1999. The time-lag between deforestation and bird extinction in tropical forest fragments. Cons Biol 13:1140–50.

Christiansen MB, Pitter E. 1997. Species loss in a forest bird community near Lagoa Santa in southeastern Brazil. Biol Cons 80:23–32.

Collar NJ, Crosby MJ, Stattersfield AJ. 1994. Birds to watch 2. Cambridge UK: BirdLife International.

Corlett RT, Turner IM. 1997. Long-term survival in tropical forest remnants in Singapore and Hong Kong. In: Laurance WL, Bierregaard RO Jr (eds). Tropical forest remnants. Chicago IL: Univ Chicago Pr. p 333–45

Costanza R, d'Arge R, de Groot R, Farber S, Grasso M, Hannon B, Limburg K, Naeem S, O'Neill RV, Pareuelo J, Raskin RG, Sutton P, van den Belt M. 1996. The value of the world's ecosystem services and natural capital. Nature 387:253–60.

Csuti B, Polasky S, Williams PH, Pressey RL, Camm JD, Kershaw M, Kiester AR, Downs B, Hamilton R, Huso M, Sahr K. 1997. A comparison of reserve selection algorithms using data on terrestrial vertebrates in Oregon. Biol Cons 80:83–97.

Curnutt J, Lockwood J, Luh H-K, Nott P. Russell G. 1993. Hotspots and species diversity. Nature 367:326–7.

Daniels R JR, Joshi NV, Gadgil M. 1990. Changes in the bird fauna of Uttara Kannada, India, in relation to changes in land use over the past century. Biol Cons 52:37–48.

Diamond JM. 1972. Biogeographic kinetics: estimation of relaxation times for avifaunas of southwest Pacific islands. Proc Natl Acad Sci USA 69:3199–3203.

Diamond J M, Bishop KD, van Balen SV. 1987. Bird survival in an isolated Javan woodland: island or mirror? Cons Biol 1: 132–42.

Ehrlich PR, Ehrlich AH. 1981. Extinction: the causes and consequences of the disappearance of species. New York NY: Random.

Ehrlich PR. 1994. Energy use and biodiversity loss. Phil Trans R Soc Lond B 344:99–104.

Gaston KJ. 1994. Rarity. London UK: Chapman & Hall.

Howard PC, Viskanic P, Davenport TRB, Kigenyi FW, Baltzer M, Dickinson CJ, Lwanga JS, Matthews RA, Balmford A. 1998. Complementarity and the use of indicator groups for reserve selection in Uganda. Nature 394:472–5.

Hughes JB, Daily GC, Ehrlich PR. 1997. Population diversity: its extent and extinction. Science 280:689–92.

Hughes JB, Daily GC, Ehrlich PR. 2000. The loss of population diversity and why it matters. In: Raven PH, Williams T (eds). Nature and human society: the quest for a sustainable world. Washington DC: National Academy Press. p 71–83.

Kattan GH, Alvarez-Lopez H, Giraldo M. 1994. Forest fragmentation and bird extinctions: San Antonio eighty years later. Cons Biol 8:138–146.

Klicka J, Zink RM. 1997. The importance of recent ice ages in speciation: a failed paradigm. Science 277:1666–9.

Laurance WF, Bierregaard Jr RO, Gascon C, Didham RK, Smith AP, Lynam AJ, Viana VM, Lovejoy TE, Sieving KE, Sites Jr JW, Andersen M, Tocher MD, Kramer EA, Restrepo C, Moritz C. 1997. Tropical forest fragmentation: synthesis of a diverse and dynamic discipline. In: Laurence WL, Bierregaard Jr RO (eds). Tropical forest fragments. Chicago IL: Univ Chicago Pr. p 502–14.

Leakey R, Lewin R. 1996. The Sixth Extinction. New York NY: Doubleday.

Lombard AT. 1995. The problems with multi-species conservation: do hot spots, ideal reserves and existing reserves coincide? S Afr J Zool 30:145–63.

Mace GM. 1994. Classifying threatened species: means and ends. Phil Trans R Soc Lond B 344:91–7.

Manne LL, Brooks TM, Pimm SL. 1999. Relative risk of extinction of passerine birds on continents and islands. Nature 399:258–61.

May RM. 2000. The dimensions of life on earth. In: Raven PH, Williams T (eds). Nature and human society: the quest for a sustainable world. Washington DC: National Academy Press. p 30–45.

May RM, Lawton JH, Stork NE. 1995. Assessing extinction rates. In: Lawton JH, May RM (eds). Extinction Rates. Oxford UK: Oxford Univ Pr. p. 1–24.

Mayer AL, Pimm SL. 1998. Integrating endangered species protection and ecosystem management: the Cape Sable Seaside-sparrow as a case study. In: Mace GM, Balmford A, Ginsberg JR (eds). Conservation in a changing world. Cambridge UK: Cambridge Univ Pr. p 53–68.

McKinney ML, Lockwood JL, Daniel F. 1996. Does ecosystem and evolutionary stability include rare species? Palaeogeogr Palaeoclim Palaeoecol 127:191–207.

Myers N. 2000. The meaning of biodiversity loss. In: Raven PH, Williams T (eds). Nature and human society: the quest for a sustainable world. Washington DC: National Academy Press. p 63–70.

Myers N. 1979. The sinking ark: a new look at the problem of disappearing species. London UK: Pergammon Pr.

Myers N. 1988. Threatened biotas: "hot spots" in tropical forests. Environmentalist 8:1–20.

Myers N. 1990. The biodiversity challenge: expanded hot-spots analysis. Environmentalist 10:243–56.

Myers N. 1992. The primary source: tropical forests and our future. New York NY: WW Norton.

Newmark WD. 1996. Insularization of Tanzanian parks and the local extinction of large mammals. Cons Biol 10:1549–56.

Norton B. 1988. Commodity, amenity and morality. In: Wilson EO, Peter FM (eds). Biodiversity. Washington DC: National Acad Pr. p 200–5.

Pimm, SL. 1991. The balance of nature? Chicago IL: Univ Chicago Pr.

Pimm SL, Askins RA. 1995. Forest losses predict bird extinction in eastern North America. Proc Natl Acad Sci USA 92:9343–7.

Pimm SL, Lawton JH. 1998. Planning for biodiversity. Science 2279:2068–9.

Pimm SL, Moulton MP, Justice LJ. 1994. Bird extinctions in the central Pacific. Phil Trans R Soc Lond B 344:27–33.

Pimm SL, Russell GJ, Gittleman JL, Brooks TM. 1995. The future of biodiversity. Science 269:347–50.

Prendegast JR, Quinn RM, Lawton JH, Eversham BC, Gibbons DW. 1993. Rare species, the coincidence of diversity hot spots and coincidence strategies. Nature 365:335–7.

Pressey RL, Humphries CJ, Margules CR, Vane-Wright RI, Williams PH. 1993. Beyond opportunism: key principles for systematic reserve selection. Trends Ecol Evol 8:124–8.

Preston FW. 1962. The canonical distribution of commonness and rarity. Parts 1 and 2. Ecology 43:185–215, 410–32.

Raup DM. 1979. Size of the Permo-Triassic bottleneck and its evolutionary implications. Science 206:217–8.

Raven PH. 1988. Our diminishing tropical forests. In: Wilson EO, Peter FM (eds). Biodiversity. Washington DC: National Acad Pr. p 119–122.

Reid WV. 1992. How many species will there be? In: Whitmore TC, Sayer JA (eds). Tropical deforestation and species extinction. London UK: Chapman & Hall. p 55–73.

Reid WV. 1998. Biodiversity hotspots. Trends Ecol Evol 13:275–80.

Renjifo LM. 1998. Changes in the avifauna of an Andean region following eight decades and a half of forest fragmentation. In: North American Ornithological Conference, 6–12 April 1998. CD-ROM. St. Louis.

Robinson WD. 1999. Long-term changes in the avifauna of Barro Colorado Island, Panama, a tropical forest isolate. Cons Biol 13:85–97.

Rosenzweig ML. 1995. Species diversity in space and time. Cambridge UK: Cambridge Univ Pr.

Russell GJ, Brooks TM,. McKinney L, Anderson CG. 1998. Present and future taxonomic selectivity in bird and mammal extinctions. Cons Biol 12:1365–76.

Simon JL. 1986. Disappearing species, deforestation and data. New Sci. 110(May):60–3.

Smith FDM., May RM, Pellew R, Johnson TH, Walter KR. 1993. Estimating extinction rates. Nature 364:494–6.

Soulé ME, Wilcox BA, Holtby C. 1979. Benign neglect: a model of faunal collapse in the game reserves of East Africa. Biol Cons 15:259–72.

Steadman D. 1995. Prehistoric extinctions of Pacific island birds: biodiversity meets zooarcheology. Science 267:1123–31.

Steadman D. 1997. Human-caused extinction of birds. In: Reaka-Kudla HML, Wilson DE, Wilson EO (eds). Biodiversity II. Washington DC: Joseph Henry Pr. p 139–61.

Stork NE. 1997. Measuring global biodiversity and its decline. In: Reaka-Kudla HML, Wilson DE, Wilson EO (eds). Biodiversity II. Washington DC: Joseph Henry Pr. p 41–68.

Terborgh J. 1975. Faunal equilibria and the design of wildlife preserves. In: Golley FB, Medina E (eds). Tropical ecological systems. New York NY: Springer Verlag. p 369–80.

Terborgh J, Lopez L, Tello J, Yu D, Bruni AR. 1997 Transitory states in relaxing ecosystems of land bridge islands. In: Laurence WF, Bierregaard Jr RO (eds). Tropical forest remnants. Chicago: Univ Chicago Pr. p 256–74.

TNC [The Nature Conservancy]. 1996. TNC priorities for conservation: 1996 annual report card for U.S. plant and animal species. Arlington VA: The Nature Conservancy.

Turner IM. 1996. Species loss in fragments of tropical rain forest: a review of the evidence. J Appl Ecol 33:200–9.

Van Jaarsveld AS, Freitag S, Chown SL, Muller C, Koch S, Hull H, Bellemy C, Krüger M, Endrödy-Younga S, Mansell MW, Scholtz CH. 1998. Biodiversity assessment and conservation strategies. Science 279:2106–8.

Wilson EO. 2000. The creation of biodiversity. In: Raven PH, Williams T (eds). Nature and human society: the quest for a sustainable world. Washington DC: National Academy Press. p 22–9.

Wilson EO. 1992. The diversity of life. Boston MA: Belknap Harvard.

THE MEANING OF BIODIVERSITY LOSS

NORMAN MYERS
Upper Meadow, Old Road, Headington,
Oxford OX3 8SZ United Kingdom

INTRODUCTION

By THE TIME we convene for the third National Forum on Biodiversity in 2007, we may have lost 1 million of Earth's putative 10 million species, counting all extinctions since the start of the biotic crisis a half-century ago. As with many other natural resources and their environmental services, we shall probably not understand the full consequences of biodiversity's decline until we, or rather our descendants, are obliged to learn by strictly empirical means. The loss could turn out to be greater than today's best theoretical models are likely to suggest. Meantime, we must peer into a clouded future and discern as best we can the "meaning" of this loss.

The world is increasingly subject to the dictates of the marketplace. Whether one likes this or not, it is a fact of biodiversity's life. So this paper deals largely with commercial and economic values of biodiversity as expressed through the marketplace or shadow prices. Many other values are at stake and are unamenable to even the most ingenious proxy pricing. It is surely the case, too, that there are many other values that we are simply not yet aware of. This paper's findings should be viewed strictly as a minimalist assessment.

THE ROLE OF POPULATIONS

Biodiversity is generally taken to comprise not only species but also units of species plus ecological processes. Those units, or populations, are more important than is sometimes supposed. Earth's 10 million species feature a rough total of 2.2 billion populations, and we are losing these populations at a rate of 43,000 per day—proportionately far faster than we are losing species (Ehrlich and Daily 1993; Hughes and others 1997). That is important because it is populations rather than species that supply us with the myriad environmental services (known ecosystem services) that support our lifestyles, if not our very survival.

Key question: suppose that in the foreseeable future we lose 50% of all species and the surviving species lose 90% of their populations. Which will carry the greater consequences for the environmental stability of the biosphere? Which will be the most adverse for ecosystem services and environmental stability, whether at local, regional, or global levels?

Some evolutionary biologists believe that speciation and other forms of origination (plus novelty, innovation, and the like) often stem from core populations; other scientists think that they derive primarily from peripheral populations. In support of the second viewpoint is the notion that populations in border zones of a species's distribution often contain a greater amount of genetic variability and are therefore best able to respond to the environmental pressures that might well arise in greatest measure at the limit of the species's distribution. Is it not in this border zone, then, that speciation processes are most likely to arise and develop? Or are the populations that are most "productive" in an evolutionary sense more likely to lie in the heartland of a species's distribution? Is there any substantive evidence from the palaeontological past to indicate which has been the most frequent and productive response? Or is it a case of both together? Or conceivably neither? Could it be that the richest resources for natural selection occur in the heartland zone but that natural selection pressures are greatest in the peripheries?

However we view these uncertainties, it is certain that many species have already lost many of their populations. Consider the case of wheat. In 1996, the crop flourished across an expanse of more than 240 million hectares, with a rough average of 2 million stalks per hectare. Wheat plants totaled almost 500 trillion individuals—probably a record. (In comparison, consider that 1 trillion seconds equals about 32,000 years.) As a species, then, wheat is the opposite of endangered. But because of a protracted breeding trend toward genetic uniformity, the crop has lost the great bulk of its populations and most of its genetic variability. In extensive sectors of wheat's original range where wild strains have all but disappeared, there is virtual "wipeout" of endemic genetic diversity. Of Greece's native wheats, 95% have become extinct; and in Turkey and extensive sectors of the Middle East, wild progenitors find sanctuary from grazing animals only in graveyards and castle ruins. As for wheat germplasm collections, they were described more than a dozen years ago as "completely inadequate"—and that was without considering such future threats as macropollution in the form of acid rain and enhanced UV-B radiation.

In the rest of this paper, we consider species, these being the most recognizable components of biodiversity.

UTILITARIAN BENEFITS OF SPECIES

Conservation biologists increasingly face the question, What is biodiversity good for? Naive as it might sound to some, it is a valid question. There is no longer enough room for a complete stock of biodiversity on an overcrowded planet with almost 6 billion humans and their multifarious activities, let alone a projected two-thirds increase in human numbers and a several-times increase in human activities within the next half-century. So biodiversity must stake its claims for living space in competition with other causes. Generally speaking, biodiversity must urge the merits of its cause through what it contributes to human welfare, preferably in the way that most appeals to political leaders and the general public, namely, in economic terms. This is a strictly anthropocentric approach, and limited as it might seem, it reflects how the world (although not the planet) works.

There are two categories of economic contributions: material goods and environmental services. The first has been frequently and widely (Baskin 1997; Daily 1997; Ehrlich 1992; Myers 1982) documented, principally in the form of new and improved foods, medicines, and raw materials for industry and sources of bio-energy. The second has been far less documented even though it was identified as unusually important 2 decades ago (Westman 1977) and even though its total value is far greater than that of the first (Bishop 1993; Ehrlich 1992; Risser 1995). The main reason for this lacuna is that scientists find it much harder to demonstrate the precise nature of the services, and it is still harder to quantify them economically. Whereas the benefits of material goods tend to accrue to individuals, often producers or consumers in the marketplace, the values of environmental services generally pertain to society; hence, they mostly remain unmarketed (Brown and others 1993).

Material goods

From morning coffee to evening nightcap, we benefit in our daily lives from our fellow species. Without recognizing it, we use hundreds of products each day that owe their origin to wild plants and animals. Conservationists can well proclaim that by saving the lives of wild species, we might be saving our own. Yet we enjoy the manifold benefits of biodiversity's genetic library after scientists have intensively investigated only one in 100 of Earth's 250,000 plant species and a far smaller proportion of the millions of animal species.

Regrettably, there is not space here to do more than cite a few economic evaluations to demonstrate the utilitarian clout at issue. Consider crop-plant germplasm. Wheat and corn germplasm collected in developing countries by the International Maize and Wheat Improvement Center near Mexico City benefits industrialized countries to the tune of $2.7 billion per year. In Italy, wheat germ pasta contributes $300 million per year to the pasta industry. In Australia, grain varieties have boosted annual harvests by as much as $2.2 billion between 1974 and 1990. One-fifth of the value of the billion-dollar US rice crop is attributed to genetic infusions (Evanson 1991). As for new foods, North American stores now feature all manner of exotic vegetables and fruits; from 1970 to 1985, the number of items available doubled to more than 130, and in some instances to as many as 250. By the middle-1980s, specialty produce, mostly from Asia and Latin America, had become a $200-million-a-year business in the United States alone (Vietmeyer 1986).

The cumulative commercial value of plant-based medicines in developed nations is estimated to amount to $500 billion[1] during the 1990s (McNeely and others 1993; Principe 1997). Two anticancer drugs from the rosy periwinkle generate sales totaling more than $250 million per year in the United States alone, and all plant-derived anticancer drugs combined save around 30,000 lives in the United States each year (Principe 1997). According to the National Cancer Institute, tropical forests alone could well contain 20 plants with materials for several additional superstar anticancer drugs (Douros and Suffness 1980).

A number of analysts have attempted an economic assessment of tropical forest plants' overall potential worth, not just for anticancer purposes. Estimates range from $420 billion (Pearce and Puroshothaman 1993) to $900 billion (Gentry 1993; Mendelsohn and Balick 1995).

Suppose that until the year 2050 we will witness the extinction every 2 years of one plant species with medicinal potential. The cumulative retail-market loss from each such extinction will amount to $12 billion for the United States alone (Principe 1997).

Environmental Services

Species supply us with entire suites of environmental services, which can be defined as functional attributes of natural ecosystems that are beneficial to humankind (Baskin 1997; Daily 1997). They include generating and maintaining soils, converting solar energy into plant tissue, sustaining hydrological cycles, storing and cycling essential nutrients (notably through nitrogen fixation), supplying clean air and water, absorbing and detoxifying pollutants, decomposing wastes, pollinating crops and other plants, controlling pests, running biogeochemical cycles (of such vital elements as carbon, nitrogen, phosphorus, and sulfur), controlling the gaseous mix of the atmosphere (which helps to determine climate), and regulating weather and climates (both macroclimates and microclimates). In addition, biodiversity provides sites for research, recreation, tourism, and inspiration.

However, it is far from true that all forms of biodiversity can contribute to all environmental services or that similar forms of biodiversity can perform similar tasks with similar efficiency. How far do environmental services depend on biodiversity itself? Recent research suggests that they are highly resilient in the face of some loss of species, and they can keep on supplying their services even in highly modified states. A sugar cane plantation might be more efficient at producing organic material than the natural vegetation that it replaced, and a tree farm might be more capable of fixing atmospheric carbon than a natural forest. At the same time, many natural ecosystems with low biodiversity, such as tropical freshwater swamps, have a high capacity to fix carbon.

---

[1] On the basis on an average of $50 billion per year ($15 billion in the United States alone).

Similarly, the services supplied by one form of biodiversity in one locality might not necessarily be supplied by a similar form of biodiversity in another locality. Just because a wetland on the Louisiana coast performs a particular suite of functions, we cannot assume that a wetland on the Georgia coast will perform the same functions—still less an inland wetland in Massachusetts or California and even less a montane wetland in Sweden or a forest wetland in Thailand. Services tend to be site-specific. That makes it much more difficult for conservation biologists to demonstrate the intrinsic value of wetlands or any other biotopes.

Biodiversity plays two critical roles. It provides the biospheric medium for energy and material flows, which in turn provide ecosystems with their functional properties; and it supports and fosters ecosystem resilience (Ehrlich and Roughgarden 1987; Schulze and Mooney 1994). The latter attribute could turn out to be the leading service supplied by biodiversity insofar as all other services appear to depend on it to a sizable degree (Perring and others 1995). As biodiversity is depleted, there is often—not always—a decline in the integrity of ecosystem processes that supply environmental services.

Environmental services are so abundant and diverse that we cannot do more here than look at an illustrative selection. First, consider biotas as carbon sinks. The value of carbon storage in tropical forests as a counter to global warming is around $1,000–3,500/ha per year, depending on the type of forest (Brown and Pearce 1994). The value of the carbon-storage service supplied by Brazilian Amazonia is estimated to be some $46 billion (Guttierez and Pearce 1992). It has been further estimated that replacing the carbon-storage function of all tropical forests could well cost $3.7–25 trillion (Panayotou and Ashton 1992).

Next, note the role of biodiversity in protecting soil cover. Excessive runoff from denuded catchments causes soil erosion and siltation in valleyland watercourses. Siltation of only reservoirs costs the global economy some $6 billion per year in lost hydropower and irrigation water. In the last 200 years, the average topsoil depth in the United States has declined from 23 cm to 15 cm; this costs the average American consumer around $300 per year through loss of nutrients and water, with total annual costs (including degradation of watershed systems; pollution of soils, water, and air; and other off-farm problems) to the United States of $44 billion. Worldwide costs of soil erosion are around $400 billion per year (Pimentel and others 1995).

Consider, too, the important but little-recognized services performed by wetlands. These services include a supply of freshwater for household needs, sewage treatment, cleansing of industrial wastes, habitats for commercial and sport fisheries, recreation sites, and storm protection (Mitsch and Gosselink 1993). Their economic value can be sizable. Louisiana wetlands are estimated to be worth $6,000–16,000/ha with an 8% discount rate, or $22,500–42,500/ha with a 3% discount rate. At the lowest value, the current annual rate of loss of these wetlands is levying costs of about $600,000/km^2 per year; at the largest value, $4.4 million/km^2 (late 1980s values). Marshlands near Boston are valued at $72,000/ha per year solely on the basis of their role in reducing flood damage (Hair 1988).

About one-third of the human diet depends on insect-pollinated vegetables, legumes, and fruits. At least 40 crops in the U.S are completely dependent on insect pollination with a marketplace value of $30 billion (Pimentel and others 1992).

Finally, note the vital part played by biodiversity in the fast-growing sector of ecotourism. Each year, people taking nature-related trips contribute to the national incomes of the countries concerned a sum estimated to be at least $500 billion, perhaps twice as much (Eagles and others 1993). Much of these ecotourists' enjoyment reflects the animal life that they encounter. In the late 1970s, each individual lion in Kenya's Amboseli Park produced $27,000 per year in tourist revenues, and an elephant herd produced $610,000 per year (Western and Henry 1979); today's figures would be much higher with many more tourists in the park. In 1994, whale-watching in 65 countries and dependent territories attracted 5.4 million viewers and generated tourism revenues of $504 million, with annual rates of increase of over 10% and almost 17%, respectively. A pod of 16 Bryde's whales at Ogata in Japan would, according to conservative estimates, produce at least $41 million from whale-watchers over the next 15 years (and be left alive), whereas if killed (as a one-shot affair) they would generate only $4.3 million (Hoyt 1995). In 1970, ecotourism in Costa Rica's Monteverde Cloud Forest Reserve generated revenues of $4.5 million, or $1,250/ha, to be compared with $30–100/ha for land outside the reserve (Tobias and Mendelshn 1991). Florida's coral reefs are estimated to generate $1.6 billion per year in tourism revenues (Adams 1995).

OVERALL FINDINGS

A team of ecologists and economists has recently attempted a comprehensive evaluation of all the goods and services stemming from biodiversity. They offer a preliminary and exploratory total of $33 trillion per year (Constanza and others 1997), compared with a global GNP of $28 trillion. Thus, the world's gross natural product is in the same league as the world's gross national product and probably exceeds it.

Consider, too, Biosphere 2, the technosphere in the Arizona desert with its semisuccessful life-support systems for eight Biospherians over a period of 2 years. The cost was about $150 million, or $9 million per person per year. The same services are provided to the rest of us by natural processes at no cost. But if we were charged at the rate levied by Biosphere 2, the total bill for all Earthospherians today would come to $3 quintillion (Avise 1994).

CONCLUSION

The biggest challenge of all is to determine a comprehensive answer to the question, What is biodiversity good for? At present lamentable rates of research and analysis, we might eventually find responses to that question only by discovering what has been lost after much biodiversity has been eliminated, with its goods and services.

Conservation biologists should feel more inclined to simply reject the question, What is biodiversity good for? We shall not have anywhere near a sufficient answer within a timeframe to persuade political leaders, policy-makers, and the public (let alone the professional skeptics). Rather, we should invoke the uniqueness and irreversibility arguments and throw the burden of proof on the doubters, requiring them to demonstrate that biodiversity is generally worth so little that it can be dispensed with if human welfare demands as much, through, for example, agricultural encroachment on wildland habitats. True, there is vast uncertainty about what biodiversity contributes to the human cause. But because of the asymmetry of evaluation, the doubters are effectively saying that they are *completely* certain that we, and our descendants for millions of years (until evolution restores the loss), can manage well enough without large quantities of biodiversity.

ACKNOWLEDGMENT

This paper was written with financial support from a Pew fellowship in conservation and environment.

REFERENCES

Adams J. 1995. Ecotourism: conservation tool or threat? Washington DC: World Wildlife Fund-US.

Avise JC. 1994. The real message from Biosphere 2. Conserv Biol 8:327–9.

Baskin Y. 1997. The work of nature. Washington DC: Island Pr.

Bishop RC. 1993. Economic efficiency, sustainability and biodiversity. Ambio 22:69–73.

Brown K, Pearce DW. 1994. The economics of project appraisal and the environment. In: Weiss J (ed). London UK: Edward Elgar. p 102–23.

Brown K, Pearce D, Perrings C, Swanson T. 1993. Economics and the conservation of global biological diversity. Washington DC: Global Environment Facility, The World Bank.

Buchmann S, Nabhan GP. 1996. The forgotten pollinators. Washington DC: Island Pr.

Costanza R, d'Arge R. De Groot R, Farber, S, Grasso M, Hannon B, Limburg K, Naeem S, O'Niell RV, Paruelo J, Raskin RG, Sutton P, van den Belt M. 1997. The value of the world's ecosystem services and natural capital. Nature 387:253–60.

Daily GC (ed). 1997. Nature's services: societal dependence on natural ecosystems. Washington DC: Island Pr.

Damania AB (ed). 1993. Biodiversity and wheat improvement. Chichester UK: Wiley.

Douros JD, Suffness M. 1980. The National Cancer Institute's natural products antineoplastic development program. In: Carter SK, Sakuri Y (eds). Recent results in cancer research. Bethesda MD: National Cancer Inst. 70:21–44.

Eagles PF, Buse JSD, Huengaard GT (ed). 1993. The Ecotourism Society annotated bibliography. North Bennetton VT: The Ecotourism Soc.

Ehrlich PR, Daily GC. 1993. Population, extinction and saving biodiversity. Ambio 22:64–8.

Ehrlich PR, Ehrlich AH. 1992. The value of biodiversity. Ambio 21: 219–26.

Ehrlich PR, Roughgarden J. 1987. The science of ecology. New York NY: MacMillan.

Evanson RE. 1991. Genetic resources: assessing economic value. In: Vincent JR, Crawford EW, Hoehn J (eds). Valuing environmental benefits in developing economies. East Lansing MI: Michigan State Univ Pr. p 169–81.

Gentry A. 1993. Tropical forest biodiversity and the potential for new medicinal plants. In: Kinghorn AD, Balandrin MF (eds). Chemical medicinal agents from plants. Washington DC: American Humane Soc. p 13–24.

Guttierez B, Pearce DW. 1992. Estimating the environmental benefits of the Amazon forest: an international evaluation exercise. London UK: CSERGE, University Coll.

Hair JD. 1988. The economics of conserving wetlands: a widening circle. Washington DC: National Wildlife Fed.

Hoyt E. 1995. The worldwide value and extent of whale watching. Bath UK: Whale and Dolphin Conservation Soc.

Hughes JB, Daily GC, Ehrlich PR. 1997. Population diversity: its extent and extinction. Science 278:689–92.

McNeely J and others (eds). 1993. Biodiversity prospecting: using genetic resources for sustainable development. Washington DC: World Resources Inst.

Mendelsohn R, Balick MJ. 1995. The value of undiscovered pharmaceuticals in tropical forests. Econ Bot 49: 223–8.

Mitsch WJ, Gosselink JG. 1993. Wetlands. New York NY: Reinhold.

Myers N. 1983. A wealth of wild species: storehouse for human welfare. Boulder CO: Westview Pr.

Panayotou T, Ashton PS. 1992. Not by timber alone. Washington DC: Island Pr.

Pearce D, Puroshothaman S. 1993. Protecting biological diversity: The economic value of pharmaceutical plants. London UK: Center for Social and Economic Research into the Global Environment, University Coll.

Perring BC, Maler KG, Folke C, Holling CS, Jansson BO (eds). 1995. Biodiversity loss: ecological and economic issues. Cambridge UK: Cambridge Univ Pr.

Pimentel D, Harvey C, Resosudarmo P, Sinclair K, Kurz D, McNair M, Crist S, Shpritz L, Fitton L, Saffouri R, Blair R. 1995. Environmental and economic costs of soil erosion and conservation benefits. Science 267:1117–22.

Pimentel D, Stachow U, Takacs DA, Brubaker HW, Dumas AR, Meaney JJ, O'Niel JAS, Onsi DE, Carzilius DB. 1992. Conserving biological diversity in agriculture/forestry systems. Bioscience 42:354–62.

Principe P. 1997. Monetizing the pharmacological benefit of plants. In: Balick MJ, Elisabetsky W, Laird S (eds). Tropical forest medical resources and the conservation of biodiversity. New York NY: Columbia Univ Pr. p 191–218.

Risser PG. 1995. Biodiversity and ecosystem function. Cons Biol 9:742–6.

Schulze ED, Mooney HA (eds). 1994. Function (study edition). New York NY: Springer-Verlag.

Tobias D, Mendelsohn R. 1991. Valuing ecotourism in a tropical rainforest reserve. Ambio 20:91–3.

Vietmeyer ND. 1986. Lesser-known plants of potential use in agriculture and forestry. Science 232:1379–84.

Western D, Henry W. 1979. Economics and conservation in third world national parks. BioScience 29:414–8.

Westman WE. 1977. How much are nature's services worth? Science 197:960–4.

THE LOSS OF POPULATION DIVERSITY
AND WHY IT MATTERS

JENNIFER B. HUGHES
GRETCHEN C. DAILY
PAUL R. EHRLICH
Department of Biological Sciences, Stanford University,
Stanford, CA 94305-5020

BIODIVERSITY ENCOMPASSES variation at all levels of biological organization, including individuals, populations, species, and ecosystems (Wilson 1988), yet much of the current scientific and public concern over the extinction crisis is focused on the loss of species. The rate of species extinction, however, reflects only one aspect of the loss of biodiversity and its consequences. What if no further species became extinct, but every nonhuman species suddenly were reduced to a single, minimal population? Although global species diversity would remain unchanged, the planet would be largely devoid of life, and civilization as we know it would collapse. This is because many of the benefits that biodiversity confers on humanity are delivered locally, through populations of species. This extreme scenario highlights the idea that species, although important, are not the only dimension of biodiversity that we should be concerned about losing.

In this paper, we examine the consequences of the gradual extinction of populations that is occurring today. First, we discuss the importance of populations to humanity. Then, we present estimates of population diversity, that is, the number of populations on Earth. Finally, we make a preliminary attempt to evaluate the rate of populations extinction.

WHAT IS A POPULATION?

Populations are geographical entities within a species, usually distinguished ecologically or genetically (Ehrlich and Daily 1993). The ecological entity is a demographic unit—a group of individuals whose population dynamics are not influenced substantially by migration from nearby conspecific groups; that is, the fluctuations in the size of the population of one group are independent of those of other groups (Brown and Ehrlich 1980). The genetic entity is a Mendelian population (Sinnott and others 1950), defined here as a genetically distinguishable group of individuals that evolves independently of other groups. Demographic units may be Mendelian populations and vice versa, but the two are not necessarily congruent. As with species, both kinds of populations exist as parts of continua in space, rather than as clear, discrete units.

We adopted the Mendelian-population definition for our estimates of population diversity and its extinction rate for two reasons. First, the Mendelian definition directly includes the genetic variation between groups of individuals, and, as discussed below, this variation is of great importance to humanity. Second, we found that quantitative data on genetic-population structure was more comparable across species and investigations than were quantitative data on demographic-population structure.

THE IMPORTANCE OF POPULATIONS

Why should one be concerned about the extinction of populations? Much has been written about the ethical and practical reasons for halting the species-extinction crisis that is driven by human activities (Ehrlich and Ehrlich 1981; Ehrlich and Wilson 1991; Myers 1979; Wilson 1992). Certainly, we agree with these reasons, yet simply arguing for saving species obscures an essential link between biodiversity and human welfare. Ultimately, most of the benefits that biodiversity confers on humanity are delivered through populations. These benefits include aesthetic enjoyment, discovery and improvement of pharmaceuticals and agricultural crops, species conservation, replenishment of stocks of economically valuable species, and, perhaps most important, delivery of ecosystem services.

Aesthetic Value

Natural ecosystems are composed of populations of organisms, their physical environments, and the interactions between them. As such systems are disrupted or destroyed, people's enjoyment of their ambience and the aesthetic values of their component populations (for example, birds, butterflies, reef fishes, flowering plants, and shade trees) is diminished. In addition, the total aesthetic value of individual species declines as their populations disappear, although the aesthetic value of "rarity" may partially, and somewhat paradoxically, compensate for this loss. For instance, although wild populations of grizzly bears and remnants of old-growth redwood forest exist in the United States, the total aesthetic benefit conferred on Americans by watching a grizzly cub play or by hiking in a cathedral-like, old-growth redwood forest is relatively small because few people can experience them firsthand.

Genetic Value

Much of the genetic diversity in species exists as genetic differences between populations. For an average animal species, 25–30% of its total genetic variability is due to differences between populations. In an average outcrossing plant species, 10–20% of its genetic variability occurs among populations, whereas a selfing plant species exhibits about 50% of its genetic variability among populations (Hammond 1995). One result of this differentiation is that populations of the same species may produce different types or quantities of defensive chemicals (Dolinger and others 1973; Goméz-Pompa and others 1972; Hwang and Lindroth 1997), compounds that may have medicinal value.

An example of how genetic variation among populations is important to pharmaceuticals is the story behind the development of penicillin. The successful development of penicillin as a therapeutic drug did not occur until 15 years after Alexander Fleming's discovery of the compound in common bread mold. One reason for this delay was a worldwide search to find a strain (that is, a population) of the mold that produced greater quantities of penicillin than the original strain produced (Dowling 1977).

Population diversity among wild relatives of crops also supplies critical genetic material to agricultural strains. Genetically uniform strains of the world's three major crops (wheat, rice, and maize) are planted widely; as a result, large fractions of the harvest can be threatened at one time by a new disease or pest (Plucknett and others 1987). Thousands of strains, or populations, of wild relatives of crops may need to be tested until one is found that carries the desired genetic resistance that can be used to protect the crop. For example, when the grassy stunt virus emerged as a serious threat to the rice crop in Southeast Asia in the late 1960s and 1970s, an extensive search for resistant varieties of rice was conducted at the gene bank of the International Rice Research Institute. Five thousand accessions from populations all over the world and 1,000 breeding lines were screened. Only one accession of a wild rice collected in India was found to resist the virus (Plucknett and others 1987). Genetic variation in wild populations of crop species also will be crucial in providing genetic material to sustain yields with changing growing conditions, especially climate (Daily and Ehrlich 1990).

Species-Conservation Value

By definition, populations are essential to the conservation of species diversity, and the number and size of populations influence the probability of persistence of the entire species. Migrants between populations can prevent the local extinction of a species by contributing critical individuals when numbers are low (the rescue effect) (Brown and Kodric-Brown 1977) or by supplying the genetic variation needed to adapt to changing environmental conditions (Lande 1988). If local extinction does occur, individuals from other populations can recolonize the area. The threat of rapid global climatic change makes the safety net of population diversity for species even more important; a species that has many populations is more likely to include individuals that are genetically suited to new conditions than is a species that has only one or a few populations (Kareiva and others 1993).

Direct Economic Value

Destruction of populations of an economically valuable species not only increases the probability that the species will become extinct in the near future but also may decrease the species' harvest level. In the short term, as populations are exterminated, fewer will remain to be harvested; in the longer term, when a species is composed of a metapopulation, the stock levels of the remaining populations also may decline (Pulliam 1988). The reduction of these economically important species often has direct consequences for local peoples. For instance, overharvesting of oceanic fish stocks and the resulting decline in yields lead to loss of income to fishermen and loss of an important source of protein for much of the human population (Kaufman and Dayton 1997; Peterson and Lubchenco 1997; Safina 1995).

Ecosystem-Service Value

Perhaps the most important benefit that populations confer on humanity is ecosystem services. Ecosystem services include natural processes, such as purification of air and water, detoxification and decomposition of waste, generation and maintenance of soil fertility, pollination of crops and natural vegetation, and control of pests (Daily 1997). These services are provided by populations, and population diversity (that is, the number of populations) at global, regional, and local levels affects the provisioning of ecosystem services. (The size and density of populations also influence the provisioning of ecosystem services. These dimensions will be discussed later. For now, we simply address numbers of populations.)

Greater global population diversity probably enhances the delivery of global ecosystem services, such as regulation of biogeochemical cycles and stabilization of climate (Alexander and others 1997). The larger the area that remains under natural tree cover in the Canadian taiga, the greater the amount of carbon stored there. Although deforestation in this region might not result in the extinction of any species, a large-scale loss of tree populations would influence the balance of greenhouse gases in the atmosphere worldwide (Woodwell and others 1983).

For many ecosystem services, however, global numbers of populations are not as important as regional population diversity. In other words, for these services, it is not only necessary that many populations exist somewhere in the world but also that they exist within the region of interest. These services include, for instance, mitigation of floods and droughts by forests and purification of water by forests and wetlands (Ewel 1997; Myers 1997). Loss of these services occurs when forests and wetlands are destroyed in a region, regardless of the continued existence of their component species elsewhere. New York City provides an excellent example of the value of regional population diversity. The city was famed for its pure water, which came from the Catskill Mountains, 100 miles to the north. For most of the city's history, natural purification processes, which are carried out by populations of soil organisms and plants, were sufficient to cleanse the water, but in recent years, land development and associated human activities reduced the efficacy of these processes. In 1996, city water officials floated an environmental bond issue to purchase land, freeze development on other lands, and subsidize the improvement of septic tanks in the water-supply area. It is hoped that these actions will restore and safeguard the local populations that filter and purify the water. If so, an investment of $1 billion in natural purification services will have saved city taxpayers $6–$8 billion, the additional avoided cost (over 10 years) of building a water-treatment plant (Chichilnisky and Heal 1998).

Regional population diversity is also necessary for control of pests. The importance of populations that serve a pest-control function is illustrated dramatically when an organism is transplanted to a new environment that lacks populations of predators capable of keeping it from becoming a pest. The importation of the prickly pear (*Cactus opuntia*) into Australia by early settlers is a classic case. Apparently originally intended as an ornamental plant, in the absence of its normal predators the cactus spread over vast areas. It occupied some 25 million hectares in New South Wales and Queensland, and half the area was covered so densely that the land could not be used for farming or ranching. The costs of poisoning or removing the cactus were more than the land was worth. The problem was solved eventually by importing a moth that is a voracious cactus-eater from the South American homeland of the opuntia. Once regional populations of that moth, *Cactoblastis cactorum*, were established, the cactus was decimated and the problem was solved. Although the cactus still can be found in Australia, it occurs only in scattered clumps since natural pest control has been re-established (Ehrlich 1986).

Pollinators are critical to agriculture, and the decline of regional populations of native pollinators, chiefly as a result of pesticides and destruction of habitat, has not gone unnoticed (Buchmann and Nabhan 1996). For more than 60 crops planted in the United States, farmers are forced to pay keepers of the European honeybee to transport their hives to the fields or orchards that require pollinating. Hiring beekeepers costs farmers more than $60 million a year and the federal government more than $80 million in subsidies, and these numbers are still increasing because of growing problems in the beekeeping industry (disease and hybridization with the aggressive Africanized honeybee) (Nabhan and Buchmann 1997).

Population diversity at a particular location (that is, local species diversity) also affects ecosystem functioning and thus the delivery of ecosystem services (Chapin and others 1997). In greenhouse and field experiments, plant productivity has been found to increase with species diversity (Naeem and others 1994). The stability of plant productivity also has been linked with greater richness of species. More diverse grassland plots seem to be more resistant to drought and grazing disturbances than less diverse plots (McNaughton 1977, Frank and McNaughton 1991; Tilman and Downing 1994). Thus, it appears that local population diversity is closely coupled to local ecosystem functioning.

Because regional and global services are performed by an aggregate of local ecosystems, the consequences of a reduction in local population diversity probably will extend beyond the local ecosystem. In other words, the loss of populations from one location, which alters the functioning of the local ecosystem, may in turn affect the delivery of larger-scale services. For example, the global carbon cycle may be influenced not only by the total number of tree populations on the planet but also by the diversity of populations at many locations.

One important question that remains to be resolved is the extent to which "weedy" species, spreading into and establishing populations in areas where native populations have been extirpated, can continue to supply ecosystem services. For such services as pest control, evidence is abundant that such compensation will be rare. The cotton disaster in the Cañete Valley in Peru is a classic example. Populations of natural enemies of potential cotton pests were destroyed by repeated, heavy applications of pesticides, and no weedy species moved in to assume the role of the natural predators. As a result, numerous obscure organisms became pests and destroyed the cotton crop (Barducci 1972).

For other services, such as flood control and soil retention, the potential for substitution by weeds, at least in the short term, sometimes may be high. In many cases, however, we are largely ignorant of the ability of weeds to maintain services over the long run. Furthermore, the capacity for large-scale technological substitution of ecosystem services appears limited (Ehrlich and Mooney 1983). The Biosphere 2 project, a materially closed, human-made ecosystem, is a case in point. Despite hundreds of millions of dollars invested in development and operating costs, scientists failed to engineer a system that could support eight people with food, air, and water for 2 years (Cohen and Tilman 1996; see also Daily 2000). That venture dramatically illustrated that we do not know yet how to replicate the life-support services that the mix of populations in natural ecosystems provides for free.

THE EXTENT OF POPULATION DIVERSITY

Given the numerous reasons to be concerned about the fate of population diversity, we recently attempted to quantify the extent of that diversity and the rate of its loss. In this section and the next, we give an overview of these calculations (for further details, see Hughes and others 1997). Again, we define population diversity as the number of populations on the planet; another aspect of population diversity is the degree of divergence among populations, but we do not consider that aspect here.

Many of the difficulties that plague attempts to estimate species diversity also hinder an estimation of population diversity. The debate over definitions of species has persisted for decades (for example, Coyne and others 1988; Dobzhansky 1935; Ehrlich 1961; Masters and Spencer 1989; Mayr 1940 and 1969), and defining a population is no simpler. Also, the small fraction of species cataloged so far (approximately 1.75 million species of 10 million or more [Hammond 1995]) represents a regionally and taxonomically biased view of the planet's biodiversity. These problems are inherent in estimates of species diversity and are inevitably present in estimates of population diversity as well. For instance, as with most estimates of species, our population estimate is restricted to eukaryotes, because information on the diversity of bacteria and viruses is almost nonexistent, although the diversity is probably enormous. Nonetheless, just as approximations of species diversity have been made despite these difficulties, enough information exists to allow us to make a preliminary evaluation of biodiversity at the level of the population.

Our method of estimating global population diversity involved three steps. First, we reviewed the literature on population differentiation for a broad range of taxa and estimated the average number of populations per unit area for a series of species. Then we calculated the average size of the range of a species with a sample of available species range maps. The product of the resulting two numbers yielded an approximation of the average number of populations per species. Finally, we multiplied that number by the total number of species to arrive at the number of populations on Earth.

We searched 15 journals published from 1980 to 1995 for genetic studies on population differentiation, reading more than 400 articles and finding 81 that provided appropriate data for our calculations. We were able to estimate the number of populations per unit area for 82 species. Most of the species were vertebrates (n = 35), followed by plants (n = 23), arthropods (n = 19), mollusks (n = 4), and one flatworm (platyhelminth).

To quantify the number of populations of a species per unit area, we determined whether the sampling locations described in the articles were in separate populations or were within a single population. If statistically significant differentiation between localities was reported in the paper, we considered all the localities to be separate populations. We then calculated the number of populations per unit area as the number of sampling locations divided by the extent of the entire sampling area. If the researchers did not find significant differentiation between the localities, we assumed that they had sampled from within one population and that the size of a population was the size of the sampling area. Many studies found an intermediate amount of differentiation. For instance, in some studies, a significant difference was found only between two clusters of sites. In these cases, we assumed that there were two populations within the sampling area. This procedure yielded a conservative estimate of one population per 10,000 km² for an average species.

What are some problems with this evaluation of populations per unit area? First is the taxonomic bias mentioned above. Arthropods make up about 65% of the planet's species, and birds account for probably less than 0.01% (Hammond 1995). In our data on population structure, however, arthropods accounted for only 20% of the species, whereas birds accounted for more than 11%. Second, the evaluation of population differentiation for an average species is limited by the sampling intensity of each study. In other words, the estimate is probably conservative, since in many cases additional sampling in the study area may have revealed further differentiation. Finally, the molecular markers chosen may not always reveal notable differences between groups (for example, Legge and others 1996), again making the estimate on the conservative side.

To estimate the average range of a species, we digitized more than 2,400 species range maps from guidebooks for birds, mammals, fishes, and butterflies from a number of geographical regions. Equally weighting the four taxonomic groups, the mean size of the range of a species is 2.6 million square kilometers. Averaging the range size estimates of the largest group, the arthropods (here just butterflies), led to a range of 2.2 million square kilometers per species. These numbers are quite similar, so we conservatively used the lower number, 2.2 million square kilometers, as our estimate of the average size of the range of a species.

This evaluation of the average size of the range of a species is the most probable source of inflation in our estimate of population diversity. The shaded areas on distribution maps virtually always encompass unsuitable habitats, where populations do not occur (Gaston 1994). Also, the majority of sources we used were limited to temperate regions, even though it is estimated that two-thirds of species diversity exists in the tropics (Raven 1983). This misrepresentation also may inflate the population estimates because, in some taxa, the sizes of species ranges tend to increase toward the poles (Pagel and others 1991; Rapoport 1982).

One aspect of our method may compensate somewhat for these biases, however. The sources we used restricted their species range maps to one continent, so the full range of intercontinental species was not taken into account. Therefore, we may have underestimated considerably the size of the range of some species, such as birds that have Holarctic ranges.

The product of the estimates of the average populations of a species per unit area and the average size of the range of a species was an average of 220 populations per species. Using three published calculations of global numbers of species (5, 14, and 30 million from, respectively, Hammond 1995, Raven 1985, and Erwin 1982), we arrived at three estimates of the total number of populations: 1.1, 3.1, and 6.6 billion populations.

POPULATION EXTINCTION

In presenting the methods of our estimation of the current rate of population extinction, it is useful to begin with a summary of how species extinction rates usually are assessed. Estimates are derived largely from species-area relationships and from the rate habitat loss due to deforestation (Lawton and May 1995; Wilson 1992). The most commonly used species-area model is $S = cA_z$, in which S is the number of species, c and z are constants estimated from empirical studies, and A is the area where the species are found (Rosenzweig 1995; see also Pimm and Brooks this volume). This relationship between area (size of the habitat) and number of species is illustrated in figure 1. The graph reveals a convenient rule of thumb: a 90% decrease in area of habitat should result in roughly a 50% decrease in species diversity.

By applying estimates of rates of tropical deforestation to this model, one can approximate the rate of species extinction in tropical forests. With a very conservative estimate of tropical deforestation of 0.8% per year, the rate of extinction of tropical forest species is predicted to lie between 0.1% and 0.3% each year, depending on the value of z used in the species-area model. If we assume that 14 million species exist globally and that two-thirds of all species exist in tropical forests, species diversity in tropical forests is declining by roughly 9,000–26,000 species per year, or 1–3 species per hour (this last calculation was reported incorrectly in Hughes and others 1997).

No comparable work relates numbers of populations to area of habitat. Although a wide range of relationships could be justified, depending on the spatial and time scales considered, in the absence of information we used the simplest and most intuitive, namely, that changes in population numbers and area correspond in a roughly one-to-one fashion in ecological time. That is, when 90% of an area is destroyed, about 90% of the populations in the original area are exterminated (figure 1). The basis of the difference between the population-area relationship and the species-area relationship is the size of the unit. When a population is destroyed, other populations of the species still may exist elsewhere. Thus, initially the population-loss curve in figure 1 is steeper than the species-loss curve. Eventually, however, when the last populations are destroyed, all the species become extinct as well, and the curves converge.

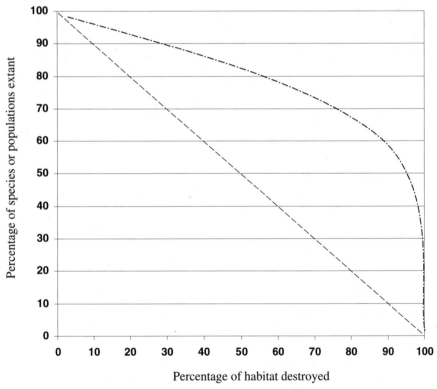

FIGURE 1   Predicted species-area and populations-area relationships. The species curve $(- \bullet -)$ is $S = cA_z$, where S is the number of species, A is the area size, and c and z are constants. Here, $z = 0.30$, so a 90% decrease in area corresponds to a 50% decrease in species diversity. In contrast, the population curve is linear, so a 90% decrease in area corresponds to a 90% decrease in population diversity.

If, indeed, a one-to-one population-area relationship exists, the rate of population extinction in tropical forests is estimated at 0.8% per year, directly proportional to the rate of habitat loss. Using our mid-range estimate of global population diversity (3.1 billion populations) and assuming that two-thirds of all populations exist in tropical forests (simply because species are distributed in this way), we estimate that 16 million populations per year, or roughly 1,800 per hour, are being exterminated in tropical forests alone. This is an absolute rate of 3 orders of magnitude higher and a percentage rate 3–8 times higher than conservative estimates of species extinction.

BIODIVERSITY AT DIFFERENT LEVELS

An investigation of population diversity does not complete the picture of biodiversity. Much remains to be explored at other levels, such as genetic, individual, and ecosystem levels, all of which are tightly interrelated. Little is known about how these different levels of biodiversity relate to ecosystem functioning. For example, for any given population, the number of individuals, the genetic variation between individuals, and the area occupied may affect the delivery of ecosystem services and other benefits provided by that population. The number of blue spruce trees may be important for global services, whereas the density of the trees may be critical for regional flood control. Similarly, although cougars exist in the San Francisco Bay area, the number of individuals is so low that numbers of local deer remain largely unchecked by these natural predators.

The effect of humans on natural areas is so extensive that every level of organization of biodiversity is threatened, even ecosystem diversity. In North America, for instance, the World Wildlife Fund estimates that 32 of a total of 116 ecoregions (that is, ecosystem types) in North America are critically threatened (Ricketts and others 1999). The consequences of the extinction of entire ecosystem types are not known, but the effect could be far-reaching if the particular assemblages of species are important for the delivery of some ecosystem services. In other words, the destruction of ecosystem types not only may result in the loss of the populations and species contained within them, but also may result in the loss of unique processes that are generated by certain combinations of species.

CONCLUSIONS

The crisis of biodiversity is more severe than species extinction rates alone would suggest: Population extinction is occurring at a rate that is 3 orders of magnitude higher than the rate of species extinction. The rapid loss of population diversity means the loss of the benefits described above and, in particular, the loss of the life-support systems on which humanity relies. Thus, the destruction and degradation of habitat and the decline of populations are of great concern even when they do not endanger species globally.

This conclusion has direct implications for both conservation biologists and policy-makers. Biologists must emphasize to the public and policy-makers the importance to humanity of all levels of biodiversity, instead of simply species diversity. This shift will require that biologists stress the functional benefits of biodiversity rather than relying only on the charismatic appeal of individual species. The most important message for policy-makers is one that ecologists long have recognized: Preservation of habitat is crucial for the preservation of biodiversity and the life-support systems that maintain human civilization. The current legislative focus on species conservation neglects crucial dimensions of biodiversity. To protect the benefits that humanity derives from biodiversity, an Endangered Biodiversity Act would be more appropriate than an Endangered Species Act.

Policy-makers also should be putting major effort into developing the field of restoration ecology (Ehrlich and Daily 1993). Unlike species, populations often can be re-established in a relatively short time, and they sometimes evolve with substantial genetic differences from the source populations (Johnston and Selander 1971). Thus, it may be possible to alleviate some of the effects of population extinction, but funds are needed to encourage this line of research.

Finally, the development of an economic-accounting system that internalizes the values of ecosystem goods and services (Costanza and Folke 1997; Goulder and Kennedy 1997) seems critical for the implementation of these policies.

ACKNOWLEDGMENTS

We thank Carol Boggs, Anne Ehrlich, Jessica Hellman, Claire Kremen, John-O Niles, and Taylor Ricketts for many constructive comments. This work has been supported by Peter and Helen Bing, the Pew Charitable Trusts, the Winslow Foundation, and the late LuEsther Mertz.

REFERENCES AND RECOMMENDED READING

Alexander SE, Schneider SH, Lagerquist K. 1997. The interaction of climate and life. In: Daily GC (ed). Nature's services: societal dependence on natural ecosystems. Washington DC: Island Pr. p 71–92.

Barducci TB. 1972. Ecological consequences of pesticides used for the control of cotton insects in Cañete Valley, Peru. In: Farvar MT, Milton JP (eds). The careless technology: ecology and international development. Garden City NY: Natural History Pr. p 423–38.

Brown L, Ehrlich PR. 1980. Population biology of the checkerspot butterfly Euphydryas chalcedona: structure of the Jasper Ridge colony. Oecologia 47:239–51.

Brown JH, Kodric-Brown A. 1977. Turnover rates in insular biogeography: effect of immigration on extinction. Ecology 58:445–9.

Buchmann SL, Nabhan GP. 1996. The forgotten pollinators. Washington DC: Island Pr.

Chapin III FS, Walker BH, Hobbs RJ, Hooper DU, Lawton JH, Sala OE, Tilman D. 1977. Biotic control over the functioning of ecosystems. Science 277:500–4.

Chichilnisky G, Heal G. 1998. Economic returns from the biosphere. Nature 391:629–630.

Cohen JE, Tilman D. 1966. Biosphere 2 and biodiversity: the lessons so far. Science 274:1150–1.

Costanza R, Folke C. 1997. Valuing ecosystem services with efficiency, fairness, and sustainability as goals. In: Daily GC (ed). Nature's services: societal dependence on natural ecosystems. Washington DC: Island Pr. p 49–68.

Coyne JA, Orr HA, Futuyma DJ. 1988. Do we need a new species concept? Syst Zool 37:190–200.

Daily, GC. 2000. Countryside biogeography and the provision of ecosystem services. In: Raven PH, Williams T (eds). Nature and human society: the quest for a sustainable world. Washington DC: National Academy Press. p 104–13.

Daily GC (ed). 1997. Nature's services: societal dependence on natural ecosystems. Washington DC: Island Pr. 392 p.

Daily GC, Ehrlich PR. 1990. An exploratory model of the impact of rapid climate change on the world food situation. Proc Royal Soc London 241:232–44.

Dobzhansky T. 1935. A critique of the species concept in biology. Philos Sci 2:344–355.

Dolinger PM, Ehrlich PR, Fitch WL, Breedlove DE. 1973. Alkaloid and predation patterns in Colorado lupine populations. Oecologia 13:191–204.

Dowling HF. 1977. Fighting infection. Cambridge MA: Harvard Univ Pr. 339 p.

Ehrlich PR, Daily GC. 1993. Population extinction and saving biodiversity. Ambio 22:64–8.

Ehrlich PR, Ehrlich AH. 1981. Extinction. New York NY: Ballantine Books.

Ehrlich PR, Mooney HA. 1983. Extinction, substitution, and ecosystem services. BioScience 33:248–54.

Ehrlich PR, Wilson EO. 1991. Biodiversity studies and science policy. Science 253:758–62.

Ehrlich PR. 1961. Has the biological species concept outlived its usefulness? Syst Zool 10:167–76.

Ehrlich PR. 1986. The Machinery of Nature. New York NY: Simon and Schuster.

Erwin TL. 1982. Tropical forests: their richness in Coleoptera and other arthropod species. Coleopt Bull 36:74–82.

Ewel KC. 1997. Water quality improvement by wetlands. In: Daily GC (ed). Nature's services: societal dependence on natural ecosystems. Washington DC: Island Pr. p 329–344.

Frank DA, McNaughton SJ. 1991. Stability increases with diversity in plant communities: empirical evidence from the 1988 Yellowstone drought. Oikos 62:660–662.

Gaston KJ. 1994. Rarity. London UK: Chapman and Hall.

Global biodiversity assessment. 1995. Cambridge UK: Cambridge Univ Pr. p 113–38.

Goméz-Pompa A, Vásquez-Yanes C, Guevara S. 1972. The tropical rainforest: a nonrenewable resource. Science 177:762–5.

Goulder LH, Kennedy D. 1997. Valuing ecosystem services: philosophical bases and empirical methods. In: Daily GC (ed). Nature's services: societal dependence on natural ecosystems. Washington DC: Island Pr. p 23–47.

Hammond PM. 1995. The current magnitude of biodiversity. In: Heywood VH (ed). Global biodiversity assessment. Cambridge UK: Cambridge Univ Pr. p 113–38.

Hughes JB, Daily GC, Ehrlich PR. 1997. Population diversity: its extent and extinction. Science 278: 689–692.

Hwang S-Y, Lindroth RL. 1997. Clonal variation in foliar chemistry of aspen: effects on gypsy moths and forest tent caterpillars. Oecologia 111:99–108.

Johnston RF, Selander RK. 1971. Evolution in the house sparrow. 11. Adaptive differentiation in North American populations. Evolution 25:1–28.

Kareiva PM, Kingsolver JG, Huey RB (eds). 1993. Biotic interactions and global change. Sunderland MA: Sinauer. 559 p.

Kaufman L, Dayton P. 1997. Impacts of marine resource extraction on ecosystem services. In: Daily GC (ed). Nature's services: societal dependence on natural ecosystems. Washington DC: Island Pr. p 275–93.

Lande R. 1988. Genetics and demography in biological conservation. Science 241:1455–60.

Lawton JH, May RM (eds). 1995. Extinction rates. Oxford UK: Oxford Univ Pr. 233 p.

Legge JT, Roush R, Desalle R, Vogler AP, May B. 1996. Genetic criteria for establishing evolutionary significant units in Cryan's buckmoth. Cons Biol 10:85–98.

Masters JC, Spencer HG. 1989. Why we need a new genetic species concept. Syst Zool 30:270–9.

Mayr E. 1940. Systematics and the origin of species. New York NY: Columbia Univ Pr.

Mayr E. 1969. The biological meaning of species. Biol J Linnean Soc 1:311–20.

McNaughton SJ. 1977. Diversity and stability of ecological communities: a comment on the role of empiricism in ecology. Amer Natur 111: 515–25.

Myers N. 1997. Biodiversity's genetic library. In: Daily GC (ed). Nature's services: societal dependence on natural ecosystems. Washington DC: Island Pr. p 255–73.

Myers N. 1979. The sinking ark. Oxford UK: Pergamon Pr.

Nabhan GP, Buchmann SL. 1997. Services provided by pollinators. In: Daily GC (ed). Nature's services: societal dependence on natural ecosystems. Washington DC: Island Pr. p 133–50.

Naeem S, Thompson LJ, Lawler SP, Lawton JH, Woodfin RM. 1994. Declining biodiversity can alter the performance of ecosystems. Nature 368:734–7.

Pagel MD, May RM, Collie AR. 1991. Ecological aspects of the geographical distribution and diversity of mammalian species. Amer Natur 137:791–815.

Peterson CH, Lubchenco J. 1997. Marine ecosystem services. In: Daily GC (ed). Nature's services: societal dependence on natural ecosystems. Washington DC: Island Pr. p 177–94.

Plucknett DL, Smith NJH, Williams JT, Anishetty NM. 1987. Gene banks and the world's food. Princeton NJ: Princeton Univ Pr.

Pulliam HR. 1988. Sources, sinks, and population regulation. Amer Natur 132:652–61.

Rapoport EH. 1982. Aerography: geographical strategies of species. Oxford UK: Pergamon.

Raven PH. 1985. Disappearing species: a global tragedy. Futurist 19:8–14.

Raven PH. 1983. The challenge of tropical biology. Bull Ecol Soc Amer Spring: 4–12.

Ricketts T, Dinerstein E, Olson D, Loucks C, Eichbaum W, Kavanagh K, Hedao P, Hurley P, Carney K, Abell R, Walters S. 1999. Terrestrial ecoregions of North America. A conservation assessment. Washington DC: Island Pr. 485 p.

Rosenzweig ML. 1995. Species diversity in space and time. Cambridge UK: Cambridge Univ Pr. 436 p.

Safina C. 1995. The world's imperiled fish. Sci Amer 273:46–53.

Sinnott EW, Dunn LK, Dobzhansky T. 1950. Principles of genetics. New York NY: McGraw-Hill.

Tilman D, Downing JA. 1994. Biodiversity and stability in grasslands. Nature 367:363–5.

Wilson EO (ed). 1988. Biodiversity. Washington DC: National Acad Pr. 521 p.

Wilson EO. 1992. Diversity of Life. Cambridge MA: Belknap.

Woodwell GM, Hobbie JE, Houghton RA, Melillo JM, Moore B, Peterson BJ, Shaver GR. 1983. Global deforestation: contribution to atmospheric carbon dioxide. Science 222:1081–6.

KEEPING A FINGER ON THE PULSE
OF MARINE BIODIVERSITY:
HOW HEALTHY IS IT?

JERRY R. SCHUBEL
New England Aquarium, Central Wharf, Boston, MA 02110-3399
CHERYL ANN BUTMAN
Department of Applied Ocean Physics and Engineering, Woods Hole Oceanographic Institution,
Mailstop 12, Woods Hole, MA 02543

"If all the beasts were gone, man would die of great loneliness of spirit."

Chief Sealth of the Duwamish Tribe
in a letter to President Franklin Pierce, dated 1855

SETTING THE STAGE

THE WORLD OCEAN stretches from pole to pole, covers 71% of Earth, and represents more than 99% of the planet's total biosphere volume, or living space. The ocean is nature's ultimate womb. Most scientists believe that life originated there. It is composed of a rich mosaic of habitats large and small, from some of the seemingly most homogeneous and remote, such as the deep-sea floor, to some of the most heterogeneous and accessible, such as the vibrant coral reefs.

The world ocean is the greatest repository of biodiversity at the second highest level of taxonomic organization, the phylum. This level distinguishes organisms according to their basic body plans. Sponges and chordates (which include humans), for example, constitute separate phyla within the animal kingdom. The sequence of phyla roughly reflects the trend in evolution. Of Earth's 33 animal phyla, 28 are found in the ocean, and only 11 are found on land. Moreover, 13 of the animal phyla are endemic (native and restricted) to the marine environment, whereas only one animal phylum is endemic to land: the ancient slug-like Onychophora. At the lowest level of taxonomic organization, the species, biodiversity apparently is much higher on land than in the sea. Yet most of the world ocean is still unsampled, and many of the species collected are still unidentified.

As ocean exploration continues, large numbers of new entries into the library of marine biodiversity are expected at virtually every level of taxonomic organization, particularly at the species level.

The relentless discovery of multitudes of new species, from microorganisms to vertebrates, has driven a revolution in marine taxonomy—the identification of species. New techniques, including flow cytometry (described later) and molecular tools involving gene sequencing, now replace or augment traditional, morphology-based methods for classifying and identifying a wide variety of marine organisms. Some new taxonomic capabilities have found immediate applications in conservation. With molecular techniques, for example, some whale species now can be identified solely from the meat that is sold in the marketplace; this facilitates enforcement of restrictions on the hunting of threatened or endangered species.

In the sea, as on land, the greatest threats to biodiversity come from one species, *Homo sapiens*. The imprint of humans is found throughout the world ocean, but it is most evident and poses by far the greatest threat along its margins—in the coastal zone. Still, documentation of changes in biodiversity caused by human activities is limited by the difficulty in sampling most ocean habitats and identifying the organisms collected. That is, assessing the health of marine biodiversity is seriously hindered because, in most cases, we cannot even take its pulse.

There are several excellent books on the general topic of marine biodiversity, in addition to numerous scholarly articles in the burgeoning primary literature (see list at end). Nearly all were written by scientists for scientists or for scientifically literate readers. As a complement to that literature, this paper describes for the layperson a general picture of biodiversity in the world ocean, how humans are altering it, and the threats that loom on the horizon. We also suggest some elements critical to any integrated management plan for minimizing human threats to marine biodiversity. Thus, this paper celebrates the diversity of marine life at all levels, laments the threats to it, and summons humankind to rise to the challenge of its conservation.

GENERAL CHARACTERISTICS

Exploring the Sea

Most people encounter the world ocean only at its margins and experience only a few meters of its depth. Yet the continental shelf, the shallowest region of the sea floor, constitutes a mere 7% of the ocean's area. Of the whole sea floor (about 300 million square kilometers), 83% is more than 1,000 m below the surface and constitutes a zone known as the deep sea. Deep-sea environments are far more difficult to sample and characterize than even the most remote terrestrial habitats. In fact, it is only within the last 50 years that scientists have been able to observe directly, sample, and experiment in untethered "manned" submersibles, like the deep-sea submergence vehicle *Alvin* (figure 1). But there is a lot of catching-up to do; the deep sea is still the most undersampled marine environment.

FIGURE 1   Deep-sea submergence vehicle *Alvin* being launched from its 274-ft support ship, research vessel *Atlantis*. The three-person submersible can dive to almost 5,000 m, enabling it to reach 86% of the world ocean floor. *Alvin* typically makes 150–200 research dives each year. It was commissioned in 1964, is owned by the US Navy, and is operated by the Woods Hole Oceanographic Institution as a national oceanographic facility. (Photo credit: Rod Catanach.)

Most marine research is done relatively blind from surface research vessels, with nets used to sample the water column and grabs and corers to sample the sea floor. In areas of thousands to tens of thousands of square kilometers, typically less than a square kilometer of the sea has been sampled. That is equivalent to characterizing the entire fauna and flora in a backyard from a sample the size of the head of a pin! Moreover, planktonic (free-floating) and nektonic (free-swimming) marine organisms are constantly in motion, so temporal and spatial variations are easily confounded.

Because of the sampling issues, knowledge of different groups of marine organisms is strongly conditioned by their habitats, their mobility, and the scale of their distributions. In general, distributions of easily accessible, larger, and relatively sedentary organisms (for example, intertidal barnacles and mussels) are better documented than those of inaccessible, smaller, and highly mobile life forms (for example, zooplankton and deep-sea fishes). Microorganisms, which occur in virtually every marine habitat, are by far the most undersampled and undercharacterized. Whales are the largest animals in the world, yet because certain species (for example, blue and sperm whales) are so mobile and can dive deeper than 800 m, there are few accurate estimates of their population sizes, let alone knowledge of their dynamics. Until the 1960s, population sizes were estimated entirely from visual observations made from whaling vessels. Photographic identification studies and tag-recapture estimates now provide more accurate information on some coastal species (for example, humpback and right whales), but population sizes and movements of off-shore species remain largely a mystery.

The Aqueous Medium

Marine organisms are enveloped by water, whereas terrestrial organisms are enveloped by air, and the differences between these two fluids—water and air—account for many of the differences in life found in the marine and terrestrial environments. Water is 1,000 times denser than air, and the difference in density has a number of important consequences. First, water acts as a thermostat, buffering against rapid and large changes in temperature not only within the fluid, but for the entire planet. Second, water provides buoyancy that counteracts gravity, reduces the need for physical supporting structures, and facilitates vertical mobility of animals. Third, the much greater kinetic energy (that associated with motion) of water relative to air virtually precludes the existence in the sea of large, rigid, stationary organisms on the scale of trees. And fourth, the much greater dissolving power of water (than of the atmosphere) provides a relatively rich nutritional soup that enables marine plants to receive all their nourishment directly from the enveloping fluid; in contrast, most terrestrial plants require both the atmosphere and the soil to obtain water and nutrients. With respect to optical clarity for photosynthesis, however, the air wins hands down.

The Diversity of Habitats

On land, there are millions of distinct and fixed habitats spanning a large range in size. Some terrestrial habitats are as small as or even smaller than a single tree in a rain forest that supports numerous highly endemic species, and some are as large as the Serengeti plains, stretching for hundreds of kilometers. In the sea, particularly in the water column of the open ocean and over vast expanses of the deep-sea floor, the number of distinctly different habitats is comparatively small. The spatial extent of these habitats is large, however, and marine habitats are intimately connected via the motion and mixing of the fluid medium; thus, endemism is much rarer in the sea than on land. What is surprising is the degree of biological heterogeneity—the biodiversity—that abounds in this seemingly homogeneous seawater medium.

In the sea, organisms have evolved in response to variables other than physical space—variables that might have no terrestrial analogues. In the open ocean, water circulation patterns can create discrete habitats. In the cold coastal waters of the western North Atlantic, for example, are "warm-core rings" that have pinched off of the swift, northward-flowing Gulf Stream (figure 2). Likewise, semienclosed pockets of cold coastal water—cold-core rings—can spin off into the Gulf Stream. Because different species assemblages are associated with different water masses, regional biodiversity is enhanced by these water-mass intrusions.

In the immense sedimentary plains of the deep-sea floor, there is an extraordinary diversity of animals, perhaps rivaling the biodiversity of tropical rain forests. Organisms living on or in the relatively thin layer of sea-floor sediments—the

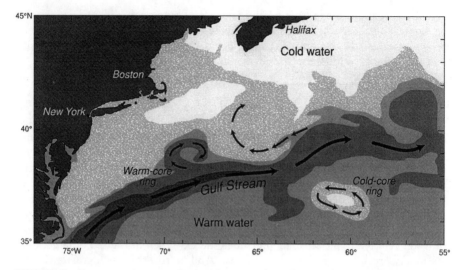

FIGURE 2   Summertime sea-surface temperature from Cape Hatteras to Nova Scotia in the western North Atlantic. The drawing was made from a satellite image taken on June 13, 1997. Arrows indicate the direction of water flow (currents) as determined from drifter studies. (Original satellite image used courtesy of the Ocean Remote Sensing Group of the Johns Hopkins University Applied Physics Laboratory; satellite image mosaic overlaid with drifter data was constructed by Dick Limeburner.)

benthos—have evolved largely in response to the highly limited and unpredictable food supply in the deep ocean and thus have remarkable adaptations for exploiting ephemeral and patchily distributed organic matter. In shallow water, benthic organisms experience more habitat variability over small spatial scales than organisms in overlying waters (figure 3). There is also less direct connectivity among habitats in the benthic zone than in the pelagic zone (open water). Thus, most benthic organisms have planktonic larvae (meroplankton) that can expand species distributions over larger areas, provide some insurance against local catastrophic events, and recolonize areas where populations have been eliminated by human activities. The greater habitat diversity and lower connectivity in the benthic zone results in species diversity much greater than that in the water column. Moreover, within the world ocean, the largest number of phyla are represented in the benthos.

In contrast with the seemingly featureless sedimentary sea floor, coral reefs scream and shout with habitat complexity. Coral reefs are, in fact, analogous to rain forests in that the most conspicuous habitat in the reefs is provided by living organisms—the corals themselves. Corals create the underlying structure for the reefs and provide attachment sites for many invertebrates and protection for numerous fishes. Moreover, because corals contain their primary producers—tiny algae called zooanthellae that live in the coral tissue—they have a built-in food source. Coral reefs are believed by many marine-biodiversity experts to be the repositories of the greatest biodiversity in the world ocean if one scales them for size, that is, species per unit area. The other contender for this distinction is the deep-sea floor.

Diversity and Ecosystem Function

Marine ecosystems are knitted together by relationships among organisms, particularly by who eats whom. In the water column, the food web involves encounters between freely moving predators and prey. Chance, random processes, and adaptation to survive for long periods without food are important driving forces in many more marine than land ecosystems.

Ecosystems have functional attributes—such as the capacity to capture, store, and transfer energy and nutrients—and they also contribute to societal needs. Estuarine ecosystems, for example, tend to have relatively low biodiversity but high productivity, and they contain many commercially important fish and shellfish species. In contrast, coral-reef ecosystems have high biodiversity but low productivity and are now exploited largely for ecotourism.

Some species are more important to the functioning of an ecosystem than others. If they are removed, their roles might be lost, leaving a hole in the food web,

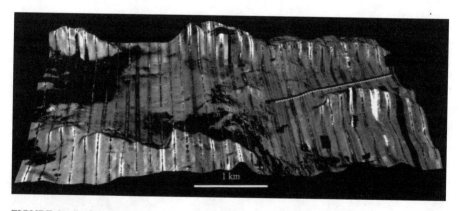

FIGURE 3   Reflectivity of bottom sediments in Massachusetts Bay, just off shore of Boston (see figure 2), superimposed on the bathymetry. Measurements were made with a remote-sensing technique—sidescan sonar—where the travel time of sound between the ship and the sea floor yields information on the texture of the sediments. Areas of boulders, represented by the lightest tone, were typically found at the crests of the small, submerged hills. The intermediate gray tone is sand containing various amounts of gravel. The darkest areas are fine-grained (muddy) sediments, often in the depressions between hills. Note that the mottled region in the center of the mosaic shows 10- to 100-m-scale patches of sands and muds. Different communities of organisms occur in different sediment types, and such intricate heterogeneity in sediment texture results in enhanced local biodiversity. The strips represent the ship track and are about 150 m apart. The dotted line is the location of a new wastewater outfall diffuser, which is being installed deep within the sea floor. The outfall pipe starts at the shore and extends northeast. (Sidescan mosaic courtesy of Brad Butman of the US Geological Survey.)

such as a gap in energy transfer, in nutrient cycling, or in some other crucial function. Most critical are the "keystone" species. If one of these is removed—by overexploitation or by a natural or human-assisted disaster—the ecosystem changes dramatically. Sea otters are a keystone species of eastern Pacific kelp communities. Sea otters eat sea urchins, which in turn eat kelp. When sea otters were hunted to near extinction along the US West Coast, urchin populations exploded and devoured the kelps, thus turning magnificent, highly diverse kelp forests into featureless sandflats known as "urchin barrens".

Few marine ecological studies have identified keystone species and other critical relationships between species composition and ecosystem function. Without such knowledge, human activities that involve broad-scale removal of species or alteration of habitat can easily and inadvertently tip the delicate ecological balance, sometimes with disastrous consequences. Overfishing has already greatly diminished most of the large predators of the open ocean; the repercussions have reverberated throughout the food web (figure 4). Moreover, industrial-scale fishing has recently begun to focus on the slow-growing fishes of the largely unexplored deep sea, with unknown consequences. Researchers do not know which or how many of the commercially hunted fishes, shellfishes, or kelps are keystone

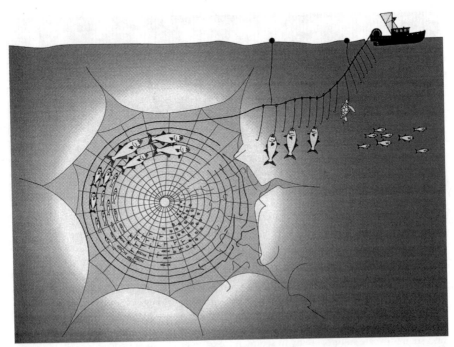

FIGURE 4   Human activities, such as longline fishing for top predators, can potentially affect the prey, the prey of the prey, and so on, contributing to the deterioration of the food web. Longlining also results in by-kill, the incidental deaths of nontargeted species, such as marine turtles and sea birds.

species, which or how many have functional equivalents, and whether and how their losses might unravel an ecosystem's intricate web.

THE MONTAGE OF MARINE BIODIVERSITY

In this section, we reveal some of the major components of the montage of marine biodiversity and describe some characteristic properties and processes that contribute to the maintenance of the marine fabric of nature. Examples are chosen to be illustrative; it is impossible to be comprehensive. We close with some observations about the general patterns of marine biodiversity around the world.

Primary Producers

There are many similarities between terrestrial and marine environments. Both depend ultimately on photosynthesis for nearly all their energy. Thus, it always starts with plants. Photosynthetic organisms in both environments produce organic material used as food by herbivorous animals, which are preyed on by carnivorous animals to form complex food webs. A primary difference between the terrestrial and marine environments is that in the ocean, most of the plants are microscopic cells floating in the water. These cells are known as phytoplankton.

The Greek root of the word plankton is *planktos*, which means "wanderer". And phytoplankton roam the high seas. But for these plant cells to photosynthesize, they must remain in the upper sunlit layer—the euphotic zone. This layer varies in thickness depending on geographic location, but it rarely exceeds 200 m in the open ocean and can be 10 m or less in coastal areas. Staying within the euphotic zone can be challenging for phytoplankton. Gravity pulls them down, and other physical processes, such as convergences (downward-directed currents), transport them away from the light. Phytoplankton have remarkable adaptations that counteract those forces, such as bubbles of fatty material and elaborate spines that reduce sinking.

Collecting and identifying phytoplankton is demanding, to say the least. The cells are so small (0.1–100 μm) and fragile that they are damaged or destroyed by conventional sampling with nets and filters; this makes identification under a microscope laborious or impossible. In contrast, flow cytometry—a new nonintrusive technique for counting and identifying phytoplankton—uses a laser to discern the fluorescence characteristics, size, and shape of each cell. In the late 1980s, the shipboard use of this technique led to the discovery of a new group of marine photosynthetic bacteria, the prochlorophytes. These tiny (0.2–2.0 μm) organisms account for up to 40% of all chlorophyll (the major photosynthetic pigment) in some regions of the ocean.

At least one-third of the annual global carbon fixation occurs in the sea. A substantial portion of this carbon is fixed by cells less than 1 μm in size because they are so numerous. Some estimates suggest that the annual fixation of carbon by phytoplankton smaller than 5 μm is similar to that by the world's rain forests. In addition to the importance of phytoplankton to carbon fixation, they can affect marine chemistry by taking up and releasing nutrients. Furthermore, because different phytoplankton species are consumed by different animals, conserving

phytoplankton diversity is critical to conserving the functioning of marine ecosystems.

Whereas phytoplankton account for more than 95% of oceanic primary productivity, the larger multicellular plants—the macroalgae—seagrasses, and marsh
grasses—might be the major producers in some near-shore regions. The kelp forests of temperate, rocky shores host a diversity of animal life, from shellfish to sea
otters. Floating mats of brown algae harbor unique animal communities in the
open-ocean region of the Sargasso Sea. And seagrasses and marsh plants, occupying critical coastal habitat, stabilize sediments and support distinctive populations of small fishes and invertebrates.

Because of light limitations, all marine plants attached to the bottom are restricted to shallow coastal areas. These ocean habitats are the most directly susceptible to human effects. Because they are close to the shore, attached plants
are especially vulnerable to human activities in the sea, such as dredging and sewage disposal, and on land, such as agriculture and urbanization. Entry of suspended sediments and nutrients into the sea from these activities often greatly
reduces water clarity (limiting photosynthesis) and degrades habitat (for example,
for attachment).

The biology of the sea is driven principally by primary productivity that occurs
within a region that accounts for less than 1% of the total volume of the world
ocean. Most life in the remaining 99% of the ocean volume depends on food
produced within the thin upper layer—food that is either preyed on directly at
the surface or scavenged at great depth. Alternative pathways to energy production include sulfur-reducing bacteria that obtain their energy from chemical
sources (as opposed to the sun) through a process called chemosynthesis. These
primary producers are largely symbiotic, living in tissues of other organisms. Although chemosynthesis produces a small fraction of the sea's total primary productivity, it contributes to marine biodiversity by extending the living space of and
facilitating a rich variety of microhabitats.

Primary Consumers

Most plant food is much smaller in the sea than on land, and so too are most
marine primary consumers. The grazers—the zooplankton ("wandering animals")—are small animals that spend either all (holoplankton) or a portion (meroplankton) of their lives in the plankton. Meroplankton are the larval stages of
such animals as clams, snails, worms, crabs, and flatfish that, as adults, live on or
in the bottom. Meroplankton generally bear no resemblance to the adult form
and, like the other plankton, are highly adapted to a suspended existence. Holo-
planktonic species dominate the zooplankton in numbers and biomass. They exhibit an extraordinary diversity of form and function, ranging in size from the tiny
copepods (hundreds to thousands of micrometers) to the larger jellyfish (millime-
ters to a few meters).

Nearly all major groups of animals on Earth, except insects, are represented in
the zooplankton. Whereas insects, as a group, contain three-fourths of all known
animal species on the planet and one-half of all known animal and plant species
combined, copepods, the dominant animal group within the zooplankton, have

more individuals than any other group of animals on Earth, except perhaps round-worms. There are an estimated 1 quintillion (10^{18}) copepods in the sea—1,000–1,000,000 individuals under each square meter of sea surface!

Although phytoplankton are largely restricted to the upper sunlit layer, zoop-lankton can extend their realm vertically by migrating. This behavior enhances their living space and provides them wider access to food and more opportunity to escape from predators. Zooplankton are intermediaries in the food web. They eat the primary producers—phytoplankton—and are eaten by the secondary con-sumers—finfish, shellfish, and some whales. Most fish and marine mammals can-not eat phytoplankton directly, but need the zooplankton to repackage them into larger and more nutritious food (like energy bars). Remarkably, however, tiny zooplankton are the major food source for baleen whales, the largest animals on Earth. The mouths of these huge whales contain large comb-like filters called a baleen (figure 5). This sievelike structure retains zooplankton-sized organisms that are slurped off by the whale's tongue. There is no terrestrial analogue for this highly size-disparate food chain, which is more extreme than if an elephant fed on a diet of ants!

There are also benthic grazers—a wide variety of suspension-feeding clams, worms, and other invertebrates that extend feeding structures into the water col-

FIGURE 5   Baleen whale, with typical zooplankton prey shown in the insert. The great-est size disparity between a predator (whale) and prey (zooplankton) occurs in the marine environment. (Photo credit: Paul Erickson.)

umn, where they actively or passively collect plankton. Because these animals process such large amounts of water and are essentially glued in place on the bottom, they are particularly susceptible to human activities that result in the deposition of particles or pollutants on the sea floor.

Secondary Consumers

Although predatory fish—the predominant secondary consumers—swim the entire world ocean, they eat and concentrate in particular areas during important stages of their life cycles, such as reproduction and early development as larvae and juveniles. Thus, for example, salmon return to spawn in the rivers of their births, and eels congregate for mass spawnings in the Sargasso Sea. Protected near-shore areas, such as estuaries and mangrove swamps, are important nurseries for a wide variety of larval and juvenile fishes. Fish can feed at several levels of the food chain, the size of the fish being a relatively good predictor of the size of the prey. Thus, small fish species and larval and juvenile fish eat zooplankton, and bigger fish eat smaller fish.

The coastal ocean has afforded great opportunity for diversification in fish species, providing a wide range of habitats, prey types, and nursery areas. The number of species (13,000) of coastal marine fishes could be more than 10 times greater than the number of species (1,200) of true oceanic fishes—those which spend their entire life cycles in the open ocean. Fish move much more than the food that they prey on. Fish mobility results in cosmopolitan species distributions and blurs the boundaries of biodiversity patterns at any given time.

Marine Mammals and Turtles

Although marine mammals are a relatively small group—only some 119 species—their endearing characteristics have generated a great deal of conservation attention. The ancestors of marine mammals left the land and began their return to the ocean as early as 50 million years ago, evolving a range of diversity greater than what land mammals had left behind.

Whales, manatees, and dugongs are fully adapted to their aquatic habitats, but marine carnivores—sea otters, polar bears, seals, sea lions, and walrus—divide their lives between land and sea. Although the diversity of many animal groups, both on land and in the sea, increases in habitats nearer the equator (for example, table 1), seals, sea lions, and walrus show the reverse pattern. Only the rarest species, the monk seal, occurs and breeds in the tropics. Toothed whales are the most diverse group of marine mammals. On the basis of size alone, it might be thought that all marine mammals would have been collected, identified, and described centuries ago. Yet, for the cryptic and difficult-to-study beaked whales, seven new species were described in this century, the most recent one in 1991!

Marine mammals are characterized by their radical anatomical, physiological, and behavioral adaptations to life in the ocean. Manatees and dugongs are marine herbivores; they depend on rich growths of aquatic plants. Other marine mammals are consummate divers, with thick layers of blubber, fur, streamlined bodies, fins, flippers, and other modifications that allow them to remain below the surface for long periods and at extreme depths. Dives of sperm whales and el-

TABLE 1   Some General Patterns of Marine Biodiversity

1. The midwaters of the open ocean appear to be the least heterogeneous environment on the planet and might have the lowest biodiversity.

2. The greatest biodiversity in the ocean occurs either in coral reefs or on the deep-sea floor. (Most scientists give the nod to coral reefs.)

3. Some exotic, low-stress environments, such as hydrothermal vents and deep-sea trenches, are characterized by relatively low species diversity.

4. In the waters of the open ocean, biodiversity increases from the North Pole to the tropics, but evidence of such a gradient in the Southern Hemisphere is far less convincing. The Arctic is younger and has less diversity and lower endemism than the Antarctic.

5. Biodiversity is higher near coasts than in the open ocean because of the greater hetereogeneity of coastal habitats.

6. The highest levels of biodiversity in most marine systems are found under conditions of very low, but not the lowest, primary productivity.

7. The strongest gradients (zones of most rapid change) of species diversity occur along gradients of primary productivity.

8. Losses of marine diversity are much higher near the coast than in the open sea because of the pervasiveness of anthropogenic effects in coastal habitats.

9. Biodiversity is higher in benthic than in pelagic systems. The small-scale heterogeneity of the deep-sea floor was grossly underestimated until the last 2 decades.

10. Most phyla in the world ocean occur in the archetypal habitat—the bottom sediments where life on Earth is believed to have originated.

ephant seals, for example, to depths of more than 1,000 m for over an hour have been recorded.

The seven species of sea turtles also spend little time at the surface and cover huge distances on their migratory routes. The largest sea turtle, the leatherback, is warm-bodied, enabling it to live in waters from Venezuela to Newfoundland. Coming ashore to breed on tropical beaches every 2–3 years, these turtles can weigh up to 1,500 lb (680 kg) and dive to more than 600 m.

Some Patterns of Marine Biodiversity and Their Causes

Ptolemy once observed that it is the role of the scientist "to tell the most plausible story that saves the facts." This charge is difficult when the facts are few and several stories could "save" them equally well. That is often the situation in attempts to generalize about patterns in marine biodiversity—for example, geographically, with depth, or across taxa—from existing data. Keeping in mind that marine biodiversity is grossly undersampled and underdescribed, in table 1 we list some of the general patterns in diversity that are fairly clear.

THE ORIGIN OF DIVERSITY

In most of this paper, we write about biodiversity as a function of natural communities. Another, fundamental way of looking at biodiversity is as the outcome of evolutionary processes: the creation and loss of species. Today's biodiversity is only a small sample of the creatures that have come and gone over evolutionary time (Jeffries 1997). The fossil record contains the traces of plants and animals far different from those now occupying ocean habitats.

New species arise from changes in the genetic makeup of subpopulations. In sexually reproducing organisms, one species can become two when subpopulations of the original species become reproductively isolated and thus do not freely exchange genes. If their genetic makeups then diverge so much that the two populations can no longer interbreed, they become separate species. Geography—continents, islands, submerged mountain ranges, deep-sea canyons, seamounts—can isolate populations. Reproductive isolation can also lead to speciation when the breeding seasons or mating behaviors of two subpopulations become sufficiently different.

Physical structure and changes thereof provide opportunities for isolation of populations. Gyres (alterations in oceanic circulation) and land bridges can impose barriers to interbreeding. For example, the Isthmus of Panama separated the Atlantic and Pacific Oceans, isolating many populations that formerly mixed. On the two sides of the isthmus are many common species, but some populations have diverged enough to become separate species.

The Gulf of California formed only 6 million years ago. It is not part of the circulation pattern of the eastern Pacific, so not only do populations of such organisms as sardines, which occur in both water masses, fluctuate independently of one another, but the gulf has endemic species closely related to similar species in the Pacific. For example, the vaquita, an endangered marine mammal closely related to the harbor porpoises of the Pacific and Atlantic, is endemic to the Gulf.

Small marine populations, especially in the nekton, might be less likely to become isolated than terrestrial or freshwater populations of similar size, because of the more "open" nature of marine systems. For example, freshwater covers only 1% of Earth's surface but accounts for 40% of the 23,000 species of fish.

ANTHROPOGENIC THREATS AND EFFECTS

Anthropogenic threats to the biodiversity of the world ocean are in five major categories: overexploitation of resources, pollution, habitat alteration, introduction of exotic species, and global climate change. The first four categories include threats that are both historical and current. Threats to marine life from global climate change are imminent. Marine biodiversity can be affected by a single threat or several threats, sometimes with devastating and unknown consequences (figure 6). The vast oyster reefs of the Chesapeake Bay, for example, that once filtered the estuary's entire volume every week, now filter it only once a year because of stock depletion due to overfishing, pollution, habitat alteration, and disease.

If we were to give the world ocean the equivalent of a physical examination to determine its fitness, a strong, rhythmic heartbeat would represent the healthy system that was characteristic of prehuman times—and indeed typical of most areas until several thousand years ago. Human activities, however, have caused

FIGURE 6   The potential effect of a very remote human activity—whaling—on deep-sea biodiversity. The flesh of falling whale carcasses provides food for deep-sea organisms, and the skeletons support a chemoautotrophic-based food chain similar to that in hydrothermal vent and seep areas (indicated as minivolcanoes billowing puffs of smoke). Thus, whale skeletons might be critical stepping stones for hydrothermal-vent communities (depicted on the left). Whale carcasses were released from whaling vessels in densities and geographic patterns that were probably distinctly different from natural whale falls. In fact, after the early 1900s, the supply of human-killed whales to the sea floor decreased dramatically because almost the entire whale, including the skeleton, was used for various products (depicted on the right). Before and after the 1900s, the number of whales in the ocean was reduced dramatically; this potentially diminished (and in some cases stopped) the supply of whale-carcass stepping stones for hydrothermal vent and other chemoautotrophic-based communities in the deep sea. (From Butman and Carlton [1995]; used with permission from the American Geophysical Union.)

a rapid deterioration in the health of many marine ecosystems. Understanding the history of effects on marine biodiversity from the four current anthropogenic threats is important for developing strategies to minimize future effects. Maintaining healthy marine ecosystems requires constant vigilance; keeping a finger on the pulse of marine biodiversity could save its life and the critically important ecosystem services that it provides.

Overexploitation of Resources

Overfishing has dramatically reduced the stocks of many, perhaps most, of the preferred edible fish and shellfish species in the world ocean and led, for example, to recent closures of the so-called inexhaustible great fishing banks, such as Georges Bank and the Grand Banks. Entire marine ecosystems have been severely, perhaps irreversibly, altered because of overexploitation of top carnivores and grazers. The herbivorous green turtle population in the Caribbean, for example, likely numbered 60–300 million individuals in the 17th century, before the exploration of the New World. The current population numbers only in the tens of thousands—a reduction of more than 99%! Now, several centuries later, we can only speculate on the vast changes in the natural marine ecosystem that resulted from this dramatic decline in perhaps the largest marine reptilian population in the Caribbean—a decline attributable almost entirely to human hunting.

Human hunting (of fish, shellfish, vertebrates, reptiles, and birds) and collecting (of seaweeds, sea urchins, shells, and corals) has removed or nearly removed ecologically important species from otherwise balanced food webs and has had substantial indirect effects, including by-catch and by-kill (the incidental take of nontargeted species), such as the hooking of sea turtles and albatross on longlines used to fish for tuna and swordfish (figure 4); destruction or disturbance of habitat, such as critical sea-floor habitat of benthic invertebrates by shellfish dredges and bottom-fish trawls; and genetic changes, such as the regional hunting to local extinction of some whale species, which decreases the total genetic material in the species's gene pools.

Pollution

For over a century, the coastal ocean has been assaulted with large quantities of various municipal, industrial, agricultural, and human wastes. For example, chemical pollutants have caused tumors and diseases in fish and shellfish and have affected reproduction in seabirds (DDT made pelican eggshells so thin that they broke when the birds sat on them); oil spills have resulted in local mass deaths of organisms at virtually every link in the food chain; agricultural fertilizers have killed coral reefs by stimulating the growth of seaweeds that overgrow them; and nutrient enrichment in estuaries has stimulated large algal blooms that sometimes lead to the consumption of most of, or all, the oxygen in the water column and deaths of immobile organisms. Although coastal habitats will continue to receive most of the human-derived wastes, the deep sea has been proposed as an additional dumping ground, especially for radioactive material. Dumping one of the most dangerous waste materials in the least-studied marine environment with, arguably, the highest biodiversity on Earth should be reason for great concern.

Habitat Alteration

Coastal habitats have been decisively altered to accommodate the "needs" of human society. For example, large portions of wetlands and salt marshes have been eliminated by dredging, filling, and diking to create new fastlands (dry land), and large areas of mangrove swamps have been altered to create shrimp ponds for aquaculture. Salt marshes and mangroves are highly productive marine systems that serve as nursery grounds for young fish. Seawalls, jetties, and groins, by design, alter the natural currents and thus can affect transport of organisms in the water or organisms that depend on it for food or respiration. Mining (upland and coastal) and deforestation cause erosion on land that ends up in the sea. Increased suspended sediment negatively affects such organisms as corals, which require clear water for photosynthesis by their symbiotic zooanthellae. Clear waters once characterized virtually all tropical areas, but no more.

Introduction of Exotic Species

A startling array and number of marine organisms have been transported around the world ocean by humans, principally in the ballast water of ships. Water is pumped into a ship's hold in one port to stabilize its load and then pumped out in another, sometimes halfway around the globe. Introduced species can outcompete and even eliminate local species. Several introductions have changed entire ecosystems. The European zebra mussel introduced into the Great Lakes led to economic losses of hundreds of millions of dollars per year, and the carnivorous American comb jellyfish introduced into the Black and Azov Seas caused declines in the zooplankton biomass of up to 90% and resulted in large declines in the anchovy fishery (anchovies eat zooplankton).

Global Climate Change

For decades, human activities have been generating compounds that rise into the atmosphere and destroy the ozone that shields the planet from the sun's ultraviolet radiation (UV). If such activities continue, marine organisms might suffer because phytoplankton, zooplankton, fish, corals, and benthic organisms experience harmful effects from biologically damaging ultraviolet radiation (UV-B). Global warming caused by enhancement of the "greenhouse effect" (wherein such gases as CO_2 and CH_4, generated by human activities, prevent the escape of heat radiating from Earth) is expected to cause a substantial increase in sea level and alter ocean circulation. Adaptation typically occurs over very long periods—thousands to millions of years. Thus, marine biodiversity could be seriously affected if organisms cannot adapt to human-accelerated global climate changes that take place over decades or perhaps a century.

COASTAL VULNERABILITY

The most vulnerable parts of the sea are the coastal areas, the focal point of most human activities that threaten marine biodiversity. Assaults on the coastal ocean have been relentless and, in many parts of the world, are still increasing in magnitude, persistence, and area affected. The cause is continuing human population growth, which is disproportionately faster in coastal areas.

Throughout the United States and the world, about 50% of the human population inhabits a narrow fringe around the periphery of the continents, a coastal region that is only 100 km wide. That percentage and the absolute numbers of human beings are increasing each year, and society as a whole is experiencing a new phenomenon—the emergence along the coasts of "mega-cities", cities with populations over 10 million (table 2). Moreover, most mega-cities are now in the developing world. Because the developing world is concentrated in tropical and subtropical areas, mega-cities occur in coastal regions that have the highest ma-

TABLE 2   The Growth of the World's Mega-Cities (Data from United Nations Population Fund)

Year	Number of 15 Largest Cities with Populations over 10 Million	Number of 15 Largest Cities in Developed World	Population of 15th Largest City, millions
1950	1 (New York City)	11	3.3
1970	3 (Tokyo, NYC, Shanghai)	8	6.7
1994	14 (Tokyo topped 26 million)	4	9.8
2015[a]	15 (7 to exceed 20 million)	1	15

[a] Projected

rine biodiversity.   Large human populations inevitably result in large effects on the natural environment, so all biodiversity—marine and terrestrial—in these regions is at high risk.

Cities and countries in the developing world do not yet have the infrastructure to deal with the environmentally unforgiving consequences of highly localized human populations, such as overexploitation of resources, habitat alteration, and pollution.   Thus, wave after wave of unprocessed or uncontained human, municipal, industrial, and agricultural wastes will continue to travel across the land-sea interface unless steps are taken quickly to forestall this situation.

Clearly, humans have been having large, negative effects on marine biodiversity. Luckily, they also can do something about it.   It is ridiculous to ask humans, animals with a right to life on this planet, to have no impact on the environment. Virtually every other species has some impact.   In fact, herein lies the origin of "ecology"—the study of relationships between organisms and their environment. But what members of Homo sapiens have over other species is the right to choose their impacts and to minimize those which must occur.   The foundation for such decisions is in knowledge of natural patterns of biodiversity and the processes that maintain them.   Such data are, however, meager, at best, for most organisms in most environments in the sea.

CONSERVATION STRATEGIES

Entire issues of journals and several books have been devoted to strategies and tactics for conserving marine biodiversity at all levels. Our purpose here is much more modest: to identify some critical concepts that form a foundation on which to build any comprehensive, integrated, and sustainable initiative to conserve marine biodiversity—concepts that derive principally from the distinctiveness of marine, compared with terrestrial, systems (table 3)—and to emphasize the importance of raising public awareness and understanding of the need to conserve marine biodiversity.

There is no comprehensive, coherent, integrated plan for conserving the world's marine biodiversity.   Development of such a plan will require far greater knowledge than exists today and far greater cooperation across multiple jurisdictions

TABLE 3   Some Critical Concepts That Form a Foundation on Which to Build Any Comprehensive, Integrated, and Sustainable Management Plan to Conserve Marine Biodiversity

Processes that generate biodiversity do not always occur in the places where the diversity is observed. For example, the greatest biodiversity in the open-ocean water column occurs at depths of around 1,000 m, but the processes responsible for it occur at much higher in the water column—in the euphotic zone.

Scales—temporal and spatial—of processes in the sea, especially in open-ocean systems, are larger than the terrestrial realm, and many of the strategies developed for conservation of biodiversity on land might be ineffective in the sea. Moreover, the coupling of open-ocean and coastal systems and of coastal water quality with land-use activities argues for integrated management plans that cross land-ocean boundaries.

Habitat protection is the most effective way to conserve biodiversity. Protection must be provided not just for isolated habitats, but for an interlocking mosaic of habitats that make up landscape (seascape) diversity. That is particularly critical in coastal areas—the ecotone that represents the transition from land to water. Present management and conservation strategies place too much emphasis on unusual habitats, on species of special human interest, and on economics and too little emphasis on biodiversity itself.

Productivity is an important consideration in setting priorities among regions of marine biodiversity to conserve. Many regions with high biodiversity have low productivity; likewise, some important regions have low diversity but high productivity. The latter areas—salt marshes, estuaries, and upwelling regions—are often highly exploited by humans.

Fragility and resilience are important criteria for establishing habitat protection. Some of, but not all, the marine ecosystems at risk are fragile. Many estuaries are exploited and polluted but resilient—up to a point. The danger is of failure to act until a system's ability to recover has been exceeded—at least on time scales important to humans.

Rarity is a relative term in describing the "abundance" of organisms. Moreover, it is critical to distinguish between concentration and total number. Many marine species are "rare" in that the numbers of individuals per unit volume of seawater (their concentrations) are small. For example, 80–90% of pelagic species are consistently "rare". But because of the enormous size of their habitat, the number of organisms per pelagic species is large if integrated over the world ocean. Ecological theory developed for terrestrial environments suggests that organisms that are rare—in terms of concentration—play a negligible role in ecosystem functioning, but such theory might not apply to the sea.

Marine protected areas should be established for regions that present essential ecological conditions and promote critical ecological processes. Sanctuaries and reserves are but one component of an overall strategy for conserving marine biodiversity. The size of protected areas required for a substantial impact on conserving marine biodiversity, particularly for highly migratory species, is much greater than for the terrestrial environment. Criteria for establishing the size of protected areas and for assessing their effectiveness must be developed in advance.

Population management and habitat protection should be integrated approaches to conserving biodiversity, particularly of migratory and cosmopolitan animals, such as some of the large predatory fishes and marine birds. Although there has been some progress recently, marine animals, like terrestrial animals, are too often "protected" on the basis of quotas or the conservation of isolated habitat in the form of small sanctuaries; these approaches thus fall short of appreciating the large ambit—the living space—of many marine organisms.

Greater public awareness is critical in achieving and sustaining success in conserving marine biodiversity. We have failed to capture the public's attention on the importance of the loss of biodiversity, and the increasing disconnect of people from nature exacerbates the problem. Because most people see the marine environment, interact with it, and have their greatest impact on it at the coast, this is the place to focus our attention.

than has ever occurred. And it all starts with goals. Goals for conserving marine biodiversity should be stipulated both in terms of values and uses important to society, with measurable and understandable performance indicators, and in terms of the fundamental value of biology. Furthermore, the effective development and execution of any conservation initiative requires a serious reconnection of people with nature. The best opportunity for broad-scale public education and involvement in conserving marine biodiversity is at the coast, where people are most likely to experience and appreciate the sea. The International Year of the Ocean, 1998, was an excellent starting point for global participation. One vehicle for raising public awareness could be the international network of aquariums, which draw more than 200 million visitors each year and where specific local and regional marine-biodiversity issues can be placed in a global context. Whatever the tactics, they must be developed now, lest our children and theirs know not the magnificent beauty and bounty of the sea.

CLOSING COMMENTS

The world ocean is experiencing substantial and startling losses of biodiversity. Arguments about whether coral reefs or rain forests support the greatest diversity are silly and dangerous; they divert attention from the real issues. Conservation of the planet's biodiversity—marine and terrestrial—is critically important. We choose the word *conservation* advisedly. Biodiversity cannot be preserved; it can and must be protected or conserved. Evolution and extinction are natural processes. But now, for the first time in the planet's history, one species—ours—has demonstrated its capacity to destroy large numbers of other species and their habitats. Never before has one species had such a profound, pervasive, and pernicious effect on so many others. Ironically, the other creatures with which we share this planet would be far better off in the absence of "intelligent life".

ACKNOWLEDGMENTS

We thank Bob Beardsley, Brad Butman, Just Cebrian, Mark Chandler, Gregory Early, Scott Kraus, Ken Mallory, Dan Pearlman, Carl Safina, Vicke Starczak, Gregory Stone, and particularly Carolyn Levi, for ideas, information, and suggestions for improving this paper. Jayne Doucette and Jack Cook did the graphic illustrations, for which we are grateful. We thank Paul Erickson for the slide show that was part of the oral presentation. C.A. Butman was supported by a Pew Charitable Trusts Fellowship in Conservation and the Environment. This is Contribution 9620 from the Woods Hole Oceanographic Institution.

SUGGESTIONS FOR FURTHER READING

Angel MV. 1993. Biodiversity of the pelagic ocean. Cons Bio 7:760–72.
Butman CA, Carlton JT. 1995. Marine biodiversity: some important issues, opportunities and critical research needs. Rev Geophys Suppl:1201–9.

Chandler M, Kaufman L, Mulsow S. 1996. Human impact, biodiversity and ecosystem processes in the open ocean. In: Mooney HA, Cushman JH, Medina E, Slaad OE, Schulze ED (eds). Functional roles of biodiversity: a global perspective. New York NY: J Wiley. p 431–74.

Dobson AP. 1996. Conservation and biodiversity. New York NY: Sci Amer Lib. p 264.

Huston MA. 1994. Biological diversity: the coexistence of species on changing landscapes. New York NY: Cambridge Univ Pr. p 681.

Jackson JBC. 1997. Reefs since Columbus. Coral Reefs 16 Suppl:S23–S32.

Jeffries MJ. 1997. Biodiversity and conservation. New York NY: Routledge. p 208.

NRC [National Research Council]. 1995. Understanding marine biodiversity: a research agenda for the nation. Washington DC: National Acad Pr. p 114.

NRC [National Research Council]. 1996. Stemming the tide: controlling introductions of non-indigenous species by ships' ballast water. Washington DC: National Acad Pr. p 141.

Norse EA (ed). 1993. Global marine biodiversity: a strategy for building conservation into decision making. Washington DC: Island Pr. p 383.

Oceanus. 1995. Marine biodiversity I. Fall/Winter.

Oceanus. 1996. Marine biodiversity II. Spring/Summer.

Safina C. 1995. The world's imperiled fish. Sci Amer 273:46–53.

Perlman DL, Adelson G. 1997. Biodiversity: exploring values and priorities in conservation. Malden MA: Blackwell. p 182.

Perrings C, Maler KG, Folke C, Hollings CS, Jansson BO (eds). 1995. Biodiversity loss: economic and ecological issues. New York NY: Cambridge Univ Pr.

Peterson M (ed). 1992. Diversity of oceanic life: an evaluative review. Washington DC: The Center for Strategic and Information Studies. p 108.

Reaka-Kudla ML, Wilson DE, Wilson EO (eds). 1997. Biodiversity II: understanding and protecting our biological resources. Washington DC: Joseph Henry Pr. p 551.

Wilson EO (ed). 1988. Biodiversity. Washington DC: National Acad Pr. p 521.

COUNTRYSIDE BIOGEOGRAPHY AND
THE PROVISION OF ECOSYSTEM SERVICES·

GRETCHEN C. DAILY

Department of Biological Sciences, Stanford University,
Stanford, CA 94305-5020

HUMANITY HAS become a dominant force on Earth, altering important characteristics of the atmosphere, oceans, and terrestrial systems. One of the many consequences of these alterations is the extinction of populations and species, which
is projected to drive biodiversity to its lowest level since humanity came into being (Ehrlich and Ehrlich 1981; Wilson 1992). A crucial set of policy questions is
when, where, and how to direct societal activities to soften or reverse their effect
on biodiversity.

In addressing these questions, one is immediately confronted with a set of tradeoffs in the allocation of resources (such as land and water) to competing uses, to
competing individuals and groups of people, and ultimately to competing value
systems. These tradeoffs are becoming increasingly vexing from both ethical and
practical perspectives. They involve our most important ideals (such as ensuring
a prosperous future for our children), our oldest tensions (such as between individual and societal interests), and sometimes our bloodiest tendencies (such as
using genocide as a convenient way of gaining control over resources). Society is
poorly equipped to handle these tradeoffs, and they are appearing everywhere; the
well-being of current and future generations hinges on how the tradeoffs are dealt
with.

The short-term benefits of alteration of habitats, the primary cause of loss of
biodiversity, are typically clear and allow relatively small groups of immediate
beneficiaries to exert great influence on the political process in favor of short-term
exploitation. In contrast, the arguments for conservation tend to be diverse and
difficult to measure, and the benefits of any single decision about conservation are

diffused over very large numbers of people. The arguments for conservation typically are drawn from any of four distinct lines of reasoning: ethical, aesthetic, direct economic, and indirect economic (Heywood 1995; Hughes and others 2000; Ehrlich and Ehrlich 1992). Ethical reasoning involves the conviction that, as the dominant species on the planet, humanity has the responsibility of stewardship toward "The Creation," its only known living companions in the universe. This moral responsibility exists independent of the perceived value of nonhuman organisms to human well-being. The other three classes of argument rest on the benefits that humanity derives from other organisms, which I collectively refer to here as "ecosystem services."

The reasons for stemming the loss of biodiversity thus range in character from the intangible, the spiritual and philosophical, to the purely anthropocentric and pragmatic (for a nice overview, see Goulder and Kennedy 1997). One might say they span the spectrum from things that make life worth living to things that make life possible at all. Clearly, both ends of the spectrum are important, although the significance ascribed to each varies considerably with social context and understanding. Lack of public understanding of societal dependence on natural ecosystems is a major hindrance to the implementation of policies needed to bring the human economy into balance with the capacity of Earth's life-support systems to sustain it.

The purpose of this paper is to explain this dependence briefly, to describe how recognition of it can help resolve the tradeoffs that society now faces, and to indicate where society could invest profitably in broadening and deepening the scientific understanding of ecosystem services. First, I briefly characterize ecosystem services in biophysical and economic terms. Then, I indicate how the concept provides a framework that, if supported with appropriate institutions and policies, allows us to incorporate ecosystem-service values into decision-making. Finally, I turn to a key underlying biological issue: the capacity of human-dominated landscapes to support biodiversity and sustain ecosystem services. My emphasis throughout is on the anthropocentric and pragmatic.

LIFE ON THE MOON

Society derives a wide array of life-support benefits from biodiversity and the natural ecosystems within which it exists. These benefits are captured in the term "ecosystem services", the conditions and processes through which natural ecosystems, and the species that are a part of them, sustain and fulfill human life (Daily 1997; Holdren and Ehrlich 1974). These services include the production of ecosystem goods, such as seafood, timber, forage, and many pharmaceuticals, which represent an important and familiar part of the economy.

Perhaps the easiest way to appreciate the importance of biodiversity in supplying life-support goods and services is by way of a thought experiment that removes the familiar backdrop of Earth. Imagine trying to set up a happy life on the moon. Assume for the sake of argument that the moon miraculously already had some of the basic conditions for supporting human life, such as an atmosphere, a climate, and a physical soil structure similar to those on Earth. After packing one's possessions and coaxing one's family and friends into coming along, the big question would be, Which of Earth's millions of species would be needed to make the sterile moonscape habitable?

One could choose first from among all the species used directly for food, drink, spices, fiber, timber, pharmaceuticals, and industrial products, such as waxes, rubber, and oils. Even if one were selective, this list could amount to hundreds or even thousands of species. And one would not have begun considering the species needed to support those used directly: the bacteria, fungi, and invertebrates that recycle wastes and help make soil fertile; the insects, bats, and birds that pollinate flowers; and the herbaceous plants, shrubs, and trees that hold soil in place, nourish animals, and help control the gaseous composition of the atmosphere that influences Earth's climate. No one knows exactly how many or which combinations of species would be required to support human life. So, rather than listing individual species, one would have to list instead the life-support services required by the lunar colony and try to choose groups of species able to perform them. A partial list of such services includes the following (Daily 1997):

- production of a wide variety of ecosystem goods;
- purification of air and water;
- mitigation of flood and drought;
- detoxification and decomposition of wastes and cycling of nutrients;
- generation and preservation of soils and renewal of their fertility;
- pollination of crops and natural vegetation;
- dispersal of seeds;
- control of the vast majority of agricultural pests;
- maintenance of biodiversity;
- protection from the sun's harmful ultraviolet rays;
- partial stabilization of climate;
- moderation of weather extremes and their effects; and
- provision of aesthetic beauty and intellectual stimulation that lift the human spirit.

The closest attempt to carry out this experiment here on Earth was the first Biosphere 2 "mission" (Cohen and Tilman 1996). A facility was constructed on 3.15 acres in Arizona that sealed off its inhabitants as much as possible from the outside world; eight people were meant to live inside for 2 years without the transfer of materials in or out. The experimenters had to decide which species to use to populate the closed ecosystem. They moved in tons of soil (with its huge abundance and variety of little-known fungi, arthropods, worms, and microorganisms), added numerous other animals and plants, and fueled the system with sunlight (through transparent walls) and electricity (at an annual cost of about $1 million). Biosphere 2 featured agricultural land and elements of a variety of natural ecosystems, such as forest, savanna, desert, and even a miniature ocean.

In spite of an investment of $200 million in the design, construction, and operation of this model Earth, it proved impossible to supply the material and physical needs of the eight "Biospherians" for the intended stay. Many unexpected and unpleasant problems arose, including a drop in the concentration of oxygen from 21% to 14%, a level normally found at an elevation of 17,500 ft; skyrocketing concentrations of carbon dioxide with large daily and seasonal fluctuations; high concentrations of nitrous oxide to the point where brain functioning can be impaired; extinction of 19 of 25 vertebrate species; extinction of all pollinators (thereby dooming most of the plant species to eventual extinction); population explosions of aggressive vines and crazy ants; and failure of water-purification systems.

The basic conclusion from this experiment is that there is no demonstrated alternative to maintaining the viability of "Biosphere 1," Earth (Cohen and Tilman 1996). Ecosystem services operate on such a grand scale and in such intricate and little-explored ways that most could not be replaced by technological means (Ehrlich and Mooney 1983). They existed for millions or billions of years before humanity evolved, making them easy to take for granted and hard to imagine disrupting beyond repair. Yet escalating effects of human activities on natural ecosystems now imperil the delivery of these services. The primary threats are changes in the uses of lands, causing loss of biodiversity and facilitating biotic invasion, and synergisms of these with alteration of biogeochemical cycles, release of toxic substances, possible rapid change of climate, and depletion of stratospheric ozone (Daily 1997b).

MANAGEMENT OF NATURAL CAPITAL

Maintaining Earth as a suitable habitat for *Homo sapiens* will require society to begin to recognize natural ecosystems and their biodiversity as capital assets, which, if properly managed, will yield a flow of benefits over time. Relative to physical capital (buildings, equipment, and so on), human capital (skills, knowledge, health, and so on, embodied in the labor force), and financial capital, natural capital is poorly understood, little valued, scarcely monitored, and undergoing rapid depletion. Sustainable management of ecosystem services will require a systematic characterization of the services, in biophysical, economic, and other terms along with the development of financial mechanisms and policy institutions to provide the means of monitoring and safeguarding them.

Characterization involves an explicit cataloging of important services on a variety of scales. In other words, which ecosystems supply what services? For a given location, which are supplied locally, which are imported, and which are exported? Characterization also involves finding answers to other questions (Costanza and Folke 1997; Daily 1997c; Holdren 1991), such as, what is the effect of alternative human activities on the supply of services?

The administration of New York City first considered replacing its natural water-purification system (the Catskill Mountains) with a filtration plant but found that it would cost an estimated $6–8 billion in capital plus $300 million per year to operate. The high costs prompted investigation of an alternative solution, namely restoring and safeguarding the natural purification services of the Catskills. That would involve purchasing land in and around the watershed to protect it and subsidizing several changes on privately owned land: upgrading sewage-treatment plants; improving practices on dairy farms and undertaking "environmentally sound" economic development. The total cost of this option was estimated at about $1.5 billion (Revkin 1997).

Thus, New York City had a choice of investing in $6–8 billion in physical capital or $1.5 billion in natural capital. It chose the latter option, raising an environmental bond issue to fund its implementation. This financial mechanism captured the important economic and public-health values of a natural asset (the watershed) and distributed them to those assuming the responsibilities of stewardship for the asset and its services.

The Catskills supply many other valuable services, such as control of flooding, sequestration of carbon, conservation of biodiversity, and, perhaps above all, beauty, serenity, and spiritual inspiration. Moreover, these services benefit others besides consumers of water in New York City. It would be absurd to try to express the full value of the ecosystem services provided by the Catskills in dollars. In this case, fortunately, there was no reason to try: even a lower estimate of the value of the natural asset was sufficient to induce adopting a policy of conservation.

The challenge is to extend this model to other geographic locations and to other services. The US Environmental Protection Agency recently estimated that treating, storing, and delivering safe drinking water to the United States without taking this approach would require an investment in physical capital of $138.4 billion over the next 20 years. More than 140 municipalities in the United States now are considering watershed protection, an option that aligns market forces with the environment, as a more cost-effective option than building artificial treatment facilities (The Trust for the Public Land 1997). Indeed, interest is growing worldwide in adopting watershed conservation. Rio de Janeiro and Buenos Aires, for example, are investigating this option; both have highly threatened watersheds of enormous biotic value (Chichilnisky and Heal 1998).

Extending this model to other services requires that an ecosystem meet two conditions. First, it must supply at least one good or service to which a commercial value can be attached. Second, some of that value must be appropriable by the steward of the ecosystem (Chichilnisky and Heal 1998). Public goods and services are difficult to privatize: if provided for one, they are provided for all, so their providers typically cannot appropriate all the value of the good or service. Natural water purification is a public service, but access to the resulting high-quality water is excludable; thus, the case of a watershed works by bundling a public service with a private good. Private capital could be mobilized in this cause to the benefit of both individual investors and society at large (Chichilnisky and Heal 1998).

In principle, this approach could be made to work for other ecosystem goods and services, such as for realizing and safeguarding biodiversity, ecotourism, and carbon-sequestration values. With appropriate institutional support (such as that needed for the management of common property resources), mechanisms for safeguarding sources of flood control, pollination, and pest-control services also may be developed. This is an important subject for further interdisciplinary investigation by persons from academe, government, and the private sector.

COUNTRYSIDE BIOGEOGRAPHY

Attaining the ultimate goal of sustainably managing natural capital will require a deeper understanding of the relative effects of alternative activities on biodiversity and ecosystem services. A key question is, Where do critical thresholds lie in the relationships between the condition and extent of an ecosystem and the quality of the services that it supplies? Let us explore this issue from the perspective of the modification of ecosystems by agricultural activities.

Food production is arguably humanity's most important activity. It is also the most important proximate cause of the loss of biodiversity worldwide, involving major direct and indirect effects, including conversion of natural habitat to agricultural use, facilitation of biotic invasion through trade (thereby increasing the rate of introduction of exotic species) and alteration of habitat (thereby increasing the susceptibility of native communities to invasion), and application of chemical fertilizers and pesticides.

In the face of such effects, the fates of organisms that once made their homes in unbroken expanses of natural habitat range along a broad continuum. At one end is the decline of population to local and eventually global extinction; at the other end is expansion into human-controlled landscapes. Biologists have paid considerable attention to the status of the biotas of fragments of natural habitat, such as forest patches, and comparably little attention (outside the context of pest management) to the organisms that occupy the highly disturbed matrix in which those fragments occur. One reason for this emphasis is undoubtedly the crisis nature of extinction: given the justified panic to save remaining natural habitat, it is taking some time to appreciate a complementary opportunity, namely, to enhance the hospitality of agricultural landscapes for biodiversity. The emphasis traces to other factors, including the prominence of the theories of island biogeography and the island paradigm in conservation biology; the assumption that a very small fraction of species is capable of persisting outside of "islands" of natural habitat, that is, in human-controlled habitats; and the frequent (although often subconscious) projection of disdain for humanity's destruction of natural habitat onto the organisms that profit from it.

The organisms that can take advantage of countryside, rural and suburban landscapes devoted primarily to human activities, deserve more attention for a series of reasons. First, it is unlikely that many large, relatively undisturbed tracts of natural habitat will remain in the face of projected growth in the size, food needs, and environmental effects of the human population. Second, the potential for conserving many species might rest on preserving or enhancing some aspects of rural landscapes that contain remnants of native habitat in lieu of protecting large tracts of undisturbed habitat, which is generally much less feasible socioeconomically. Third, the supply of some important ecosystem services—such as pest control, pollination, and water purification—will depend in many instances on the biodiversity that occurs locally, in the vicinity of human habitation, in countryside habitats. Finally, a growing interest in restoration also will require comparing the conservation value of alternative sites for the establishment and succession of desired community assemblages.

Countryside biogeography is the study of the diversity, abundance, conservation, and restoration of biodiversity in rural and other human-dominated landscapes. Broad issues in this area pertain to the future course, societal consequences, and appropriate policy responses to the mass extinction currently under way. They include the following sorts of questions.

• What is the relationship between levels of agricultural intensification and biodiversity in countryside landscapes? Measures of agricultural intensification include the frequency distribution of clearing sizes, the ratio of clearing to hedgerow areas, the spatial configuration and relative coverage of native and human-dominated habitats within the countryside landscape, the diversity of crops under cultivation, modification of the hydrological cycle, and the levels and types of chemical fertilizers and pesticides applied.

• Which species traits confer an advantage for survival in the face of tropical deforestation and other major alterations of habitat?

• Are these traits distributed randomly across taxa, or are some groups of organisms especially resistant and others especially prone to extinction? In other words, will the current episode of extinction nip off the buds of the evolutionary tree of life relatively uniformly, or will it eliminate some major limbs, dramatically reshaping the future diversity and evolution of life?

• Can simple mathematical theory be developed to predict patterns of persistence of biodiversity in countryside landscapes?

• How accurately can patterns of biodiversity in countryside habitats be predicted on the basis of remotely sensed information on land use (for example, with images from satellites)?

• How effectively can countryside biotas perform ecosystem services?

• What practical measures can be taken to enhance the capacity of countryside habitats to sustain biodiversity and ecosystem services as well as human activities?

This is not the place for a comprehensive review of work addressing those issues. I offer instead a few illustrative findings to date:

• In Europe, more than 50% of the land area with high conservation value is under low-intensity farming. Examples of these habitat types include blanket bog, northern Atlantic wet heath, lowland hay meadow, heather moorlands, wood pasture, alpine pasture, and nonirrigated cereal steppe (Bignal and McCracken 1996). Intensification of farming practices in recent decades has resulted in declining populations of many species of birds throughout Europe. For instance, nine of the 11 species of waders listed in the Red Data Book that occur in Sweden are seriously threatened by changes in farming practices there (Johansson and Blomqvist 1996).

• Recent studies are beginning to illuminate the strength and type of biotic control over the functioning of ecosystems (Chapin and others 1997). Greater richness of species can enhance the stability of the ecosystem. In species-poor plots of grassland in Minnesota, for example, a severe drought caused a reduction in productivity of more than 90% from predrought levels, whereas productivity in species-rich plots was reduced by 50% (Tilman 1994). Alterations in habitat that change the functional diversity and composition of plant species appear especially likely to have major effects on various properties of ecosystems (Hooper and Vitousek 1997; Tilman and others 1997).

- In the vicinity of Las Cruces in southern Costa Rica, a significant fraction of the native avian species appear to be persisting, at least temporarily, in open countryside habitats in a mixed-agricultural landscape that retains 27% of its once-continuous forest cover. Of possible original totals in the 33 species of birds under consideration, it appears that 1–9% have become extinct locally, 50–54% are restricted to habitats of forested countryside, and 36–40% occur in habitats of open countryside that are as far as 6 km from the nearest large tracts (at least 200 hectares) of forest (Daily and others in review).

- Some systems of cultivation used in coffee production appear to have high potential for conserving birds and other elements of the native biota. Systems of cultivation that use shade trees, plantations with tall canopy cover, diverse stratification, little pruning, and low levels of insecticides are especially rich in birds, including both resident and neotropical migrant species (Greenberg and others 1997). Strikingly high abundances of arthropods and richness of species have also been found. For example, fogging of shade trees with pyrethrins in a Costa Rican coffee plantation in formerly upland-rainforest habitat yielded a richness of coleopteran and hymenopteran species comparable with that of samples from trees in upland rainforests in Peru and Brazil (Perfecto and others 1996).

- Nocturnality might confer an advantage of dispersal and possibly of survival in the face of tropical deforestation. Surveys of the diversity of diurnal birds and butterflies and nocturnal beetles and moths in forested patches reveal the classic island biogeographic pattern for birds and butterflies (in other words, fewer in smaller patches) but similarly high diversities of moths and beetles among forested patches of all sizes (0.1–225 hectares). A possible mechanism explaining this apparent advantage is that typically the movement of nocturnal species occurs when the conditions of thermal, humidity, and solar radiation are similar between native forest and cleared areas; during the day, the hot, dry, and bright conditions in open areas might impede dispersal seriously for many organisms (Daily and Ehrlich 1996).

Ideally, further effort in empirical and theoretical research on issues of countryside biogeography eventually will allow us to predict patterns of biodiversity in human-dominated landscapes worldwide (White and others 1997). This would be an important step toward characterizing and monitoring the effects of humans on ecosystems and the services they supply.

CONCLUSIONS

The human population and its standards of living are maintained by a steady depletion of natural capital assets, including renewable-resource stocks and waste sinks that, if they were safeguarded, could sustain a flow of ecosystem goods and services through time. In our collective behavior, there is little recognition or systematic accounting, let alone nurturing, of these critical capital assets. Tremendous payoff could result from further research on managing Earth's life-support systems. Such research should be oriented toward developing the following:

• a broader and deeper understanding of the functioning of Earth's life-support systems and the effects of humanity on them, especially in countryside habitats;
• systematic accounting and monitoring of the condition of these systems;
• ways of quantifying the importance of ecosystems at the margin, from biophysical, economic, and cultural (aesthetic and spiritual) perspectives, that is, ways of determining, for instance, how much importance should be attached to the preservation or destruction of the next unit of habitat;
• ways of incorporating these values into a framework for decision-making; and
• ways of creating appropriate institutions and policies to allow the individuals or societies that safeguard life-support systems for the public good to realize the value of their stewardship.

In the market-driven culture that prevails today, the concept of ecosystem services offers a new way to approach actions of conservation by confronting market forces on their own terms. This concept has promise because it integrates biophysical and social dimensions of managing the biosphere; it offers rational, practical solutions to tradeoffs in allocation of resources to competing uses and people; and it is adaptable to different economic and cultural circumstances. Similarly, countryside biogeography can reveal new strategies for preserving biodiversity and ecosystem services in the context of some of humanity's most important activities. Nevertheless, these frameworks are just two tools to complement the many others required for protecting biodiversity (Raven 1990; Raven and Wilson 1992). In our quest to safeguard the systems that make life possible, it is critical that we not lose sight of what makes life worth living.

ACKNOWLEDGMENTS

I am grateful for insightful comments from Scott Daily, Michael Dalton, Paul Ehrlich, Geoffrey Heal, and Jennifer Hughes. This work was supported by the generosity of Peter and Helen Bing, the Pew Charitable Trusts, and the Winslow Foundation.

REFERENCES

Bignal EM, McCracken DI. 1996. Low-intensity fanning systems in the conservation of the countryside. J Appl Ecol 33:413–24.
Chapin FS III, Walker BH, Hobbs RJ, Hooper DU, Lawton JH, Sala OE, Tilman D. 1997. Biotic control over the functioning of ecosystems. Science 277:500–4.
Chichilnisky G, Heal G. Forthcoming. Securitizing the biosphere. Nature.
Chichilnisky G, Heal G. 1998. Economic returns from the biosphere. Nature 391:629–30
Cohen JE, Tilman D. 1996. Biosphere 2 and biodiversity: the lessons so far. Science 274:1150–1.

Costanza R, Folke C. 1997. Valuing ecosystem services with efficiency, fairness, and sustainability as goals. In: Daily GC (ed). Nature's services: societal dependence on natural ecosystems. Washington DC: Island Pr.

Daily GC (ed). 1997. Nature's services: societal dependence on natural ecosystems. Washington DC: Island Pr.

Daily GC, Ehrlich PR, Sanchez-Azofeifa GA. Countryside biogeography: utilization of human-dominated habitats by the avifauna of southern Costa Rica. forthcoming.

Daily GC, Ehrlich PR. 1996. Nocturnality and species survival. Proc Natl Acad Sci 93:11709–12.

Daily GC. 1997. Introduction: What are ecosystem services? In: Daily GC (ed). Nature's services: societal dependence on natural ecosystems. Washington DC: Island Pr. p 1–10.

Daily GC. 1997. Valuing and safeguarding Earth's life-support systems. In: Daily GC (ed). Nature's services: societal dependence on natural ecosystems. Washington DC: Island Pr. p 365–74.

Ehrlich PR, Ehrlich AH. 1981. Extinction: the causes and consequences of the disappearance of species. New York NY: Random House.

Ehrlich PR, Ehrlich AH. 1992. The value of biodiversity. Ambio 21: 219–26.

Ehrlich PR, Mooney HM. 1983. Extinction, substitution and ecosystem services. BioScience 33: 248–54.

Goulder LH, Kennedy D. 1997. Valuing ecosystem services: philosophical bases and empirical methods. In: Daily GC (ed.). Nature's services: societal dependence on natural ecosystems. Washington DC: Island Pr. p 23–47.

Greenberg R, Bichier P, Cruz Angon A, Reitsma R. 1997. Bird populations in shade and sun coffee plantations in Central Guatemala. Conserv Biol 11:448–59.

Greenberg R, Bichier P, Sterling J. Forthcoming. Bird populations and planted shade coffee plantations of eastern Chiapas. Biotropica.

Heywood VH (ed). 1995. Global biodiversity assessment. Cambridge UK: Cambridge Univ Pr.

Holdren J. 1991. Report of the planning meeting on ecological effects of human activities. Unpublished mimeo. National Research Council, Irvine CA.

Holdren JP, Ehrlich PR. 1974. Human population and the global environment. Amer Sci 62:282–92.

Hooper DU, Vitousek PM. 1997. The effects of plant composition and diversity on ecosystem processes. Science 277:1302–5.

Hughes JB, Daily GC, Ehrlich PR. 2000. The loss of population diversity and why it matters. In: Raven PH, Williams T (eds). Nature and human society: the quest for a sustainable world. Washington DC: National Academy Press. p 71–83.

Johansson OC, Blomqvist D. 1996. Habitat selection and diet of lapwing Vanellus vanellus chicks on coastal farmland in SW Sweden. J Appl Ecol 33:1030–40.

Perfecto I, Rice RA, Greenberg R, Van der Voort ME. 1996. Shade coffee: a disappearing refuge for biodiversity. BioScience 46:598–608.

Perrings C. 1995. The economic value of biodiversity. In: Heywood VH (ed). Global biodiversity assessment. Cambridge UK: Cambridge Univ Pr. p 823–914.

Raven PH, Wilson EO. 1992. A fifty-year plan for biodiversity surveys. Science 258:1099–100.

Raven PH. 1990. The politics of preserving biodiversity. BioScience 40:769–74.

Revkin AC. 1997. Billion-dollar plan to clean New York City water at its source. The New York Times, New York. 31 August 1997, p 3. The Trust for Public Land. 1997. Protecting the source. San Francisco CA: The Trust for Public Land. 28 p.

Tilman D, Downing JA. 1994. Biodiversity and stability of grasslands. Nature 367:363–5.

Tilman D, Knops J, Wedin D, Reich P, Ritchie M, Siemann E. 1997. The influence of functional diversity and composition on ecosystem processes. Science 277:1300–2.

Vitousek PM, Mooney HA, Lubchenco J, Melillo JM. 1997. Human domination of Earth's ecosystems. Science 277:494–9.

White D, Minotti PG, Barczak MJ, Sifneos JC, Freemark KE, Santelmann MV, Steinitz CF, Kiester AR, Preston EM. 1997. Assessing risks to biodiversity from future landscape change. Conserv Biol 11:349–60.

Wilson EO. The diversity of life. 1992. Cambridge MA: Harvard Univ Pr.

P A R T

2

LESS WELL-KNOWN
INDIVIDUAL FORMS
OF LIFE

MICROBIAL DIVERSITY
AND THE BIOSPHERE

NORMAN R. PACE

Departments of Plant and Microbial Biology and Molecular and Cell Biology,
University of California, Berkeley, CA
(Current address: Department of Molecular, Cellular, and
Developmental Biology,
University of Colorado, Boulder, CO 80309-0347)

INTRODUCTION

MICROORGANISMS OCCUPY a peculiar place in the human view of life. They receive little attention in our general texts of biology. They are largely ignored by most professional biologists and are virtually unknown to the public except in the contexts of disease and rot. Yet the workings of the biosphere depend absolutely on the activities of the microbial world (Madigan and others 1996). And a large bulk of global biomass is microbial (Whitman and others 1998). Our texts articulate biodiversity in terms of large organisms: insects usually top the count of species. Yet if we squeeze out any insect and examine its contents under the microscope, we find hundreds or thousands of distinct and unidentified microbial species. A handful of soil contains billions of microorganisms, of so many types that accurate numbers remain unknown. At most only a few of these microorganisms would be known to us; only about 5,000 noneukaryotic organisms have been formally described (Bull and others 1992) in contrast with the half-million described insect species. We know little about microbial biology, a part of biology that looms large in the sustenance of life on this planet.

The reason for our poor understanding of the microbial world lies in the fact that microorganisms are tiny, individually invisible to the eye. The mere existence of microbial life was recognized only relatively recently in history, about 300 years ago, with Leeuwenhoek's invention of the microscope. Even under the microscope, however, the simple structures of most microorganisms, usually nondescript rods and spheres, prevented their classification by morphology, through which large organisms had always been related to one another. It was not until the late 19th century and the development of pure-culture techniques that microorganisms could be studied as individual types and characterized to some extent, mainly by nutritional criteria. However, the pure-culture approach to the study of the microbial world seriously constrained the view of microbial diversity because most microorganisms defy cultivation by standard methods. Moreover, the morphological and nutritional criteria used to describe microorganisms failed to provide a natural taxonomy, ordered according to evolutionary relationships. Molecular tools and a perspective based on gene sequences are now alleviating these constraints to some extent. Even the early results are changing our perception of microbial diversity.

A SEQUENCE-BASED MAP OF BIODIVERSITY

Before the development of sequence-based methods, it was impossible to know the evolutionary relationships connecting all of life and thereby to draw a universal evolutionary tree. Whittaker, in 1969, just as the molecular methods began to develop, summarized evolutionary thought in the context of the "five kingdoms" of life: animals, plants, fungi, protists ("protozoa"), and monera (bacteria) (Whittaker 1969). There also was thought to be a higher, seemingly more fundamental taxonomic distinction between eukaryotes, organisms that contain nuclear membranes, and prokaryotes, predecessors of eukaryotes that lack nuclear membranes (Chatton 1937). Those two categories were considered independent and coherent groups. The main evolutionary diversity of life on Earth, four of the five traditional taxonomic kingdoms, was believed to lie among the eukaryotes, particularly the multicellular forms. These still-pervasive notions had never been tested, however, and they proved to be incorrect.

The breakthrough that called previous beliefs into question and brought order to microbial, indeed biological, diversity emerged with the determination of molecular sequences and the concept that sequences could be used to relate organisms (Schwartz and Dayhoff 1978; Zuckerkandl and Pauling 1965). The incisive formulation was reached by Carl Woese, who, by comparing ribosomal RNA (rRNA) sequences, established a molecular sequence-based phylogenetic tree that could be used to relate all organisms and reconstruct the history of life (Woese 1987; Woese and Fox 1977). Woese articulated the now-recognized three primary lines of evolutionary descent, termed "urkingdoms" or "domains": Eucarya (eukaryotes), Bacteria (initially called eubacteria), and Archaea (initially called archaebacteria) (Woese and others 1990).

Figure 1 is a current phylogenetic tree based on small-subunit (SSU) rRNA sequences of the organisms represented. The construction of such a tree is conceptually simple (Swofford and others 1996). Pairs of rRNA sequences from different organisms are aligned, and the differences are counted and considered to be some measure of "evolutionary distance" between the organisms. There is no consideration of the passage of time, only of change in nucleotide sequence. Pairwise differences between many organisms can be used to infer phylogenetic trees, maps that represent the evolutionary paths leading to the modern-day sequences.

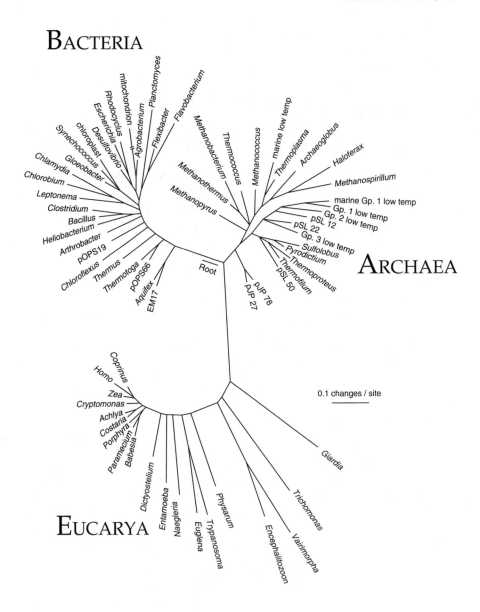

FIGURE 1 Universal phylogenetic tree based on small-subunit ribosomal RNA sequences. Sixty-four rRNA sequences representative of all known phylogenetic domains were aligned, and a tree was produced with FASTDNAML (Barns and others 1996; Maidak and others 1997). That tree was modified to the composite one shown by trimming lineages and adjusting branchpoints to incorporate results of other analyses. The scale bar corresponds to 0.1 change per nucleotide. (From Pace 1997, reprinted with permission.)

The tree in figure 1 is largely congruent with trees made by using any molecule in the nucleic acid-based, information-processing system of cells. But phylogenetic trees based on metabolic genes, those involved in manipulation of small molecules and in interaction with the environment, commonly do not concur with the rRNA-based version; see Doolittle and Brown (1994), Palmer (1997), and Woese (1998) for reviews and discussions of phylogenetic results with different molecules. Incongruities in phylogenetic trees made with different molecules can reflect lateral transfers or even the intermixing of genomes in the course of evolution. Some metabolic archaeal genes, for instance, appear much more highly related to specific bacterial versions than to their eucaryal homologues; other archaeal genes seem decidedly eukaryotic; still others are unique. Nonetheless, recently determined sequences of archaeal genomes show clearly that the evolutionary lineage of Archaea is independent of both Eucarya and Bacteria (Bult and others 1996; Smith and others 1997).

INTERPRETING THE MOLECULAR TREE OF LIFE

"Evolutionary distance" in the type of phylogenetic tree shown in figure 1, the extent of sequence change, is read along line segments. The tree can be considered a rough map of the evolution of the genetic core of the cellular lineages that led to the modern organisms (sequences) included in the tree. The time of occurrence of evolutionary events cannot be extracted reliably from phylogenetic trees, despite common attempts to do so. Time cannot be accurately correlated with sequence change, because the evolutionary clock is not constant in different lineages (Woese 1987). This disparity is evidenced in figure 1 by the fact that lines leading to the different reference organisms are not all the same length; these different lineages have experienced different extents of sequence change. Nonetheless, the order of occurrence of branchings in the trees can be interpreted as a genealogy, and intriguing insights into the evolution of cells are emerging.

A sobering aspect of large-scale phylogenetic trees like that shown in figure 1 is the graphic recognition that most of our legacy in biological science, historically based on large organisms, has focused on a narrow slice of biological diversity. Thus, we see that animals (represented in figure 1 by *Homo*), plants (*Zea*), and fungi (*Coprinus*) constitute small and peripheral branches of even eukaryotic cellular diversity. If the animals, plants, and fungi are taken to make up taxonomic "kingdoms", we must recognize as kingdoms at least a dozen other eukaryotic groups, all microbial, with at least as much independent evolutionary history as that which separates the three traditional eukaryotic kingdoms. The taxonomic term *kingdom* has no molecular definition. I use it to indicate main lines of radiation in the particular domain; 14 such "kingdom-level" lines are associated with the eucaryal line of descent in figure 1 (see also Sogin 1994).

The rRNA and other molecular data solidly confirm the notion stemming from the last century that the major organelles of eukaryotes—mitochondria and chloroplasts—are derived from bacterial symbionts that have undergone specialization through coevolution with the host cell. Sequence comparisons establish mitochondria as representatives of Proteobacteria (the group in figure 1 including *Es-*

cherichia and *Agrobacterium*) and chloroplasts as derived from cyanobacteria (*Synechococcus* and *Gloeobacter* in figure 1) (Sapp 1994). Thus, all respiratory and photosynthetic capacity of eukaryotic cells was obtained from bacterial symbionts; the "endosymbiont hypothesis" for the origin of organelles is no longer hypothesis but well-grounded fact. The nuclear component of the modern eukaroytic cell did not derive from one of the other two lineages, however. The rRNA and other molecular trees show decisively that the eukaryotic *nuclear* line of descent extends as deeply into the history of life as do the bacterial and archaeal lineages. The mitochondrion and chloroplast came in relatively late. This late evolution is evidenced by the fact that mitochondria and chloroplasts diverged from free-living organisms that branched peripherally in molecular trees. Moreover, the most deeply divergent eukaryotes lack even mitochondria (Cavalier-Smith 1993). These latter organisms, little studied but sometimes troublesome creatures—such as *Giardia*, *Trichomonas*, and *Vairimorpha*—nonetheless contain at least a few bacterium-type genes (Bui and others 1996; Germot and others 1996; Roger and others 1996). That might be evidence of an earlier mitochondrial symbiosis with Eucarya that was lost (Palmer 1997) or perhaps other symbiotic or gene-transfer events between the evolutionary domains.

The root of the universal tree in figure 1, the point of origin of the modern lineages, cannot be established by using sequences of only one type of molecule. However, recent phylogenetic studies of gene families that originated before the last common ancestor of the three domains have positioned the root of the universal tree deep on the bacterial line (Doolittle and Brown 1994). Therefore, Eucarya and Archaea had a common history that excluded the descendants of the bacterial line. The period of evolutionary history shared by Eucarya and Archaea was an important time in the evolution of cells during which the refinement of the primordial information-processing mechanisms occurred. Thus, modern representatives of Eucarya and Archaea share many properties that differ from bacterial cells in fundamental ways. One example of similarities and differences is in the nature of the transcription machinery. The RNA polymerases of Eucarya and Archaea resemble each other in subunit composition and sequence far more than either resembles the bacterial type of polymerase. Moreover, whereas all bacterial cells use sigma factors to regulate the initiation of transcription, eucaryal and archaeal cells use TATA-binding proteins (Marsh and others 1994; Rowlands and others 1994).

THE METABOLIC DIVERSITY OF LIFE

The molecular-phylogenetic perspective, as depicted in figure 1, is a reference framework within which to describe microbial diversity; the sequences of genes can be used to identify organisms. That is an important concept for microbial biology. It is not possible to describe microorganisms as is traditional with large organisms, through their morphological properties. To be sure, some microorganisms are intricate and beautiful under the microscope, but mainly they are relatively unfeatured at the resolution of routine microscopy. Therefore, to distinguish different types of microorganisms, early microbiologists turned to metabolic properties of the organisms, such as their sources of carbon, nitrogen, and energy. Microbial taxonomy accumulated as anecdotal descriptions of metabolically and morphologically distinct types of organisms that were essentially unrelatable. Molecular phylogeny now provides a framework within which we can relate organisms objectively and through which we can interpret the evolutionary flow of the metabolic machineries that constitute microbial diversity.

Laboratory studies of microbial metabolism have focused mainly on such organisms as *Escherichia coli* and *Bacillus subtilis*. In the broad sense, such organisms metabolize much as animals do; we are all "organotrophs," using reduced organic compounds for energy and carbon. Organotrophy is not the prevalent form of metabolism in the environment, however. Autotrophic metabolism—fixation of CO_2 to reduced organic compounds—must necessarily contribute to a greater biomass than the organotrophic metabolism that it supports (a principle long appreciated by ecologists). Energy for fixing CO_2 is gathered in two ways: "phototrophy" (photosynthesis) and "lithotrophy" (coupling the oxidation of reduced inorganic compounds—such as H_2, H_2S, and ferrous iron—to the reduction of a chemical oxidant, a terminal electron acceptor, such as oxygen, nitrate, sulfate, sulfur, and CO_2). Thus, metabolic diversity can be generalized in terms of organotroph or autotroph, phototroph or lithotroph, and the nature of the electron donor and acceptor.

The phylogenetic patterns of types of carbon and energy metabolism among different organisms do not necessarily follow the evolutionary pattern of rRNA (figure 1). Presumably, that is because of past lateral transfers of metabolic genes and larger-scale symbiotic fusions. Nonetheless, domain-level tendencies might speak to the ancestral nature of the three domains of life (Kandler 1993). The perspective, here, is limited mainly to Archaea and Bacteria. Such broad generalities cannot yet be assessed for the Eucarya, because so little is known about the metabolic breadth of the domain and the properties of the most deeply divergent lineages. There is considerable information about one pole of eukaryotic diversity—that represented by animals, plants, and fungi. We know little about the other pole—the amitochondriate organisms that spun off the main eucaryal line early in evolution (Sogin 1994). The known instances of such lineages—represented by *Trichomonas*, *Giardia*, and *Vairimorpha* in figure 1—are primarily pathogens. Pathogenicity in humans is a rare trait among the rest of eukaryotes and bacteria, and no archaeal pathogen is known. That correlation might indicate that nonpathogenic, deeply divergent eukaryotes are abundant in the environment but not yet detected. They should be sought in anaerobic ecosystems, possibly coupled metabolically to other organisms. A driving theme of the eucaryal line seems to be the establishment of physical symbiosis with other organisms. Beyond that, the general metabolism of the rudimentary eukaryotic cell seems simple and based on fermentative organotrophy. By virtue of symbiotic partners, however, eukaryotes are able to take on phototrophic or lithotrophic lifestyles and to respire using the electron-acceptor oxygen (Smith and Douglas 1987).

Symbiotic microorganisms commonly confer the lithotrophic way of life even on animals, although this was only recently recognized. The 2-m-long submarine vent tubeworm *Riftia pachyptila*, for instance, lives in the vicinity of sea-floor hydrothermal vents and metabolizes H_2S and CO_2 by means of sulfide-oxidizing, CO_2-fixing, bacterial symbionts (Tunnicliffe 1992). This invertebrate and metabolically similar ones might contribute substantially to primary productivity in the ocean (Kates and others 1993; Lutz and others 1994). It is not necessary to go to (from our perspective) unusual places, such as ocean-floor vents, to encounter equally fascinating H_2S-dependent eukaryotes (Fenchel and Finlay 1995). Under foot at the ocean beach, for example, microbial respiration of seawater sulfate creates an H_2S-rich ecosystem populated by little-known creatures, such as *Kentrophoros*, a flat, gulletless ciliate that under the microscope appears fuzzy because it cultivates a crop of sulfide-oxidizing bacteria on its outer surface (Fenchel and Finlay 1989); the bacteria are ingested by endocytosis and thereby provide nutrition for *Kentrophoros*. In other anaerobic environments, methanogens, members of Archaea, live intracellularly with eukaryotes and serve as metabolic hydrogen sinks (Embley and Finlay 1994). Still other symbioses based on inorganic energy sources are all around us and are little explored for their diversity of microbial life (Fenchel and Finlay 1995).

Many lithotrophic but comparatively few organotrophic representatives of Archaea have been obtained in pure culture (Kates and others 1993). There are primarily two metabolic themes, both relying on the use of hydrogen as a main energy source. Among the known members of Euryarchaeota, one of the two archaeal kingdoms known through cultivated organisms, the main electron acceptors are CO_2 and the product CH_4, "natural gas." Most of the CH_4 encountered in the outer few kilometers of Earth's crust or on the surface is determined by isotopic analysis to be the product of methanogenic archaea communities, past and present. Such organisms probably constitute a huge component of global biomass. They certainly offer an inexhaustible source of renewable energy to humankind.

The general metabolic theme of the other established kingdom of Archaea, Crenarchaeota, also is the oxidation of H_2, but with a sulfur compound as the terminal electron acceptor. All the cultivated representatives of Crenarchaeota also are thermophiles. Consequently, such organisms have been referred to as thermoacidophilic or hyperthermophilic archaeons; some grow at the highest known temperatures for life, up to 113°C in the case of *Pyrolobus fumaris* (Stetter 1995). These crenarchaeotes might seem bizarre to us, capable as they are of thriving at temperatures sometimes above the usual boiling point of water on a diet of H_2, CO_2, and S and exhaling H_2S. Yet, in terms of the molecular structures of the basic cellular machineries, these creatures resemble eukaryotes far more closely than either resembles our gut bacterium *E. coli* (Marsh and others 1994).

The metabolic diversity of microorganisms is usually couched in terms of the use of complex organic compounds. From that standpoint and on the basis of cultivated organisms, metabolic diversity seems to have flowered mainly among the Bacteria. Even here, however, reliance on organic nutrients probably was not ancestral. The most deeply branching of the cultured bacterial lineages, represented by *Aquifex* and *Thermotoga* in figure 1, are basically lithotrophs that use H_2 as an energy source and such electron acceptors as sulfur compounds (*Ther-*

motoga) or low concentrations of O_2 (*Aquifex*) (Pitulle and others 1994). Cultivated organisms from these deeply branching bacterial lineages also are all thermophilic and thus share two important physiological attributes with the deeply branching and slowly evolving Archaea; a H_2-based energy source and growth at high temperatures. That coincidence suggests that the last common ancestor of all life also metabolized H_2 for energy at high temperatures; this inference is consistent with current notions regarding the origin of life—that it came to be in the geothermal setting at high temperature (Pace 1991).

Chlorophyll-based photosynthesis was a bacterial invention. It seems to have appeared well after the establishment of the bacterial line of descent at or before the divergence of the line in figure 1 leading to *Chloroflexus*, a photosynthetic genus (Pierson 1993), and after the deeper divergences, such as those leading to *Aquifex* and *Thermotoga*, which are not known to have photosynthetic representatives. Most bacterial photosynthesis is anaerobic, however. Oxygenic photosynthesis, the water-based photosynthetic mechanism that produces the powerful electron acceptor O_2, arose only in the kingdom-level lineage of cyanobacteria. This invention changed the surface of Earth profoundly and is conventionally thought to be the basis, directly or indirectly, of most present-day biomass.

Anaerobic photosynthesis is widely distributed in the late-branching bacterial kingdoms. The more ancient theme of lithotrophy, metabolism of inorganic compounds, is also widely distributed phylogenetically, intermixed with organotrophic organisms. The pattern suggests that organotrophy arose many times from otherwise photosynthetic or lithotrophic organisms. Indeed, many instances of bacteria can switch between these modes of nutrition, carrying out photosynthesis in the light and lithotrophy or organotrophy in the dark. Particularly among bacteria, the type of energy metabolism seems highly volatile in evolution; bacteria that are closely related by molecular criteria can display strikingly different phenotypes when assessed in the laboratory through the nature of their carbon and energy metabolism. In the relatively closely related "gamma subgroup" of the kingdom of Proteobacteria (delineated by the genus *Escherichia* in figure 1), for instance, we find the phenotypically disparate organisms *E. coli* (organotroph), *Chromatium vinosum* (H_2S-based phototroph), and the symbiont of the tubeworm *R. pachyptila* (H_2S-based symbiont). The superficial metabolic diversity of these types of bacteria belies their underlying close evolutionary relatedness, giving no hint of the close similarities of their basic machineries. The versatility of Bacteria makes the metabolic machineries of Archaea and Eucarya seem comparatively monotonous. As the sequences of diverse genomes are compared, it will be possible to map the flow of metabolic genes onto the rRNA-based tree and see how metabolic diversity has been molded through evolution.

The molecular perspective gives us more than just a glimpse of the evolutionary past; it also brings a new future to the discipline of microbial biology. Because the molecular-phylogenetic identifications are based on sequence, not metabolic properties, microorganisms can be identified without the requirement for cultivation. Consequently, all the sequence-based techniques of molecular biology can be applied to the study of natural microbial ecosystems, heretofore little known with regard to organismal makeup.

A SEQUENCE-BASED GLIMPSE OF BIODIVERSITY IN THE ENVIRONMENT

Knowledge of microorganisms in the environment has depended mainly on studies of pure cultures in the laboratory. Rarely are microorganisms so captured, however. Studies of several types of environments estimate that more than 99% of organisms seen microscopically are not cultivated with routine techniques (Amann and others 1995). With the sequence-based taxonomic framework of molecular trees, only a gene sequence, not a functioning cell, is required to iden- tify an organism in terms of its phylogenetic type. The occurrence of phyloge- netic types of organisms, "phylotypes," and their distributions in natural commu- nities can be surveyed by sequencing rRNA genes obtained from DNA isolated directly from the environment. A molecular-phylogenetic assessment of an un- cultivated organism can provide insight into many of its properties through com- parison with its relatives. Analysis of microbial ecosystems in this way is more than a taxonomic exercise in that the sequences provide experimental tools, such as molecular hybridization probes, that can be used to identify, monitor, and study the microbial inhabitants of natural ecosystems (Amann and others 1995; Hugen- holtz and Pace 1996; Pace and others 1985).

Every nucleic acid-based study of natural microbial ecosystems so far performed has uncovered novel types of rRNA sequences, often representing major new lin- eages only distantly related to known ones. The discovery of rRNA sequences in the environment that diverge more deeply in phylogenetic trees than those of cultivated organisms is particularly noteworthy. It means that the divergent or- ganisms recognized by rRNA sequence are potentially more different from known organisms in the lineage than the known organisms are from one another. The deepest divergences in both Bacteria and Archaea were first discovered in rRNA- based surveys of communities associated with hot springs in Yellowstone National Park (See Hugenholtz and others 1998, for review).

The gene-based studies of organisms in the environment have substantially expanded our view of the extent of microbial diversity, reflected in new branches in phylogenetic trees. Figure 2 shows a diagrammatic tree of known bacterial di- versity. When Woese first summarized the phylogeny of the phylogenetic domain Bacteria, he could articulate about 12 main phylogenetic groups. These groups have been called "phyla," "kingdoms," or "phylogenetic divisions"; I use the lat- ter term. The number of recognized bacterial phylogenetic divisions has ex- panded now to about 36 (figure 2). About one-third of these divisions, indicated by the outlined wedges in figure 2, have no known cultivated representative and were detected only by rRNA gene-based studies of environmental organisms. Some of the most abundant organisms in the biosphere fall into these divisions with no cultured examples. Their abundance identifies such organisms as wor- thy of future study (Hugenholtz and others 1998). Environmental surveys of rRNA genes also have expanded the known diversity of Archaea and revealed that such organisms, previously thought restricted to "extreme" environments (from the human standpoint), in fact are ubiquitous. Crenarchaeota, for in- stance, all of whose cultured representatives are thermophiles, is revealed by the

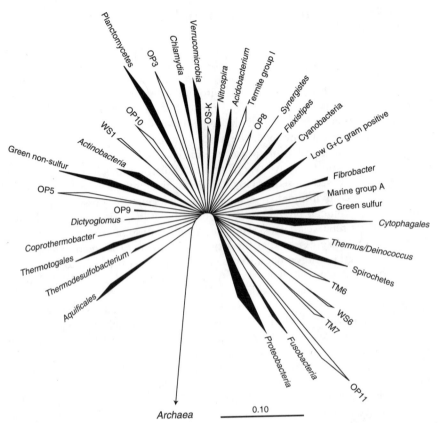

FIGURE 2   Evolutionary-distance tree of the bacterial domain showing currently recognized divisions and putative (candidate) divisions. The tree was constructed with the ARB software package (with the Lane mask and Olsen rate-corrected neighbor-joining options) and a sequence database modified from the March 1997 ARB database release (Strunk and others 1998). Division-level groupings of two or more sequences are depicted as wedges. The depth of a wedge reflects the branching depth of the representatives selected for a particular division. Divisions that have cultivated representatives are shown in black; divisions represented only by environmental sequences are shown in outline. The scale bar indicates 0.1 change per nucleotide. The aligned, unmasked datasets used for this figure are available from http://crab2.berkeley.edu/~pacelab/176.htm. (From Hugenholtz and others 1998, reprinted with permission.)

molecular studies to be abundant in the marine environment and in soils (see Pace 1997, for review).

MICROBIAL DIVERSITY AND THE LIMITS OF THE BIOSPHERE

Textbooks generally portray only a part of the global distribution of life—the part that is immediately dependent on either the harvesting of sunlight or the metabolism of the decay products of photosynthesis. The molecular phylogenetic record shows, however, that lithotrophic metabolism preceded and is more widespread phylogenetically and geographically than either phototrophy or organotrophy. The lithotrophic biosphere potentially extends kilometers into the Earth's crust, an essentially unknown realm (Ghiorse 1997). These considerations suggest that lithotrophy contributes far more to the biomass of Earth than currently thought. If so, where is it?

Part of the lithotrophic biomass is in microhabitats all around us, usually away from light and O_2. It is not necessary to look far to find such environments: the rumens of cattle and the guts of termites and humans, for example, are important sources of CH_4, a signature of hydrogen metabolism. Most life that depends on inorganic energy metabolism, however, probably is in little-known environments, according to poorly understood geochemistry. The oceans, for instance, cover 70% of Earth's surface to an average depth of 4 km. Most life in the ocean is microbial, and the metabolic patterns of such organisms are not understood. Does the occurrence of a large standing crop of low-temperature crenarchaeotes, potentially H_2 oxidizers, indicate an unsuspected, lithotrophy-based food chain in the oceans? Another little-studied environment with global importance is the deep subsurface (Fredrickson and Onstott 1996; Gold 1992; Lovely 1995). There is increasing evidence that the Earth's crust is shot through with biomass wherever the physical conditions permit. Metabolism of H_2 is a dominant theme among organisms isolated from geothermal settings or deep aquifers (Pedersen 1993; Stevens 1997). H_2 is generated readily by abiotic mechanisms, such as interaction of water with iron-bearing basalt, the main stuff of Earth's crust. Consequently, a food source is unlikely to be limiting in most subterranean environments; it is likely to be the oxidant, the terminal electron acceptor, that limits growth. Nonetheless, it seems possible that much, perhaps most, of the biomass on Earth is subterranean, a biological world based on lithotrophy. Although the metabolic rate of this subterranean biosphere is likely to be far lower than in the more dynamic, photic environment, life is likely to be as pervasive in occurrence, and perhaps in cellular diversity, as we experience on the surface.

The opportunities for discovery of new organisms and development of resources based on microbial diversity are greater than ever before. Molecular sequences have finally given microbial biologists a way to define their subjects—through molecular phylogeny. The sequences also are the basis of the tools that will allow microbial biologists to explore the distribution and roles of the organisms in the environment. Microbial biology can now be a whole science and can study the organism in the ecosystem.

ACKNOWLEDGMENTS

I thank Sydney Kustu, Gary Olsen, and Carl Woese for helpful comments on the manuscript and Sue Barns and Phil Hugenholtz for assistance with figures. Research in my laboratory is supported by grants from the National Science Foundation and the National Institutes of Health. This article is based on an earlier one (Pace 1997).

REFERENCES

Amann RI, Ludwig W, Schleifer KH. 1995. Phylogenetic identification and in situ detection of individual microbial cells without cultivation. Microbiol Rev 59(1):143–69.

Barns, SM, Delwiche CF, Palmer JD, Pace NR. 1996. Perspectives on archaeal diversity, thermophily, and monophyly from environmental rRNA sequences. Proc Natl Acad Sci USA 93:9188–93.

Bui ETN, Bradley PJ, Johnson PJ. 1996. A common evolutionary origin for mitochondria and hydrogenosomes. Proc Natl Acad Sci USA 93:9651–6.

Bull AT, Goodfellow M, Slater JH. 1992. Biodiversity as a source of innovation in biotechnology. Ann Rev Microbiol 46:219–52.

Bult CJ, White G, Olsen J, Zhou L, Fleischmann RD, Sutton GG, Blake JA, Fitzgerald LM, Clayton RA, Gocayne JD. 1996. Complete genome sequence of the methanogenic aracheon, Methanococcus jannaschii. Science 273:1058–73.

Cavalier-Smith T. 1993. Kingdom Protozoa and its 18 phyla. Microbiol Rev 57:953–94.

Chatton E. 1937. Titres et travoux scientifiques. Sottano Italy: Sete.

Doolittle WF, Brown JR. 1994. Tempo, mode, the progenote and the universal route. Proc Natl Acad Sci USA 91:6721–8.

Embley TM, Finlay BJ. 1994. The use of small subunit rRNA sequences to unravel the relationships between anaerobic ciliates and their methanogen endosymbionts. Microbiology 140:225–35.

Fenchel T, Finlay BJ. 1989. Kentrophoros: a mouthless ciliate with a symbiotic kitchen garden. Ophelia 30:75–93.

Fenchel T, Finlay BJ. 1995. Ecology and evolution in anoxic worlds. In: May RM, Harvey PH (eds). Oxford UK: Oxford Univ Pr. p 276.

Fredrickson JK, Onstott TC. 1996. Microbes deep inside the earth. Sci Amer 275:68–73.

Germot A, Philippe H, le Guyader H. 1996. Presence of a mitochondrial-type 70-kDa heat shock protein in Trichomonas vaginalis suggests a very early mitochondrial endosymbiosis in eukaryotes. Proc Natl Acad Sci USA 93:14614–7.

Ghiorse WC. 1997. Subterranean life. Science 275:789–90.

Gold T. 1992. The deep hot biosphere. Proc Natl Acad Sci USA 89:6045.

Hugenholtz P, Goebel BM, Pace NR. 1998. Impact of culture-independent studies on the emerging phylogenetic view of bacterial diversity. J Bacteriol 180(18):4765–74.

Hugenholtz P, Pace NR. 1996. Identifying microbial diversity in the natural environment: a molecular phylogenetic approach. Trends Biotech 14:190–7.

Kandler O. 1993. The early diversification of life. In: Bengtson S (ed). Early life on earth. Nobel Symp 84. New York NY: Columbia Univ Pr. p 152–60.

Kates M, Kushner DJ, Matheson AT (eds). 1993. The biochemistry of Archaea (Archaebacteria). Amsterdam Netherlands: Elsevier. 582 p.

Lovely D, Chapelle F. 1995. Deep subsurface microbial processes. Rev Geophys 33:365–81.

Lutz RA, Shank TM, Fornari DJ, Haymon RM, Lilley MD, Von Damm KL, Desbruyeres D. 1994. Rapid growth at deep-sea vents. Nature 371:663–4.

Madigan MT, Martinko JM, Parker J. 1996. Brock biology of microorganisms. Upper Saddle River NJ: Prentice Hall. 986 p.

Maidak BL, Olsen GJ, Larsen N, Overbeek R, McCaughey MJ, Woese CR. 1997. The RDP (ribosomal database project). Nucleic Acids Res 25:109–10.

Marsh TL, Reich CI, Whitelock RB, Olsen GJ. 1994. Transcription factor IID in the Archaea: sequences in the Thermococcus celer genome would encode a product closely related to the TATA-binding protein of eukaryotes. Proc Natl Acad Sci USA 91:4180–4.

Pace NR. 1991. Origin of life—facing up to the physical setting. Cell 65:531–3.

Pace NR. 1997. A molecular view of microbial diversity and the biosphere. Science 276:734–40.

Pace NR., Stahl DA, Lane DJ, Olsen GJ. 1985. Analyzing natural microbial populations by rRNA sequences. ASM News 51:4–12.

Palmer JD. 1997. Organelle genomes: going, going, gone! Science 275:790–1.

Pedersen K. 1993. The deep subterranean biosphere. Earth Sci Rev 34:243–60.

Pierson BK. 1993. The emergence, diversification, and role of photosynthetic eubacteria: early life on earth. New York NY: Columbia Univ Pr.

Pitulle C, Yang Y, Marchiani M, Moore ERB, Siefert JL, Arangno M, Jurtshuk P, Fox GE. 1994. Phylogenetic position of the genus Hydrogenobacter. Int J Syst Bacteriol 44:620–6.

Roger AJ, Clark CG, Doolittle WF. 1996. A possible mitochondrial gene in the early-branching amitochondriate protist Trichomonas vaginalis. Proc Natl Acad Sci USA 93:14618–22.

Rowlands T, Baumann P, Jackson SP. 1994. The TATA-binding protein: a general transcription factor in eukaryotes and archaebacteria. Science 264:1326–9.

Sapp J. 1994. Evolution by association: a history of symbiosis. New York NY: Oxford Univ Pr. 255 p.

Schwartz RM, Dayhoff MO. 1978. Origins of prokaryotes, eukaryotes, mitochondria, and chloroplasts. Science 199:395–403.

Smith DC, Douglas AE. 1987. The biology of symbiosis. In: Willis AJ, Sleigh MA (eds). London UK: Edward Arnold. 302 p.

Smith DR, Doucette-Stamm LA, Deloughery C, Lee H, Dubois J, Aldredge T, Bashirzadeh R, Blakely D, Cook R, Gilbert K and others. 1997. Complete genome sequence of Methanobacterium thermoautotrophicum deltaH: functional analysis and comparative genomics. J Bacteriol 179(22):7135–55.

Sogin ML. 1994. The origin of eukaryotes and evolution into major kingdoms. In Bengtson S (ed). Early life on earth. Nobel Symp No 84. New York NY: Columbia Univ Pr. p 181–92.

Stetter KO. 1995. Microbial life in hyperthermal environments. ASM News 61(6):285–290.

Stevens TO. 1997. Lithoautotrophic microbial ecosystems in the subsurface. FEMS Microbiol Rev 20(3–4):327–37.

Strunk O, Gross O, Reichel B, May M, Hermann S, Struckmann N, Nonhoff B, Lenke M, Vilbig A, Ludwig T, Bode A, Schleifer KH, Ludwig W. 1998. ARB: a software environment for sequence data. Submitted to Nucleic Acids Res. http://www.mikro.biologie.tu-meunchen.de.

Swofford DL, Olsen GJ, Waddell PJ, Hillis DM. 1996. Phylogenetic inference. In: Hillis DM, Moritz C, Mable BK (eds). Molecular systematics, 2nd ed. Sunderland MA: Sinauer. p 407–514.

Tunnicliffe V. 1992. Hydrothermal-vent communities of the deep sea. Amer. Sci 80:336–49.

Whitman WB, Coleman DC, Wiebe WJ. 1998. Prokaryotes: the unseen majority. Proc Natl Acad Sci USA 95:6578–83.

Whittaker RH. 1969. New concepts of kingdoms of organisms. Science 163:150–60.

Woese C. 1998. The universal ancestor. Proc Natl Acad Sci USA 95:6854–9.

Woese CR. 1987. Bacterial evolution. Microbiol Rev 51(2):221–71.

Woese CR, Fox GE. 1977. Phylogenetic structure of the prokaryotic domain: the primary kingdoms. Proc Natl Acad Sci USA 74:5088–90.

Woese CR, Kandler O, Wheelis ML. 1990. Towards a natural system of organisms: proposal for the domains Archaea, Bacteria, and Eukarya. Proc Natl Acad Sci USA 87:4576–9.

Zuckerkandl E, Pauling L. 1965. Molecules as documents of evolutionary history. J Theor Biol 8:357–66.

BIODIVERSITY, CLASSIFICATION, AND NUMBERS OF SPECIES OF PROTISTS

JOHN O. CORLISS
P.O. Box 2729, Bala Cynwyd, PA 19004

CLEAR RECOGNITION of the great evolutionary gulf between the prokaryotes (essentially bacteria) and the eukaryotes (all other organisms) nearly four decades ago led to numerous studies that preoccupied research biologists' time for some years. But in the 1970s, attention began to refocus on the equally important specific fields of eukaryogenesis (the evolutionary appearance of cells above the bacterial level) and the phylogenetic origin of multicellular-multitissued organisms themselves, with the recognition that filling the gap between bacteria and animals/plants seemed to require some intermediate level of organismic organization. The hypothetical "gap-fillers"—to the surprise, perhaps, of many experimental biologists but not of field and taxonomic protozoologists and phycologists—turned out to be represented by the largely unicellular eukaryotic microorganisms, a huge assemblage (tens of thousands of species) of widespread but often poorly known forms that now can collectively be called the protists. Thus dawned the interdisciplinary field of protistology, arbitrarily said to have reached a recognizable state in about 1975 (Corliss 1986, 1987). Vast improvements in cytological techniques, including kinds of electron microscopy (Patterson 1994) and the advent of molecular methods (Cavalier-Smith 1995), have since aided greatly in the expansion of such investigations exploiting what may be termed the protist perspective.

Biodiversity, quite new itself as a term and concept in biology, is often linked with conservation in people's minds, and the organisms involved are typically the highly visible plants and animals now living on Earth. The protists—that is, the generally unicellular and microscopic algae and protozoa and the lower fungi—are, like the bacteria, cosmopolitan and ubiquitous; but the healthy abundance of many of these microorganisms is absolutely necessary for maintenance of a sustainable world. Their roles at the base of the food chain and in nutrient recycling are known to be of the highest importance, and their potential in treating diseases is under study. Their roles in the preservation, not to mention the (past) evolution, of *other* organisms have been and are indeed indispensable. In contrast, some of the parasitic species are highly virulent to their hosts and thus can have disastrous effects on human populations, food crops, and domesticated animals. Today's unpredicted increase in appearance of opportunistic protistan parasites in AIDS patients is an example of our need to understand these organisms better. Only recently have all the points such as those mentioned above begun to become appreciated (Andersen 1992, 1998; Colwell 1997; Corliss 1989b, 1991, 1998a; Finlay 1998; Finlay and Esteban 1998; Hawksworth and Colwell 1992; John 1994; Norton and others 1996; Patterson and Sogin 1993; Sogin and Hinkle 1997; Vickerman 1992, 1998). But it is increasingly clear that much further work is required to assess the multiple roles of protists in natural ecosystems.

To speak quantitatively about the numbers of known protistan species, a main aim of this paper, we must first have some idea of the qualitative nature of protists: what are they, and how can they be defined and classified? A further question—what are the probable evolutionary and phylogenetic interrelationships of what to most people is the rather large number of separate high-level taxa commonly recognized as containing protists?—is mostly well beyond discussion in the present brief essay, although it obviously affects the classification of the organisms concerned. For often detailed treatments of major aspects of the last question, the reader is referred to Coombs and others (1998), Hausmann and Hülsman (1996), Hülsmann and Hausmann (1994), Karpov (1990), Katz (1998), Knoll (1992), Kuźnicki and Walne (1993), Lipscomb and others (1998), Patterson (1994), Schlegel (1998), Sleigh (1995), Sogin and others (1996) and the many pertinent references within those works.

WHAT IS A PROTIST?

Even defining the term *protist* is somewhat controversial, so I shall offer only a broad and general description here, attempting to make clear their essential uniquenesses, in combination, as a great and diverse assemblage of organisms on Earth. Recent comments on this difficult question have appeared in works by Andersen (1998), Cavalier-Smith (1993b, 1998a), Corliss (1994a, 1998b), Hausmann and Hülsmann (1996), Margulis (1996), Patterson (1994), Vickerman (1998).

Protists, typically and mostly, are single-celled, microscopic eukaryotic organisms, occasionally forming a single tissue that can lead to large body size (for example, in some multicellular brown algae). As cells, they may have one to several nuclei; and various other organelles are always present in their cytoplasm. They represent, in general, a structural grade between the bacteria or prokaryotes and the so-called higher eukaryotes. Although eukaryotic themselves, protists do not have multicelled organs or true vascular systems, and ordinarily they do not show complex developmental or embryonic stages in their life cycles (ontogeny). Whereas the ancestors of some contemporary protistan groups very likely gave rise to lines leading to such recognized kingdoms as Fungi, Animalia, and Plantae, others—as far as we can surmise at this time—have retained their protistan nature, evolutionarily speaking. It is reasonable to assume that extinction of protistan groups occurred often during past millennia, although the fossil record to date has not been very helpful in this respect (Lipps 1993; Tappan 1980). Sexuality is not recognized in species of many taxa; asexual division is the most common mode of reproduction and allows stability of an adapted genotype.

Overall distribution today of these lower eukaryotes is cosmopolitan; nutritional and locomotive modes are many, and there are amazing structural and functional adaptations (Hausmann and Hülsmann 1996). The single-celled species are wholly independent organisms: the two terms—*cell* and *organism*—are thus not mutually exclusive descriptors (Corliss 1989a; Hausmann and Bradbury 1996). Major habitats of free-living forms include soils and bodies of freshwater and salt water; and ectosymbiotic and endosymbiotic species are found in association with numerous animal and some plant species and even other protists. Some parasitic forms are highly pathogenic (some malarial species of the genus *Plasmodium* are most notable), with hosts that include humans. Useful species (from the human perspective) include the many involved in essential food chains, in nutrient turnover in lakes and seas, in functioning as bioindicators or biomonitors of pollution and potentially as biocontrol agents, in serving as ideal cells in a multitude of biomedical and medical research projects, and in their direct roles in the petroleum, food, medicinal, agricultural, aquacultural, and other commercial industries. It has been said that 40% of global photosynthesis (carbon fixation and oxygen production) is contributed by algae, and the abundant diatoms alone are responsible for nearly half that (Andersen 1992, 1998).

HISTORICAL BACKGROUND ON PROTISTAN TAXONOMY

We need to understand background information—however brief—on the overall classification of species now known as protists to appreciate their present status. More than a century ago, Ernst Haeckel (1866, 1878; and see Aescht 1998) and a few others proposed that these "lower forms" on the ladder of life should be considered as members of a distinct third kingdom, alongside the established kingdoms of Animalia and Plantae. To shorten a lengthy tale (Corliss 1998c; Lipscomb 1991; Ragan 1997; Rothschild 1989), such ideas, for various involved reasons, did not succeed for a long time, although refined and resurrected by such notable workers as Copeland (1956) and Whittaker (1969).

When the evolutionist-geneticist-microbiologist Margulis (1974, 1988; Margulis and Schwartz 1982; Margulis and others 1990) came on the scene, her forceful arguments stimulated a great deal of research in cell and evolutionary biology. Her undiluted enthusiasm convinced many a formerly reluctant biologist to appreciate the wisdom and particularly the convenience and pedagogical usefulness of a five-kingdom arrangement for all living organisms: Monera (or Bacteria), Protista (or Protoctista), Fungi, Plantae, and Animalia. A neo-Haeckelian system seemed to have become established for all time, although battles were (and are) incessantly waged over the internal composition of the "new" kingdom: what and how many algal divisions and protozoan phyla, for example, are to be subsumed under that heading? What was definitely and irrevocably clear was that protozoa should no longer be treated as "mini-animals" and that most algae must no longer be considered to be merely "mini-plants" (Corliss 1983).

In recent years, considerable evidence has indicated that some major lines of protists share closer relationships with other kingdoms (an outstanding example is green algae with land plants) than they do with formerly neighboring protistan taxa. Such revelations threaten the stability of the whole five-kingdom concept—but this is a complex subject largely beyond further consideration here. However, it is still often convenient and appropriate (as in this paper) to treat the many protistan groups as a *single* great assemblage, although using a lowercase "p" and writing of "the protists" rather than of "a kingdom Protista". Incidentally, another problem, not due extended discussion here but deserving mention, is the nomenclatural matter of Protista and Protoctista (the latter is properly pronounced "proto-tista", in that the "c" is silent in this combination). Arguments for both names exist in the literature, but I believe that today the consensus among research protistologists favors the shorter name; and there is no rule of nomenclature that obliges one to treat the longer word as having any official priority (Corliss 1990, 1994a).

Alternatives to accepting a single formal kingdom Protista have recently been reviewed (Corliss 1994a,b, 1998c; see also Cavalier-Smith 1998a); listing them should suffice here to give the reader an appreciation of possible choices that exist. One is to recognize *no* separate high-level taxon for the protists, considering them overall to represent but an evolutionary grade or level of cellular organization (between bacteria and the higher eukaryotes) and thus sidestepping a number of high-level taxonomic problems. A second is to view groups of protists as simply mostly *independent* evolutionary lines or *lineages*, again leaving aside attempts to define high-level taxonomic interrelationships among such lines; cladistically derived phylogenetic trees (such as those constructed from collected molecular biological data) often support such a choice. Finally, to avoid a single, perhaps highly artificial kingdom for the diverse protist assemblages, some workers have proposed the assignment of these organisms to *multiple* kingdoms of eukaryotes. Such kingdoms can number from five or six to 18–20 or even more. Some might be composed solely of protists; others might contain various protistan taxa but comprise predominantly taxa of existing major kingdoms of multicellular organisms (such as the Plantae, Animalia, or Fungi).

CLASSIFICATIONAL FRAMEWORK FOR MAJOR GROUPS OF PROTISTS

Although discussion of the evolutionary or phylogenetic relationships of the diverse high-level protistan groups is beyond the scope of the present paper, a taxonomic framework of some sort is necessary for clarity in the treatment of their nature or composition, including numbers and inventories of species. Only naming or identifying major assemblages will make possible our recognizing, comparing, and retrieving information about the different groups (Mayr 1997, 1998), many of which, in the case of the protists, have been known (under a variety of names) for scores of years.

Adopting Cavalier-Smith's (1998a) five eukaryotic kingdoms and their names, and using a constellation-of-characters (Corliss 1976) approach as a basis for their taxonomic separateness, I have assigned some 14 phyla of protists to PROTO-ZOA, 11 to CHROMISTA, six to PLANTAE, and two each to FUNGI and ANIMALIA; see table 1. Thus, I am suggesting that some 35 eukaryotic phyla are required to contain the protists overall—a welcome reduction from the 45 of 15 years ago (Corliss 1984). A very brief description of the taxonomic composition of the kingdoms involved is appropriate here because even the better-known ones might no longer embrace the same phyletic taxa as in years past. A link with the classical systems, at both nomenclatural and taxonomic levels, is needed if we are to understand the present locations and interrelationships of the diverse protistan forms implicated.

PROTOZOA (literally meaning "first animals") traditionally has embraced species belonging not only to the phyla listed in Protozoa in table 1 but also to other major taxa no longer included there, most notably Cryptomonada, Haptomonada, Labyrinthomorpha, Opalinata, and some lower taxa of Bicosoecae, Chrysophyta, and Dictyochae; I consider these seven phyla to belong to the kingdom Chromista. The phyla Choanozoa, Myxozoa, and Microspora were also treated as protozoan taxa in the past. The first two of these are now assigned to Animalia, and the third to Fungi (see table 1). Even a few well-studied genera of Chlorophyta and Prasinophyta, phyla now both in Plantae in table 1, have been steadfastly embraced by protozoologists in their classification schemes. So, interestingly, the present *kingdom* Protozoa is considerably more restricted, more refined, and thus more meaningful than the former *phylum* Protozoa of the literature (see discussions in Cavalier-Smith 1993b; Corliss 1994a).

CHROMISTA never existed in former times as such, before Cavalier-Smith's (1981) proposal of this particular name (see also Cavalier-Smith 1986, 1989,

TABLE 1   Protistan Phyla (Arranged Alphabetically) Assigned to Eukaryotic Kingdoms (with Indication of the 10 Largest and the 13 Smallest Groups)

Kingdom	Included Protistan Phyla
PROTOZOA	[a]Apicomplexa, [b]Archamoebae, [a]Ciliophora, [a]Dinozoa, Euglenozoa, [a]Foraminifera, [b]Heliozoa, [b]Metamonada, Mycetozoa, [b]Neomonada, Parabasala, [b]Percolozoa, [a]Radiozoa, [a]Rhizopoda
CHROMISTA	[b]Bicosoecae, Chrysophyta, [b]Cryptomonada, [a]Diatomae, [b]Dictyochae, Haptomonada, [b]Labyrinthomorpha, Opalinata, Phaeophyta, Pseudofungi, [b]Raphidophyta
PLANTAE	[a]Charophyta, [a]Chlorophyta, [b]Glaucophyta, Prasinophyta, [a]Rhodophyta, [b]Ulvophyta
FUNGI	Chytridiomycota, Microspora
ANIMALIA	[b]Choanozoa, Myxozoa

[a] Phyla containing more than 3,000 species.
[b] Phyla containing only 300 or fewer species.

1997b, 1998a). Despite its containing some former protozoan groups (see above), it is now largely "algal" in composition, including such major well- and long-known, mainly photosynthetic groups as Chrysophyta, Diatomae, Phaeophyta, and Raphidophyta (with their many well-known classes). Furthermore, botanists have generally claimed Bicosoecae and Dictyochae at the same time that zoologists were considering them to be "first animals". In much of its composition, the kingdom Chromista of table 1 resembles the rather similar assemblage widely·known today as the stramenopiles (Patterson 1989, 1994; Sogin and Hinkle 1997). Both circumscribed groups, chromists and stramenopiles, contain predominantly species of the old and large heterokont algal assemblage of the past botanical and phycological literature (see historical reviews in Corliss 1984, 1994a).

PLANTAE, a kingdom for scores of years, has conventionally been composed not only of the bryophytes, pteridophytes, and higher aquatic and terrestrial species (gymnosperms, angiosperms, and so on), but also traditionally of all the so-called algae, ranging from the prokaryotic cyanobacteria (blue-green algae) through the algal classes and divisions (or phyla) listed in this paper under various kingdoms, not to mention all fungal and even bacterial taxa. Here (see Corliss 1994a, 1998c), I have restricted the plant kingdom to the vascular (multicellular) photosynthetic eukaryotes plus four phyla of green algae, one of red algae, and one for the enigmatic glaucophytes (table 1).

FUNGI, separated from Plantae by various workers during the last 40–50 years (recently more vigorously; see Barr 1992), is sometimes persistently considered basically as a "plant" group. It has long included phyla of the so-called higher fungi (Ascomycota, Zygomycota, and Basidiomycota), but it has conventionally also laid claim to various lower fungi, including diverse kinds of slime molds (now under Protozoa or Chromista; see table 1) and the water molds or so-called motile zoosporic fungi (the chytrids, which are true fungi, and members of the Pseudofungi, which are quite different in many taxonomic characteristics and now assignable as heterokontic algae to Chromista). Very recent molecular studies add the curious and possibly ancient "protozoan" group of Microspora (the microsporidians), minute intracellular parasites with unique spores, to the Fungi (Canning 1998; Edlind 1998; Keeling and McFadden 1998).

ANIMALIA, long recognized as the haven for numerous invertebrate and vertebrate phyla, is little affected by protistan studies and reclassifications. However, the always-enigmatic Myxozoa (protozoan myxosporidians of the literature) are now thought to be animals of some sort (Anderson 1998; Cavalier-Smith 1998a; Corliss 1998b; Schlegel and others 1996; Siddall and others 1995; Smothers and others 1994). One might add to the kingdom, as I have controversially done (Corliss 1998c), the choanoflagellates, definitely considered a link to the sponges of Animalia and now to the Fungi as well (Cavalier-Smith 1998a,b).

OBSTACLES TO DERIVING RELIABLE ESTIMATES OF NUMBERS OF PROTISTS

There are many reasons why our knowledge of the kinds and numbers of protistan species generally lags far behind that for numerous other groups of organisms; these deserve brief mention here. In general, and above all, their tremendous diversity—combined with their often microscopic size, cosmopolitan nature, lack of overt sexual processes, and no helpful fossil record—renders precise study of the morphology, taxonomy, and evolution of most of them very difficult. Furthermore, protists for scores of years have been described by a motley array of naturalists, zoologists, botanists, mycologists, cell biologists, ecologists, limnologists, microscopists, parasitologists, and, more recently, geneticists and evolutionary and molecular biologists—persons with highly diverse backgrounds and conceptual outlooks and, in many cases, without rigorous taxonomic training or even proper awareness of the relevant systematic literature in protistology. Other, more specific reasons for our continuing ignorance and uncertainty about numbers of protists in existence include the following (sometimes overlapping) factors:

• The lack of a universal definition of a species among largely asexual eukaryotic microorganisms. There are few guidelines to assist taxonomic protistologists in their choice from a veritable smorgasbord of kinds (some overlapping) of species in the biological literature: morphological, phenetic, nominal, ecological, cryptic, endemic, taxonomic, parataxonomic, biological, asexual, sexual, genetic (sibling or syngenic), molecular, and chimaeric. The morphospecies concept seems reliable for numerous protists (Finlay and others 1996). But what are the criteria for recognition of the separateness of presumably closely related species? And to what extent does polymorphism (common and often striking in many groups of protists) complicate the problem, not to mention the historical acquisition of some endocytoplasmic inclusions or organelles by engulfment of (or invasion by) "foreign" microorganisms in eons past (Bardele 1997; Cavalier-Smith and Lee 1985; Gray 1992; Margulis 1993, 1996; Sapp 1994; Taylor 1987a)?

• An abundance of nomenclatural problems, exacerbated by lack of clarity in recognizing boundaries at the species and much higher taxonomic levels (Corliss 1993). Different protists have inadvertently been given identical names, and the same protist might have been described independently under different names: this has happened especially in cases of the so-called ambiregnal protists (Corliss 1984, 1986, 1995; Patterson 1986b; Taylor and others 1986). Names of species not accepted by later revisers (for example, of a genus or family) fall into synonymy with the oldest name available (rule of priority); but the senior synonym itself could be associated with an inadequately described organism. Problems are compounded by lumpers and splitters (Corliss 1976) in taxonomic protistology. And, in recent years, with the seemingly constant shifting about in the assignment of various groups to individual higher taxa, anyone tabulating species must be careful not to count the same organism twice under different headings in the ever-changing scheme of higher protistan classification (Corliss 1998b).

• The justification of new species continually being described in the literature on practically all protistan taxonomic groups (see the *Zoological Record* and relevant botanical and algal lists and monographs). The lack of appropriate techniques of study in the past might often have been the cause of proliferation of what are now deemed unnecessary or unacceptable species; but, today, is it the availability of improved cytological, biochemical, and molecular methods that fuels the persistent description of new forms? Yet, as habitats and niches in diverse geographic and ecological areas (including new host species for symbiotic forms) are more thoroughly explored (perhaps even for the first time), is it not reasonable to anticipate the finding of at least *some* novel species of algae and protozoa? In general, taxonomic protistologists are in widespread agreement that this *is* inevitably the case, while they understandably bewail the shortage of trained students and funds to investigate such habitats (Andersen 1992; Vickerman 1992). For the large protozoan phylum Ciliophora, in particular, Finlay and others (1996) argue that the great majority of the free-living, free-swimming, phagotrophic forms from freshwater and salt water habitats probably have been discovered and described already and that the number of these considered acceptable reaches only a few thousand. Foissner (1998), in contrast, claims that hundreds of additional species of ciliates inhabiting such edaphic habitats as diverse soils, with perhaps as many as 75% of them not living elsewhere, have been largely and unfairly neglected.

• The different ways in which different workers categorize the areas covered in their own studies or reviews. For example, members of most protistan high-level taxa might be thought of, by some, as falling into only three major groupings, with scant attention to overlappings: free-living species, symbiotic-parasitic forms, and fossilized species (of extinct *or* contemporary taxa). But the extent to which "symbiotic" and "free-living" forms can coincide is often blurred; consider the cases of mutualistic and commensalistic forms versus "true" endoparasites and ectoparasites, or even those of symphorionts (basically independent organisms merely carried about by nonspecific "hosts"). Other investigators might divide protistan groups on the basis of their being found in different major ecosystems: freshwater, marine, estuarine, or terrestrial habitats. Numerous workers emphasize what seems to be the preference of their organisms for specific geographic areas, raising the problem of endemism versus cosmopolitanism in recognition of "new" species. Still another popular general categorization highlights modes of nutrition: autotrophs (via photosynthetic pigments) versus heterotrophs (phagotrophs and osmotrophs or saprotrophs), with bacteria and algal protists serving as the most commonly engulfed prey organisms. Unfortunately, authors have sometimes not specified the limits or boundaries used in arriving at their particular "total numbers" of species assignable to a given higher taxon.

Is it any wonder that few published works make reliable overall estimates of the total numbers of species of protists? Keeping in mind the difficulties mentioned above, I am attempting here to overcome most such obstacles and arrive as objectively as possible at reasonably accurate figures of known protists as of 1998.

NUMBERS OF PROTISTS IN MAJOR GROUPS

In the following sections, I purposely arrange major taxa of protists under widely known "tried and true" top-level *conventional* headings, using vernacular titles for such broad categories—"protozoa," "algae," "fungi," "plants," and "animals." In each section, the formal names of protistan phyla acceptable to me (see table 1) are given in boldface, often with an indication of synonymous names and with cross-references, as needed, to the kingdom (names are in all capital letters for easy recognition) in which a given phylum, in my opinion, seems best assigned today. The final location of a phylum might not match the title of the section; and for some groups, the reader is referred to fuller treatment under one of the other sections.

Sources of data leading to my estimated numbers have been many, beginning with those cited, directly or indirectly, in my own first publications on the subject (Corliss 1982, 1984). All such figures have naturally required considerable updating to take into account new species descriptions and to accommodate revisions in which former species might have been rejected. Too numerous to list here have been the useful taxonomic monographs, books, compendia, and authoritative individual papers that I have consulted. But I should mention the most helpful single modern source of information, the volume edited by Margulis and others (1990), a prodigious work that contains 36 scholarly chapters contributed by some 60 specialists on the diverse high-level taxa of protists. For some groups, our knowledge of numbers is still frustratingly fragmentary. Also distressful is the continuing instability of the exact composition of various higher taxa involved in overall protistan megasystematics, which makes exact placement of some implicated genera and their species difficult. Generic names that are representative of particular taxa, incidentally, are generally not included in the present paper, because of space limitations, illustrative of diversity though they would be. For the interested reader some 1,100 of them have recently appeared elsewhere (Corliss 1994a; and see many more in specialized phycological and protozoological textbooks and in Lee and others 1985, Margulis and others 1990, Margulis and Schwarz 1998, Parker 1982, and Tappan 1980, although genera might be quite differently classified at the highest levels in such works).

Conventional Protozoan Phyla

The taxa below follow the usual arrangement commonly found in well-known biological and more-specialized protozoological textbooks. That is, forms mostly amoeboid, although also including some amoeboflagellates, with pseudopodia of various kinds (the old rhizopod and actinopod "sarcodinids") make up the first grouping; the numerous taxa whose species are predominantly biflagellated or multiflagellated (both pigmented and nonpigmented arrays, roughly the "phytoflagellates" and "zooflagellates," respectively, of old) come next; spore-forming parasitic taxa (sporozoa and the former "cnidosporidian" groups) are then treated; and finally the ciliates, a large collection of species that represents one of the most circumscribed and noncontroversial protistan taxa of all, are mentioned.

Names given *first* (and in boldface type) follow those used in table 1; but explanations and brief descriptions plus major synonymous names are supplied when deemed helpful. Note that, although some two dozen phyletically named taxa are considered below as, in effect, *conventionally* known "protozoan-like groups," nearly half have been reassigned to kingdoms *other than* the PROTOZOA of the present paper, as pointed out in appropriate places.

Archamoebae (synonym Karyoblastea): The pelobionts (such as *Pelomyxa*, a free-living, freshwater, benthic "giant amoeba" reaching 5 mm in diameter), embracing some five or six genera if the parasitic *Entamoeba* is also accepted here. Not long ago, these amitochondriate protists were placed in a separate kingdom, the "ARCHEZOA," along with the **Metamonada** and the **Parabasala** (see below). Although descriptions of quite a few species have appeared in the literature, there is now wide agreement that the number of acceptable ones is probably less than 12. The conservative figure that I am using as a total here is 10.

Neomonada: A group of often small, free-living, marine heterotrophic flagellates and amoeboflagellates (Cavalier-Smith 1998b), still ill-defined, many formerly in Cavalier-Smith's (1993a, 1997a) phylum "Opalozoa". Depending on the workers involved, the number can range from a dozen or two to several score (including some of the "unassignable" forms of Patterson and Zölffel 1991); many genera are monotypic (that is, they have only a single species). At this time, I estimate 30 as a possible total number of valid species here.

Rhizopoda (synonym Amoebozoa, in part, of Corliss 1984): Predominantly typical amoeboid forms, including ones with tests, shells, or thecae, but some small heterotrophic flagellates here as well (see Patterson and Zölffel 1991). Some workers put the enigmatic algal *Chlorarachnion* here; others, 40 or more species of plasmodiophorans (endoparasitic slime molds). Separation from the following phyletic group is not always clear. There are at least 5,000 species, with some to be dropped (for example, hundreds of poorly described testaceous amoebae might be rejected by future workers), but predictably with many new "small naked amoebae" awaiting discovery (Vickerman 1992). A few fossil forms—and possibly 250 symbiotic species—have been described.

Mycetozoa (synonyms Eumycetozoa and Myxomycetes): A "lower fungal" plasmodial slime mold group containing both cellular and "acellular" species. Some plasmodia can be longer than 3 m. Exact boundaries are uncertain (see remark under **Rhizopoda**, above). Some 800–900 species are assigned here, with probably more to be moved in from other taxa and still others to be found and described as new. Possibly a few fossils and a number of symbiotic forms belong here as well (for example, the necrotrophic plasmodiophorans, the soil protists infecting cabbage and other plants).

Foraminifera (synonym Granuloreticulosea): The foraminifers in the broadest sense (Lee and Anderson 1991). Perhaps as many as 45,000 species have been described, with nearly 40,000 as fossilized forms (many represent extinct lines and make up the "globigerine ooze" on ocean floors and are invaluable in dating strata for the petroleum industry). The diameter of some extinct fossil shells or tests may reach 15 cm; of living extant species, up to 6.5 cm. No end of new forams is in sight, although some workers question the taxonomic significance of some minor differences in morphology of the calcareous test. A few taxonomists include 15 genera of xenophyophorans (body diameters, up to 25 cm) and 12 genera of komokiaceans here.

Labyrinthomorpha: Net slime molds of mycologists; unique parasitic forms (for example, on eelgrass); also some saprotrophic on dead tissues. These are now better placed in the kingdom CHROMISTA than in PROTOZOA or FUNGI (Cavalier-Smith 1998a). The group includes labyrinthuleans proper plus thraustochytriaceans. It totals about 50 species.

Heliozoa: Mostly a freshwater group of the classical "actinopod sarcodinids". The amazing marine *Sticholonche zanclea*, a single species formerly considered to make up the separate heliozoan class Taxopoda, is perhaps better assigned to membership in the next phylum (**Radiozoa**, below). Some 180 species have been described as heliozoa, but only about 100 might be acceptable to today's specialists on the group, which itself could still be a polyphyletic taxon (Smith and Patterson 1986).

Radiozoa (synonym Radiolaria): Spherical marine planktonic "actinopods", producers of great depths of "radiolarian ooze" on ocean floors. There are three major subgroups, of which the first two are closer taxonomically to each other than to the third (all are sometimes treated as separate phyla): Acantharia, 500 species (possibly only half valid), of which no fossils have been described; Polycystina, 10,000 species (possibly only half valid), nearly 75% of which are found as fossils; Phaeodaria, without the endosymbionts found in preceding groups, 1,150 species (possibly only 60% valid), few of which have been found as fossils.

Percolozoa: Small heterotrophic flagellates or amoeboflagellates; a considerably smaller group than when originally circumscribed (Cavalier-Smith 1993b). Some former heterolobosean genera are here, some "unassignable" forms of Patterson and Zölffel (1991), and, controversially, the ciliate-turned-flagellate (Lipscomb and Corliss 1982; Patterson and Brugerolle 1988) *Stephanopogon*. There are more than 100 species.

Bicosoecae: Small nonpigmented heterotrophic flagellates, some colonial. This group has long been claimed by protozoologists, but see the treatment under "Algae," below. It is assigned to the kingdom CHROMISTA.

Dictyochae: Silicoflagellates, some known from the fossil record; long claimed by protozoologists; but see the treatment under "Algae," below. It is assigned to the kingdom CHROMISTA.

Cryptomonada: Mostly pigmented species, although many are heterotrophic. The group has long been claimed by protozoologists, but see the treatment under "Algae," below. It is assigned to the kingdom CHROMISTA.

Haptomonada: Pigmented, but claimed also by protozoologists. It is treated here under "Algae," below. It is assigned to the kingdom CHROMISTA.

Opalinata (synonyms Protociliata, Paraflagellata): Protozoologists' well-known opalinid parasites (in the strictest sense) plus *Karotomorpha* and *Proteromonas* (Delvinquier and Patterson 1992; Patterson 1986a). More than 400 species are reported in the literature, but many of the opalinids described are from ill-fixed material; perhaps only 200 are acceptable as valid today. The group was in PRO-TOZOA and is now assigned to the kingdom CHROMISTA (Cavalier-Smith 1998a,b).

Euglenozoa: Two principal subgroups (Triemer and Farmer 1991; Vickerman and others 1991): The Euglenophyta of the algal literature, more than 1000 species, mainly free-living, freshwater, and photosynthetic, although also phagotrophic, colorless, and some symbiotic-parasitic (and rare fossil) forms are known; and the Kinetoplastidea of the protozoological-parasitological literature, more than 600 species, ranging from pathogenic blood and tissue parasites of human beings (trypanosomatids) to free-living, freshwater or salt-water biflagellated species (bodonids).

Dinozoa (synonyms Dinoflagellata, Pyrrhophyta, and Peridinea + Syndinea): A major group of unique biflagellated protists, the dinoflagellates, long claimed by both phycologists and protozoologists. The pigmented species, some also heterotrophic, are a major component of marine plankton, but 10% occur in freshwater habitats; about 50% of the species are nonpigmented; some dinos are thecate and some colonial. About half the described species have been found as fossils, exclusive of 400 genera of acritarchs (a fossil group also assigned here by some workers) but including the small taxa of ebriideans and ellobiophyceans. Some species are important symbionts of other organisms; others exhibit toxic blooms (for example, red tides) with direct and indirect effect on humans. There is a distinct taxonomic subdivision of osmotrophic, endosymbiotic forms in diverse marine hosts. A primitive group might now include the nonpigmented former apicomplexan (see below) parasite *Perkinsus* (Siddall and others 1997). There are some 4,500 species, with perhaps nearly 2,500 as fossils of some extant but mostly extinct forms (Fensome and others 1993; Taylor 1987b).

Metamonada (synonym "polymastigotes," in part): Biflagellated to multiflagellated forms, typically gut parasites of diverse hosts (from insects to humans), allegedly (with the following phylum) primitive protists (Vickerman and others 1991). They have no mitochondria but hydrogenosomes (latest review, Müller 1998). There are about 300 species, but some are in need of restudy.

Parabasala (synonym "polymastigotes," in part): Mostly parasitic multiflagellated forms (called trichomonads and hypermastigotes), amitochondriate, and with striking parabasal (Golgi) apparatus. They share enough characteristics with the above phylum (**Metamonada**) to be joined with it (and the **Archamoebae**) under the one-time kingdom "ARCHEZOA" of Cavalier-Smith (1993b, 1998a, and references therein). The group has more than 400 species, some in need of restudy; doubtless more will be found, especially in the inadequately explored insect (woodroach) digestive tract.

Choanozoa (synonyms Choanoflagellata, Craspedophyceae): Planktonic (mostly marine) nonpigmented "collar-flagellates" with a single smooth anterior flagellum, often stalked or loricate. The group was placed in the protozoan phylum **Neomonada** by Cavalier-Smith (1998b) and tentatively transferred to the kingdom ANIMALIA by Corliss (1998c) and here. There are about 150 species.

Apicomplexa: Popular name for what is essentially the still-valid "**Sporozoa**" of old (Ellis and others 1998). An "apical complex" is made visible only by electron microscopy. The species are all symbiotic in a great variety of hosts, many as harmful endoparasites (Levine 1988; Perkins 1991). They include some of the smallest protists (intracellular forms with diameters less than 1 μm), although others can be up to 10 mm long. The major subgroups are gregarines (some large), coccidians (*Toxoplasma* and others in humans), and haematozoeans (malarial organisms and others). *Perkinsus* has been transferred to the phylum **Dinozoa** (see above). The "**Ascetospora**" of the literature is tentatively placed here. There are more than 5,000 species, some questionable today because of inadequate past accounts; but parasitologists predict numerous yet-to-be-described species. Levine (1973) once estimated, on the basis of potential combinations of numbers of sporocysts and sporozoites in the oocyst (which represent important differentiating taxonomic characters), that there could be, hypothetically, more than a million species in the second sporozoan subgroup (the coccidians) alone!

Microspora (synonym Microsporidia): A highly unusual group, with very small spores (diameters less than 1 μm) containing a complex extrusome and with a chitinous cell wall. The group consists of obligate intracellular parasites found in other protists, insects, fishes, and, opportunistically, human AIDS patients. Unicellular forms long considered as protozoa, they are here placed in the kingdom FUNGI on the basis of recent molecular findings (see citations in a preceding section of this paper). There are more than 800 species.

Myxozoa (synonyms Myxosporidia and Myxospora): Formerly grouped with **Microspora** as "cnidosporidians". These are histozoic or coelozoic parasites, mainly of cold-blooded vertebrates (the cause of great economic losses in the commercial fish industry). They have valved multicellular spores with polar capsules that include extrusible filaments. They were long considered as protozoa but here are placed in the kingdom ANIMALIA mainly on the basis of molecular data (see citations in a preceding section of this paper). There are more than 1,200 species. Some species in invertebrates, formerly assigned to independent status in a (second) major class, Actinomyxidea, are now being identified as simply *stages* in the life cycle of well-known myxosporidian fish parasites (Kent and others 1994).

Ciliophora (synonym Heterokaryota): Multiciliated (usually), colorless (with exceptions), relatively large cilioprotists (general range, 10–500 μm; a few up to 5,000 mm; and some colonies up to 15 cm in diameter). They exhibit nuclear dualism (the heterokaryotic condition—two kinds of nuclei, macronuclei and micronuclei; see Raikov 1996 for latest review), and are often phagotrophs, free-living in widely diverse habitats, although many groups are symbiotic-parasitic (including *Balantidium* in humans) or symphoriontic (the latter usually stalked). This is a large phylum (ranking fifth among all protists, behind diatoms, forams, charophytes, and radiolarians), with 8–10 classes and many orders. The total number of species is often said to be at least 8,000, including about 200 fossil forms (all of tintinnids) and an estimated 2,600 symbiotic species, with many more presumably awaiting discovery (Corliss 1979; Lynn and Corliss 1991; Small and Lynn 1985). But more conservative figures have recently been offered by Finlay and others (1996), who estimate a maximum of 4,300 for their pragmatic "morphospecies" of cosmopolitan free-living, phagotrophic forms primarily from major freshwater and salt-water habitats (calculating this to be 70% of all ciliates, including symbiotic species) and who suggest that careful taxonomic revisions might reduce their number to about 3,000. The matter is controversial (there might be many more valid soil-dwelling species than is often appreciated: see Foissner 1998).

Additional *"protozoan"* **groups:** Treated in the following section as conventional green or golden-brown "algal" taxa, are three phyla from within which selected (mostly motile) subgroups have long been of interest to protozoologists: the **Chlorophyta** (the volvocine line) and the **Prasinophyta** (both assigned here to the kingdom PLANTAE) and the **Chrysophyta**, assigned to the kingdom CHROMISTA. Treated in the later section on conventional "fungal" phyla are members of the **Chytridiomycota**, a number of species of which have been routinely included in protozoological textbooks. But that phylum, in its entirety, is placed in the kingdom FUNGI in this paper. Finally, the **"Ascetospora"** or "Haplosporidea" of both old and more recent literature (for example, Cavalier-Smith 1993b; Corliss 1994a) is tentatively placed within the **Apicomplexa** (above) here, on the basis of reasoning found in Cavalier-Smith (1998a).

Conventional Algal Phyla

Deliberately omitted here is further mention of the *prokaryotic* (cyanobacterial) divisions or classes of algae, the "Cyanophyta" or blue-green algae, with some 2,000 species, and the **"Prochlorophyta"** with fewer than six. Many botanists and phycologists recognize three "true" major broad algal assemblages, the red algae, the green algae, and the chromophyte algae. The latter vast group (containing numerous classes, depending on the author) has been known by a variety of names, including Chromobionta, Heterokontae, Heterokontophyta, and even Chrysophyta (in its broadest usage). Andersen (1992), whose summarizing table on numbers of algal species overall has been especially helpful to me (see also John 1994; Norton and others 1996), considered those three diverse assemblages to be taxonomically and phylogenetically "the major algal lineages," with four additional "minor lineages" (dinoflagellates, euglenophytes, cryptophytes, and glaucophytes) listed in his table below his classes of chromophytes.

Here I recognize some 16 eukaryotic algal groups, at phylum (division) rank, eight of which I assign to the kingdom CHROMISTA, six to PLANTAE, and two to PROTOZOA (**Euglenozoa, Dinozoa:** see above). The chromistan phyla contain the majority of the species broadly classified as "chromophyte algae" by botanists, and most of their species are pigmented (that is, carry out photosynthesis). My order of presentation below more or less follows the conventional arrangement used by many phycologists. Space does not permit specific mention of the names of many often traditionally well-recognized algal taxa usually treated (and here generally so retained) at the level of *class* or below, not as divisions or phyla. Their numbers of *species* have not been left out of my overall count but are included, as appropriate, in totals given for such large, all-embracing phyla or divisions as the **Chlorophyta** and the **Chrysophyta**. Some former classes have been elevated to phyletic status, or their names have here been considered more or less synonymous with preferred different names for the higher rank of phylum. "Taxonomic inflation", bringing about a concomitant increase in names, has been as inevitably rampant, in recent years, among the "algal" protists as among their "protozoan" and "fungal" counterparts—perhaps a consequence of our increasingly precise methods of study and analysis of the systematics and phylogeny of these highly diverse eukaryotic microorganisms (Corliss 1998b).

Rhodophyta: Nonflagellated, mostly marine, macroalgae (red seaweeds), but some minute unicells as well, and some parasitic species. Meter-long multicellular parenchymatous forms appear along rocky shores. The two principal classes or subgroups, Bangiophyceae and Florideophyceae, each have several or many orders. Species encrusted with $CaCO_3$ fossilize well. Red algae are a source of commercially valuable agar and of maerl, widely used as a fertilizer. This taxon (containing some of the oldest fossil algae known) has been given very high independent ranking taxonomically or, as here, has been assigned a unique place in the kingdom PLANTAE. There are well over 5,000 species (about 750 as fossils), and more than 100 species have been described as parasites of other red algae.

Glaucophyta (synonym Glaucocystophyta): A small algal group, all with cyanelles, all freshwater, and most biflagellated. They are placed in or near **Rhodophyta** by many workers; here, they are tentatively assigned to separate phyletic status in the kingdom PLANTAE. Depending on the number of accepted genera, the species counts range from a few to about 15.

Prasinophyta (synonym Micromonadophyceae): Grass-green scaly algae, freshwater, mostly small (and possibly primitive) biflagellated unicells. One of the tiniest free-living protists belongs to the picoplanktonic genus *Micromonas* (diameter, 1 μm). These species are assigned to the kingdom PLANTAE with other green algae. Some 400 species have been described (but perhaps only half that number are fully acceptable). About 100 have been found as fossils (some of which were originally identified as acritarchs; see the comment under **Dinozoa**, above).

Chlorophyta: *The* green algae of the botanical literature, mostly unicells or colonial in freshwater, many nonmotile. The celebrated "zoochlorellae" (symbionts of many ciliates) of the classical protozoological literature (and see Reisser 1992) belong here. There are many separate classes or orders; some phycologists conservatively include here members of some of the other phyla described below (such as **Ulvophyta**). This evolutionarily important and ecologically widespread group is assigned to the kingdom PLANTAE. It contains perhaps more than

3,500 species, depending on the inclusiveness of the phylum and thus on the workers making the counts.

Ulvophyta: Includes macroscopic seaweeds from tropical marine waters. They are sessile, with coenocytic or multicellular thalli. The group is assigned to the kingdom PLANTAE and contains at least 300 species, a few of which are fossils.

Charophyta (synonyms Conjugatophyceae, Gamophyceae, and Zygonemato-phyceae, and others): Mostly (including the ubiquitous desmids) unicellular or filamentous in freshwater, vegetative stage nonflagellated, and with conjugation often involving amoeboid gametes. Larger forms—far fewer in species—are placed in a separate class, which includes the well-known stoneworts, with macroscopic thalli typically scale-covered. Several charophyte characteristics are clearly reminiscent of land plants, their evolutionary descendants. This group of green algae is assigned to the kingdom PLANTAE. It has some 12,000 species, but about 9,000 are desmids alone (of which half are of uncertain validity); stoneworts number fewer than 400 species, about 300 of which have been found only as fossils.

Dinozoa: Long (and still) claimed as algae, but treated in this paper as PRO-TOZOA (see preceding section).

Euglenozoa: Long (and still) claimed as algae, but treated in this paper as PROTOZOA (see preceding section).

Bicosoecae: Freshwater and marine nonpigmented flagellates, some with loricae. These were formerly placed within **Chrysophyta**. The group is assigned to the kingdom CHROMISTA and contains about 40 species.

Dictyochae (synonym Dictyochophyceae): Silicoflagellates, formerly in **Chrysophyta**, with a number of fossil marine forms. *Dictyocha* is the only genus with extant species (as the phylum is restricted here). The group is assigned to the kingdom CHROMISTA. It has fewer than 12 species (excluding *Actinomonas* and *Pedinella* and close relatives that are placed here by some phycologists).

Cryptomonada (synonym Cryptophyta): Well-known freshwater and marine mostly pigmented biflagellated protists, some phagotrophic, some endosymbiotic. The group is controversially assigned to the kingdom CHROMISTA but *not* within the large heterokontic moiety. There are about 200 species.

Haptomonada (synonyms Coccolithophora, Haptophyta, and Prymnesiophyta): Yellow-brown algae, typically marine flagellates with unique haptonema arising between a pair of polar flagella and a body usually covered with layers of scales, some known as coccoliths. The group is controversially assigned to the kingdom CHROMISTA but *not* within the large heterokontic moiety. Some 500 living and 1,200 fossil species have been described; the celebrated white cliffs of Dover are composed mostly of coccoliths.

Chrysophyta: The golden-brown algae, numerous freshwater species, some with delicate loricae, and many producing a unique statospore as a resting stage. The group is assigned to the kingdom CHROMISTA as a major phylum of the heterokontic moiety there. There are perhaps more than 1,500 species, several classes of which are given independent phyletic status by some workers. The total includes about 250 fossil forms.

Diatomae (synonyms Bacillariophyta, Diatomea, and Diatomophyceae): The diatoms. "Bacillariophyceae" is the most popular name used for this taxonomic group. They are yellow-brown unicells, widespread planktonic and benthic forms in salt-water amd today especially freshwater habitats and are also found in moist soil; a few are endosymbionts of the protozoan foraminifers (Lee 1992). They are nonflagellated in the vegetative stage. Diatoms have the characteristic two-valved siliceous test or frustule, which is readily fossilizable and the main component of commercially useful diatomite ("diatomaceous earth"). The group is assigned to the kingdom CHROMISTA. The number of recorded forms has apparently reached 100,000, including fossils of both extinct and extant forms, according to Round and others (1990), or even 200,000 according to Mann and Droop (1996). But some conservative phycologists have estimated that only about 25% (or less) might be acceptable as truly separate extant species. The ratio of living to fossil forms has been given as 2:3. Some diatom specialists (personal communications acknowledged in Andersen 1992 and Norton and others 1996; see also John 1994) predict that the "real" (potentially describable) number of species of these highly abundant and very important autotrophic protists might reach an amazing total of 10,000,000!

Raphidophyta (synonym Chloromonadophyceae and inappropriately known as "the chloromonads", but *Chloromonas* is a genus of the green algal phylum **Chlorophyta** in the kingdom PLANTAE): Small group of yellow-green algae, from freshwater and salt-water habitats, formerly placed in **Chrysophyta** by some workers. The group is assigned to the kingdom CHROMISTA. It has fewer than three dozen species.

Phaeophyta (synonyms Fucophyceae and Melanophyceae): The brown heterokont algae, with multicellular filaments or thalli. These are large seaweeds (kelp) of intertidal or subtidal habitats, gigantic protists reaching lengths of up to 60 m. Many are of commercial value, directly as food or as sources of alginates, fertilizers, vitamins, and minerals. The group, sometimes closely linked to the **Chrysophyta**, is assigned to the kingdom CHROMISTA. It has more than 1,600 species, a few described as fossils and a few as symbionts on other algae or seagrass.

Conventional Fungal Phyla

Under consideration in this paper are *only* the basically unicellular "fungus-like" protists, not the long-accepted "higher" fungal taxa. As implied in earlier sections, botanists (that is, mycologists) formerly claimed many protozoan-protistan groups as "lower" fungi, particularly the slime molds (including the labyrinthulids and plasmodiophorans) and the zoosporic taxa. In recent years, there has been growing acceptance of removal from the kingdom FUNGI not only of the various slime molds but also of two of the three flagellated (independently motile) groups (see **Pseudofungi**, below), leaving only the chytrids as true (although unicellular and flagellated) fungi.

Myxomycetes (synonym Myxomycota): See account under the "protozoan" phylum **Mycetozoa**, above. The group is assigned to the kingdom PROTOZOA.

Labyrinthomorpha: See the group, by the same name, above (with other "protozoan" phyla). But the group is assigned in this paper to the kingdom CHROMISTA.

Pseudofungi (synonyms, at least in part, Mastigomycetes, Oomycota, Phycomycetes, and Pseudomycota): Zoosporic protists separable into two subphyletic zoosporic taxa, the Oomycetes (synonym Oomycota) and the Hyphochytriomycetes (synonyms Hyphochytrea and Hyphochytridiomycota) of the literature. Both groups are assigned to the kingdom CHROMISTA. These small but numerous freshwater "water molds," whose zoospores have an anteriorly projecting flagellum bearing mastigonemes, parasitize hosts ranging from other protists and aquatic plants to fishes and, via the soil, grapes and potatoes. Many species are also saprotrophic on detritus and dead tissues in aqueous and terrestrial habitats. More than 800 species of oomycetes have been described, although some are now considered doubtful; about two dozen species are known from the second taxon.

Chytridiomycota: A third taxon of zoosporic protists, but with posteriorly projecting smooth flagellum (no mastigonemes) and taxonomically remaining in the kingdom FUNGI. They have several fungal characteristics, including chitinous cell walls in their hyphal stage, although they are basically unicellular. They are symbionts or saprobes in soil and freshwater habitats (Powell 1993); a few, treated as protozoa in past years, are found in the digestive tract of horses and ruminants. Some 900 species have been described.

Microspora: As pointed out above (under conventional "protozoan" phyla), only very recently have true fungal affinities been discovered for these tiny intracellular parasites presumably of ancient phylogenetic origin. See the protozoan section (above) for data on the group; but recall that it now properly belongs here in the kingdom FUNGI as the second phylum of fungal protists (the first being the **Chytridiomycota**, see immediately above).

Conventional Plant Phyla

Considering here only the basically unicellular (or multicellular but not truly multitissued) "lower" plants, I hardly need to point out that formerly *all* "algal" taxa, including some claimed also by protozoologists, were treated as members of the kingdom PLANTAE. Because all fungi were under this banner, too, it follows that the "lower" fungi, most groups of which are now considered to be members of the totally protistan kingdoms PROTOZOA and CHROMISTA, were also formerly claimed by botanists as plants. Today, I assign or retain essentially only

the red and (some of) the green algal groups as protistan assemblages in the kingdom PLANTAE (see conventional "algal" section, above).

Conventional Animal Phyla

Considering the protozoa as basically unicellular organisms, it is common
knowledge that they were long treated as animals, as a single phylum (or eventually at best a subkingdom) of the kingdom ANIMALIA. It follows that *all* subtaxa
of such protozoan protists were considered taxonomically as microscopic "first"
animals. Some algal groups were also included, mostly under the title of "Phyto-
mastigina" or "Phytomastigophora", as well as two phyla (the chytrids and
microsporidians—see above) now treated as true fungi. In this paper, I append
only two protistan phyla to the kingdom ANIMALIA (see above, under conventional "protozoan" phyla), namely the **Choanozoa** (controversially) and the
Myxozoa.

SOME SUMMARIZING OBSERVATIONS

Using (with appropriate caution) data given on the preceding pages, we can
draw several conclusions concerning total numbers of species of protists (see also
table 2). A grand total of at least 213,000 species, distributed among the 35 phyla
recognized in this paper, have been described in the literature to date. Interestingly enough, about 113,000 of these are fossil forms. Five of the 18 phyla known
to have any fossils at all contain 98% of the known fossil protists; these groups,
in order of richness in fossil species, are the diatoms, foraminifers, radiozoa, dinoflagellates, and haptomonads. In fact, the diatoms and forams alone are responsible for 90% of them. Still, fossil forms also represent an important percentage
of the species of some of the smaller phyla (for example, 15–25% of chrysophytes,
prasinophytes, and rhodophytes).

Among the extant contemporary forms, numbering some 100,000 species, only
about 14% can be labeled as symbionts in the broadest sense, ranging from
symphoriontic, commensalistic, and mutualistic forms to obligate ectoparasites and
endoparasites (with the latter including some highly pathogenic microorganisms)
on and in all kinds of protistan, plant, fungal, and animal hosts. Free-living
species would thus seem to outnumber greatly the symbiotic forms. The percentage figure given above, however, is somewhat misleading. If, in addition to fossils, we also leave to one side the huge number (40,000) of *non*fossil diatoms
(many controversial anyway?), the roughly 14,000 symbiotic species become nearly
one-fourth (23%) of all other extant protists.

Incidentally, 95% of the 14,000 symbiotic species are members solely of the 10
following phyla: **Apicomplexa** (all), **Ciliophora** (one-third), **Myxozoa** (all), **Chytridiomycota** (all), **Pseudofungi** (all), **Microspora** (all), **Metamonada** plus **Parabasala** (essentially all), **Euglenozoa** (some euglenids plus essentially all trypanosomatids), and **Opalinata** (all). But the remaining 5% include scattered important
species found among dinoflagellates, cryptophytes, chlorophytes, and rhodophytes,
with the majority pigmented; among amoebae, mycetozoa, and amoeboflagellates;
and among members of various other usually smaller protistan groups.

TABLE 2 Numbers of Described Species of Protists per Major Taxonomic Group[a]

Kingdoms and Their Phyla	Total Reported Species[b]	Percent Fossils	Percent Symbionts[c]	Additional Notes and Comments
PROTOZOA				
Archamoebae	10	None	Few	Amitochondriate, primitive(?)
Neomonada	30	None	None	Also presumably primitive group
Rhizopoda	5,000	Few	5%?	Uncertain taxonomic boundaries
Mycetozoa	900	Few	20%?	Uncertain taxonomic boundaries
Foraminifera	45,000	89%	Few	Some species uncertain
Heliozoa	180	None	None	Uncertain boundaries and species
Radiozoa	11,650	65%	None	Many species uncertain?
Percolozoa	100	None	Few	Uncertain taxonomic boundaries
Euglenozoa	1,600	Rare	45%	Most symbionts kinetoplastideans
Dinozoa	4,500	55%	2–3%	Symbionts few but important
Metamonada	300	None	>95%	Nearly all intestinal parasites
Parabasala	400	None	>95%	Nearly all intestinal parasites
Apicomplexa	5,000	None	100%	No nonparasitic form identifiable?
Ciliophora	8,000	2.5%	33%	Only few parasitic in primates
CHROMISTA				
Labyrinthomorpha	50	None	100%	Some saprotrophic on dead tissue
Pseudofungi	850	None	100%	Some saprotrophic on dead tissue
Bicosoecae	40	None	Few	Nonpigmented heterotrophs
Dictyochae	10	Few	None	The silicoflagellates
Cryptomonada	200	None	Few	Nonheterokontic group
Haptomonada	1,700	70%	None	Prymnesiophytes of phycologists
Opalinata	400	None	100%	Many species uncertain?
Chrysophyta	1,500	15%?	None	Uncertain taxonomic boundaries
Diatomae	100,000	60%?	Few	Total number highly uncertain
Raphidophyta	35	None	None	Needs more study
Phaeophyta	1,600	Few	Few	Kelp, giants among protists!
PLANTAE				
Rhodophyta	5,000	15%	2%?	Allegedly evolutionarily ancient
Glaucophyta	15	None	None	Uncertain taxonomic status
Prasinophyta	400	25%	None	Some species uncertain
Chlorophyta	3,500	Few	Few	Uncertain taxonomic boundaries
Ulvophyta	300	Few	None	Uncertain taxonomic boundaries
Charophyta	12,000	2.5%	None	Includes the numerous desmids
FUNGI				
Chytridiomycota	900	Rare	100%	Some saprotrophic on dead tissue
Microspora	800	None	100%	Formerly protozoan protists
ANIMALIA				
Choanozoa	150	None	None	Formerly protozoan protists
Myxozoa	1,200	None	100%	Formerly protozoan protists

[a] Classification as in table 1, but with phyla differently ordered.
[b] Numbers given are generally, but not always, already more or less accepted by, or assumed to be acceptable to, various specialists on the implicated groups.
[c] Includes all relationships with hosts, from protists as benign epibionts to pathogenic endoparasites.

Additional calculable totals and other comments can be found in table 2, where the phyla of the five kingdoms are in a different arrangement from that found in table 1, in keeping more closely with the order of their presentation on the preceding pages.

Briefly, the totals (here rounded off) per kingdom of *all* species of *protists* contained therein (be they fossilized; free-living or symbiotic; autotrophic, heterotrophic, or mixotrophic; benthic or planktonic; from aquatic or terrestrial habitats; and so on) are as follows: PROTOZOA (as restricted in this paper), 82,700 species; CHROMISTA (with its mixture of many traditional algal phyla and some others), 106,400 (but 94% of these are diatoms); PLANTAE (six algal phyla), 21,200; FUNGI (two phyla), 1,700; and ANIMALIA (two phyla), 1,350.

With respect to *described* species versus putatively *valid* or *acceptable* species, I have despaired of solving all such problems here. In my calculations (and in table 2), I have generally used numbers from the first category—that of described forms—on the basis of the original literature (or reliable second-hand sources). For the great majority of protistan phyla, there has seldom been to date a significant difference between the two sets of figures, so I have not cited the latter numbers in this paper. However, there are two striking examples of disparity or discrepancy between the numbers—described versus acceptable species—in the cases of **Diatomae** and **Radiozoa.** Of the 100,000 (or more!) diatom species (extant and extinct) allegedly established in the literature, are as few as 10,000–12,000 the maximal number acceptable to many phycologists today? Or are authors of the lower figure excluding fossil (and some other) forms from their estimates without clearly informing their readers of the fact? For the radiozoa, are about half the 11,650 described species now to be considered by protozoologists to be invalid or uncertain? Or do some papers on the subject seem confusing only to the unsophisticated reader? I suggest that specialists, not generalists like me, should discuss and eventually solve or at least clarify such serious problems to everyone's satisfaction.

Whereas there is little doubt that many species of protists have not been carefully enough described in a comparative way (and thus really are "lumpable") and that endemism has been overused as a basis for newness (including that old parasitological dictum, "A new host means a new species"), is it possible that only a relatively few truly new species remain undetected in the largely unexplored biomes of eukaryotic microorganisms?

On the basis of personal communication with many protistologists, I am obliged to draw the conclusion that, for numerous groups, vast numbers of unique protists *do* await description. Perhaps we have only scratched the surface regarding the biodiversity of these organisms. Thus, with rare exception, I have not attempted to include estimates of the *probable* numbers of species assignable to the phyla described to date.

OUTLOOK AND GOALS FOR THE FUTURE

The roles of protists in natural ecosystems are, in a general way, beginning to be appreciated, but they are hardly yet understood to a very helpful degree, one applicable to humankind's many environmental challenges. Awareness of their potential is only the first step in the process of getting to know them better. Several major needs are becoming clear, as exposed very briefly below:

• The biodiversity of protistan groups must be studied in greater depth. That is, we need to understand their distribution—and functions—on a global scale to focus on their diverse interactions with other organisms in a wide variety of habitats. To investigate their ecology, we must improve our knowledge of their taxonomy (and vice versa: Corliss 1992). More-thorough comparative studies need to be carried out, with use of the most precise sampling and cytological techniques now available.

• Reaching widespread agreement on the nature of a *protistan species* is imperative. If we do not understand the dimensions of a species definition, taxonomic and nomenclatural problems will continue to plague our progress. And we can hardly prepare inventories without knowing the identity of our material in considerable depth.

• More-extensive work on the phylogeny of the protists will throw light on their evolutionary relationships with the prokaryotic bacteria and with the other eukaryotes, the latter assemblages all supposedly having had protistan origins. An interdisciplinary approach thus needs to continue to be taken in studying protists because of the value of viewing major problems from different points of view. Cladistic trees and taxonomic classification systems must be refined and become more supportive of one another.

• Practical reasons for studying many protistan groups more intensely are related to their direct and indirect effects on human welfare, ranging from their basic food-chain involvement (nutrient and mineral recycling), their roles in agriculture and aquaculture, and their commercial, medicinal, and biomonitoring uses to their being causative agents of major diseases.

• Clearly, more financial support is needed for protistological research, for teaching and training more students and technicians, for maintenance and expansion of culture collections and gene banks, and for preparing appropriate inventories or censuses of species numbers. All these activities are necessary for determining future avenues worthy of exploration in the vast field of protistan biodiversity.

REFERENCES

Aescht E (ed). 1998. Welträtsel und Lebenswunder: Ernst Haeckel—Werk, Wirkung und Folgen. Stapfia 56:1–506.

Andersen RA. 1992. Diversity of eukaryotic algae. Biod Cons 1:267–92.

Andersen RA. 1998. What to do with protists? Austral Syst Bot 11:185–201.

Anderson CL. 1998. Phylogenetic relationships of the Myxozoa. In: Coombs GH, Vickerman K, Sleigh MA, Warren A (eds). Evolutionary relationships among protozoa. Dordrecht The Netherlands: Kluwer Acad Publ. p 341–50.

Bardele CF. 1997. On the symbiotic origin of protists, their diversity, and their pivotal role in teaching systematic biology. Ital J Zool 64:107–13.

Barr DJS. 1992. Evolution and kingdoms of organisms from the perspective of a mycologist. Mycologia 84:1–11.

Canning EU. 1998. Evolutionary relationships of microsporidia. In: Coombs GH, Vickerman K, Sleigh MA, Warren A (eds). Evolutionary relationships among protozoa. Dordrecht The Netherlands: Kluwer Acad Publ. p 77–90.

Cavalier-Smith T. 1981. Eukaryote kingdoms: seven or nine? BioSystems 14:461–81.

Cavalier-Smith T. 1986. The kingdom Chromista: origin and systematics. Prog Phycol Res 4:309–47.

Cavalier-Smith T. 1989. The kingdom Chromista. In: Green JC, Leadbeater JSC, Diver WL (eds). The chromophyte algae: problems and perspectives. Oxford UK: Clarendon Pr. p 381–407.

Cavalier-Smith T. 1993a. The protozoan phylum Opalozoa. J Euk Microbiol 40:609–15.

Cavalier-Smith T. 1993b. Kingdom Protozoa and its 18 phyla. Microbiol Rev 57:953–94.

Cavalier-Smith T. 1995. Evolutionary protistology comes of age: biodiversity and molecular cell biology. Arch Protistenk 145:145–54.

Cavalier-Smith T. 1997a. Amoeboflagellates and mitochondrial cristae in eukaryote evolution: megasystematics of the new protozoan subkingdoms Eozoa and Neozoa. Arch Protistenk 147:237–58.

Cavalier-Smith T. 1997b. Sagenista and Bigyra, two phyla of heterotrophic heterokont chromists. Arch Protistenk 148:253–67.

Cavalier-Smith T. 1998a. A revised six-kingdom system of life. Biol Revs 73:203–66.

Cavalier-Smith T. 1998b. Neomonada and the origin of animals and fungi. In: Coombs GH, Vickerman K, Sleigh MA, Warren A (eds). Evolutionary relationships among protozoa. Dordrecht The Netherlands: Kluwer Acad Publ. p 375–407.

Cavalier-Smith T, Lee JJ. 1985. Protozoa as hosts for endosymbioses and the conversion of symbionts into organelles. J Protozool 32:376–9.

Colwell RR. 1997. Microbial biodiversity and biotechnology. In: Reaka-Kudla ML, Wilson DE, Wilson EO (eds). Biodiversity II: understanding and protecting our biological resources. Washington DC: Joseph Henry Pr. p 279–87.

Coombs GH, Vickerman K, Sleigh MA, Warren A (eds). 1998. Evolutionary relationships among protozoa. Dordrecht The Netherlands: Kluwer Acad Publ.

Copeland HF. 1956. The classification of lower organisms. Palo Alto CA: Pacific Bk.

Corliss JO. 1976. On lumpers and splitters of higher taxa in ciliate systematics. Trans Amer Microsc Soc 95:430–42.

Corliss JO. 1979. The ciliated protozoa: characterization, classification, and guide to the literature. 2nd ed. Oxford UK, New York NY: Pergamon.

Corliss JO. 1982. Numbers of species comprising the phyletic groups assignable to the kingdom Protista. (Abstr.) J Protozool 29:499.

Corliss JO. 1983. Consequences of creating new kingdoms of organisms. BioScience 33:314–8.

Corliss JO. 1984. The kingdom Protista and its 45 phyla. BioSystems 17:87–126.

Corliss JO. 1986. Progress in protistology during the first decade following reemergence of the field as a respectable interdisciplinary area in modern biological research. Prog Protistol 1:11–63.

Corliss JO. 1987. Protistan phylogeny and eukaryogenesis. Int Rev Cytol 100:319–70.

Corliss JO. 1989a. The protozoon and the cell: a brief twentieth-century overview. J Hist Biol 22:307–23.

Corliss JO. 1989b. Protistan diversity and origins of multicellular/multitissued organisms. Bull Zool 56:227–34.

Corliss JO. 1990. Toward a nomenclatural protist perspective. In: Margulis L, Corliss JO, Melkonian M, Chapman DJ (eds). Handbook of Protoctista. Boston MA: Jones and Bartlett.

Corliss JO. 1991. Introduction to the protozoa. In: Harrison FW, Corliss JO (eds). Microscopic anatomy of invertebrates. New York NY: Wiley-Liss. 1:1–12.

Corliss JO. 1992. The interface between taxonomy and ecology in modern studies on the protists. Acta Protozool 31:1–9.

Corliss JO. 1993. Should there be a separate code of nomenclature for the protists? BioSystems 28:1–14.

Corliss JO. 1994a. An interim utilitarian ("user-friendly") hierarchical classification and characterization of the protists. Acta Protozool 33:1–51.

Corliss JO. 1994b. The place of the protists in the microbial world. USFCC Newsl 24(3):1–6.

Corliss JO. 1995. The ambiregnal protists and the codes of nomenclature: a brief review of the problem and of proposed solutions. Bull Zool Nomencl 52:11–7.

Corliss JO. 1998a. The protists deserve attention: what are the outlets providing it? Protist 149:3–6.

Corliss JO. 1998b. Classification of protozoa and protists: the current status. In: Coombs GH, Vickerman K, Sleigh MA, Warren A (eds). Evolutionary relationships among protozoa. Dordrecht The Netherlands: Kluwer Acad Publ. p 409–47.

Corliss JO. 1998c. Haeckel's kingdom Protista and current concepts in systematic protistology. Stapfia 56:85–104.

Delvinquier BLJ, Patterson DJ. 1992. The opalines. In: Kreier JP, Baker JR (eds). Parasitic protozoa. 2nd ed. New York NY and London UK: Academic Pr. 3:247–325.

Edlind TD. 1998. Phylogenetics of protozoan tubulin with reference to the amitochondriate eukaryotes. In: Coombs GH, Vickerman K, Sleigh MA, Warren A (eds). Evolutionary relationships among protozoa. Dordrecht The Netherlands: Kluwer Acad Publ. p 91–108.

Ellis JT, Morrison DA, Jeffries AC. 1998. The phylum Apicomplexa: an update on the molecular phylogeny. In: Coombs GH, Vickerman K, Sleigh MA, Warren A (eds). Evolutionary relationships among protozoa. Dordrecht The Netherlands: Kluwer Acad Publ. p 255–74.

Fensome RA, Taylor FJR, Norris G, Sarjeant WAS, Wharton DI, Williams GL. 1993. A classification of living and fossil dinoflagellates. Micropaleontology Special Publication No 7. New York NY: Micropaleontology Pr, American Museum of Natural History.

Finlay BJ. 1998. The global diversity of protozoa and other small species. Int J Parasitol 28:29–48.

Finlay BJ, Corliss JO, Esteban G, Fenchel T. 1996. Biodiversity at the microbial level: the number of free-living ciliates in the biosphere. Quart Rev Biol 71:221–37.

Finlay BJ, Esteban GF. 1998. Freshwater protozoa: biodiversity and ecological function. Biod Cons 7:1163–86.

Foissner W. 1998. An updated compilation of world soil ciliates (Protozoa, Ciliophora), with ecological notes, new records, and descriptions of new species. Europ J Protistol 34:195–235.

Gray MW. 1992. The endosymbiont hypothesis revisited. Int Rev Cytol 141:233–357.

Haeckel E. 1866. Generelle Morphologie der Organismen... 2 vols. Berlin Germany: G Reimer.

Haeckel E. 1878. Das Protistenreich. Leipzig Germany: Günther.

Hausmann K, Bradbury PC (eds). 1996. Ciliates: cells as organisms. Stuttgart Germany: G Fischer.

Hausmann K, Hülsmann N. 1996. Protozoology. 2nd ed. Stuttgart Germany and New York NY: Georg Thieme.

Hawksworth DL, Colwell RR. 1992. Microbial Diversity 2: biodiversity amongst microorganisms and its relevance. Biod Cons 1:221–6.

Hülsmann N, Hausmann K. 1994. Towards a new perspective in protozoan evolution. Europ J Protistol 30:365–371.

John DM. 1994. Biodiversity and conservation: an algal perspective. The Phycologist 38:3–15.

Karpov SA. 1990. System of Protista. Mezh Tipograf, OMPi, Omsk. (in Russian)

Katz LA. 1998. Changing perspectives on the origin of eukaryotes. Trends Ecol Evol 13:493–7.

Keeling PJ, McFadden GI. 1998. Origins of microsporidia. Trends Microbiol 6:19–23.

Kent ML, Margolis L, Corliss JO. 1994. The demise of a class of protists: taxonomic and nomenclatural revisions proposed for the protist phylum Myxozoa Grassé, 1970. Can J Zool 72:932–7.

Knoll AH. 1992. The early evolution of eukaryotes: a geological perspective. Science 256:622–7.

Kuźnicki L, Walne PL. 1993. Protistan evolution and phylogeny: current controversies. Acta Protozool 32:135–40.

Lee JJ. 1992. Taxonomy of algae symbiotic in foraminifera. In: Reisser W (ed). Algae and symbioses: plants, animals, fungi, viruses, interactions explored. Bristol UK: Biopress Ltd. p 79–92.

Lee JJ, Anderson OR (eds). 1991. Biology of Foraminifera. London UK: Acad Pr.

Lee JJ, Hutner SH, Bovee EC (eds). 1985. An illustrated guide to the protozoa. Lawrence KS: Society of Protozoologists.

Levine ND. 1973. Protozoan parasites of domestic animals and of man. 2nd ed. Minneapolis MN: Burgess.

Levine ND. 1988. The protozoan phylum Apicomplexa. 2 vols. Boca Raton FL: CRC Pr.

Lipps JH (ed). 1993. Fossil prokaryotes and protists. Boston MA: Blackwell Scientific.

Lipscomb DL. 1991. Broad classification: the kingdoms and the protozoa. In: Kreier JP, Baker JR (eds). Parasitic protozoa. 2nd ed. New York NY and London UK: Academic Pr. 1:81–136.

Lipscomb DL, Corliss JO. 1982. Stephanopogon, a phylogenetically important "ciliate," shown by ultrastructural studies to be a flagellate. Science 215:303–4.

Lipscomb DL, Farris JS, Källersjö M, Tehler A. 1998. Support, ribosomal sequences and the phylogeny of the eukaryotes. Cladistics 14:303–38.

Lynn DH, Corliss JO. 1991. Ciliophora. In: Harrison FW, Corliss JO (eds). Microscopic anatomy of invertebrates. New York NY: Wiley-Liss. 1:333–467.

Lynn DH, Small EB. 1997. A revised classification of the phylum Ciliophora Doflein, 1901. Rev Soc Mex Hist Nat 47:65–78.

Mann DG, Droop SJM. 1996. Biodiversity, biogeography and conservation of diatoms. Hydrobiologia 336:19–32.

Margulis L. 1974. Five-kingdom classification and the origin and evolution of cells. Evol Biol 7:45–78.

Margulis L. 1988. Systematics: the view from the origin and early evolution of life. Secession of the protoctista from the animal and plant kingdoms. In: Hawksworth D, Davies RG (eds). Prospects in systematics. Oxford UK: Clarendon Pr. p 430–43.

Margulis L. 1993. Symbiosis in cell evolution. 2nd ed. San Francisco CA: WH Freeman.

Margulis L. 1996. Archaeal-eubacterial mergers in the origin of Eukarya: phylogenetic classification of life. Proc Natl Acad Sci USA 93:1071–6.

Margulis L, Corliss JO, Melkonian M, Chapman DJ (eds). 1990. Handbook of Protoctista. Boston MA: Jones and Bartlett.

Margulis L, Schwartz KV. 1982. Five kingdoms: an illustrated guide to the phyla of life on earth. 1st ed. San Francisco CA and New York NY: WH Freeman.

Margulis L, Schwartz KV. 1998. Five kingdoms: an illustrated guide to the phyla of life on earth. 3rd ed. New York NY: WH Freeman.

Mayr E. 1997. This is biology: the science of the living world. Cambridge MA: Belknap Pr.

Mayr E. 1998. Two empires or three? Proc Natl Acad Sci USA 95:9720–3.

Müller M. 1998. Enzymes and compartmentation of core energy metabolism of anaerobic protists—a special case in eukaryotic evolution? In: Coombs GH, Vickerman K, Sleigh MA, Warren A (eds). Evolutionary relationships among protozoa. Dordrecht The Netherlands: Kluwer Acad Publ. p. 109–32.

Norton TA, Melkonian M, Andersen RA. 1996. Algal biodiversity. Phycologia 35:308–26.

Parker SP (ed). 1982. Synopsis and classification of living organisms. 2 vols. New York NY: McGraw-Hill.

Patterson DJ. 1986a. The fine structure of Opalina ranarum (family Opalinidae): opalinid phylogeny and classification. Protistologica 21 (year 1985):413–28.

Patterson DJ. 1986b. Some problems of ambiregnal taxonomy and a possible solution. Symp Biol Hung 33:87–93.

Patterson DJ. 1989. Stramenopiles: chromophytes from a protistan perspective. In: Green JC, Leadbeater BSC, Diver WI (eds). The chromophyte algae: problems and perspectives. Oxford UK: Clarendon Pr. p 357–79.

Patterson DJ. 1994. Protozoa: evolution and systematics. In: Hausmann K, Hülsmann N (eds). Progress in protozoology (Proceedings of the IX International Congress of Protozoology, Berlin, 1993). Stuttgart Germany: G Fischer. p 1–14.

Patterson DJ, Brugerolle G. 1988. The ultrastructural identity of Stephanopogon apogon and the relatedness of the genus to other kinds of protists. Europ J Prostistol 23:279–90.

Patterson DJ, Sogin ML. 1993. Eukaryote origins and protistan diversity. In: Hartman H, Matsuno K (eds). The origin and evolution of the cell. Singapore: World Scientific Publ Co. p 13–46.

Patterson DJ, Zölffel M. 1991. Heterotrophic flagellates of uncertain taxonomic position. In: Patterson DJ, Larsen J (eds). The biology of free-living heterotrophic flagellates. Oxford UK: Clarendon Pr. p. 427–76.

Perkins FO. 1991. "Sporozoa": Apicomplexa, Microsporidia, Haplosporidia, Paramyxea, Myxosporidia, and Actinosporidia. In: Harrison FW, Corliss JO (eds). Microscopic anatomy of invertebrates. New York NY: Wiley-Liss. 1:261–331.

Powell MJ. 1993. Looking at mycology with a Janus face: a glimpse at chytridiomycetes active in the environment. Mycologia 85:1–20.

Ragan MA. 1997. A third kingdom of eukaryotic life: history of an idea. Arch Protistenk 148:225–43.

Raikov IB. 1996. Nuclei in ciliates. In: Hausmann K, Bradbury PC (eds). Ciliates: cells as organisms. Jena Germany: G Fischer. p 221–42.

Reisser W (ed). 1992. Algae and symbioses: plants, animals, fungi, viruses, interactions explored. Bristol UK: Biopress Ltd.

Rothschild LJ. 1989. Protozoa, Protista, Protoctista: what's in a name? J Hist Biol 22:277–305.

Round FE, Crawford RM, Mann DG. 1990. The diatoms: biology & morphology of the genera. Cambridge UK: Cambridge Univ Pr.

Sapp J. 1994. Evolution by association: a history of symbiosis. New York NY: OUP Inc.

Schlegel M. 1998. Diversität und Phylogenie der Protisten—aufgedeckt mit molekularen Merkmalen. Stapfia 56:105–18.

Schlegel M, Lom J, Stechmann A, Bernhard D, Leipe D, Dycova I, Sogin ML. 1996. Phylogenetic analysis of complete small subunit ribosomal RNA coding region of *Myxidium lieberkuehni:* evidence that Myxozoa are Metazoa and related to the Bilateria. Arch Protistenk 147:1–9.

Siddall ME, Martin DS, Bridge D, Desser SS, Cone DK. 1995. The demise of a phylum of protists: phylogeny of Myxozoa and other parasitic Cnidaria. J Parasit 81:961–7.

Siddall ME, Reece KS, Graves JE, Burreson EM. 1997. 'Total evidence' rejects the inclusion of *Perkinsus* species in the phylum Apicomplexa. Parasitology 115:165–76.

Sleigh MA. 1995. Progress in understanding the phylogeny of flagellates. Cytology 37:985–1009.

Small EB, Lynn DH. 1985. Phylum Ciliophora Doflein, 1901. In: Lee JJ, Hutner SH, Bovee EC (eds). An illustrated guide to the protozoa. Lawrence KS: Society of Protozoologists. p 393–575.

Smith R, Patterson DJ. 1986. Analyses of heliozoan interrelationships: an example of the potentials and limitations of ultrastructural approaches to the study of protistan phylogeny. Proc R Soc Lond 227B:325–66.

Smothers JF, von Dohlen CD, Smith Jr LH, Spall RD. 1994. Molecular evidence that the myxozoan protists are metazoans. Science 265:1719–21.

Sogin ML, Hinkle G. 1997. Common measures for studies of biodiversity: molecular phylogeny in the eukaryotic microbial world. In: Reaka-Kudla ML, Wilson DE, Wilson EO (eds). Biodiversity II: understanding and protecting our biological resources. Washington DC: Joseph Henry Pr. p 109–22.

Sogin ML, Morrison HG, Hinkle G, Silberman JD. 1996. Ancestral relationships of the major eukaryotic lineages. Microbiologia SEM 12:17–28.

Tappan H. 1980. The paleobiology of plant protists. San Francisco CA: WH Freeman.

Taylor FJR. 1987a. An overview of the status of evolutionary cell symbiosis theories. In: Lee JJ, Frederick JF (eds). Endocytobiology II. Annals NY Acad Sci 503:1–16.

Taylor FJR (ed). 1987b. The biology of dinoflagellates. Oxford UK: Blackwell Scientific.

Taylor FJR, Sarjeant WAS, Fensome RA, Williams GL. 1986. Proposals to standardize the nomenclature in flagellate groups currently treated by both the botanical and zoological codes of nomenclature. Taxon 35:890–6.

Triemer RE, Farmer MA. 1991. An ultrastructural comparison of the mitotic apparatus, feeding apparatus, flagellar apparatus and cytoskeleton in euglenoids and kinetoplastids. Protoplasma 164:91–104.

Vickerman K. 1992. The diversity and ecological significance of Protozoa. Biod Cons 1:334–41.

Vickerman K. 1998. Revolution among the Protozoa. In: Coombs GH, Vickerman K, Sleigh MA, Warren A (eds). Evolutionary relationships among protozoa. Dordrecht: Kluwer Acad Publ. p 1–24.

Vickerman K, Brugerolle G, Mignot J-P. 1991. Mastigophora. In: Harrison FW, Corliss JO. 1991. Microscopic anatomy of invertebrates. New York NY: Wiley-Liss. 1:13–159.

Whittaker RH. 1969. New concepts of kingdoms of organisms. Science 163:150–60.

ESTIMATING THE EXTENT OF
FUNGAL DIVERSITY IN THE TROPICS

K.D. HYDE
W.H. HO
Department of Ecology and Biodiversity, The University of Hong Kong,
Pokfulam Road, Hong Kong

J.E. TAYLOR
Department of Plant Pathology, University of Stellenbosch,
Private Bag X1, Matieland 7602, South Africa

D.L. HAWKSWORTH
MycoNova, 114 Finchley Lane, Hendon, London NW4 1DG, UK

INTRODUCTION

WITH THE RAPID global destruction of tropical habitats, many people—including conservationists, research scientists, and those wishing to use biodiversity—are beginning to recognize that we should find out what we are destroying before it is too late. Tropical deforestation has become the crucible of today's extinction crisis (Wildman 1997), but we should not forget that many other habitats are under threat.

But why in particular do we need to measure fungal diversity? Why do we even want to know which fungi are present in an ecosystem? Why not just measure their isozyme activity or use molecular techniques to indicate fungal presence? Mycologists have robust answers to such probing questions (Hawksworth 1991, 1993, 1998; Hyde 1996a,b; Lodge and others 1996), but conservationists and ecologists, let alone the broader public and politicians, are rarely appropriately briefed. Fungi are important in biological control, in medicine, in biotechnology, in bioactive novel compounds, in decomposition, in nutrient cycling, as actual and potential food resources, in enzyme and organic compound production, and in pollution monitoring. Few other organisms can boast such a successful record of usefulness to humanity! Four of the most important classes of life-saving pharmaceuticals known are produced by fungi: penicillin from *Penicillium chrysogenum*, cephalosporins from *Acremonium chrysogenum*, cyclosporin from *Tolypocladium niveum*, and lovastatin from *Aspergillus terreus* (Rossman 1997). Fungi are used in biotechnological processes or in the production of novel compounds, and they have a huge potential for use in the pharmaceutical and health-care industries (Fox 1993; Nisbet and Fox 1991; Rossman 1997; Wildman 1997). Fungi can also cause huge losses of food in storage and substantial disease in crop plants in the field. Furthermore, because of their integral role in ecosystem processes—for example, in nutrient cycling, plant growth, as a food source, and in their sensitivity to air pollution and perturbation—fungi (including lichen-forming species) are ideal organisms for measuring and monitoring biodiversity (Rossman 1994). Fungi have proved to be important, and it is up to mycologists to raise awareness of them among the wider public and politicians. Each mycologist has been challenged to devote a part of his or her working time to this task (Hawksworth 1995).

NUMBERS OF FUNGI

There are several estimates of the numbers of fungi (Cannon 1997a; Hawksworth 1991, 1993), and a working figure of 1–1.5 million species is now generally accepted (Hammond 1992; Heywood 1995; Rossman 1997). Several lines of evidence point to a similar figure, but it can be derived by comparing the number of fungi known in all habitats in a single geographical area (the British Isles) with the number of native and naturalized plant species in the same area (Hawksworth 1991). The resulting ratio of six fungi to each plant in an area, extrapolation to a conservative 270,000 global vascular plants, and the use of some allowances yielded a global total of about 1.5 million species of fungi. That figure contrasts markedly with the 72,000–100,000 species known (Hawksworth 1995; Hawksworth and Rossman 1997), and a few authors have argued that 1.5 million is too high (Aptroot 1997; May 1994); however, skepticism is based largely on a lack of familiarity with fungal distributions and host specificity and on the lack of detailed studies in the tropics. Recent studies in the tropics have found a magnitude of novelty that tends to support the figure of 1.5 million (Fröhlich and Hyde 1999; Hawksworth 1998; Hawksworth and Rossman 1997).

The extent to which new species are found varies among different systematic and ecological groups. For example, on the basis of the results of a monographic treatment of the saprobic ascomycete genus Didymosphaeria, Aptroot (1997) estimated that there were only 20,000–40,000 nonlichenized ascomycetes in the world. However, his estimate was based on the assumption that only seven of the 550 taxa classified in Didymosphaeria actually belong to that genus (Aptroot 1995). He considered this a general trend in ascomycete systematics; although it might be for many long unrevised genera, a realistic figure has been used in totaling the world's described fungi (Hawksworth and others 1983, 1995). In other genera, the opposite trend is occurring. Oxydothis previously had 27 species names, but the number was increased to 42 after publication of the monograph of Hyde (1994), and a further 23 species have since been found on palms in Australia, Brunei, Ecuador, and Hong Kong (Fröhlich 1997). Aptroot's assumptions are also based on wide species concepts. How can we be sure that fungi with a wide host and biogeographical range and with varied structure are of the same species in the absence of inoculation experiments, incompatibility tests, and molecular data?

There are many potential pitfalls in endeavoring to extrapolate from limited datasets. For instance, a study of phyllachoraceous taxa in Australia and later extensive collecting across the continent increased the number of species known in the country only from 103 to 109 (Pearce and others 1997); extrapolation would provide lower estimates of fungi. But a study of palms in north Queensland identified 202 ascomycete taxa, of which eight genera and 95 species were new to science (Fröhlich and others 1997); extrapolation of these figures would provide much higher estimates of fungi. The original estimate of 1.5 million fungi (Hawksworth 1991) endeavored to account for numerous variables, and recent data from various sources (Cannon 1997; Fröhlich and others 1997a; Hawksworth 1993, 1998; Hyde 1995, 1996a) all point to the figure of 1.5 million as, if anything, conservative.

The number of vascular plants in the United Kingdom is about 2,089, and the number of fungi (including lichen-forming species) is estimated at 12,000 (Hawksworth 1991). Hong Kong, an island smaller than the Isle of Wight or Vancouver Island, has more than 1,700 vascular plants; if the ratio of six fungi to each plant species holds, there are more than 10,000 fungi in Hong Kong. We know of fewer than 500 species (or 5% of 10,000) of fungi in Hong Kong, but some plants have already been shown to support a large number of fungi, many of which are host-specific or family-specific (Fröhlich 1997; Fröhlich and Hyde 1999; Taylor 1997). Those findings indicate that a ratio of 1:6 for vascular plants to fungi might be low, at least in the tropics.

MEASURING FUNGAL BIODIVERSITY

Why Should We Measure Fungal Diversity Rapidly?

The necessity for immediate assessments, new research, and rapid monitoring methods for measuring biodiversity is undisputed, and many of the general recommendations made also apply to fungi and other microorganisms (Burley and Gauld 1997). The ideal way to measure fungal diversity would be an all-taxa biodiversity inventory (Janzen and Hallwachs 1993) for fungi—an all-mycota biodiversity inventory (AMBI), as discussed further below. Because of the difficulty in detecting many fungi and because of their diverse nature, there is an urgent need for all fungi in one geographical region to be identified (Rossman 1994). This would provide basic data against which the results of other external and internal studies could be measured. However, because of the diverse ecologies, seasonality, sporadic findings, and so on, such a survey would take teams of specialists decades. There are at least 31 separate fungal niches in a tropical forest, almost all of which need different techniques and specialists to inventory (Hawksworth and others 1997).

Inasmuch as total inventories will always be impractical (except for a few sites), alternative methods for estimating fungal biodiversity and preparing environmental impact assessments are essential. If they can be developed, there will no longer be any reason for fungi not to be used widely in environmental monitoring, impact assessments, and ecological research. Indeed, because of their sheer diversity and niche specificity, fungi could prove to be especially valuable as indicators of different kinds of environmental changes and ecological processes.

An All-Mycota Biodiversity Inventory?

The need for an AMBI is undeniable. The scientific benefits would be immense with respect to providing a dataset against which to test hypotheses on species richness and host specificity. Permanent plots or otherwise circumscribed sites need to be established to initiate such an inventory. The most intensively inventoried sites for fungi in the world are two in the United Kingdom: Esher Common in Surrey and Slapton Key National Nature Reserve in Devon. Each has around 2,500 species recorded after several decades of work, but neither has been completely surveyed—species are still being added, some niches have not been sampled, and the two sites have only about one-third of the recorded species in common despite many similarities in the plants of the two sites. The true number of fungi in the two UK sites, both of which have been intensively affected by human influences, could well be around 3,000. Whatever the total in those disturbed temperate sites, the richness in pristine tropical forests can be expected to be much greater because of the much larger numbers of potential host plants and insects. No sites in the tropics have yet been studied to a comparable depth; one was contemplated in Costa Rica but has since been abandoned, and some are now being planned in Brunei and Taiwan.

Biodiversity measurement in one plot or site is complicated by the need to sample large numbers of habitats and by the diversity of the fungi encountered. Most of the fungi collected can be predicted to be new to science, and their identification only to genus or family level might be possible (Hawksworth and others 1997; Hyde and Hawksworth 1997). In the case of the Guanacaste project, an estimated 50,000 fungi were probably present, of which around 35,000 could be expected to be new to science (Cannon 1996a). Once some site inventories are complete, protocols for accurate measurement of fungal diversity can be developed and tested in them. However, because of the problems mentioned above, it is unlikely that such results will be available within the next 20 years. In the interim, we must develop the best protocols we can on the basis of existing knowledge.

Alternative Approaches to Inventorying and Monitoring

What is our best way forward? The problems associated with selecting target genera or families or specific habitats as a measure of biodiversity have been discussed (Burley and Gauld 1996). The best approach is thought to be to integrate target groups and specific habitats. Carefully selected permanent plots (selected to incorporate a high degree of plant and habitat diversity) would be established under the auspices of local scientists. The plant species within a plot would be identified and labeled if possible. Mycological inventory could then be carried out over a period of years with input from appropriate specialists. The larger basidiomycetes (for example, polypores), ascomycetes (for example, *Xylaria*), and some biological groups of fungi (for example, entomophagous fungi, freshwater fungi, and lichen-forming fungi) could be collected and identified (spatially and temporally) over the whole plot, inasmuch as their numbers would be manageable. The microfungi could be investigated in smaller plots or individual host trees (Hyde and Hawksworth 1997).

Microhabitat Predictors

Because of the difficulties likely to be encountered if we choose to use predictor sets of fungi as a measure of species richness, another approach could be to select microhabitat predictors. Specific microhabitats in an area would be chosen, and a measure of the diversity of microfungi in those microhabitats would be made according to standard protocols. Random collections of leaf litter followed by isolation according to standard techniques might give us a good measure of overall diversity. If this were used with standardized isolations from random soil samples, estimation of fungal endophyte numbers in an endemic tree species, estimation of aerofungi and lichenized fungi on bark or leaves, and collections of *Xylaria* species, we might have a tangible, albeit qualitative, estimate of the actual diversity in a given region.

The microhabitat components chosen for such an approach to the estimation of biodiversity could vary from habitat to habitat, at this stage; we have no data on which microhabitats would be most representative of microfungal species richness. In the absence of an AMBI, a specific research effort on selected microhabitats is needed to assess which ones would yield the best indications of species richness, to prepare standard protocols for these, and to test the protocols for effectiveness, reproducibility, and ease of application. Five to eight years of coordinated effort across forest regions would be required to allow the identification of suitable predictor microhabitats and then to provide methods for rapid evaluations of fungal diversity in different regions.

Rapid Biodiversity Assessment

In rapid biodiversity assessment or RBA (Beattie and others 1993), numbers of fungi would be estimated without identification as to named species but by sorting them into recognizable, similar species units based on morphological similarities (Cannon 1997b; Hyde 1997; Hyde and Hawksworth 1997). Trained biodiversity technicians ("paratechnicians") are required to sort specimens into recognizable taxonomic units (RTUs). It has been demonstrated that RTU estimates of spiders, ants, polychaetes, and mosses made by biodiversity technicians can be close enough to formal taxonomic estimates of species richness to be useful for RBA (Oliver and Beattie 1993), and we see no reason why this should not be tried with fungi.

The idea of applying RBA in mycology has been viewed optimistically by several authors (Cannon 1997b; Hyde 1997; Hyde and Hawksworth 1997), but it is not clear that workable reproducible protocols can be developed. High priority is now attached to the production of protocols, which can be tested by paratechnicians, revised, and widely promulgated, as addressed in more detail elsewhere (Will-Wolf and others 1999).

ASSESSMENT WITH MOLECULAR TECHNIQUES

The use of molecular techniques in estimating fungal biodiversity has been mentioned as a possibility (Cannon 1997b, 1999; Liew and others 1998) but not used. These techniques have been used to estimate bacterial species, including species that cannot be grown in culture (Tiedje and Zhou 1996), and theoretically they can be applied to fungi, although within-species genotype variation and the low proportion of fungi on which any sequence data are available, they pose particular problems in interpretation. Molecular techniques can be used to access litter and soil samples; although they are still tedious at the cloning stage, the rapid development of automated sequencing machines and computer generation of phylogenetic trees is making them increasingly feasible (Liew and others 1998).

DIVERSITY ASSESSMENT WITH IMAGE ANALYSIS

Computerized image analysis has been successfully developed to identify high-profile groups of fungi, such as airborne species, that might cause allergic responses in humans and trigger asthma (Benyon and others 1997). It could be feasible to develop identification by computerized image analysis for other groups of fungi, such as soil, litter, or mosaics of lichen-forming fungi on leaves or bark. However, computerized analysis is expensive to develop, and the method is unlikely to provide an alternative for wide-scale fungal assessments soon.

SELECTED GROUPS FOR RAPID BIODIVERSITY ASSESSMENT

Recognizing the problems of inventorying all the fungi present in an area and the limitations of various other approaches as reviewed above, we discuss here some candidate groups for use in RBA.

Macromycetes

The large basidiomycetes are probably the easiest group of fungi to record in biodiversity surveys because they are conspicuous, easy to collect, and generally easily identified as to genus. Further separation at the species level can be carried out on site even if the fungi cannot be given an existing species name; they can be given numbers (for example, *Coprinus* sp.) (Hyde 1997). After a spell of rain, the fruiting bodies of macromycetes will flourish in most habitats, but it must be remembered that this is only a representative sample of those actually present; long-term studies over many years are needed to approach a full survey of larger fungi in a site, as demonstrated by studies in Malaysia and Puerto Rico in particular (Hawksworth 1993). Over a period of 7 weeks, sites in Tai Po Kau Nature Reserve and on the University of Hong Kong campus were visited during the wet season. Representative collections of all macromycetes were made on each visit and sorted into recognizable species units. The cumulative numbers of basidiomycetes at both sites indicated that the total numbers of recognizable species units had not been established. Numerous visits to each site are therefore required to obtain best estimates of species numbers.

Xylariaceae

The Xylariaceae are a large family of ascomycetes, most of which are relatively conspicuous. They are particularly well represented in the tropics, although their identification at the species level might require good access to the literature, and some genera are rich in undescribed species in the tropics. They can easily be spotted in the field, occur on the forest floor, or sprout from dead stumps, logs, and branches; because of the robust nature of most species, they require minimal care in handling. A short visit to a site can generate large numbers of xylariaceous taxa. Identification to genus, species, or other recognizable units is relatively easy for paratechnicians. Because the fruiting bodies of these fungi are tough and persistent, they can provide a better comparative measure of fungal diversity than is the case with the more ephemeral, larger basidiomycetes.

Lichen-forming fungi

The lichen-forming fungi in tropical regions are confined mainly to the bark of trees and leaves. Their development depends heavily on light penetration, and in dense tropical forests most will be in the canopy layers. If they can be assessed, lichens can be especially attractive for RBA because of their perennial nature and variations in shape and color. In numerous cases, lichens have been surveyed by schoolchildren as a part of studies of air-pollution patterns, including one in Hong Kong (Thrower 1980). In tropical forests, some groups whose spores or other propagules are large or that for other reasons can be dispersed over only short distances (for example, Thelotremataceae) act as indicators of forests with long histories of ecological continuity; in Thailand, lichens on bark have been related to fire histories (Wolseley and others 1995).

The value of lichens living on leaves in the tropics as indicators of habitat disturbance has been demonstrated by a series of elegant studies in Costa Rica (Lücking 1997). The species forming mosaics on leaf surfaces lend themselves to being counted by eye and with a 10x hand lens by nonspecialists, so they can generate comparable data if similar trees and canopy-sampling strategies are used.

Lichens are now widely used in site assessments in temperate forests (Rose 1992) and merit parallel attention in the tropics. Lichens not only act as indicators of air pollutants and habitat disturbance themselves. Because a wide range of invertebrates feed on or are camouflaged to resemble lichens and provide hiding and breeding places for insects sought by insectivorous birds, sites with a high lichen diversity will also be rich in other dependent organism groups.

Endophytes

Endophytes are fungi or bacteria that for all or part of their life cycle live in tissues of living plants and cause unapparent and asymptomatic infections entirely in the plant tissues but cause no disease symptoms (Wilson 1995). There have been many papers on endophyte associations, mainly from temperate countries, but with some attention paid to tropical habitats (Dreyfuss and Petrini 1984; Fisher and others 1993; Rodrigues 1994; Rodrigues and Petrini 1997; Rodrigues and Samuels 1990). It is now believed that all plants have associated endophytes and that their foliage holds a reservoir of fungi, which can be easily recovered and isolated into culture. Isolation materials are widely available and inexpensive, and plant material is easily sampled and transported. Simple standardized protocols can be constructed to ensure comparability between samples, although allowances must be made for host specificity, and sampling is ideally restricted to particular kinds of trees.

Some endophytes have been shown to be organ- or tissue-specific, so sampling different parts of the plant (Fisher and Petrini 1988, 1990; Petrini and Fisher 1988) and varying the preparations and media used for their recovery yield different assemblages (Chapela and Boddy 1988; Fisher and others 1993; Petrini and others 1992; Pfenning 1997). With the exception of some fungi, such as xylariaceous anamorphs and some species of coprophilous fungi, endophytes are seldom recovered from soil or decaying vegetation (Bills and Polishook 1992).

J.E. Taylor has studied the endophytes and saprobes associated with the Chinese palm *Trachycarpus fortunei*, saprobes associated with Australian endemic *Archontophoenix alexandrae*, and the pantropical *Cocos nucifera*—in and outside the natural biogeographic range of the former two species. Standardized sampling is needed at all the sites, and several sites at each location were investigated. Sampling was undertaken at the same time of the year (depending on seasonality and precipitation) to obtain comparable results. The results generated by both the endophytic and saprobic studies indicate a decrease in species numbers on palms when they are outside their natural range, unless they are in equivalent habitats with—in the case of palms, for instance—a source of fungi from other palm hosts.

SELECTED HABITATS FOR RAPID BIODIVERSITY ASSESSMENT

As a complement to selecting particular groups of fungi to subject to RBA, we suggest that particular habitats be examined in addition to selected groups.

Palms

Palms are an integral part of most tropical forests and so are a valuable host group for comparisons of the fungi present. The fronds and stems of palms are robust, long-lived, and available for colonization by fungi over a relatively long period.

Investigations of palm pathogens, saprobes, and endophytes have revealed a high diversity of palm microfungi, mainly ascomycetes and related mitosporic fungi (Fröhlich 1992; Hyde 1992, 1993, 1994; Hyde and others 1997; Rodrigues 1994; Rodrigues and Petrini 1997). Fröhlich was intrigued by the seemingly limitless microfungal species that could be found on a single palm species in a given patch of forest and investigated the number of species that could be supported by a single host tree. An individual palm tree contains many distinct microhabitats: trunks, stems, roots, frond blades, petioles, inflorescences, fruits, seeds, and assorted appendages such as flagella and spines; these tissues vary in attractiveness to different fungi with age and health. To sample the mycota completely, it would be necessary to examine the following habitats separately:

- all the living palm surfaces, especially the frond blade (phylloplane), for lichens;
- any diseased areas, such as leafspots or frond tips, for pathogens;
- the surfaces and interior of all the senescing and dead tissues for saprobes;
- living tissue collected in the field and incubated in the laboratory for latent pathogens and saprobes;
- the interior of all the healthy, fleshy organs, including the roots, for endophytes; and
- the root surfaces and interior for mycorrhiza.

Studies of the fungal saprobe numbers by Fröhlich and Hyde (1999) indicate that 172 species of saprobes occurred on three fan palms (*Licuala* sp.) in Brunei Darussalam (sampled three times over $1^1/_2$ years), and 100 species of saprobes occurred on three fan palms (*Licuala ramsayi*) in Australia (sampled once). Palm saprobes could be a useful target group for biodiversity assessment; substantial data can be collected with minimal fieldwork.

Bamboo

Bamboo is also a good substrate for biodiversity assessment in the tropics because it is relatively common. In a preliminary study, a *Bambusa* sp. and *Dendrocalamus* sp. were collected in Tai Po Kau Country Park, Hong Kong, and on Mt. Makiling, Los Baños, in the Philippines. One decaying culm in each of three replicated clumps was cut down and chopped into pieces measuring about 25 × 3 cm. Twenty pieces were randomly selected and taken to the laboratory, where they were incubated and kept moist for 1–2 weeks. Each piece was microscopically examined, fungi were recorded, and the number of species on each host at each site was recorded.

The bamboo on Mt. Makiling was found to support 114 species, and that in Tai Po Kau Park 101 species. The hosts had different mycota, and the use of bamboo in RBA therefore seems likely to be effective.

Pandanus

Pandanus leaves are another good substrate for microfungi, particularly hyphomycetes. Collection is relatively simple and involves a pair of secateurs and thick protective gloves. Material can be collected dry or after rain. It should be returned to the laboratory and incubated for a few days. The hyphomycetes present on the material should sporulate quickly and can be identified to provide a measure of fungal diversity. Ascomycetes on *Pandanus*, bamboo, and palms can deteriorate after several days of incubation. If the material is allowed to air dry after a week of incubation, however, this arrests the deterioration of the ascomycetes and allows storage for long periods if necessary.

Freshwater fungi

Fungi flourish on submerged decaying plant material in freshwater; over 300 species of hyphomycetes (Goh and Hyde 1996) and about 300 species of ascomycetes (Shearer 1993) have been recorded. The number of new taxa is increasing (Goh and Hyde 1996). Thomas (1996) defines freshwater fungi as any species that rely on free freshwater for all or part of their life cycle. The richest fungal assemblages occur in average-size, more or less clean, well-aerated forest streams and rivulets with fairly turbulent water (Subramanian 1983).

The common sampling techniques involve collecting substrates—such as foam, water, and submerged plant debris—and examining them microscopically either directly or after incubation in moist or water aeration chambers. Plating is also common but is more labor-intensive and time-consuming and is not appropriate for rapid assessment. Foam filtration and water filtration are convenient and widely adopted. Foam and water samples usually contain conidia of numerous "Ingoldian fungi" with branched or coiled conidia that are separable microscopically without special training.

In contrast with the above methods, which yield mostly freshwater hyphomycetes, the incubation of wood samples from freshwater in moist chambers reveals diverse ascomycetes. Wong (1997) listed 363 species of freshwater ascomycetes, among which 303 were recorded on submerged wood, 18 on submerged bamboo, 40 on submerged leaves and two in foam samples. Hyde, Ho, Tsui, and Ranghoo, University of Hong Kong (pers. comm.), also noted that an extremely rich ascomycete biota occurred in tropical lakes and rivers. Examination of a good collection of wood samples by a mycologist takes about a month, in contrast with 3–5 days needed for a foam or water sample. However, it does reveal another important group of freshwater fungi, and it is therefore recommended for biodiversity assessment in freshwater habitats.

Pathogens

Plant pathogens might prove useful for estimating biodiversity. Collection will involve wandering around a site and collecting diseased leaves, which can be taken to the laboratory and examined. It is important that the collectors have a trained eye, but if this is the case it is possible to estimate diversity of plant pathogens in the field without laboratory examination. Many diseases are host-specific, and different fungal pathogens on a given plant usually differ in the symptoms that they cause.

It is rare to find leafspots on rain-forest plants, particularly palms and *Pandanus* species, and tar spots of phyllachoraceous taxa are also rare. However, large numbers of pathogens occur in gardens, nurseries, or monocultured crops.

Other Habitats for Rapid Biodiversity Assessment

The habitats suggested above are those we have worked on, and they have proved to be excellent sources of fungal diversity. Many others could perform equally as surrogates for biodiversity measurement of, for example, entomophagous fungi in the rain forests of Thailand (Hywel-Jones 1997) or leaf-litter fungi in Costa Rica (Bills and Polishook 1995). There are also ways of standardizing techniques for isolating soil fungi (Cannon 1996b). Disturbed forests harbor fewer rare species in the soil than undisturbed forest and so might be a good indicator of fungal diversity (Pfenning 1997). A concerted effort by mycologists is now needed to try to develop these target groups and microhabitat predictors. A given subset of target groups or microhabitat predictors is unlikely to work for all habitats, so a folio of basic methods must be selected to match local needs. However, until an all-mycota biodiversity inventory can be carried out, we would be wise to develop these methods to obtain estimates of fungal diversity.

MYCODIVERSITY TECHNICIANS

It is unlikely that trained mycologists will always be available for or have the time to devote to measuring fungal diversity in a given habitat, but it can be possible to use mycodiversity technicians (Hyde 1997; Hyde and Hawksworth 1997). Mycodiversity technicians (a kind of "parataxonomist") are not formally trained, but rather undergo minimal training to help in the task of biodiversity assessment. Their value can be exemplified by the use of students to measure endophyte diversity in one species of *Pandanus* and one of *Livistona* in Hong Kong and of summer students to measure larger basidiomycetes in two plots in Hong Kong.

In an experiment carried out with 54 students in Hong Kong, *Livistona chinensis* (a nonnative naturalized palm) and *Pandanus furcatus* were sampled for endophytes from the same piece of secondary woodland. The sampling and time-tabling were carried out as follows: *Livistona chinensis* (27 students divided into eight groups), eight plants sampled with 32 sampling units per individual (mature and immature leaves only); *Pandanus furcatus* (27 students divided into eight groups), eight plants sampled with 16 sampling units per individual (mature and immature leaves only). Far more fungi were recovered from *P. furcatus* in the pilot studies, so the number of sampling units had to be limited to a manageable amount. An alternative method would be to sample more individuals of a single host.

The general process was as follows:

Week 1 Collection of plant material and surface sterilization of samples.
Weeks 2-5 No formal practical classes, but a visit by several students twice a week to check each group's samples for growth of endophytes and subbing onto individual plates and for recording results.
Week 6 Sorting of fungi into morphospecies and checking each culture for sporulation identification attempted as far as possible.
Week 7 Presentation of results and discussion.

Although some fungi would sporulate after 7 weeks, for the purposes of this practical class it was necessary to limit the number of weeks devoted to the study.

Several skilled demonstrators were necessary to assist with running the practical work, especially sorting the isolates into morphospecies and identifying them. In addition, the results were entered onto a database for the students, and the results were presented in a form suitable for statistical analysis. Assistance for 2 hours per week was also necessary for the intervening weeks when the students carried out subculturing. Recording of data was performed accurately, and there were few errors in the final dataset. The only technical problem was in numbering the individual isolates recovered by each group of students; this problem can be circumvented by allotting each group a series of numbers—1–60, 61–120, and so on—or different prefixes, such as A1–A60. The study was labor-intensive and required considerable effort and input by the assisting demonstrators and technicians. However, the advantages were that this applied approach enabled students to investigate a fairly difficult concept and to carry out real scientific investigation on previously unstudied hosts. The students became proficient in sterile techniques, recording of results, and data analysis. The practical class can be carried out in later years on a variety of hosts, giving each group the chance to undertake a first study of endophytes from a specific host plant. Alternatively, technicians familiar with surface sterilization techniques and recording of data could undertake the labor-intensive parts of the work, leaving the identification to the trained mycologists.

In a separate experiment, we used four students over the summer break to compare basidiomycete diversity in a plot at Tai Po Kau Nature reserve with that in one on the Hong Kong University campus. The group first measured out a 1-ha plot and visited the site weekly for 7 weeks. During each visit, the students would walk through the plots, along paths parallel to one side of the plot; the paths were about 10 m apart. The mycodiversity technicians collected representatives of any macromycetes visible from the paths, placed them in suitable containers, and took them to the laboratory. The specimens were identified, isolated, photographed, and dried. Species that could not be identified were placed in recognizable taxonomic units and treated as above. Slide preparations were also made for future reference, and materials and slides are held in the herbarium (HKU) at the University of Hong Kong. Collections from later visits could be compared with photographs and slides from previous visits; in this way, it was possible to identify newly collected species.

Fifty-seven fungi were collected at Tai Po Kau Nature Reserve and 51 at the site at the University of Hong Kong. This indicates that the diversity of fungi was similar in the two sites. However, not all the macromycetes present would have been detected, so the cumulative number had not leveled off. It is interesting to note that more fungi were found at Tai Po Kau during the first visit (25 taxa) than at the other site (13 taxa). Although inconclusive, this pilot experiment indicates that

• mycodiversity technicians can be used in fungal diversity assessment;
• further studies are required to establish whether a single visit to a site to assess fungal diversity is representative; and
• further studies are required to establish how many visits to a site are required to collect an adequate representation of the macromycete species present.

TOWARD A SET OF PROTOCOLS FOR THE RAPID ASSESSMENT OF FUNGAL DIVERSITY

Here we propose protocols for several target groups and predictor habitats that should provide tangible estimates of fungal biodiversity in the tropics. We propose that at least six of these protocols be chosen to obtain a reasonable estimate

of biodiversity and remove bias from any individual protocol. This set of procedures is proposed to provide a starting point for diversity assessment of fungi. The
set can be tested, procedures can be added, and others can be removed, until we
establish a robust mechanism for estimating fungal diversity across a range of global habitats. For the purposes of this exercise, we assume that appropriate human resources are not available and that specialized help will be provided by
mycodiversity technicians. Most of the studies should be carried out during wet
spells.

Macromycetes

The easily visible, larger fungi are an ideal target group on which to base biodiversity estimates, as long as they are integrated with other estimates to eliminate bias. We have found that a plot of 50–100 m² can be thoroughly investigated in 2–3 hours, when all the larger fungi can be collected. This will require
walking throughout the plot along lines 10 m apart, from where most fungi can
be seen. Compartmentalized plastic fishing-tackle boxes or egg boxes are suitable
and can be used to take the samples to the laboratory. It must be wet during the
period under study, and at least 10 weekly visits should be made to each plot
under investigation. It might be necessary to estimate the diversity of the longer-
living polyspores on only one visit.

In the laboratory, untrained technicians can visually sort the material into
morphospecies by using form and color. Slides and spore prints can be prepared,
fresh specimens photographed, and single-spore isolations attempted. The specimens can then be freeze-dried or air-dried and placed in a reference collection
for future study. Total diversity must exclude duplications of the same species
collected on each visit. Over a 10-week period, it should be possible to obtain
an indicative estimate of macromycete diversity by using mycodiversity technicians
(mostly for unnamed specimens) or trained mycologists (for named specimens).

Lichen-forming Fungi

Lichens are especially attractive for use in rapid biodiversity assessment because
they are perennial and generally can be sorted into morphospecies first by eye and
then with a hand lens. Although microscopic and chemical studies might be
needed for critical determinations, they are not necessary when comparative assessments of species numbers are required. Experience with previously untrained
students has shown that 1–2 days of training is sufficient to train a mycodiversity
technician to survey these fungi. Lichens are long-lived, so only a single site visit
is necessary, although ideally it should last for 2–4 days. The same sample plot
as used for macromycetes could be surveyed, and it would be valuable to collect
data on morphospecies distinguishable on tree bark and leaves separately. Where
possible, canopy samples should be obtained from recently fallen or felled trees,
although it is often the understory rather than the exposed crowns that are richest in leaf-inhabiting species. As many as possible should be examined because
there can be variations due to light regimes and other microclimatic factors, which
will affect the development of different species.

The numbers of morphospecies can be compared directly by pooling or considering separately the datasets on bark and leaf-inhabiting species. The numbers of species to be found can be considerable. For example, studies of the crowns of 14 trees in a semideciduous tropical forest in Guyana yielded 100 lichen-forming fungi on leaves alone (Sipman 1997).

For those wishing to go further, the literature on the collection and identification of lichen-forming fungi is immense, but we recommend particularly a recent well-illustrated guide to New Zealand lichens (Malcolm and Galloway 1997). A detailed overview of lichen collection and identification is in press (Will-Wolf and others 1999).

Xylariaceae

The Xylariaceae constitute a tangible target group for biodiversity assessments because only a short period of training is needed to enable mycodiversity technicians to recognize them in the field. We have found that a plot of 50–100 m^2 can be thoroughly investigated in 2–3 hours, when all the visible xylariaceous fungi can be collected. This requires walking throughout the plot along lines 10 m apart and closely examining potential substrates, especially logs on the ground. Most of these fungi are robust and require no special handling. It is often not possible to separate them in the field; therefore, all specimens will need to be returned to the laboratory for microscopic examination. It must be wet during the period under study, and we suggest that at least five visits, 2 weeks apart, be made to each plot under investigation.

In the laboratory, mycodiversity technicians can visually sort the fungi into groups (genera) according to form and to a lesser extent color. Further separations can be made with a hand lens or a dissecting microscope. Slide preparations of spores and asci are, however, essential for species-richness assessments, and the mycodiversity technicians will need to draw and measure these structures. Gross structure, asci, and spores could be photographed from fresh specimens, and isolations into culture from single ascospores can be attempted. The specimens can then be air-dried and placed in a reference collection for possible future study. Total diversity must exclude overlap of the same species collected on each visit. Over five visits, comparative estimates of the diversity of the Xylariaceae present can be obtained by mycodiversity technicians.

Logs and Branches

Numerous dead branches occur on the floor in most tropical habitats and can provide components for rapid fungal-diversity assessments. A short period of training provides mycodiversity technicians with the skill to collect logs and examine them for fungi in the laboratory. The plot of 50–100 m^2 can be used, but in this case 20 logs can be randomly collected during a wet period and then incubated in moist chambers.

In the laboratory, mycodiversity technicians visually examine the logs with a dissecting microscope and make slides of fungi encountered. They can then sort the fungi into different groups on the basis of a minimum of taxonomic knowledge, that is, morphology, spore size, shape, septation, and color. In this way, fungi can be sorted into morphospecies. Photographs can be taken and specimens preserved or cultures attempted as for macromycetes. Total-diversity assessments must exclude overlap of the same species collected on each visit. We suggest that 20 samples is sufficient for mycodiversity technicians to provide a reasonably comparative estimate of the diversity of fungi on logs.

Endophytes

It is relatively easy for mycodiversity technicians to carry out standard procedures to recover endophytic fungi. The methods will depend on the host plant, and pilot studies need to be undertaken to develop optimal sampling and surface sterilization techniques. Surface sterilization techniques are outlined in the methodology of every paper dealing with the recovery of these fungi (Petrini 1986; Petrini and others 1992; Schulz and others 1994). The number of samples necessary to yield at least 80% of all the endophyte taxa at a single site has been estimated (Petrini and others 1992) at a maximum of 40 individuals per species and 30–40 sampling units per individual.

Although fairly labor-intensive, most of the techniques can be used by relatively unskilled technicians, or students, and results can be obtained in less than 3 months. The equipment necessary is inexpensive and widely available. Mycodiversity technicians will need to learn isolation techniques and spend some time at the microscope to separate fungi into "species units" or "morphospecies". The same or allied hosts should be chosen to eliminate differences due to host diversity. Suggested species for which we already have results are palms (Fröhlich 1997; Taylor 1997), bamboo (Umali, unpublished), and mangroves (Rodrigues and Petrini 1997). Sporulation was promoted in many of these cultures with 43–52 species identified within 32 genera; however, different strains of the same species often exhibited different cultural characteristics.

Palm Fungi

Palm rachids or petioles probably support the highest diversity of palm fungi, so we suggest these for biodiversity assessment. Numerous dead rachids can be found on the floor or attached to living palms in most tropical habitats and so are considered an ideal component of a suite of fungal-diversity assessment protocols. A short period of training will provide mycodiversity technicians with the skill to collect samples and examine them for fungi in the laboratory. The same plot of 50–100 m^2 can be used, but in this case 20 rachid samples can be "selectively" randomly collected during a wet period or a dry period. They can be examined after air drying.

In the laboratory, mycodiversity technicians examine the samples with a dissecting microscope and make slides of fungi encountered. Fungi are sorted into different groups on the basis of minimal taxonomic knowledge—spore type, spore size, shape, septation, and color. In this way fungi can be sorted into morphospecies. Photographs can be taken and specimens preserved or cultures attempted as for macromycetes. Total diversity must exclude overlap of the same species collected on each visit. We suggest that 20 samples are sufficient for microdiversity technicians to provide a comparative estimate of fungi on palms.

Bamboo Fungi

Dead bamboo culms support a high diversity of fungi, and we suggest these for biodiversity assessment. Numerous dead culms can be found on the floor or standing, and two species of bamboo can be selected and sampled. Training, sampling, examination, and interpretation are similar to those for palms, and the same plot of 50–100 m² can be used. Samples should first be examined for ascomycetes and basidiomycetes and then incubated for 14 days, after which they can be examined for other fungi.

Fungi on Pandanus

Dead *Pandanus* leaves support a high diversity of fungi, and we suggest these for biodiversity assessment. Numerous dead leaves can be found on the floor or attached to the plants, and these can be randomly collected. The type of training required and procedures to be followed are similar to those for palms and bamboos. The same plot of 50–100 m² can be used, but in this case 20 leaves can be "selectively" randomly collected during a wet period or a dry period. Leaves should first be examined for ascomycetes and basidiomycetes and then incubated for 3 days, after which they can be examined for other fungi.

Freshwater Fungi

Comparative-biodiversity studies of fungi in freshwater habitats require the use of standard methods for foam and water examination. Examination of 10 foam collections and the filtrates from two membrane-filtered (pore size, 5–8 μm) 5-L water samples from three locations in the freshwater habitat should be carried out. Foam samples are collected in separated, clean, sterilized vials and preserved with the addition of formal-acet-alcohol or stored in an icebox. In the water-filtration method, the membrane filter is stained and fixed with lactic acid cotton blue or lactic acid fuchsin. This preserves and stains the spores and renders the membrane filter semitransparent. Samples should be examined until the number of new morphospecies recorded declines to a minimum; this can be assessed by plotting a cumulative graph of the number of new taxa recorded versus the number of slides examined in foam samples or the number of filter papers from the water-filtration method examined. The direct examination of random wood samples for surface fungi could also provide a good estimate of fungal diversity.

Observation of conidia on semitransparent filter membranes might be difficult, especially with respect to the minute characters of fungal spores. The difficulty can be overcome with the use of a high-power dissecting microscope or a compound microscope with an upper light source. Conidia might also be covered by particles filtered in the filtration process; these can occlude important features.

CONCLUSIONS

Fungi are ideal organisms to work with in the field and in the laboratory. Collection requires a visit to the area under investigation and either collection of the visible fungi concerned or collection of small parts of the habitat under investigation. In the laboratory, the fungi are easy to handle and can be photo‑graphed and dried. Alternatively, isolations can be made with established tech‑niques. Most fungi grow rapidly in culture and require no complicated procedures for their study.

We have endeavored to indicate the richness of tropical fungi and compara‑tive approaches to the assessment of fungal diversity between different tropical sites. The several‑protocol approach that we recommend is essential to capture some representation of the microfungi present. This is critical because the mi‑crofungi make up the highest proportion of fungi in any ecosystem. However, if time is short, we recognize that there could be advantages in paying particular attention to macromycetes and lichen‑forming fungi in preliminary assessments.

In the exploration of fungal diversity, much attention has been focused on ob‑taining data on species richness in different systematic groups or in particular niches or substrates. Such studies have been important in vindicating hypoth‑eses regarding the richness of the world's mycota, but we believe that it is now time to focus on securing comparative data on the richness of fungi in different sites. The approaches to rapid assessment of biodiversity in fungi described here are intended both to further discussion as to the best suite of protocols to recom‑mend and more important to stimulate more work in tropical sites, even in the absence of experienced mycologists.

ACKNOWLEDGMENTS

We are grateful to P.H. Raven for encouraging us to prepare this contribution for these proceedings. Helen Leung is thanked for technical assistance.

REFERENCES

Aptroot A. 1995. A monograph of Didymosphaeria. Stud Mycol 37:1–175.

Aptroot A. 1997. Species diversity in tropical rainforest ascomycetes: lichenized versus non‑lichenized; folicolous versus corticolous. Abstr Botan 21:37–44.

Beattie AJ, Majer JD, Oliver I. 1993. Rapid biodiversity assessment: a review. In: Beattie A (ed). Rapid biodiversity assessment, proceedings of the Biodiversity Assessment Workshop. Sydney Aus‑tralia: Maquarie Univ. p 4–14.

Benyon FHL, Jones AS, Tovey ER. 1997. Image analysis differentiates spores of allergenic fungal genera and species. In: Barker RM, Barker WR (eds). Abstracts of the Joint National Conferences Adelaide '97. Adelaide Australia: Joint National Conferences Organising Committee, State Herbarium of South Australia. p 82.

Bills GF, Polishook JD. 1992. Recovery of endophytic fungi from Chamaecyparis thyroides. Sydowia 44:1–12.

Bills GF, Polishook JD. 1995. Abundance and diversity of microfungi in leaf litter of a lowland rainforest in Costa Rica. Mycologia 86:187–98.

Burley JM, Gauld I. 1996. Measuring and monitoring forest biodiversity: a commentary. In: Boyle TJB, Boontawee B (eds). Measuring and monitoring biodiversity of tropical and temperate forests. Bogor Australia: CIFOR. p 19–46.

Cannon PF. 1996a. An ATBI—how to find one and what to do with it. Inoculum 46(4):1–4.

Cannon PF. 1996b. Filamentous fungi. In: Hall GS (ed). Methods for the examination of organismal diversity in soils and sediments. Wallingford UK: CAB International. p 125–43.

Cannon PF. 1997a. Diversity of the Phyllachoraceae with special reference to the tropics. In: Hyde KD (ed). Biodiversity of tropical fungi. Hong Kong: Hong Kong Univ Pr. p 255–278.

Cannon PF. 1997b. Options and strategies for rapid assessment of fungal diversity. Biod Cons 6:669–680.

Cannon PF. 1999. Options anc constraints in rapid biodiversity assessments in natural ecosystems. Fungal Div 2:1–15.

Chapela IH, Boddy L. 1988. Fungal colonisation of attached beech branches. II. Spatial and temporal organisation of communities arising from latent invaders in bark and functional sapwood under different moisture regimes. New Phytol 110:47–57.

Dreyfuss M, Petrini O. 1994. Further investigations on the occurrence and distribution of endophytic fungi in tropical plants. Bot Helv 94:33–40.

Fisher PJ, Petrini O. 1988. Tissue specificity by fungi endophytic in Ulex europaeus. Sydowia 40:46–50.

Fisher PJ, Petrini O. 1990. A comparative study of fungal endophytes in xylem and bark of Alnus species in England and Switzerland. Mycol Res 94:313–9.

Fisher PJ, Petrini O, Sutton BC. 1993. A comparative study of fungal endophytes in leaves, xylem, and bark of Eucalyptus in Australia and England. Sydowia 45:338–45.

Fox FM. 1993. Tropical fungi: their commercial potential. In: Isaac S, Frankland JC, Watling R, Whalley AJS (eds). Aspects of tropical mycology. Cambridge UK: Cambridge Univ Pr. p 253–63.

Fröhlich J. 1992. Fungi associated with foliar diseases of palms in North Queensland. Honours thesis, Univ of Melbourne.

Fröhlich J. 1997. Biodiversity of microfungi associated with tropical palms. PhD thesis, Univ of Hong Kong.

Fröhlich J, Hyde KD. 1999. Biodiversity of palm fungi in the tropics: are global fungal diversity estimates realistic? Biod Cons 8:977-1004.

Fröhlich J, Taylor JE, Hyde KD. 1997. Biodiversity of microfungi associated with palms in the tropics. In: Barker RM, Barker WR (eds). Abstracts of the Joint National Conferences Adelaide '97. Adelaide Australia: State Herbarium of South Australia. p 84.

Goh TK, Hyde KD. 1996. Biodiversity of freshwater fungi. J Indus Microbiol 17:328–45.

Hammond PM. 1992. Species inventory. In: Broombridge B (ed). Global biodiversity. London UK: Chapman & Hall. p 17–39.

Hawksworth DL. 1991. The fungal dimension of biodiversity: magnitude, significance, and conservation. Mycol Res 95:641–55.

Hawksworth DL. 1993. The tropical fungal biota: census, pertinence, prophylaxis, and prognosis. In: Isaac S, Frankland JC, Watling R, Whalley AJS (eds). Aspects of tropical mycology. Cambridge UK: Cambridge Univ Pr. p. 265–293.

Hawksworth DL. 1995. Challenges in mycology. Mycol Res 99:127–8.

Hawksworth DL. 1998. Getting to grips with fungal diversity in the tropics. In: Chou S (ed). Frontiers in biology: the challenges of biodiversity, biotechnology, and sustainable agriculture. Taipei Taiwan: Academia Sinica.

Hawksworth DL, Kirk PM, Sutton BC, Pegler DN. 1995. Ainsworth & Bisby's dictionary of the fungi, 8th edition. Wallingford UK: CAB International. p. 616.

Hawksworth DL, Minter DW, Kinsey GC, Cannon PF. 1997. Inventorying a tropical fungal biota: intensive and extensive approaches. In: Janardhanan KK, Rajendran C, Natarajan K, Hawksworth DL (eds). Tropical mycology. Enfield NH: Science Publ. p 29–50.

Hawksworth DL, Rossman AY. 1997. Where are all the undescribed fungi? Phytopathology 87:888–91.

Hawksworth DL, Sutton BC, Ainsworth GC. 1983. Ainsworth & Bisby's dictionary of the fungi, 7th edition. Kew UK: International Mycological Inst. p 445.

Heywood VH (ed.) 1995. Global biodiversity assessment. Cambridge UK: Cambridge Univ Pr. p 1140.

Hyde KD. 1992. Fungi from decaying intertidal fronds of Nypa fruticans, including three new genera and four new species. Botanical J Linnean Soc 110:95–110.

Hyde KD. 1993. Fungi from palms V. Phomatospora nypae sp. nov. and notes on marine fungi from Nypa fruticans in Malaysia. Sydowia 45:199–203.

Hyde KD. 1994. Fungi from palms. XII. Three new intertidal ascomycetes from submerged palm fronds. Sydowia 46:257–64.

Hyde KD. 1995. Fungi from palms. XIII. The genus Oxydothis, a revision. Sydowia 46:265–314.

Hyde KD. 1996a. Biodiversity of microfungi in North Queensland. Austral Syst Bot 9:261–71.

Hyde KD. 1996b. Measuring biodiversity: diversity of fungi in the wet tropics of North Queensland. In: Boyle TJB, Boontawee B (eds). Measuring and monitoring biodiversity of tropical and temperate forests. Bogor Australia: CIFOR. p 271–286.

Hyde KD. 1997. Can we rapidly measure fungal diversity? Mycologist 11:176–8.

Hyde KD, Fröhlich J, Taylor JE. 1997. Diversity of ascomycetes on palms in the tropics. In: Hyde KD (ed). Biodiversity of tropical microfungi. Hong Kong: Hong Kong Univ Pr. p 141–56.

Hyde KD, Hawksworth DL. 1997. Measuring and monitoring the biodiversity of microfungi. In: Hyde KD (ed). Biodiversity of tropical microfungi. Hong Kong: Hong Kong Univ Pr. p 11–28.

Hywel-Jones NL. 1997. Biological diversity of invertebrate pathogenic fungi. In: Hyde KD (ed). Biodiversity of tropical microfungi. Hong Kong: Hong Kong Univ Pr. p 107–20.

Janzen DH, Hallwachs W. 1993. In: Highlights of the NSF-sponsored 'All taxa biodiversity workshop', Philadelphia PA. Bitnet, Biological Systematics Discussion list (taxacom@harvarda.bitnet) 16–18 April 93.

Liew EYC, Guo LD, Ranghoo VM, Goh TK, Hyde KD. 1998. Molecular approaches to assessing fungal diversity in the natural environment. Fung Div 1:1–20.

Lodge JD, Hawksworth DL, Ritchie BJ. 1996. Microbial diversity and tropical forest functioning. In: Orians G, Dirzo R, Cushman H (eds). Biodiversity and ecosytem processes in tropical forests. Berlin Germany: Springer. p 69–100.

Lücking R. 1997. The use of foliicolous lichens as bioindicators in the tropics, with special reference to the microclimate. Abstr Botan 21:99–116.

Malcom WM, Galloway DJ. 1997. New Zealand lichens: checklist, key, and glossary. Wellington New Zealand: Museum of New Zealand.

May M. 1994. Conceptual aspects of the quantification of the extent of biological diversity. Phil Trans R Soc London Biolog Sci 345:1–20.

Nisbet LJ, Fox FM. 1991. The importance of microbial diversity to biotechnology. In: Hawksworth DL (ed). The biodiversity of microorganisms and invertebrates: its role in sustainable agriculture. Wallingford UK: CAB International. p. 229–244.

Oliver I, Beattie AJ. 1993. A possible method for the rapid assessment of biodiversity. Cons Bio 7:562–8.

Pearce CA, Hyde KD, Taylor JE. 1997. Preliminary studies on the Australian phyllachoraceae. In: Barker RM, Barker WR (eds). Abstracts of the Joint National Conferences Adelaide '97. Adelaide Australia: Joint National Conferences Organising Committee, State Herbarium of South Australia. p. 87.

Petrini O. 1986. Taxonomy of endophytic fungi of aerial plant tissues. In: Fokkema NJ, Heuvel JVD (eds). Microbiology of the phyllosphere. Cambridge UK: Cambridge Univ Pr. p. 175–187.

Petrini O, Fisher PJ. 1988. A comparative study of fungal endophytes in xylem and whole stem of Pinus sylvestris and Fagus sylvatica. Trans Br Mycol Soc 91:233–8.

Petrini O, Sieber TN, Toti L, Vivet O. 1992. Ecology, metabolite production and substrate utilisation in endophytic fungi. Nat Tox 1:185–96.

Pfenning L. 1997. Soil and rhizosphere microfungi from Brazilian tropical forest ecosystems. In: Hyde KD (ed). Biodiversity of tropical microfungi. Hong Kong: Hong Kong Univ Pr. p 341–366.

Rodrigues KF. 1994. The foliar fungal endophytes of the Amazonian palm Euterpe oleracea. Mycologia 86:376–85.

Rodrigues KF, Petrini O. 1997. Biodiversity of endophytic fungi in tropical regions. In: Hyde KD (ed). Biodiversity of tropical microfungi. Hong Kong: Hong Kong Univ Pr. p 57–70.

Rodrigues KF, Samuels GJ. 1990. Preliminary study of endophytic fungi in a tropical palm. Mycol Res 94:827–30

Rose F. 1992. Temperate forest management: its effects on bryophyte and lichen floras and habitats. In: Bates JW, Farmer AM (eds). Bryophytes and lichens in a changing environment. Oxford UK: Clarendon Pr. p 211–233.

Rossman AY. 1994. A strategy for an all taxa inventory of fungal biodiversity. In: Peng CI, Chou CH (eds). Biodiversity and terrestrial ecosystems. Acad Sinica Mono Ser 14:169–94.

Rossman AY. 1997. Biodiversity of tropical microfungi: an overview. In: Hyde KD (ed). Biodiversity of tropical microfungi. Hong Kong: Hong Kong Univ Pr. p 1–10.

Schulz B, Wanke U, Draeger S, Aust HJ. 1994. Endophytes from herbaceous plants and shrubs: effectiveness of surface sterilization methods. Mycol Res 97:1447–50.

Shearer CA. 1993. The freshwater ascomycetes. Nova Hedwigia 56:1–33.

Sipman HJM. 1997. Observations on the foliicolous lichen and bryophyte flora in the canopy of a semi-deciduous tropical forest. Abstr Botan 21:153–61.

Subramanian CV. 1983. Hyphomycetes: taxonomy and biology. London UK: Academic Pr. p 502.

Taylor JE. 1997. Biodiversity and distribution of microfungi on palms. PhD thesis, Univ Hong Kong.

Thomas K. 1996. Fresh water fungi. In: Orchard AE (ed). Fungi of Australia. Canberra Australia: Austral Biol Res Stud 1(B):1–37.

Thrower S. 1980. Air pollution and lichens in Hong Kong. Lichenologists 12:305–11.

Tiedje JM, Zhou J. 1996. Analysis of non-culturable bacteria. In: Hall GS (ed). Methods for the examination of organismal diversity in soils and sediments. Wallingford UK: CAB International. p 54–65.

Wildman HW. 1997. Potential of tropical microfungi within the pharmaceutical industry. In: Hyde KD (ed). Biodiversity of tropical microfungi. Hong Kong: Hong Kong Univ Pr. p 29–46.

Will-Wolf S, Hawksworth DL, McCune B, Sipman H. 1999. Assessing the biodiversity of lichenized fungi. In: Mueller GM, Bill G, Rossman A, Burdsall H (eds). Measuring and monitoring biodiversity: standard methods for fungi. Washington DC: Smithsonian Inst.

Wilson D. 1995. Endophyte—the evolution of a term, and clarification of its use and definition. Oikos 73:274–6.

Wolseley PA, Moncreiff C, Aguirre-Hudson B. 1995. Lichens as indicators of environmental stability and change in the tropical forests of Thailand. Glob Ecol Biogeogr Let 4:116–23.

Wong SW. 1997. Ultrastructural and taxonomic studies of freshwater ascomyctes. PhD thesis. Univ of Hong Kong.

NEMATODES:
PERVADING THE EARTH AND LINKING ALL LIFE

J. G. BALDWIN
Department of Nematology, University of California,
1455 Boyce Hall, Riverside, CA 92521

S. A. NADLER
Department of Nematology, University of California,
588 Hutchinson Hall, Davis, CA 95616-8668

D. H. WALL
College of Natural Resources, Colorado State University,
Fort Collins, CO 80523

ABUNDANCE OF NEMATODES

THE PHYLUM NEMATODA (Nemata), known commonly as roundworms, contains the most abundant, common, and genetically diverse multicellular organisms (Lambshead 1993; Platt and Warwick 1983). Usually, these organisms are invisible to all but a few specialized scientists because most are essentially microscopic and transparent. More than 85 years ago, Cobb (1914) eloquently noted that "if all the matter in the universe except the nematodes were swept away, our world would still be dimly recognizable, and if, as disembodied spirits, we could then investigate it, we should find its mountains, hills, vales, rivers, lakes, and oceans represented by a film of nematodes." "So little do we know of this vast multitude of soil-inhabiting nematodes that the first spadeful of earth we lift is practically certain to contain kinds never seen before", and 'There exists...a greater disproportion between the known and the unknown than exists in almost any other class of organisms." With respect to Cobb's characterization of our knowledge of nematode diversity, relatively little has changed during the last 85 years. Somewhere between 500,000 and more than 100,000,000 nematode species are believed to exist on Earth (Lambshead 1993), but habitats certain to be richest in new species are mostly unexplored, and fewer than 25,000 species have been described (Andrássy 1992; Platt and Warwick 1983).

Diverse morphological, physiological, and behavioral adaptations allow nematodes to pervade nearly every habitat, but most habitat adaptations cross formal taxonomic boundaries, which arguably include two classes and 18 orders

176

(figure 1). Most nematodes are nothing like the vertebrate parasite *Ascaris* (figure 2J), which represents the phylum in introductory biology texts and laboratory dissections; rather, most are microscopic nonparasites designated microbivorous because they feed on small organisms, such as bacteria, fungi, and algae. A few are parasites of plants, vertebrates, or invertebrates; others browse on plants as herbivores or are predators of small organisms. Most microbivorous nematodes are of unnamed species, but these poorly characterized taxa play an important role in decomposition of organic matter and nutrient cycling in all ecosystems (Freckman 1988; Peterson and Luxton 1982). Nonparasitic nematodes are commonly the most abundant microinvertebrates of terrestrial (Nielsen 1949), marine-estuary sediment (Warwick and Rice 1979), and freshwater ecosystems; they are also taxonomically heterogeneous, transcending 12 orders (figure 1). Seabeds, ranging from the tropics to the Arctic, are the habitats by far richest in nematode species diversity (Boucher 1990; Lambshead 1993). Although 4,000–5,000 marine nematode species have been named and described, full surveys of marine habitats probably will reveal many millions of previously unknown species (Andrássy 1992; Hope and Murphy 1972). The natural histories of these marine nematodes are diverse and sometimes astounding. Consider for example, marine nematodes that are symbiotic with chemotropic sulfur bacteria, a nematode that

Phylum Nematoda

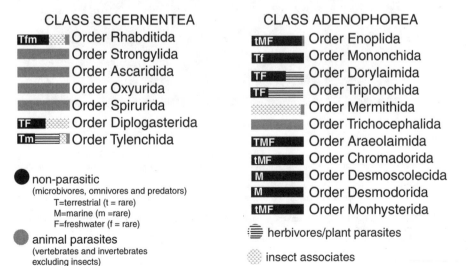

CLASS SECERNENTEA
Tfm Order Rhabditida
Order Strongylida
Order Ascaridida
Order Oxyurida
Order Spirurida
TF Order Diplogasterida
Tm Order Tylenchida

● non-parasitic
(microbivores, omnivores and predators)
T=terrestrial (t = rare)
M=marine (m =rare)
F=freshwater (f = rare)

● animal parasites
(vertebrates and invertebrates
excluding insects)

CLASS ADENOPHOREA
tMF Order Enoplida
Tf Order Mononchida
TF Order Dorylaimida
TF Order Triplonchida
Order Mermithida
Order Trichocephalida
TMF Order Araeolaimida
tMF Order Chromadorida
M Order Desmoscolecida
M Order Desmodorida
tMF Order Monhysterida

▤ herbivores/plant parasites

░ insect associates

FIGURE 1 Distribution of lifestyles and feeding habits among nematode orders. Classification system modified from Maggenti (1981) and Chabaud (1974). Validity of order Triplonchida is discussed in Decraemer (1995).

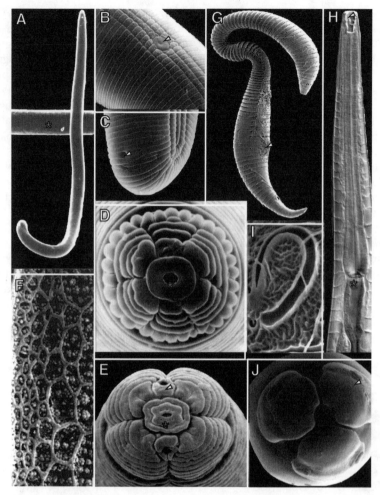

FIGURE 2   Examples of morphological diversity of surface structures of nematodes (SEM unless noted otherwise).

A. One of larger plant-parasitic nematodes, *Hoplolaimus columbus*, propped on human hair.
B. Enlargement from figure 2A showing cuticular pattern of transverse striations, lateral longitudinal lines, and phasmid sensory organ (arrow).
C. Enlargement of tail end of figure 2A showing position of anus.
D. En face view of *Hoplolaimus columbus*; during feeding, hypodermic-needle-like stylet projects from oral opening and punctures plant cell for feeding on cytoplasm.
E. En face view of plant parasitic nematode *Belonolaimus longicaudatus*; six minute openings of chemosensory receptors surround oral opening; sensory receptors, amphid (arrow), occur on each lateral side.
F. Lacelike surface pattern of new species of microbivorous nematode *Bunonema* n. sp.
G. New species of marine nematode, *Epsilonema* n. sp.; arrow indicates position of vulva.
H. Light micrograph of anterior end of new species of marine nematode; most nematodes are transparent, and oral cavity with tooth is visible; although this nematode is likely to be microbivorous, presence of tooth (arrow) might suggest that it can feed as predator; oral cavity leads to muscular esophagus, which terminates posteriorly (asterisk).
I. One of pair of chemosensory organs (amphids) on anterior end of *Paradraconema* n. sp.
J. En face view of *Ascaris suum*, large intestinal parasite of pigs. Oral opening is surrounded by pronounced papillae (arrow).

attaches its eggs to itself and guards them until they hatch, and a moderate-size nematode (4 mm long) that lacks a mouth and digestive system but is packed with symbiotic organisms (Bernard 1994). It is impossible to anticipate the richness of lifestyles that will be discovered among such underexplored habitats. Nematodes of soil and freshwater sediment are only somewhat better known; about 6,000 species in these habitats have been named, but the rate of discovery of new species in some habitats suggests that that is only a small fraction of extant species (Andrássy 1992). Although nonparasitic nematodes have great diversity in tropical and temperate soils, they characterize the biota of all soils and are even present in dry Antarctic ones (Freckman and Virginia 1997; Lawton and others 1996; Sohlenius 1980; Yeates 1980).

Many of the most thoroughly studied human parasites, being macroscopic, have been recognized since ancient times (Thorne 1961). However, parasitic taxa of economic importance, such as those of plants and vertebrates, probably include fewer than 5% of all nematode species. Nearly 2,000 species in three orders are herbivores or known to parasitize plants (figure 2A–E); a few hundred principal species are responsible for billions of dollars of crop losses annually (Barker 1994). Even in nonagricultural habitats, the impact of plant parasites can be impressive; pinewood nematodes introduced from North America are capable of killing a large pine tree in Japan in less than six weeks (Mamiya 1983; Rutherford and others 1990). Similarly, roughly 12,000 species in six major orders are known to parasitize vertebrates, but among these, only about 36 are considered to have a direct impact on human health and 300 are of veterinary importance. Most invertebrates—mainly insects—also host nematodes, the associations ranging from phoresis to obligate parasitism. Although hundreds of named nematode species in four orders are known to be associated with insects, they probably represent a small fraction of the existing species, considering insect diversity and nematode specialization on such hosts. It is very likely that this pool of largely unexplored parasites is rich in potential biocontrol agents and management tools for medical, veterinary, and agricultural insect pests.

The inadequately developed state of systematic surveys has been a major limitation in the development of a phylogenetic taxonomic system for phylum Nematoda, but such a framework is fundamental to the predictability and repeatability of all other research on the group. The problem is not unlike assembling a 1,000-piece jigsaw puzzle with 975 missing pieces! Another limitation in developing this taxonomic framework has been the historical fragmentation of nematologists into groups that study vertebrate parasites, invertebrate parasites, herbivore and plant parasites, or free-living forms in such disciplines as agriculture, parasitology, veterinary science, medicine, and ecology. As a consequence, taxonomic classifications often reflect academic specializations, rather than broad-scale nematode phylogenetic relationships. This fragmented approach is like grouping puzzle pieces by their similar shapes; in some cases (as it is with edge pieces) it might be useful; but in most cases, there is little predictive power or congruence with historical relationships. These "discipline-specific taxonomies" are now being tested and for some groups revealed as artificial by robust DNA-based phylogenetic hypotheses (Blaxter and others 1998). Fragmentation by discipline also has resulted in overemphasis on parasitic groups of obvious economic importance but with no contextual connection to the grossly understudied nonparasitic groups, which make up the majority of nematodes. Broad-based biotic inventories will transcend these previous boundaries as taxonomists with diverse specializations collaborate to investigate nematode diversity through the application of varied methods.

Most recognized nematode species have been described on the basis of a morphotype that is presumed to be unique. Such species can be synonymized if it is demonstrated that the morphotype either is not unique or representative for the taxon. Andrássy (1992) estimated that 18% of named species of terrestrial and freshwater nonparasitic nematodes are invalid because they are synonyms or because information on the species is inadequate for assessing validity; it is unclear, however, whether this estimate of "invalid species" is generally applicable to the entire phylum. A potential source of error that has been more difficult to assess is the degree to which morphotypic uniqueness is a good estimator for ontologically real species; the integration of molecular data in nematode systematics provides an independent line of evidence to help address this problem (Adams 1998; Szalanski and others 1997; Thomas and others 1997).

Current systematics practice emphasizes that discovery and description of new nematode species requires phylogenetic context for many taxonomic decisions, including reevaluation of previously described species and their relationships. Errors in estimating evolutionary history can have critical implications for over-estimating or underestimating species numbers (Adams 1998). But one advantage of phylogenetic approaches to studying species-level questions is that thorough integrated approaches to gathering character data (for example, structural, molecular, genetic, and developmental data) can promote discovery of new taxa.

GEOGRAPHIC DISTRIBUTION OF SPECIES

A measure of the distribution or degree of localization of nematode species is crucial to developing sampling strategies for estimating worldwide species richness. For example, is species abundance in the North American deserts similar to abundance in African deserts, and to what extent do the same species occur in both habitats? Because lifestyles and feeding habits of nematodes (figure 1) cover the biological spectrum, it is not surprising that species are varied in their patterns of distribution. A high degree of localization might be expected among the majority of nematodes that have low mobility and a life history that lacks a dispersal phase; these are determinants for high speciation rates (Castillo-Fernandez and Lambshead 1990). In some habitats, limited dispersal is by mechanisms that move sediment or soil, including wind or flowing water. Other species are globally distributed, because of a deep evolutionary history predating dispersal of continents (Baldwin 1992; Ferris 1979). In still other cases, nematode distribution is determined by the habitats of organisms with which they are closely associated. Many nematodes are dispersed through phoretic associations with mobile insects or birds, by anthropogenic effects (such as agriculture), or by congruence with a specific host. Such mechanisms determining geographic pattern cannot be separated from issues of isolation and speciation. Highly localized species or species restricted to a narrow habitat regime or a single host include a majority of nematodes, and these are most vulnerable to annihilation. The resources to measure the rate at which nematode species, including beneficial ones, are lost to anthropogenic effects are unavailable, but we can expect that the rate of loss is huge.

We have noted arguments that most nematodes are nonparasitic marine species (Lambshead 1993). However, in the marine environment, abundance and perhaps species geographic pattern vary somewhat with concentrations of organic matter, vertical distribution, latitude, and depth (Boucher 1990; Lambshead and others 1995). In deep seabeds and undisturbed soil systems, regardless of overall abundance, individuals of a particular species often are rare (Lambshead 1993; Grassle and Maciolek 1992; Hessler and Sanders 1967). For example, a deep-sea sample yielding 148 nematode species included only 216 individuals (Hope 1987).

Considering terrestrial and freshwater nematodes, Nicholas (1975)—citing examples from the orders Dorylaimida, Araeolaimida, and Tylenchida—argued that to a striking degree, particular genera and species occur in all parts of the world and in a variety of habitats, irrespective (within wide limits) of soil's physical and chemical factors, climate, or vegetation. We would add a range of microbivorous Rhabditida, such as *Acrobeloides nanus* and *Panagrolaimus rigidus*, which, regardless of limited sampling, are known to have distributions that include all of Europe, Australia, Asia, Africa, and North and South America (Andrássy 1984). Often such wide distribution is difficult to explain. For example, a survey of nematodes in a freshwater lake in the Galápagos, colonized during its recent history of 15,000 years, included 18 species in five orders; 16 of the species were known in other parts of the world (Abebe and Coomans 1995). Thus, mechanisms of dispersal and colonization are not fully understood. In some cases, supposed geographic limits of species are an aberration of inadequate testing. For example, recent sampling in a California desert led to the discovery of microbivorous nematodes, including species previously known only in South Africa. Such geographic limits would be difficult to explain through introductions or biogeography; it is much more likely that more-extensive surveys will demonstrate a broad distribution of these species beyond California and South Africa.

Some parasitic nematodes have broad host ranges and are distributed nearly worldwide; an example is the root knot nematode, *Meloidogyne incognita*, one of the most destructive plant pathogens of agriculture and widespread in relatively undisturbed habitats (Eisenback and Triantaphyllou 1991). Other parasitic nematodes often are highly regionalized by specific requirements of their host and habitat. For example, the citrus pathogenic variant of *Radopholus citrophilus*, which at one time nearly destroyed the Florida citrus industry, is known only in that region. This nematode's requirement for a habitat of deep sandy soils might limit its distribution (O'Bannon 1977). A closely related species, *Radopholus similis*, is distributed globally throughout the tropics, whereas most other *Radopholus* species seem to be restricted to Australia, Asia, or Africa (Huettel and Dickson 1984; O'Bannon 1977; Sher 1968). To some extent, the wide distribution of *Radopholus similis* might be affected by anthropogenic effects of agriculture, including the spread on infected corms from Asia via propagation of bananas throughout the tropics.

The cyst nematodes include a group of about six genera and 70 species, many of which have great economic importance to agriculture and diverse patterns of species distribution, determined by host specificity and coevolution, biogeography, and anthropogenic effects (Baldwin 1992; Baldwin and Mundo-Ocampo 1991). For example, the cyst nematode *Punctodera chalcoensis* is restricted to Mexico and clearly coevolved with its only host, *Zea*, including *Z. mays* (cultivated maize) and uncultivated species endemic to Mexico. The potato cyst nematodes, *Globodera rostochiensis* and *G. pallida*, were thought to be restricted to potatoes in Europe until the 1950s, when they were discovered on a shipment of potatoes from Peru. Later surveys in South America revealed that the potato cyst nematodes occur on wild plants throughout a region of the Andes, where they probably coevolved with potatoes. It is commonly believed that they were introduced to Europe with potatoes in the 1600s and later throughout many of the potato-growing regions of the world, despite rigorous regulation of shipments (Baldwin and Mundo-Ocampo 1991). Whereas a number of species of *Globodera* parasitize Solanaceae in the new world, another group of *Globodera* species seem to have coevolved with Compositae in Europe (Baldwin and Mundo-Ocampo 1991). The distribution of a wide range of cyst species can be traced from particular hosts and regions to biogeographic patterns and more recently movement of these nematodes with soil and roots associated with shipments of agricultural products (Baldwin and Mundo-Ocampo 1991; Ferris 1983, 1985; Stone 1977).

We have noted that distribution of parasites is a function of the distribution of the hosts. One general precept of animal parasitology is the expectation that a host species harbors several parasite species, some of which are probably restricted to that species of host. By extension, many nematologists who study animal parasites are not surprised when a new nematode species is described from a host species that has not been the subject of exhaustive examination. In fact, this generalization is frequently confirmed; most thoroughly investigated vertebrate hosts are likely to yield one or more novel nematode species.

The nematode parasites of domesticated hosts have been intensively studied because of their importance to agriculture. Host-parasite lists (for example, Soulsby 1982) provide some insight into the diversity of nematodes from domesticated hosts and other vertebrates. For example, bovine species have been reported to harbor more than 60 nominal nematode species representing 28 genera; these nematodes include specialists for many "habitats" within the hosts, including the digestive tract, circulatory system, respiratory system, muscles, urogenital system, skin, eyes, and body cavities. Similarly, pigs are reported to serve as hosts for 37 nominal species representing 24 genera, and equids host at least 43 nominal species representing 29 genera. It is unclear how many of the congeners reported from a single vertebrate host species are synonyms. This problem is no doubt more acute for some nematode groups than others, but recent molecular investigations show promise for investigating potential synonymies. For example, a study of nucleotide sequences indicated that three described species of vertebrate parasites (*Teladorsagia* spp.) had no detectable nucleotide differences in a rapidly evolving region of ribosomal DNA; this result was interpreted as evidence that the previously described morphotypes represented a single species (Stevenson and others

1996). Similarly, sequence-based results (Newton and others 1998) were consistent with the inference of one species in the case of two nominal (and controversial) taxa of nematode parasites (*Cooperia oncophora* and *C. surnabada*) that were previously diagnosed on the basis of structural differences; this molecular finding was also consistent with the results of cross-breeding studies (Isenstein 1971). Conversely, it is unclear how many of the taxa assigned to a single morphological taxon represent cryptic species, but again genetic methods are beginning to shed light on the nature of species complexes in nematodes (Chilton and others 1992; Chilton and others 1995).

For many nematodes of vertebrate hosts, the extent to which nominal species are regionalized is often unknown because of the absence of systematic survey data. In addition, assessing geographic distributions of nematode parasites of vertebrates is made more difficult by the fact that nematodes are typically not uniformly distributed among individuals of a host species; instead many hosts remain uninfected while a few individuals harbor many nematodes (overdispersion). Predictably, the nematode parasites of domesticated vertebrates are geographically widespread because of transport of these hosts. Exceptions to that prediction might involve vector-transmitted nematode parasites, inasmuch as hosts could be moved to regions where required vectors are not codistributed. In the case of human-mediated movement of domesticated animals, the effect on their parasites is also evident from studies of nematode population genetic structure. For example, studies of four trichostrongylid nematode species (*Ostertagia ostertagi*, *Haemonchus placei*, *H. contortus*, and *Teladorsagia circumcinta*) occurring in three domesticated hosts revealed very little genetic differentiation among geographic regions of North America; this suggests that human transport of domesticated hosts has resulted in high nematode gene flow (Blouin and others 1992; Blouin and others 1995). In contrast, the genetic structure of a trichostrongylid nematode of deer (*Mazamastrongylus odocoilei*) was much more regionalized and showed a pattern of genetic isolation by distance (Blouin and others 1995).

NUMBER OF SPECIES PER HABITAT

The vast numbers of nematode species, the diversity of their habitats (from soils and sediments to animals and plants), the paucity of surveys, and the inadequately developed state of nematode taxonomy make a global estimate of nematode species daunting. Nematodes can live in the bark of trees (pinewood nematode); as parasites of bees, lizards, and tomato plants; and even in mushrooms and earthworms. Even the smallest samples of habitat can contain hundreds of nematode species, including many unknown species. For example, Tietjen (1984) found in sampling marine sediments of the Venezuela basin that only two of 196 nematode species had been previously described. Lawton and others (1996) found over 200 nematode species in moist tropical Cameroon soils, but because of the expertise needed and the cost of identifying species, had to leave twice that many undescribed. For these reasons, the number of descriptions of nematode species is still in its infancy, and nematologists and ecologists rely on descriptions of functional groups or become specialists in the identification of particular groups of species.

Nevertheless, there is some information that we can use with caution to examine global geographic and latitudinal patterns of nematode species diversity (table 1). The few studies in tropical forest soils (Coleman 1970; Hodda and

TABLE 1 Numbers of Described or Named Nematode Species in Single Sites and Estimates of Described and Total Numbers of Extant Species in Combined Sites of a Given Habitat and Globally Across All Habitats

Number of Described/Named Species from Single Sites

Author	Habitat Type	Location	No. of Described Species
Johnson and others 1972	Forest	Indiana, US	175
Yeates 1972	Forest	Denmark	75
Lawton and others 1996	Forest	Cameroon	204
Hodda and Wanless 1994	Grassland	England	154
Freckman and Huang 1998	Grassland	Central Plains Experimental Range LTER, Colorado, US	118
Niles 1991	Agroecosystem	Indiana, US	94
Freckman and Ettema 1993	Agroecosystem	Kellogg Biological Station LTER, Michigan, US	132
Freckman and Mankau 1986	Hot desert	Nevada, US	23
Freckman and Virginia 1989	Hot desert	Jornada LTER, US	18
Freckman and Virginia 1997	Cold desert	Dry Valleys, Antarctica	3
Hope 1987	Marine sediments	East Pacific Rise	148
Dinet and Vivier 1977	Marine sediments	Bay of Biscay	316
Jensen 1988	Marine sediments	Norwegian Sea	92
Lambshead 1993	Marine sediments	San Diego Trough	116

Number of Species Per Habitat Type, Globally

Author	Habitat Type	Described Species	Total No. of Species
Andrássy 1992	Continental	5,600	
	Marine	5,450	
Lambshead 1993	Marine sediments	4,000	10^6–10^8
Hoberg 1997	Helminthic parasites of vertebrates	>12,000	

Number of Species Across All Habitats of Globe

Author	Habitat Type	Described Species	Total No. of Species
Mayr and others 1953	All habitats, globally	12,000	
May 1988	All habitats, globally	1,000,000[a]	
Barnes 1989	All habitats, globally	12,000	
Brusca and Brusca 1990	All habitats, globally	12,000	
Hammond 1992	All habitats, globally	15,000	>1,000,000
Wilson 1992	All habitats, globally	12,000	
Hawksworth 1995	All habitats, globally	25,000	100,000–1,000,000

[a] May's estimate of described species seems exceptionally high, and Hammond (1992) suggests that he might have intended this as an estimate of *actual*, rather than *described*, species.

Wanless 1994) before Lawton's study did not indicate that the tropics were foci of nematode diversity or abundance, as was found for nematode species associated with plants and animals. Temperate forests had much higher species diversity. Although table 1 shows that the number of species is similar in Indiana agroecosystems and in the Cameroon forest, species were still being identified in the Cameroon when the study ended. The results seem to support Hammond (1992), who suggested that the well-documented floristic richness of areas that are seasonally dry might not be paralleled by the richness of terrestrial invertebrates, fungi, or microorganisms. At the other extreme of diversity are the soils of the Antarctic Dry Valleys (78°S), which have only three endemic species—perhaps the lowest nematode diversity on Earth (Freckman and Virginia 1997). Surprisingly, agroecosystems, which we think of as being disturbed ecosystems, can have over 100 nematode taxa (Freckman and Ettema 1993; Yeates 1980). Yet we also know that different agronomic practices have different effects on nematode diversity, including the loss of some species. Those examples indicate the need for a more strategic effort in identifying these speciose animals.

The lack of knowledge about the total numbers of nematode species present on even the smallest scales—say, in 10 g of soil—makes estimates of total species in different habitats or even globally a monumental task. There is considerable variability in estimates that have been attempted (table 1), with estimates of total nematode species globally as low as 100,000 (GBA 1995); in contrast, Lambshead and others (1993) estimated the highest number of marine nematode species alone to be 100,000,000.

A fundamental problem encountered in estimating regional or global nematode species diversity is that many of the traditional diversity methods might not be applicable to such a group as poorly described and as varied as nematodes. For the near future, estimates of global nematode biodiversity will remain speculative. Methods often used to estimate species numbers for poorly known but speciose groups do not seem particularly appropriate for nematodes. For example, one traditional method is to extrapolate from reliable available data with ratios that are established for better-known groups, such as birds or vascular plants. For nematodes, reliable data on species numbers on even the smallest scale are available for only a small number of habitats. Furthermore, by definition, groups that are well described can be very different from such poorly described groups, so it cannot be assumed that patterns are similar in both. Other methods, such as relating first principles and processes of ecology to species number (such as body size, food web structures, trophic links, and parasite-host relationships) are similarly compromised because the data behind the relationships are based on groups that are easy to access and might not be representative of nematodes. Methods that base estimates of total species numbers on patterns in the number of species that have already been detected are likely to be uninformative because rates of discovery probably are biased by our interests and resources rather than reflecting a true insight into the ratio of discovered to undiscovered species (Hammond 1992; May 1988, 1998).

Until a major effort is launched by the scientific community, we will continue to guess at this most abundant and important group of invertebrates. One approach that shows promise in estimating nematode species numbers in different habitats, and ultimately globally, is to develop theoretical estimates based on energetics. A model is being developed to estimate the total nematode biomass that can be supported in different habitats, on the basis of the carbon available in those habitats. This information might be used to estimate maximal viable nematode population sizes and species number at each trophic level (Freckman and Moore 1998).

GAPS IN KNOWLEDGE?

Although nematodes are the most species-rich phylum of metazoa, most of the world's habitats are undersampled, and we can only begin to envision the numbers of species. Worldwide, wherever habitat is destroyed, microscopic nematode species with highly localized distribution are likely to be lost at a high rate with a cost to humankind that includes loss of biological control agents and unique natural compounds. The compounds, largely unexplored, might include anticoagulants, plant-growth regulators, and antimicrobials in bacteria dependent on nematodes (Jarosz 1996). Loss of nematode species also results in loss of opportunities to measure anthropogenic disturbance; the latter has implications for understanding and managing ecosystems. Inadequate knowledge of nematode diversity has led to introduction of destructive nematode species. Most recent surveys are biased toward the regions most accessible to the world's small and shrinking supply of nematode taxonomists, and development of a taxonomic system is biased to parasites that have obvious economic and medical effects. The greatest need is for data from representative habitats that will provide a predictive framework for testing hypotheses of global species diversity and targeting areas of strategic importance for the conservation and use of nematode species. Whereas it would be difficult to identify any region, even Europe or North America, as adequately sampled, the most undersampled habitats, relative to suspected diversity, are deep ocean sediments, wetlands, and tropical nematode-invertebrate associations. Data from expanded nematode surveys in Asia and Africa are particularly needed.

Assessment and conservation of nematode species diversity are confounded by an inadequate taxonomic and phylogenetic framework. The situation is exacerbated by the need for a new generation of nematode taxonomists with broad training in a range of classical and molecular tools in a rigorous phylogenetic context (Barker 1994; Ferris 1994; Hyman and Powers 1991; Systematics Agenda 2000 1994). Taxonomists can broaden their reach by working closely with parataxonomists, specialists trained to work with taxonomists for field collecting, processing material, and routine identifications. Furthermore, substantial financial resources are needed to support research in nematode biodiversity in general and to facilitate greater integration of nematode taxonomists throughout the world via readily accessible and well-supported taxonomic collections and electronic databases.

NUMBER OF TRAINED PEOPLE AVAILABLE

There is little argument that the number of persons trained to identify and describe new species of nematodes is woefully inadequate in light of the enormous size of the phylum and its largely underexplored state. No group of nematodes is as species-rich as marine forms. Lambshead (1993), predicting 100,000,000 species of marine nematodes, notes that even if there were only 1,000,000 species of nematodes, the task would be overwhelming; if 20 active marine nematode taxonomists worldwide were collectively describing about 200 species per year (an optimistic set of assumptions), it would take 5,000 years to describe the species! Considering broader groups of nematodes, about 30 taxonomists worldwide have the experience necessary to describe species in one or more orders of nonparasitic nematodes, including some marine taxa (SON Systematics Resources Committee 1994). The availability of taxonomists is only slightly greater for more specialized nematodes of economic importance. About 60 taxonomists worldwide work on one or more families of the orders Tylenchida, Dorylaimida, and Triplonchida, which include herbivores and plant parasites (figure 1). Taxonomy is a small portion of the overall responsibilities of these scientists; each might describe one or two species per year. For the six major orders of vertebrate parasites (figure 1), there are about 45 taxonomists worldwide. Species belonging to the four orders that include parasites or associates of other invertebrates (figure 1) have received expanded attention in recent years because of their potential as biological control agents of insect pests; but only about 25 specialists worldwide work on these nematodes. From the perspective of advancing knowledge of nematode biodiversity, more taxonomic emphasis should be placed on understanding nonparasitic taxa, particularly aquatic species.

PROBLEM-SOLVING AND APPROACHES FOR GREATER AWARENESS

Increasing public value of ecology and the environmental sciences presents nematologists with opportunities to respond to a public receptive to understanding food webs, nutrient cycling, regulation of microbial populations, natural history, and biogeography. Bernard (1994) suggests that ocean-dwelling nematodes, because of their remarkable structural and biological variation, are potential magnets for programs that illustrate their natural history. Research on the nematode model *Caenorhabditis elegans* has already been featured in PBS programs on genetics, reproduction, and developmental biology. Internet sites, such as the Tree of Life Project (D.R. Madison and W.P. Madison; http://phylogeny.arizona.edu/tree/phylogeny.html), should not be underestimated in their effect on increasing public awareness of nematodes. The cost associated with such increased public awareness is that nematologists must be willing to set aside other responsibilities to accept outreach opportunities.

Understanding of nematode biodiversity will advance most efficiently when there is a global program harnessing the world's nematology expertise for coordinated sampling and database development; the first step in such a program will be to develop a sampling strategy designed to yield maximal information on global diversity with minimal expenditure of time and human and economic resources. Because nematodes are so pervasive and integrated with all of life, a global sampling program must be integrated with appropriate complementary expertise, including soils, marine biology, entomology, botany, and vertebrate zoology.

Development of the taxonomic infrastructure in nematology requires a commitment to programs for strengthening taxonomic expertise; these include international programs for training new taxonomists and support for systematic museum collections. Specimen-based collections provide basic tools for taxonomic work as repositories of preserved specimens, preserved DNA, and supporting databases and Web sites to aid inventory and access for specialists. Among the most promising new approaches for taxonomists are molecular tools refined to increase the efficiency and accuracy of surveys through databases that link morphological and molecular species diagnostics (NSF Workshop on Systematics and Inventory of Soil Nematodes, http://www.nrel.colostate.edu/soil/home.html) (Szanlanski and others 1997; Thomas and others 1997). Surveys and species identification are inseparable from the task of reconstructing a phylogenetic framework for nematodes (Adams 1998), but the tools for developing such a framework are more powerful and promising than ever before (Blaxter and others 1998).

The sciences that support advances in our understanding of nematode biodiversity are at an important crossroads. Nematology's base of classical taxonomists—with their wealth of information on nematode structure, diagnoses, and natural history—has been seriously eroded over the last 15 years. Although molecular systematists are applying modern approaches that offer much promise to advance the discipline, the most fruitful outcomes will come from collaborative efforts of classically trained nematologists, molecular systematists, and other scientists who can apply novel tools that enhance our ability to address complex problems in biodiversity. There is a very narrow window of opportunity to train the next generation of nematode biosystematists schooled in both classical and new approaches, given that many universities have hired scientists who use reductionist approaches in preference to those who use the entire organism as the unit of study. With the reduction and dispersion of the remaining expertise, training of broad-based nematode biosystematists is more expensive because it often requires on-site work at several laboratories. Given our lack of data on the full extent of nematode diversity, it is not practical to estimate how many additional scientists are required to develop a thorough understanding of the phylum. But doubling the number of nematode biosystematists during the next two decades seems to be a conservative and necessary first step. Clearly, more emphasis should be placed on taxa that are not of direct economic importance, such as nonparasitic soil and aquatic species.

Those scientific imperatives cannot be accomplished without addressing some serious practical considerations. What agencies will provide the funding for this important research? How will the current pecking order among scientists be altered so that organismal biologists are considered to be as essential to university research programs as scientists who study fundamental processes at the molecular level? Documents like this cannot serve science if such practical problems remain unresolved or if scientists themselves do not adopt a more panoramic perspective of biology and, in the process, see fit to set priorities in research support for groups of organisms that are poorly understood and rapidly disappearing.

ACKNOWLEDGMENTS

We acknowledge support of National Science Foundation grants PEET 97-12355, DEB-93-18249, OPP 91-20123, and DEB 96-26813 and the help of Gina Adams. This work is a contribution to OPP 92-11773, the MCM LTER.

REFERENCES

Abebe E, Coomans A. 1995. Freshwater nematodes of the Galápagos. Hydrobiologica 299:1–51

Adams B. 1998. Species concepts and the evolutionary paradigm in modern nematology. J Nematol 30:1–21.

Andrássy I. 1984. Klasse Nematoda. Stuttgart West Germany: Gustav Fischer Verlag. 509 p.

Andrássy I. 1914. A short census of free-living nematodes. Fund Appl Nematol 15:187–8.

Baldwin JG. 1992. Evolution of cyst and non-cyst-forming Heteroderinae. Ann Rev Phytopath 30:271–90.

Baldwin JG, Mundo-Ocampo M. 1991. Heteroderinae, cyst and noncyst-forming nematodes. In: Nickle WR (ed). Manual of agricultural nematology. New York NY: Marcel Dekker. p 275–362.

Barker KR. 1994. Plant and soil nematodes—societal impact and focus for the future. Bioscience 44:568–569.

Barnes RD. 1989. Diversity of organisms. How much do we know? Amer Zool 29:1075–84.

Bernard EC. 1994. Nematology in the 21st century: a foray into the future. Phytopath News 28:40–1.

Blaxter ML, DeLey P, Garey JR, Liu LX, Scheldeman P, Vierstraete A, Vanfleteren JR, Mackey LY, Dorris M, Frisse LM, Vida JT, Thomas WK. 1998. A molecular evolutionary framework for the phylum Nematoda. Nature. 392:71–75.

Blouin MS, Dame JB, Tarrant CA, Courtney CH. 1992. Unusual population genetics of a parasitic nematode: mtDNA variation within and among populations. Evolution 46:470–6.

Blouin MS, Yowell CA, Courtney CH, Dame JB. 1995. Host movement and the genetic structure of populations of parasitic nematodes. Genetics 141:1007–14.

Boucher G. 1990. Pattern of nematode species diversity in temperate and tropical subtidal sediments. Mar Ecol 11:133–46.

Brusca RC, Brusca GJ. 1990. Invertebrates. Sunderland MA: Sinauer. 922 p.

Castillo-Fernandez D, Lambshead PJD. 1990. Revision of the genus Elzalia Gerlach, 1957 (Nematoda: Zyalidae) including three new species from an oil producing zone in the Gulf of Mexico, with a discussion of the sibling species problem. Bull Br Mus Nat Hist 56:63–71.

Chabaud AG. 1974. Keys to subclasses, orders and superfamilies. In: Anderson RC, Chabaud AG, Willmott S (eds). CIH keys to the nematode parasites of vertebrates. London UK: CAB Intl, Headley Brothers Ltd, The Invicta Pr No 1. p 6–17.

Chilton NB, Beveridge I, Andrews RH. 1992. Detection by allozyme electrophoresis of cryptic species of Hypodontus macropi (Nematoda: Strongyloidea). Intl J Parasitol 22:271–9.

Chilton NB, Gasser RB, Beveridge I. 1995. Differences in a ribosomal DNA sequence of morphologically indistinguishable species with the Hypodontus macropi complex (Nematoda: Strongyloidea). Intl J Parasitol 25:647–51.

Cobb NA. 1914. Nematodes and their relationships. US Dept Agric Ybk. p 457–90.

Coleman DC. 1970. Nematodes in the litter and soil of El Verde rainforest in a tropical rainforest: a study of irradiation and ecology at El Verde, Puerto Rico. Div of Technical Information, US Atomic Energy Commission. p E103–4.

Decraemer W. 1995. The family Trichodoridae: Stubby root and virus vector nematodes. London UK: Kluwer Academic. p 360.

Dinet A, Vivier MH. 1977. Le meiobenthos abyssal di Golfe de Gascoigne: Conderations sur les donnees quantitatives. Cash Biol Mar 20:109–23.

Eisenback JD, Triantaphyllou HH. 1991. Root-knot nematodes: Meloidogyne species and races. In: Nickle WR (ed). Heteroderinae, cyst and non-cyst-forming nematodes. Manual of agricultural nematology. New York: Marcel Dekker. p 191–274.

Ferris VR. 1979. Cladistic approaches in study of soil and plant parasitic nematodes. Amer Zool 19:1195–215.

Ferris VR. 1983. Phylogeny, historical biogeography, and the species concept in soil nematodes. In: Stone AR, Platt HM, Khalil LF (eds). Concepts in nematode systematics. London NY: Academic Pr. p 143–61.

Ferris VR. 1985. Evolution and biogeography of cyst-forming nematodes. OEPP/EPPO Bull 15:123–9.

Ferris VR. 1994. The future of nematode systematics. Fund Appl Nematol 17: 97–101.

Freckman DW. 1988. Bacterivorous nematodes and organic-matter decomposition. Agric Ecosys Envir 24:195–217.

Freckman DW, Ettema CH. 1993. Assessing nematode communities. In: agroecosystems varying human intervention. Agric Ecosyst Envir 45:239–61.

Freckman DW, Mankau R. 1986. Abundance, distribution, biomass, and energetics of soil nematodes in a northern Mojave desert. Pedobiologia 29:129–42.

Freckman DW, Virginia RA. 1997. Low diversity Antarctic soil nematode communities: distribution and response to abundance. Ecology 78:363–9.

Grassle JF, Maciolek NJ. 1992. Deep-sea species richness: regional and local diversity estimates from quantitative bottom samples. Amer Nat 139:313–41.

Hammond PM. 1992. Species inventory. In: Jenkins M, Swanson TM, Synge H (eds). Global biodiversity: status of the earth's living resources. London UK: Chapman &Hall. p 17–39.

Hawksworth DL. 1995. Magnitude and distribution of biodiversity. In: Heywood V (ed). Global biodiversity assessment. Cambridge UK: Cambridge Univ Pr. p 107–92.

Hessler RR, Sanders HL. 1967. Faunal diversity in the deep sea. Deep-Sea Res 14:65–78.

Hoberg EP. 1997. Phylogeny and historical reconstruction: host-parasite systems as keystones in biogeography and ecology. In: Reaka-Kudla ML, Wilson DE, Wilson EO (eds). Biodiversity II. Washington DC: Joseph Henry Pr. p 243–62.

Hodda M, Wanless FR. 1994. Nematodes from an English chalk grassland: species distributions. Nematologica 40:116–32.

Hope WD. 1987. Appendix G. In: Spiess FN, Hessler R, Wilson G, Wudert M (eds). Environmental effects of deep-sea dredging. San Diego CA: University of California.

Hope WD, Murphy DG. 1972. A taxonomic hierarchy and checklist of the genera and higher taxa of marine nematodes. Contrib Zool 307:101.

Huettel RN, Dickson DW. 1984. *Radopholus citrophilus* n. sp (Nematoda), a sibling species of *Radopholus similis*. Washington DC: Proc Helminthol Soc 51:32–5.

Hyman BC, Powers TO. 1991. Integration of molecular data with systematics of plant parasitic nematodes. Ann Rev Phytopath 29:89–107.

Isenstein RS. 1971. The polymorphic relationship of *Cooperia oncophora* (Railliet, 1898) Ransom, 1907, to *Cooperia surnabada* Antipin, 1931 (Nematoda: Trichostrongylidae). J Parasitol 57:316–9.

Jarosz J. 1996. Ecology of anti-microbials produced by bacterial associates of *Steinernema carpocapsae* and *Heterorhabditis bacteriophora*. Parasitology 112:545–52.

Jensen P. 1988. Nematode assemblages in the deep sea benthos of the Norwegian sea. Deep-Sea Res 35:1173–84.

Johnson SR, Ferris VR, Ferris JM. 1972. Nematode community structure of forest woodlots: I Relationships based on similarity coefficients of nematode species. J Nematol 4:175–82.

Lambshead PJD. 1993. Recent developments in marine benthic biodiversity research. Oceanis 19:5–24.

Lambshead PJD, Ferrero TJ, Wolf GA. 1995. Comparison of the vertical distribution of nematodes from two contrasting abyssal sites: the northeast Atlantic subject to different seasonal inputs of phytodetritus. Intl Rev Gesamt Hydrobiol 80:327–31.

Lawton JH, Bignell DE, Bloemers GF, Eggleton P, Hodda ME. 1996. Carbon flux and diversity of nematodes and termites in Cameroon forest soils. Biodiv Cons 5:261–73.

Maggenti A. 1981. General nematology. New York NY: Springer-Verlag. p 372.

Mamiya Y. 1983. Pathology of the pine wilt disease caused by *Bursaphelenchus xylophilus*. Ann Rev Phytopath 21:201–20.

May RM. 1988. How many species are there on earth? Science 241:1441–9.

Mayr E, Linsley EG, Usinger RL. 1953. Methods and principles of systematic zoology. New York NY: McGraw-Hill. p 328.

Newton LA, Chilton NB, Beveridge I, Gasser RB. 1998. Genetic evidence indicating that Cooperia surnabada and Cooperia oncophora are one species. Intl J Parasitol. 28:331–6.

Nicholas WL. 1975. The biology of free-living nematodes. Oxford UK: Clarendon Pr 215 p.

Overgaard–Nielsen JC. Studies on the soil microfauna: The soil inhabiting nematodes. Aarhus Denmark: Naturhistorist Museum. 2:1–131.

Niles RK. 1991. Relationship between agronomic management and nematode community structure. PhD dissertation. West Lafayette IN: Purdue University (Diss Abstr 91–32486). 291 p.

O'Bannon JH. 1977. Worldwide dissemination of *Radopholus similis* and its importance in crop production. J Nematol 9:16–25.

Peterson H, Luxton MA. 1982. Comparative analysis of soil fauna populations and their role in decomposition processes. Oikos 39:287–388.

Platt HM, Warwick RM. 1983. Free-living marine nematodes, Part 1: British Enoplids. Cambridge UK: Cambridge Univ Pr. 307 p

Rutherford TA, Mamiya Y, Webster JM. 1990. Nematode-induced pine wilt disease: factors influencing its occurrence and distribution. Forest Sci 36:145–55.

Sher SA. 1968. Revision of the genus *Radopholus Thorne*, 1949 (Nematoda: Tylenchoidea). Washington DC: Proc Helminthol Soc 35:219–37.

Sohlenius BA. 1980. Abundance biomass and contributions to energy flow by soil nematodes in terrestrial ecosystems. Oikos 34:186–94.

SON Systematics Resources Committee. 1994. Nematol Newsl 40:11–5.

Soulsby EJL. 1982. Helminths, arthropods, and protozoa of domesticated animals, 7th edition. London UK: Bailliére Tindall. 809 p.

Stevenson LA, Gasser RB, Chilton NB. 1996. The ITS-2 rDNA of *Teladorsagia circumcincta, T. trifurcata,* and *T. davtiani (Nematoda: Trichostrongylidae)* indicates that these taxa are one species. Intl J Parasitol 26: 1123–6.

Stone AR. 1977. Recent developments and some problems in the taxonomy of cyst-nematodes, with a classification of the Heteroderoidea. Nematologica 23:272–88.

Systematics Agenda 2000. 1994. Systematic Agenda 2000: Charting the biosphere. New York NY: American Museum of Natural History. p 20.

Szalanski AL, Sui DD, Harris TS, Powers TO. 1997. Identification of cyst nematodes of agronomic and regulatory concern with PCR-RFPL of ITS1. J Nematol 29:255–67.

Thomas WK, Vida JT, Frisse LM, Mundo M, Baldwin JG. 1997. DNA sequences from formalin-fixed nematodes: integrating molecular and morphological approaches to taxonomy. J Nematol 29:250–4.

Thorne G. 1961. Principles of nematology. New York NY: McGraw Hill. p 553.

Tietjen JH. 1984. Distribution and species diversity of deep-sea nematodes in the Venezuela Basin. Deep-Sea Res 31:119–32.

Wall DW, Huang SP. 1998. Response of the soil nematode community in a shortgrass steppe to long-term and short-term grazing. Appl Soil Ecol 9:39–44.

Warwick RM, Rice R. 1979. Ecological and metabolic studies on free living nematodes from an estuarine mudflat. Estuar Coastal Mar Sci 9:251–57.

Wilson EO. 1992. The diversity of life. Cambridge MA: Belnap Pr and Harvard Univ Pr. 424 p.

Yeates GW. 1972. Nematodes of a Danish beech forest in methods and general analyses. Oikos 23:128–89.

Yeates GW. 1980. Soil nematodes in terrestrial ecosystems. J Nematol 11:213–29.

GLOBAL DIVERSITY OF MITES

R. B. HALLIDAY
Australian National Insect Collection, CSIRO Entomology,
GPO Box 1700, Canberra Act 2601, Australia

B. M. OCONNOR
Museum of Zoology, University of Michigan,
Ann Arbor, Michigan, 48109-1079

A. S. BAKER
Department of Entomology, The Natural History Museum,
Cromwell Road, London SW7 5BD, United Kingdom

INTRODUCTION

THE MITES and ticks make up the order Acari (or Acarina) within the class Arachnida. They differ from other arachnids in that, in virtually all cases, all traces of body segmentation have disappeared. The body of a mite is divided into the gnathosoma (mouthparts) and the idiosoma. There is no recognizable head, and the structures normally associated with the head, such as eyes and brain, are incorporated into the idiosoma. There are no antennae, but in many groups the first pair of legs is long and slender and serves a sensory function. Their ancestral arachnid relatives were predatory, but the mites have diversified from this origin to the extent that they now occupy an extraordinarily diverse range of niches. Many have remained predatory, but other groups have adapted to plant feeding and scavenging on dead plant matter and, alone among the arachnids, have developed into parasites of vertebrate and invertebrate animals. They have developed a wide range of associations with other organisms, including phoretic relationships, some of which have produced morphological and behavioral adaptations in both the mites themselves and the animals that they use for transportation.

The acari are usually subdivided into seven suborders. Astigmata includes forms that occur in patchy habitats, such as dung, carrion, decaying wood, fungi, and the nests of mammals and birds; the latter have evolved into many families of vertebrate parasites, the common pests of stored products that occur in the home and in food storage, and the dust mites that are implicated in allergy and asthma. Oribatida contains essentially feeders on dead plant material and fungi; these mites play an important part in litter decomposition and soil formation, and a few are important as intermediate hosts of cestode parasites. Prostigmata is a diverse assemblage of predators, plant parasites, algal feeders, and parasites of vertebrate and invertebrate animals. Mesostigmata also includes parasites, but most of its members are predators in soil and decomposing organic material. Ixodida contains the ticks, which are exclusively parasites of vertebrates that feed by drawing blood from their hosts with specially adapted mouthparts. The members of the remaining suborders, Opilioacarida and Holothyrida, are less diverse and less well known; they occur in damp habitats, such as forest leaf litter, and are believed to be predators.

Mites are small. The smallest adults are plant parasites about 80 μm long; the largest are predators about 13 mm long. Most are 400–800 μm long. Their small size has contributed to the diversity of their life cycles and basic biology and has allowed them to exploit an extraordinary variety of niches. The taxonomic and ecological diversity of mites is accompanied by structural diversity. Some exhibit specialized adaptations of the mouthparts, from elongated attenuated chelicerae for sucking plant or animal fluids in parasitic groups to very heavy robust chewing mouthparts in families that feed on fungi or dead plant material. Associations with other organisms have produced specialized structures and organ systems used for grasping and holding—such as hypertrophied claw-like setae, modified mouthparts, and adhesive sucker plates—and modifications of the life cycle to synchronize with other species on which they depend for feeding and dispersal.

Mites are equally diverse in their modes of reproduction. Some copulate with direct sperm transfer from the male to the primary or a secondary genital aperture of the female. In others, males transfer sperm to the female with their chelicerae or deposit a spermatophore on the substrate for females to pick up. Each of those systems is associated with specific structural and behavioral adaptations, some of which are highly complex.

In the face of the tremendous diversity of this group of organisms, generalizations about taxonomic and ecological diversity are difficult. We do not attempt a comprehensive overview of all these varied phenomena here. Instead, we attempt the more modest objective of assessing the evidence of how many species of mites exist. It would be easy to make sweeping generalizations about the existence of millions of species on the basis of our acknowledged ignorance of most mite groups, but we have resisted that temptation. Instead, we present an analysis of the state of knowledge of the mite faunas of Great Britain, Australia, and North America based on the results of recent taxonomic studies of selected taxa. We use these data as the basis of an extrapolation to an estimate of the number of species that remain to be described in these regions and then, more speculatively, to an estimate of the total world fauna of mite species. Previous estimates of the number of described mite species are summarized in table 1, and our current estimate is 48,200 described species.

TABLE 1   Various Estimates of Numbers of Described Mite Species

Source	No. of Species
Wharton 1964	17,500
Levi and others 1968	20,000
Krantz 1970	30,000
Walter and others 1996	45,000
Present work	48,200

MITE FAUNA OF AUSTRALIA

We have conducted a comprehensive survey of the literature associated with the Australian mite fauna. By August 1997, about 2,700 described species of mites in Australia were known—Astigmata, 330; Oribatida, 330; Prostigmata, 1,270; Ixodida, 80; Holothyrida, 3; and Mesostigmata, 675 (Halliday 1998). We may then ask ourselves how to extrapolate this figure to derive an estimate of the size of the total fauna. One approach is suggested by the figures presented in table 2. The table lists a series of mite groups in which the Australian fauna has been the subject of modern revision and shows the numbers of species before and after revision. The totals show that the previously known fauna was multiplied by a factor of about 2.9 as a result of revision. That might lead to an expectation that the total Australian mite fauna could include about 7,800 species if the same trend is repeated in other groups.

However, we believe for several reasons that a multiplying factor of 2.9 is a serious underestimate. The water-mite survey published by Cook (1986) essentially reported the results of a single collecting expedition by a single person, and many

TABLE 2   Numbers of Described Mite Species in Selected Families before and after Modern Revisions, Australia

| Group | Revision | No. Species | | |
		Before	After	Factor
Water mites	Cook 1986	133	334	2.51
Macrochelidae	Halliday 1986a,b, 1988, 1990, 1993	25	64	2.56
Phytoseiidae	Schicha 1987	31	97	3.13
Scutacaridae	Mahunka 1967	4	19	4.75
Steganacaridae	Niedbala 1987, 1989	6	63	10.5
Ascidae	Niedbala and Colloff 1997 Walter and others 1993 Walter and Lindquist 1997 Halliday and others 1998	17	45	2.65
TOTAL		216	622	2.88

other species of water mites have been reported in Australia since then (for example, Harvey, 1987, 1989, 1990a,b,c,d, 1996). Schicha's (1987) revision of Phytoseiidae included mainly species associated with economic crop plants. The fauna associated with native plants, such as those occurring in rain forests, has only recently been studied (for example, Schicha and O'Dowd 1993) and is likely to be a rich source of new species. The study of the Ascidae reported by Halliday and others (1998) deliberately excluded many undescribed species and documented only enough species to record the presence of each genus in Australia.

It is also informative to examine the state of knowledge of some groups that are not in table 2. The known Australian fauna of Eriophyoidea includes only 53 described species of a world total of 2,884 (Armine and Stasny 1994). The United States has 635 described species of Eriophyoidea (Baker and others 1996), described from 579 host plant species. The Australian flora is very likely to include over 25,000 species of plants (George 1981), so if the same relationship between plant species and mite species numbers exists in Australia, the eriophyoid fauna of Australia might exceed 5,000 or even 10,000 species. A total of 64 species of described feather mites has been recorded from Australia. The number of bird species in Australia is about 700 (Slater 1970), and the feather mites demonstrate a high degree of host specificity (Gaud and Atyeo 1996), so the number of feather mite species might exceed 1,000. Hirschmann and Wisniewski (1993) listed some 2,000 described Uropodina species worldwide, but only 67 from Australia are known. The uropodine fauna of Australian rain forests is extremely rich and is likely to yield hundreds of species. Until the 1980s, the Australian fauna of Halacaridae had been very little studied and included fewer than 20 described species. However, a single collecting trip to a small island (Rottnest Island, WA; area, 1,900 ha) in 1991 yielded over 80 species (Bartsch 1996), most of which remain undescribed. The Australian fauna now comprises 80 described species, almost all of which were first described in the last five years. The Australian coastline of over 36,000 km is certain to yield hundreds of species of Halacaridae. About 300 described species of Tarsonemidae are known worldwide (Lindquist 1986), but only eight from Australia are recorded. Bearing all those factors in mind, we conservatively estimate that the Australian mite fauna is likely to exceed 20,000 species, more than seven times as many as the known described species.

MITE FAUNA OF GREAT BRITAIN AND IRELAND

On the basis of the checklists of Turk (1953) and Luxton (1996), the recent monograph by Hillyard (1996), and a search of the *Zoological Record*, we obtained a total of 1,740 species for the mite fauna of Great Britain and Ireland—Astigmata, 265; Oribatida, 303; Prostigmata, 675; Ixodida, 22; and Mesostigmata, 475. However, the regularity with which acari new to science or newly recorded are still being found (for example, Luxton 1996; Skorupski and Luxton 1996) is evidence that the fauna is incompletely known.

A multiplying factor of 1.49 was obtained with the method described above (table 3), and this gives a projected total fauna of 2,590 species. Examination of the taxonomic attention that the various mite taxa have received in Great Britain and Ireland, however, suggests that far more are likely to occur there. The Ixodida have been extensively studied, and it is unlikely that additional native species will be discovered (Arthur 1963; Hillyard 1996). The Oribatida have also been the subject of much taxonomic work over many years (for example, Michael 1884, 1888; Luxton 1996), but new records are still being uncovered (Luxton 1996). Some families of Mesostigmata, such as the Macrochelidae, have been reviewed more than once in the last 50 years (Evans and Browning 1956; Hyatt and Emberson 1988), but most families have not been studied in great detail. Of all the orders, the Astigmata and Prostigmata have been the least studied; apart from Green and Macquitty (1987) and Gledhill and Viets (1976), there are no modern taxonomic monographs or reviews of the British members of either taxa. More often, descriptions or records appear in studies of particular habitats, such as human dwellings and food stores (Hughes 1976) or hosts (Hyatt 1990).

Many localities and habitats remain to be comprehensively sampled. Green and Macquitty (1987) acknowledged that their monograph on the marine mites could not be regarded as a complete account of the British fauna, because so much of the coastline was unexplored for halacarids. Similarly, Gledhill (1979) anticipated additional records of freshwater mites when the hyporheic zone of superficial riverine gravels and sands was more exhaustively sampled. Surveys of terrestrial mites—such as the Eupodidae, Rhagidiidae, and Phytoseiidae (table 3)—have concentrated on the faunas of woodland soils and plants, but even such a narrow range of situations has yielded many new species and records. Such habitats as tree-hole litter, fungi, mosses, and intertidal areas are still to be thoroughly examined.

It can be expected, therefore, that surveys of the majority of mite taxa will result in large numbers of species being added to the fauna of Great Britain and Ireland, as will a close examination of the many habitats for which the mite fauna is not known in any detail.

TABLE 3 Numbers of Described Mite Species in Selected Taxa before and after Modern Revisions, British Isles

Group	Revision	No. Species		Factor
		Before	After	
Water mites	Gledhill and Viets 1976	226	273	1.21
Halacaridae	Green and Macquitty 1987	27	65	2.41
Eupodidae	Baker 1987	5	24	4.8
Rhagidiidae	Baker 1987	4	11	2.75
Parasitinae	Hyatt 1980	21	36	1.71
Pergamasus	Bhattacharyya 1963	13	32	2.46
Macrochelidae	Hyatt and Emberson 1988	23	32	1.39
Phytoseiidae	Baker pers. obs.	25	39	1.56
TOTAL		344	512	1.49

MITE FAUNA OF NORTH AMERICA

OConnor (1990) reviewed the status of the mite fauna of North America north of Mexico by using data derived from species lists maintained by systematists working with each major taxonomic group. At that time, 5,106 described species had been recorded for the region—Opilioacarida, 1; Ixodida, 83: Mesostigmata, 869; Prostigmata, 2,803; Oribatida + "Endeostigmata", 930; and Astigmata, 420. Contributors to that dataset were asked to estimate the total number of species expected in the North American fauna. The estimates were Opilioacarida, 1; Ixodida, 84; Mesostigmata, 2,827; Prostigmata, 7,977; Oribatida + "Endeostigmata", 15,300; and Astigmata, 3,611. That is a total estimated fauna of 29,800 species, which means a multiplying factor of almost six. Since those data were reported, several revisionary works and compilations have added to the fauna. Farrier and Hennessey (1993) cataloged the free-living Mesostigmata, recording over 1,300 species in "North America", including Mexico and parts of Central America. Baker and Tuttle (1994) revised the spider-mites (Tetranychidae) of the United States, and Baker and others (1996) revised the Eriophyoidea of the United States. Numerous smaller-scale revisions of families and genera have also been published, most notably for several groups of Oribatida (for example, Behan-Pelletier 1986, 1989, 1990, 1993, 1994; Norton and others 1996) and water mites (for example, Smith 1989a,b, 1990a,b, 1991a,b, 1994). The latter data provide something of a test of the earlier estimates of faunal diversity. With the method described above (see table 4), an increase factor of 3.24 was obtained for the oribatid groups recently revised and 2.65 for the water mites. The prior published estimates would predict increase factors of 16.5 for the oribatids and 1.79 for the water mites. The much lower observed increase factor for the oribatids might be explained by the fact that the recent revisions dealt primarily with faunas occurring in boreal Canada, where species diversity might be expected to be lower than

TABLE 4   Numbers of Described Mite Species in Selected Taxa before and after Modern Revisions, North America

Group	Revision	No. Species		Factor
		Before	After	
Veigaia	Hurlbutt 1984	23	27	1.17
Oribatida	Behan-Pelletier 1986, 1989, 1990, 1993, 1994	25	81	3.24
Water mites	Smith 1989a,b, 1990a,b, 1991a,b, 1992a,b, 1994	23	61	2.65
Astigmata	OConnor 1991	4	57[a]	14.25
Phytoseiidae	Cunliffe and Baker 1953	26	150	5.77
	Farrier and Hennessey 1993			
Oplitis	Hunter and Farrier 1976a,b	8	26	3.25
TOTAL		109	402	3.69

[a] Most remain undescribed.

in North America as a whole. The larger than predicted increase for the water-mite taxa might reflect the fact that many of the included groups specialize in stream and hyporheic environments, which have received much less attention than lotic environments. The data again point out the danger in extrapolating too much from limited observations.

As has been the case for other regions, North American taxa of economic and medical importance have received more attention. For example, in the first compilation of North American Phytoseiidae, Cunliffe and Baker (1953) reported 26 species. Farrier and Hennessey's 1993 catalog lists 150, for an increase factor of 5.77. Baker and Tuttle's 1994 review of the thoroughly studied North American Tetranychidae lists 218 species, of which only 12 were newly reported. The fauna of Ixodida is essentially completely known. However, many other taxa have received scant attention in North America; relatively little information is available on such major faunal elements as the free-living Prostigmata and Mesostigmata and the arthropod-associated Mesostigmata and Astigmata. Evidence of the latter includes OConnor's (1991) report of 57 species of insect-associated Astigmata collected at a single forested site in northern Michigan; of those, only four were previously described. Later collecting at the same site has yielded an additional 25 undescribed species (OConnor, unpublished data). Most taxa that have received some taxonomic treatment have not been thoroughly surveyed over the entire continent; most published treatments are at best regional. Major areas of endemism—such as California, the Pacific Northwest, the arid Southwest, and subtropical Florida—remain poorly collected for most free-living taxa.

HOW MANY SPECIES OF MITES HAVE BEEN DESCRIBED?

Various figures have been quoted for the number of mite species described worldwide (table 1). We now present a total of 48,200 nominate species (to June 1997). This figure was obtained from three main sources: an index of species at The Natural History Museum, London, which was maintained until 1977; the *Zoological Record* from 1978 on, which showed that an average of 788 new species were being described each year during that period (table 5); and, for Ixodida, Keirans (1992). The search of the *Zoological Record* from 1978 to 1996 also showed that mite species names were being synonymized at the rate of about 40 per year during that period, so this figure should be moderated slightly. Nevertheless, the figure of 48,200 should be regarded as a realistic assessment of the total number of valid mite species known worldwide.

CONCLUSIONS

It is evident from the estimates of biodiversity in the geographic regions considered above that the true number of mite species in the world fauna is much higher than 48,200. New species of Ixodida are now found only infrequently. The groups in which there are likely to be the greatest increases are the Astigmata, Mesostigmata, Oribatida, and Prostigmata (table 6). Many parts of the world have no active specialists in mite taxonomy or have not been the subject

TABLE 5 Number of New Mite Species Recorded in *Zoological Record*, 1978–1996

Volume	Year	No. New Species
115	1978	852
116	1979	766
117	1980	1135
118	1981	781
119	1982	812
120	1983	765
121	1984	988
122	1985	707
123	1987	795
124	1988	799
125	1989	878
126	1990	939
127	1991	651
128	1992	644
129	1993	701
130	1994	886
131	1995	506
132	1996	579
	Mean	788

of faunal surveys. Similarly, the mites in many habitats and associations are poorly known, such as the soils of tropical rain forests and species associated with other arthropods (Welbourn 1983). Taxonomic attention is sometimes focused on a group when its economic importance is recognized, with the resulting description of huge numbers of new species. For mites, perhaps the best example is seen in the family Phytoseiidae. In the 1950s, when the potential of some phytoseiid species as biological control agents was first noticed, 165 species were listed in the world fauna (Chant 1959). By 1994, no fewer than 1,745 species had been described (Kostiainen and Hoy 1996), and another 55 new species have been found since then. That represents an increase factor of more than 10; if applied to our figure for valid species in other groups, the factor would yield a total of over a half-million.

TABLE 6 Numbers of Species Described in the Five Largest Mite Suborders

Order	No. Species	Years
Astigmata	3,986	1864–1978
Oribatida	5,221	1864–1978
Prostigmata	16,205	1864–1978
Mesostigmata	7,580	1864–1978
Ixodida	834	1864–1997

We can suggest several strategies that will help to increase our knowledge of the world's mite fauna. It would be useful to document and publicize the existence of collections of unsorted material, such as Berlese funnel samples, so that material collected incidentally by nonacarologists and samples of no immediate interest to acarologists is not lost but is available for later study. It would also be useful if entomologists, mammologists, and ornithologists were encouraged to retain the parasitic and phoretic mites that they find, rather than discarding them, and to draw them to the attention of acarologists. The specimens should either be kept with the host or have collection data included with them if separation is necessary. But the availability of specimens is not likely to be the most important limiting factor in the progress of systematic acarology. Serious assessment of mite biodiversity will continue to be inhibited by the shortage of trained taxonomists, especially in tropical areas, where species diversity is likely to be much greater than in the areas we have examined here.

ACKNOWLEDGMENTS

We thank Emma de Boise, Department of Entomology, The Natural History Museum, London, for help in the assessment of numbers of valid mite species and Dave Walter for generously providing access to unpublished information.

REFERENCES

Armine JW, Stasny TA. 1994. Catalog of the Eriophyoidea (Acarina: Prostigmata) of the world. West Bloomfield MI: Indira Publ House.

Arthur DA. 1963. British ticks. London UK: Butterworths.

Baker AS. 1987. Systematic studies on mites of the superfamily Eupodoidea (Acari: Acariformes) based on the Fauna of the British Isles. PhD thesis. London UK: Univ of London.

Baker EW, Tuttle DM. 1994. A guide to the spider mites (Tetranychidae) of the United States. West Bloomfield, Michigan: Indira Publ House.

Baker EW, Kono T, Amrine JW, Delfinado-Baker M, Stasny T. 1996. The Eriophyoid mites of the United States. West Bloomfield MI: Indira Publ House.

Bartsch I. 1996. Halacarines (Acari: Halacaridae) from Rottnest Island, Western Australia: the genera Agauopsis Viets and Halacaropsis gen nov. Rec W Aust Mus 18:1–18.

Behan-Pelletier VM. 1986. Ceratozetidae (Acari: Oribatei) of the western North American subarctic. Can Ent 118:991–1057.

Behan-Pelletier VM. 1989. Limnozetes (Acari: Oribatida: Limnozetidae) of northeastern North America. Can Ent 121:453–506.

Behan-Pelletier VM. 1990. Redefinition of Megeremaeus (Acari: Megeremaeidae) with description of new species, and nymphs of M. montanus Higgins and Woolley. Can Ent 122:875–900.

Behan-Pelletier VM. 1993. Eremaeidae (Acari: Oribatei) of North America. Mem Ent Soc Can 168:1–193.

Behan-Pelletier VM. 1994. Mycobates (Acari: Oribatida: Mycobatidae) of America north of Mexico. Can Ent 126:1301–61.

Bhattacharyya SK. 1963. A revision of the British mites of the genus Pergamasus Berlese s. lat. (Acari: Mesostigmata). Bull Brit Mus (Nat Hist) (Zool) 11:133–242.

Chant DA. 1959. Phytoseiid mites (Acarina: Phytoseiidae): Part I. Bionomics of seven species in southeastern England. Part II; A taxonomic review of the family Phytoseiidae with descriptions of 38 new species. Can Ent Suppl 2:1–166.

Cook DR. 1986. Water mites from Australia. Mem Amer Ent Inst 40:1–568.

Cunliffe F, Baker EW. 1953. A guide to the predatory Phytoseiid mites of the United States. Pinellas Biological Laboratory Publication.

Evans GO, Browning E. 1956. British mites of the subfamily Macrochelinae Trägårdh (Gamasina, Macrochelidae). Bull Brit Mus (Nat Hist) (Zool) 4:1–55.

Farrier MH, Hennessey MK. 1993. Soil-inhabiting and free-living Mesostigmata (Acari-Parasitiformes) from North America. An annotated checklist with bibliography and index. North Carolina Agric Res Serv Tech Bull 302:1–408.

Gaud J, Atyeo WT. 1996. Feather mites of the world (Acarina, Astigmata): The supra specific taxa. Ann Mus R Afr Cent Sci Zool 277:Pt 1: 1–193, Pt II: 1–436.

George AS. 1981. Flora of Australia. Vol 1:3–24. Canberra Australia: Australian Gov Publ Serv.

Gledhill T. 1979. Some data on the freshwater mites (Hydrachnellae and Limnohalacaridae, Acari) of the British Isles and Ireland. Proc 4th Intl Congr Acarol 153–6.

Gledhill T, Viets KO. 1976. A synonymic and bibliographic check-list of the freshwater mites (Hydrachnellae and Limnohalacaridae, Acari) recorded from Great Britain and Ireland. Freshw Biol Asso Occ Publ 1:1–59.

Green J, Macquitty M. 1987. Halacarid mites. London UK: Linnean Society.

Halliday RB. 1986a. Mites of the genus Glyptholaspis Filipponi and Pegazzano (Acarina : Macrochelidae) in Australia. J Aust Ent Soc 25:71–4.

Halliday RB. 1986b. Mites of the Macrocheles glaber group in Australia (Acarina : Macrochelidae). Aust J Zool 34:733–52.

Halliday RB. 1988. The genus Holostaspella Berlese (Acarina: Macrochelidae) in Australia. J Aust Ent Soc 27:149–55.

Halliday RB. 1990. Mites of the Macrocheles muscaedomesticae group in Australia (Acarina:Macrochelidae). Invert Tax 3:407–30.

Halliday RB. 1993. Two new species of Macrocheles from Australia (Acarina:Mesostigmata: Macrochelidae). Aust Ent 20:99–106.

Halliday RB. 1998. Mites of Australia: a checklist and bibliography: Melbourne Australia: CSIRO Publishing.

Halliday RB, Walter DE, Lindquist EE. 1998. Revision of the Australian Ascidae (Acarina: Mesostigmata). Invert Tax 12:1–54.

Harvey MS. 1987. New and little-known species of the water mite genera Tartarothyas, Pseudohydryphantes and Cyclohydryphantes from Australia (Chelicerata:Actinedida:Hydryphantidae). Mem Mus Vic 48:107–22.

Harvey MS. 1989. A review of the water mite genus Australorivacarus KO Viets (Chelicerata:Actinedida : Hygrobatidae). Invert Tax 3:155–62.

Harvey MS. 1990a. A review of the water mite family Limnocharidae in Australia (Acarina). Invert Tax 3:483–93.

Harvey MS. 1990b. Pezidae, a new freshwater mite family from Australia (Acarina:Halacaroidea). Invert Tax 3:771–81.

Harvey MS. 1990c. Two new water mite genera from south-western Australia (Acarina:Aturidae:Mideopsidae). Mem Mus Vic 50:341–6.

Harvey MS. 1990d. A review of the water mite family Anisitsiellidae in Australia (Acarina). Invert Tax 3:629–46.

Harvey MS. 1996. A review of the water mite family Pionidae in Australia (Acarina:Hygrobatoidea). Rec W Aust Mus 17:361–93.

Hillyard PD. 1996. Ticks of North-west Europe. Shrewsbury UK: Field Studies Council.

Hirschmann W, Wisniewski J. 1993. Die Uropodiden der Erde. Acarologie. 40:1–466.

Hughes AM. 1976. The mites of stored food and houses. London UK: HMSO.

Hunter JE, Farrier MH. 1976a. Mites of the genus Oplitis Berlese (Acarina: Uropodidae) associated with ants (Hymenoptera: Formicidae) in the southeastern United States. Pt I. Acarologia 17:595–623.

Hunter JE, Farrier MH. 1976b. Mites of the genus Oplitis Berlese (Acarina: Uropodidae) associated with ants (Hymenoptera: Formicidae) in the southeastern United States. Pt II. Acarologia 18:20–50.

Hurlbutt HW. 1984. A study of North American Veigaia (Acarina: Mesostigmata) with comparisons of habitats of unisexual and bisexual forms. Acarologia 25:207–22.

Hyatt KH. 1980. Mites of the subfamily Parasitinae (Mesostigmata: Parasitidae) in the British Isles. Bull Br Mus (Nat Hist) (Zool) 38:237–378.

Hyatt KH. 1990. Mites associated with terrestrial beetles in the British Isles. Ent Mon Mag 126:133–47.

Hyatt KH, Emberson RM. 1988. A review of the Macrochelidae (Acari: Mesostigmata) of the British Isles. Bull Br Mus (Nat Hist) (Zool) 54:63–125.

Keirans JE. 1992. Systematics of the Ixodida (Argasidae, Ixodidae, Nuttalliellidae): an overview and some problems. In: Fivaz B, Petney T, Horak I (eds). Tick vector biology: medical and veterinary aspects. Berlin Germany: Springer-Verlag. p 1–21.

Kostiainen TS, Hoy MA,1996. The Phytoseiidae as biological control agents of pest mites and insects: a bibliography (1960–1994). Miami FL: Univ of Florida.

Krantz GW. 1970. A manual of acarology. Corvallis OR: Oregon St Univ Bookstores.

Levi HW, Levi LR, Kim HS. 1968. A guide to spiders and their kin. New York NY: Western Publ Co.

Lindquist EE. 1986. The world genera of Tarsonemidae (Acari: Heterostigmata): A morphological, phylogenetic, and systematic revision, with a reclassification of family-group taxa in the Heterostigmata. Mem Ent Soc Can 136:1–517.

Luxton M. 1996. Oribatid mites of the British Isles. A checklist and notes on biogeography (Acari, Oribatida). J Nat Hist 30:803–22.

Mahunka S. 1967. A survey of the scutacarid (Acari : Tarsonemini) fauna of Australia. Aust J Zool 15:1299–1323.

Michael AD. 1884 and 1888. British Oribatidae. London UK: Ray Society.

Niedbala W. 1987. Phthiracaroidea (Acari, Oribatida) noveaux d'Australie. Redia 70:301–75.

Niedbala W. 1989. Phthiracaroidea (Acari, Oribatida) nouveaux du royaume australien. Ann Zool 43:19–50.

Niedbala W, Colloff MJ. 1997. Euptyctime oribatid mites from Tasmanian rainforest (Acari: Oribatida). J Nat Hist 31:489–538.

Norton RA, Behan-Pelletier VM, Wang HF. 1996. The aquatic oribatid mite genus *Mucronothrus* in Canada and the western USA. (Acari: Trhypochthoniidae). Can J Zool 74:926–49.

OConnor BM. 1990. The North American Acari: current status and future projections. In: Kosztarab M, Schaefer CW (eds). Systematics of the North American insects and arachnids: status and needs. Blacksburg VA: Polytechnic Inst St Univ p 21–9.

OConnor BM. 1991. A preliminary report on the arthropod-associated astigmatid mites (Acari: Acariformes) of the Huron Mountains of Northern Michigan. Mich Academ 24:307–20.

Schicha E. 1987. Phytoseiidae of Australia and neighboring areas. West Bloomfield MI: Indira Publ House.

Schicha E, O'Dowd DJ. 1993. New Australian species of Phytoseiidae (Acarina) from leaf domatia. J Aust Ent Soc 32:297–305.

Skorupski M, Luxton M. 1996. Mites of the family Zerconidae Canestrini, 1891 (Acari: Parasitiformes) from the British Isle, with descriptions of two species. J Nat Hist 30:1815–32.

Slater P. 1970. A field guide to Australian birds. Volume 1, non-passerines. Adelaide Australia: Rigby.

Smith IM. 1989a. North American water mites of the family Momoniidae Viets (Acari: Arrenuroidea). 2. Revision of species of *Momonia* Halbert, 1906. Can Ent 121:965–87.

Smith IM. 1989b. North American water mites of the family Momoniidae Viets (Acari: Arrenuroidea). 3. Revision of species of *Stygomomonia* Szalay, 1943, subgenus *Allomomonia* Cook, 1968. Can Ent 121:989–1025.

Smith IM. 1990a. Proposal of Nudomideopsidae fam nov (Acari: Arrenuroidea) with a review of North American taxa and description of a new subgenus and species of *Nudomideopsis* Szalay, 1945. Can Ent 122:229–52.

Smith IM. 1990b. Description of two new species of *Stygameracarus* gen nov from North America, and proposal of Stygameracarinae subfam nov (Acari: Arrenuroidea: Athienemanniidae). Can Ent 122:181–90.

Smith IM. 1991a. North American water mites of the family Momoniidae Viets (Acari: Arrenuroidea). 4. Revision of species of *Stygomomonia* (*sensu stricto*) Szalay, 1943. Can Ent 123:501–58.

Smith IM. 1991b. North American water mites of the genera *Phreatobrachypoda* Cook and *Bharatalbia* Cook (Acari: Hygrobatoidea: Aturinae). Can Ent 123:465–99.

Smith IM. 1992a. North American water mites of the family Chappuisididae Motas and Tanasachi (Acari: Arrenuroidea). Can Ent 124:637–723.

Smith IM. 1992b. North American species of the genus *Chelomideopsis* Romijn (Acari: Arrenuroidea: Athienemanniidae). Can Ent 124:451–90.

Smith IM. 1994. North American species of Neomamersinae Lundblad (Acari: Hydrachnida: Limnesiidae). Can Ent 126:1131–84.

Turk FA. 1953. A synonymic catalogue of British Acari. Ann Mag Nat Hist (12)6:1–26, 81–99.

Walter DE, Lindquist EE. 1997. Australian species of Lasioseius (Acari : Mesostigmata : Ascidae); the porolosus group and other species from rainforest canopies. Invert Tax 11:525–54.

Walter DE, Halliday RB, Lindquist EE. 1993. A review of the genus Asca (Acarina : Ascidae) in Australia, with descriptions of three new leaf-inhabiting species. Invert Tax 7:1327–47.

Walter DE, Krantz GW, Lindquist EE. 1996. Acari, the mites. Tree of Life World Wide Web site http://phylogeny.arizona.edu/tree/eukaryotes/animals /arthropoda/arachnida/acari/acari.html. 5 pp. Version dated 13 December 1996.

Welbourn WC. 1983. Potential use of trombidioid and erythraeoid mites as biological control agents of insect pests. In: Hoy MA, Cunningham GL, Knutson L (eds). Biological control of pests by mites. Berkeley CA: Univ of California. p 103–40.

Wharton GW. 1964. First international congress of acarology, keynote address. In: Paillart F (ed). Proc First Intl Congr Acarol, Abbeville. p 37–43.

BIODIVERSITY OF TERRESTRIAL INVERTEBRATES IN TROPICAL AFRICA: ASSESSING THE NEEDS AND PLAN OF ACTION

SCOTT MILLER*
BARBARA GEMMILL
HANS R. HERREN
LUCIE ROGO
1 International Centre of Insect Physiology and Ecology, Box 30772, Nairobi, Kenya
MELODY ALLEN
Xerces Society, 4828 SE Hawthorne St., Portland, OR 97215

*(Current address: National Museum of Natural History, Smithsonian Institution, Washington, DC 20560-0105)

THE CHALLENGE

BIOLOGICAL RESOURCES are the basis of the prosperity of the developed world, yet the biologically rich underdeveloped nations of Africa are the economically poorest in the world. Africa's biodiversity, if conserved and developed sustainably, can be used to relieve poverty and achieve economic stability. The challenge lies in rapidly acquiring the required knowledge of the biodiversity resource: knowing what the critical species are and where they occur, obtaining information about their natural history, and establishing sustainable patterns of resource use.

Although the continent of Africa is most renowned for its highly charismatic megafauna, the greatest contributions to its biodiversity (as elsewhere) in fact lie in its other taxa, which ultimately facilitate the existence of these flagship species. Insects and other arthropods compose more than 70% of the world's fauna. Insects contribute the largest number of taxa by far to biological diversity in both Africa and the world (figure 1). Many major effects on human welfare—human and animal diseases carried by insect vectors; outbreaks of migrant pests, such as locusts and armyworms; destruction of food by plant pests; toxic residues from pesticides; and overuse and depletion of agricultural lands and adjoining forests—are problems whose answers lie well within the field of biodiversity and, more specifically, insect diversity (Hill 1997). By performing critical "service" functions within ecosystems, insects are key to the stability of ecosystems. Many insects provide a direct economic return (for example, silkworms and bees); others produce chemicals for medicinal use. Some constitute an important source of pro-

FIGURE 1 Afriscape: An imaginary landscape of the Afrotropical realm (terrestrial and freshwater) in which the size of taxa is proportional to the number of species currently known in the group it represents. Data sources include vascular plants (42,500, Groombridge 1992:66); land snails (6,000, Bruggen 1986); insects (150,000); fishes (1,800, Groombridge 1992:116); amphibians (627, Duellman 1993); reptiles (1,400, Bauer 1993); birds (1,500 Vuilleumier and Andors 1993); mammals (1,045, Cole and others 1994). Inspirational thanks to Quentin Wheeler's 1990 world speciescape. Graphic by Barbara Gemmill.

tein in the diet of rural peoples; others play predatory and parasitic roles that regulate pests (Odindo 1995). Arthropods are key in providing pollination services to both natural and human-made ecosystems.

No African invertebrate species have been documented yet as becoming extinct either directly or indirectly because of human activities during historical times, although several butterflies and lacewings might have become extinct in South Africa (Siegfried and Brooke 1995). However, invertebrates are generally so poorly known that even probable extinction is difficult to detect. With insects more than with any other taxa, we risk losing aspects of biodiversity without ever knowing their value.

THE APPROACH

This paper outlines an approach to putting terrestrial invertebrates on the agenda for conservation of biodiversity in Africa. We seek to fill two key gaps in the understanding and use of the positive aspects of insects in African biodiversity. First, almost all research on insects in tropical Africa focuses on the negative aspects of insects—for example, the problems in agriculture, forestry, livestock, and human health that are caused by less than 1% of the species of insects—and ignores the remaining 99% of insect species. Of the more than 100,000 described species of insects in the Afrotropical region, fewer than 500 species were mentioned between 1990 and 1995 in the journal *Insect Science and its Application*, one of the major African entomology journals, and 97% of the articles over this time focused on pest or other economically important species.

Most programs of biodiversity studies and conservation currently operating in tropical Africa focus on vertebrates or, secondarily, flowering plants and ignore insects, which E.O. Wilson (1987) has called "the little things that run the world" because of their key roles in ecosystem function. For example, in 1994, a survey of sets of biodiversity data available for East Africa included only 12 for insects, whereas mammals and plants had more than 50 each and birds and fish had more than 40 each (World Conservation Monitoring Centre 1994).

SOME PREMISES

Through extensive consultation, we have reviewed many programs around the world that have dealt, successfully or unsuccessfully, with similar challenges (for example, Hawksworth and Ritchie 1993; Miller 1994). By this process, we have identified some basic premises that have guided the development of our program.

First, if we are to protect biodiversity, it must have utility for human societies, and if it is to be used sustainably, it must be understood. This premise is the basis for several conservation initiatives in Costa Rica and Africa (see Janzen paper in this volume; see also Ramberg 1993 and Noss 1997). The developed-country model of protecting biodiversity in national parks is not sustainable in developing countries. Long-term protection of biodiversity depends on making it useful and valuable to the people who live amid and around it. This means that some of the biodiversity must be used to provide the means for supporting and managing the rest. Sustainable use of biodiversity requires knowing how to find what you need, understanding the implications of that use, and learning how to encourage the regeneration or recovery of the resource to support its continued use.

A second premise is that systematics provides the framework for organizing and communicating basic information about biodiversity (Janzen 1993). Thus, the involvement of the taxasphere, the international infrastructure for biological systematics, including the natural-history museums that hold most of the collections of specimens, is vital. We also expect to integrate our activities with those at smaller (for example, national or local) and larger (for example, international) levels, including BioNet International, Systematics Agenda 2000 International, and Species 2000 (Hawksworth 1997).

Finally, it is more cost-effective to use what we already know than it is to recreate basic information on biodiversity (Nielsen and West 1994; Soberon and others, 1996). An enormous body of information is theoretically available, but it is highly dispersed, extraordinarily varied in form, uncoordinated, and largely unavailable in most of Africa. Much of this information is in museum collections (Cotterill 1997). Recent developments in information technology have provided the means to achieve a coordinated information base on African insect fauna and an efficient means of disseminating that information. The task requires effective collaboration of experts and stakeholders from all aspects of the process, from the discovery through the management and use of biodiversity (World Conservation Monitoring Centre 1996).

AN ORGANIZATIONAL STRUCTURE FOR A CONTINENTWIDE BIODIVERSITY INITIATIVE

One irony of current biodiversity-conservation initiatives is that while we continually are refining our skills to document the value of the ecosystem services that biodiversity provides, few governments or legal entities are prepared to pay for the conservation of these services, which, until now, have been exploited freely by human societies. An example that is specific to our program is that no country in Africa has the resources to initiate a program of conserving insect biodiversity; the task is formidable, and the benefits are so basic and diffuse that they become lost in a sea of competing priorities. Only a highly-targeted, cost-effective program that can coordinate the resources and disseminate the benefits on a wide scale (regional or continental) can return the expected outcomes.

The leadership for such a program has been assumed by the International Centre for Insect Physiology and Ecology (ICIPE), an international institute that is situated centrally on the continent in Kenya. ICIPE is a major international institution that has more than 27 years of experience in research and monitoring of arthropods. It has developed integrated pest and vector-management techniques and biological-control strategies for insects that are disease vectors and plant pests. The institution combines research with interactive training of scientists, technicians, and farmers and herders at both national and subregional levels, and it provides training programs for graduate students from universities throughout Africa. ICIPE has memorandums of understanding with more than 20 sub-Saharan countries, and more than 30 countries worldwide have signed its charter. With an established structure in place for joint training and research with the major taxonomic and biodiversity institutions of Europe and the United States, implementation of a collaborative program would be possible without untimely delays.

THE PLAN

We have identified a mixture of projects that provide a cost-effective foundation for understanding the diversity of insects, the roles they play in natural systems, and ways to manage those interactions more effectively. Our initiative includes three main components. First, an information-management program will organize and make available a large volume of information that already exists but is not readily accessible to users. This will be coordinated with other activities that are already under way in the museum, systematics, and conservation communities and will be targeted carefully to fill key gaps. Second, a series of field projects will evaluate the use of insects as indicator organisms and will quantify their roles in ecosystem processes. In many cases, these projects will take approaches that have been successful in South Africa and the Northern Hemisphere and will apply them, with appropriate modifications, to tropical Africa. Third, training and participatory technology transfer will build on ICIPE's existing training programs, including the African Regional Postgraduate Programme in Insect Science (ARPPIS).

Information Management

An initiative for reviewing the literature and creating a database of specimens will repatriate 200 years' worth of information collected in sub-Saharan Africa and now housed in museums in the United Kingdom, France, Belgium, Germany, elsewhere in Europe, South Africa, Kenya, elsewhere in Africa, and the United States and Canada. This initiative will support individual projects and applications within the ICIPE program and will provide relevant information to a wide audience in Africa. One of the first and most important steps in managing the biodiversity of African insects is to find and organize what we already know. As we have stated, a tremendous amount of information is available but not in a cohesive and accessible form; recent developments in information technology will allow us to compile an information base on African insects; they will also allow efficient dissemination of this information (Vane-Wright 1998). Note the continuing growth of the technology from, for example, the discussion of early Internet tools in Miller (1993) to those in Helly and others (1996). As a result, gaps in our knowledge will become apparent, allowing us to establish priorities for further work. Our checklist of insects known from Africa is in progress, and an interim product is on the World Wide Web (www.icipe.org/environment/biodiversity_index.html).

Pilot Projects and Applications of Conservation Biology That Focus on Insects

A series of experiments, surveys, and applications will be designed to investigate the role of key groups of insects in the function and management of ecosystems and to provide information on the conservation and sustainable management of the insect resource. The major foci will include identification of high-priority areas for conserving biodiversity, using butterflies, fruit flies, dragonflies, and termites; impact assessments, using insects as "indicators"; and identification of the roles of insects in pollination, soil processes, and the organization of tropical-forest ecosystems.

Training and Participatory Technology Transfer

An important element in the overall program is capacity-building: producing trained technicians and scientists who will be able to implement the information-management and research tasks. In the multidisciplinary field that is inherent to conservation of biodiversity—from taxonomy to database management to field techniques—individual and institutional capacity spans every activity. This feat will be achieved through partnerships with universities, museums, advanced-research laboratories, and national institutions throughout the world. Developing an African and overseas reciprocal research exchange within Africa will ensure a permanent conduit for technology transfer. Many of the students trained through this program will become interns in museums and research centers throughout

Europe and North America, thus effecting the transfer of skills, as well as information.

Formal university training will be conducted through the ARPPIS PhD program at ICIPE, in which students undergo three years of research training. The ICIPE provides a thesis project, research facilities and supervision, and a training fellowship to support the students' maintenance, university fees, and research costs, for a total of US$30,000 per student per year. Students are registered at participating African universities, which examine the students and award them their degrees. The program has, at any one time, 20–40 students at various stages of their thesis work at ICIPE. To date, 131 scholars from 25 African countries have enrolled in the program, and 91 have graduated. The success of the ARPPIS program has stimulated the interest of universities; 18 universities have renewed their agreements with ICIPE. ICIPE is proud that after they have graduated, almost all former ARPPIS scholars have stayed in Africa to work toward solving the continent's insect-related problems. Most graduates are employed by national research systems, universities, or science-based international organizations.

Training means more than formal university training, however. Two other major activities are the enhancement of national capacities for the diffusion, adoption, and use of technology and the facilitation of the dissemination and exchange of information.

EXAMPLES OF RESEARCH ON INSECT BIODIVERSITY IN AFRICA IN SUPPORT OF SUSTAINABLE DEVELOPMENT

Recent research initiatives undertaken in Kenya have shown that basic research on the biodiversity of arthropods can contribute in substantial ways to sustainable agricultural development.

The Role of Habitat in the Agroecosystem of Maize

Losses of maize, sorghum, and other cereals caused by stemborers remains one of the biggest threats to the security of the food supply in eastern and southern Africa. Maize yields in Africa are less than half the average yield worldwide. Especially damaging is the moth *Chilo partellus* (Lepidoptera: Crambidae), an intruder from Asia that was introduced accidentally into Africa in the 1930s and has now displaced indigenous pests. This exotic species soon became infamous for causing losses of 20–80% of crops.

Native predators may play an important role in suppressing stemborer populations. Studies conducted in Kenya's Coast Province revealed that ants are the most abundant predator. The abundance and diversity of predators increased with the age of the plants and was highest at the tasseling stage of maize. An even broader—and very promising—view is being taken by looking not just at the farmer's field, but also at the surrounding environment (Khan and others 1997).

ICIPE's project on the role of wild habitat in the invasion of gramineous crops by stemborers already is yielding hard data on the benefits of preserving and managing biodiversity in small and medium-size farms. The project is developing a novel approach to pest management that uses a stimulus-deterrent ("push-pull") diversionary strategy. A better understanding of the relationship between diversity of habitat and resilience to pest challenge is being developed, as are ideas for modifying the habitat to contain this challenge.

Several plants that lower the density of stemborers by the "push-pull" strategy have been identified, resulting in higher crop yields. Especially promising in this respect are Napier grass (*Pennisetum purpureum*), Sudan grass (*Sorghum vulgare sudanens*), and molasses grass (*Melinis minutiflora*). These three important fodder grasses act as traps by "pulling" or attracting the borers and serving as reservoirs for the natural enemies of the stemborers. Furthermore, Sudan grass increases the efficiency of the natural enemies. The rate of parasitism on larvae of the spotted stemborer, *Chilo partellus*, more than tripled, from 4.8% to 18.9%, when Sudan grass was planted around maize in a field. Napier grass has its own defense mechanism against crop borers: When the larvae enter the stem, the plant produces a gum-like substance that kills the pest. Molasses grass releases volatiles that not only repel (or "push") stemborers, but also attract parasitoids. Both whole live plants of *M. minutiflora* and its volatiles were shown to attract the natural enemy of the wasp, *Cotesia sesamiae* (Hymenoptera: Braconidae). Intercropping with *M. minutiflora* increases parasitism, particularly by the larval parasitoid wasp and the pupal parasitoid *Dentichasmis busseolae* (Hymenoptera: Ichneumonidae). Analysis of the volatile oils from molasses grass shows that they contain several physiologically active compounds. Two of these inhibit egg-laying in *Chilo*, even at low concentrations. In contrast, *Chilo's* host plants (maize, sorghum, and Napier grass) have been found to contain volatile compounds, such as eugenol, that attract *Chilo* and stimulate egg-laying. Molasses grass also emits a chemical that summons the borers' natural enemies. This same substance is released by whole plants as a distress signal when they are being damaged by pests. The results of this study have opened up the new and intriguing possibility of using intact plants that have an inherent ability to release these stimuli. Such plants will be useful in ecologically based crop-protection strategies.

Commercial and Sustainable Production of Wild Insects

Wild insects long have been part of the diet of humans in Africa; termites and locusts are two highly valued food items among the arthropods. Wild insects also are husbanded for the products they create. If harvesting and use of wild insects are to be sustained with increasing population, however, they will need to be studied carefully and rationally. ICIPE currently undertakes studies of African honeybee culture and wild silk production (ICIPE 1997).

Apiculture is a traditional occupation in most African communities, but centuries-old practices of harvesting honey are inefficient and often cause the death of the colony; the aggressiveness of African honeybees has been attributed to these management practices. ICIPE is introducing improved methods of beekeeping to farmers and to women's groups, supported by research to solve the problems of queen-rearing and African honeybee aggressiveness and to improve the production of honey and other valuable hive products. Linking honey production to floral calendars can help local producers understand the direct benefit of habitat conservation.

Similarly, production of wild silk moths can provide a strong economic incentive for rural communities to adopt sound wild-land management practices as an adjunct to subsistence agriculture. Currently, silk moth larvae and pupae are harvested in bulk as a source of dietary protein, but no mechanisms exist to replenish the silkworms (the moth larvae). Techniques of sericulture (the deliberate rearing of silk moths for harvesting of the pupal cases) are unknown at the village level, yet at least three species of moth that yield high-quality wild silk have been identified. ICIPE has undertaken a project to develop methods of sericulture that are appropriate for Africa and that also will assist in conserving the valuable wild species of moths. The interest shown by authorities from national parks and by communities in East Africa is proof of the timeliness of this project and augurs well for the future of a strong conservation industry built around wild silk moths.

CONCLUSION

We expect that the foundation of knowledge and trained personnel that will be generated by this new initiative will enable sophisticated strategies of ecological monitoring and applications of sustainable development that draw on the strengths of the resource base of African arthropods. In a continent that until now has been remarkable for the coexistence of a rich and varied wildlife with human societies, we are challenged to direct development along lines that also foster the coexistence with the ubiquitous but less-noticed aspects of biodiversity, such as arthropods. Because these aspects most directly impinge on human welfare, the success of biodiversity conservation may depend on how well we meet this challenge. In the largely intact, undeveloped landscapes of Africa, we still have a tremendous chance to conserve the fine fabric and delicate linkages of nature in and with human development if only we can document its existence and importance before we have lost it.

ACKNOWLEDGMENTS

We thank all the colleagues and institutions, too numerous to name here, who have helped in the development of our ideas and plans. The government of Norway provided seed funding for the development phase of this initiative. We also acknowledge the seminal role played by biological-survey and information-management activities in Australia (ERIN), Costa Rica (INBio), and Mexico (CONABIO) in developing and testing many of the ideas that we and others have been able to build from.

REFERENCES

Bauer AM. 1993. African-South American relationships: a perspective from the Reptilia. In: Goldblatt P (ed). Biological relationships between Africa and South America. New Haven CT and London UK: Yale Univ Pr. p 244–88.

Bruggen AC van. 1986. Aspects of the diversity of the land molluscs of the Afrotropical region. Revue Zool Afr 100:29–45.

Cole FR, Reeder DM, Wilson DE. 1994. A synopsis of distribution patterns and the conservation of mammal species. J Mammalogy 75:266–76.

Cotterill FPD. 1997. The second Alexandrian tragedy, and the fundamental relationship between biological collections and scientific knowledge. In: Nudds JR, Pettitt CW (eds). The value and valuation of natural science collections: proceedings of the international conference, Manchester 1995. London UK: Geological Soc. p 227–41.

Duellman WE. 1993. Amphibians in Africa and South America: evolutionary history and ecological comparisons. In: Goldblatt P (ed). Biological relationships between Africa and South America. New Haven CT and London UK: Yale Univ Pr. p 200–43.

Eldredge LG, Miller SE. 1998. Numbers of Hawaiian species: supplement 3, with notes on fossil species. Bishop Museum Occasional Papers 55:3–15.

Groombridge B (ed). 1992. Global biodiversity: status of the Earth's living resources. London UK: Chapman & Hall.

Hawksworth DL. 1997. Biosystematics: meeting the demand. Biol Internat 35:21–4.

Hawksworth DL, Ritchie JM. 1993. Biodiversity and biosystematic priorities: microorganisms and invertebrates. Wallingford UK: CAB International. 120p.

Helly J, Case T, Davis F, Levin S, Michener W (eds). 1996. The state of computational ecology. San Diego Supercomputer Center, San Diego. 20p. [also at http://www.sdsc.edu/compeco_workshop/report/helly_publication.html].

Hill DS. 1997. The economic importance of insects. London UK: Chapman & Hall.

ICIPE [International Centre for Insect Physiology and Ecology]. 1997. 1996/1997 annual report. Nairobi Kenya: ICIPE.

Janzen DH. 1993. Taxonomy: universal and essential infrastructure for development and management of tropical wildland biodiversity. In: Sandlund OT, Schei PJ (eds). Proceedings of the Norway/UNEP expert conference on biodiversity. Trondheim Norway: Directorate for Nature Management and Norwegian Institute for Nature Research. p 100–13.

Khan ZR, Ampong-Nyarko K, Chiliswa P, Hassanali A, Kimani S, Lwande W, Overholt WA, Pickett JA, Smart LE, Wadhams LJ, Woodcock CM. 1997. Intercropping increases parasitism of pests. Nature 388:631–2.

Miller SE. 1993. The information age and agricultural entomology. Bull Entomol Res 83:471–4. [Also at http://www.bishop.hawaii.org/bishop/HBS/BER.html]

Miller SE. 1994. Development of world identification services. In: Hawksworth DL (ed). The identification and characterization of pest organisms. Wallingford UK: CAB International. p 69–80.

Nielsen ES, West JG. 1994. Biodiversity research and biological collections: transfer of information. In: Forey PL, Humphries CJ, Vane-Wright RI (eds). Systematics and conservation evaluation. Oxford UK: Clarendon Pr. p 101–21.

Noss AJ. 1997. Challenges to nature conservation with community development in central African forests. Oryx 31:180–8.

Odindo MO (ed). 1995. Beneficial African insects: a renewable natural resource: Proceedings of the 10th meeting and scientific conference of the African Association of Insect Scientists. Nairobi Kenya: African Association of Insect Scientists. iii + 251p.

Ramberg L. 1993. African communities in conservation: a humanistic perspective. J Afr Zool 107:5–18.

Siegfried WR, Brooke RK. 1995. Anthropogenic extinctions in the terrestrial biota of the Afrotropical region in the last 500,000 years. J Afr Zool 109:5–14.

Soberon J, Llorente J, Benitz H. 1996. An international view of national biological surveys. Ann Missouri Bot Gard 83:562–73.

Vane-Wright RI. 1997. African lepidopterology at the millennium. Metamorph Suppl 3:11–27.

Vuilleumier F, Andors AV. 1993. Avian biological relationships between Africa and South America. In: Goldblatt P (ed). Biological relationships between Africa and South America. New Haven CT and London UK: Yale Univ Pr. p 289–328.

Wheeler Q. 1990. Insect diversity and cladistic constraints. Ann Ent Soc Amer 83:1031–47.

Wilson EO. 1987. The little things that run the world (the importance and conservation of invertebrates). Cons Biol 1:344–6.

World Conservation Monitoring Centre. 1994. Availability of biodiversity information for East Africa, [computer disk]. Dar es Salaam Tanzania: FAO.

World Conservation Monitoring Centre. 1996. Guide to information management in the context of the Convention on Biological Diversity. Nairobi Kenya: UNEP.

GLOBAL DIVERSITY OF INSECTS:
THE PROBLEMS OF ESTIMATING NUMBERS

EBBE S. NIELSEN
LAURENCE A. MOUND
Australian National Insect Collection, CSIRO Entomology,
GPO Box 1700, Canberra ACT 2601, Australia

THE CLASS INSECTA is the most species-rich of all major groups of living multicellular organisms. Any meaningful assessment of the diversity of life on earth depends on estimates of both the number of named insect species and the number of insect species that are living but are yet unnamed or even undiscovered.

Common sense might suggest that the number of described species would be a statistic that science would have available. However, no single compilation exists of the names of described insect species, so the total number remains a matter of conjecture. Indeed, for most groups of insects, apart from the Diptera (Evenhuis 1989), published lists of species names are not readily available, despite a recent surge of interest in computer listings. The production of lists of described taxa should have high priority for insect taxonomic science, whether for a local fauna, such as the Lepidoptera of Australia (Nielsen and others 1996), or for the worldwide fauna of a particular group, such as Geometridae (Scoble 1999). Such lists provide some measure of what has been achieved at a given time. More important, they can be a means of stimulating further studies and of attracting research funding in other aspects of biology (Mound 1998). However, within the taxonomic community, tradition remains biased toward the production of scholarly nomenclatural catalogs, with details of type material that are useful only to other specialist taxonomists. Our use of the term checklist implies a product that can be used as the starting point for investigations into biological diversity by the biological and conservation community in general.

As a result of the lack of checklists, available estimates of the numbers of described species often differ widely. Indeed, at times it is difficult to understand precisely what an author means by a figure for the number of species in an insect group. The numbers may represent all published species-group names; only technically available species-group names (excluding, for example, *nomina nuda*); currently accepted valid species, excluding synonyms but with or without names of subspecies; or the estimated currently extant species, including undescribed or even undiscovered species. In recent accounts, the estimated number of named species varies from 751,000 to 950,000, and the estimated number of living species ranges from 1 million to 100 million (Hammond 1994). The numbers we quote in table 1 presumably suffer from similar problems, but we have attempted to clarify the situation whenever possible.

TABLE 1   Our Estimates of Numbers of Named and Living Species of Insects

Order	Estimated Total Named	Estimated Total Species
Collembola	7,213	50,000+
Protura	300	1,000
Diplura	659	1,500
Archaeognatha	300	1,000
Thysanura	370	500
Ephemeroptera	2,000	4,000
Odonata	4,870	5,500
Plecoptera	2,000	3,000
Blattodea	4,000	5,000
Isoptera	1,900	2,300
Mantodea	1,600	2,000
Grylloblattodea	20	30
Dermaptera	1,300	3,000
Orthoptera	12,500	20,000
Phasmatodea	2,500	3,000
Embioptera	200	2,000
Zoraptera	30	50
Psocoptera	3,500	5,000
Phthiraptera	3,000	5,000
Hemiptera	85,600	190,000
Thysanoptera	5,000	10,000
Megaloptera	300	500
Raphidioptera	200	200
Neuroptera	5,000	7,000
Coleoptera	350,000	850,000
Strepsiptera	530	700
Mecoptera	500	700
Siphonaptera	2,200	2,500
Diptera	99,000	150,000+
Trichoptera	7,000	12,000
Lepidoptera	146,500	400,000
Hymenoptera	115,000	230,000+
TOTALS	ca 865,000	ca 2,000,000+

A further problem is that "species" are not comparable units throughout the Insecta, thus at times rendering comparisons potentially misleading (Vane-Wright 1992). Although the "species" is the most commonly used unit of biodiversity, organismal diversity cannot be measured objectively solely by differences in the number of species (Hawksworth and Kalin-Arroyo 1995). It is widely agreed that the species number is the most important measure that we have, but we cannot regard it as a standard unit in any statistical sense. The existence of sibling species of Diptera that are distinguishable only through examination of their chromosomes has been well known for many years. In other groups of insects, DNA methods are increasingly demonstrating genetic differences that many authors interpret as evidence of different species. In some ways, we face the same problem that plagued Alfred Russel Wallace and Charles Darwin, in attempting to distinguish units within biological systems that sometimes appear to exhibit almost continuous variation. The peaks of ecological and evolutionary adaptation that we call species can modulate and move in space and time in response to varying pressures of selection. In conservation, it is this ability to change and adapt that we need to protect, not merely the units that we use to measure diversity.

The estimates we give of the number of named species, particularly the total diversity within each order of insects, clearly depend heavily on the bibliographic efficiency and practical experience of individual taxonomic specialists. May (1990) expressed concern that no full published list exists of all described taxa of insects. However, no individual working taxonomist has a particular requirement for such a list. Moreover, most taxonomists who work on insects in the major orders study only a small subset of any major group and thus have little requirement for even a checklist of all the available names within an order or major family. Given that some genera of insects, for example within the Ichneumonidae and Staphylinidae, include more than 1,000 species each, it is scarcely surprising that individual taxonomists have not had the resources to produce or maintain such massive checklists.

Estimates of the possible number of living insect species originate essentially from two sources. One source is the few taxonomists who have experience with very large collections, usually coupled with field experience in areas of high biological diversity. In this case, the data will have been produced haphazardly and over a long time, albeit on a wide front, and the estimate is based on the frequency with which novelties appear in collections. The second source is ecologists who are interested in estimates of species richness. In this case, the data come from intensive sampling of restricted areas over a restricted period followed by extrapolation of these numbers into unsampled areas. Not surprisingly, these techniques yield rather different estimates. The first, which is essentially a species-accumulation curve, is related to the acquisition policy of institutes and the distribution patterns of species. This method will underestimate the total number of species through any failure to score fully the many species in large genera that are difficult to distinguish because they are represented only by single individuals. The second method is concerned with the numbers of species that can be found at a single point, and any assumptions of local endemicity or host specificity when extrapolating from these data will tend to overestimate the total number.

Gaston (1991a) pointed out that few of his taxonomic colleagues who had experience with tropical diversity considered it likely that the group in which they specialized would prove to be larger than the currently described subset by orders of magnitude. Similarly, Hodkinson and Casson (1991) produced a figure for the total worldwide insect fauna of 1.87–2.49 million species on the basis of large collections of Hemiptera made in Sulawesi. In contrast, ecological estimates—such as those by Erwin (1982), Stork (1988, 1993), Kitching (1990), and Recher and Majer (1996)—imply that the world's insect fauna is 30 or more times that of the currently described subset. Current evidence from the major museum collections of sorted and labeled insect species, whether described or undescribed, does not support these larger estimates. Insect taxonomists generally concur that, although there may be as many as 5 million species of insects in the world, there are probably fewer than 10 million.

The suggestion that urgent efforts be made to describe all the world's species of insects leads us to a further series of issues. Even an estimate of 5 million species implies logistical demands that far exceed available resources. Mound (1998) pointed out that the practical problems involved in describing such very large numbers of species have never been considered seriously. These problems include communicating the information to other scientists, the effect on library budgets of a further 8 million pages of descriptions for the minimum of 4 million new species, and the effect of all the new insect material on museum budgets. Wilson (1985) expressed a more positive viewpoint by saying that describing a large proportion of the world's fauna is feasible but with the caveat that this possibility exists only if the priorities of human society change substantially from producing armaments to protecting the biosphere.

Some biologists assert that the study of highly diverse biological systems must be preceded by description of the many species that make up such systems. This is not entirely true, as indicated by the extensive karyotypic studies by M. J. D. White (1982) on species of Australian morabine grasshoppers that even now are undescribed. Similarly, Robinson and Nielsen (1989) gave an account of the Australian fauna of tineid moths, despite the fact that half the 380 known species remain formally unnamed. In both those instances, the species are sorted, labeled, and available for study in the Australian National Insect Collection. The importance of a major collection is the quantity and quality of information that it can contain, including distribution patterns in time and space and biological details, such as host plants and parasites. This information can be made available to biologists and conservation workers, even if not all the taxa are formally named. We certainly are not suggesting a moratorium on describing species of insects, but we suggest that greater thought be given to the question of what benefits will be obtained by describing a much larger fraction of the world's insect fauna.

The question of how science should respond to the problem of such a vast number of undescribed insect species is complex. Gaston (1994) pointed out that although most insect species are tropical, most taxonomic effort continues to be applied to temperate faunas. Mound (1998) indicated that science budgets in tropical countries will need to take a greater share of this burden of description in the future, but emphasized that more appropriate responses need to be considered than the ad hoc description of large numbers of species. The interesting scientific problem lies not in the description of all the species, but in why so many species exist. We need to describe formally only the species that we require for our analyses of biologically diverse systems, whether these analyses are ecological or systematic. The activity of describing species is sometimes advocated as providing the building blocks for the rest of biology. However, ad hoc description of new taxa is like the unplanned production of building blocks in the hope that one day they may find a place in our biological building. Can we not find a more rational and effective use of our resources for such a gigantic task?

Gaston (1991b) made the point that better data on the total numbers of species could be obtained by conducting detailed studies of the numbers of both described and undescribed species from a number of specified sites; that is, data should be collected purposefully, with particular objectives. Similarly, Longino (1994) has pointed out the advantages to be gained from a sampling program that has specific long-term objectives. Again, Mound (1998) pointed out that when descriptive taxonomy is incorporated into focused interdisciplinary projects on particular systems or groups, then the whole subject is enriched by data from other biological disciplines. Detailed sampling and interdisciplinary studies then have the objective of facilitating comparisons between sites, seasons, and habitats and thus are relevant to a wider community of scientists. More important, such an approach is based on the view of faunas as dynamic systems, in which processes can be studied, rather than as static systems, in which units need to be described.

As taxonomists ourselves, we find that the absence of an accurate figure for the total number of living insect species does not limit our studies of patterns in structural, behavioral, and geographic diversity. We continue to describe new species when this is relevant to our exploration of interesting patterns in nature, not as part of any program to provide names for the entire insect fauna. Far more important to us are the problems of the origin of insect diversity and of how to maintain this diversity in a rapidly changing world. In this context, we emphasize again the importance of well-curated museum collections and effective access to the information they contain (Nielsen and West 1994), because these tangible and available records of biodiversity facilitate the comparisons between sites and seasons that are valuable to the rest of society.

Table 1 summarizes the number of named species and the estimated total number of species that we consider valid, and table 1 is a brief discussion of each order according to various authors. Our estimated numbers are those we believe to be most accurate, given our current knowledge.

TABLE 2   Various Authors' Inventories of the Insect Orders

Order	Reference	Comments
Collembola	Hopkins 1996	Estimated more than 50,000 species worldwide.
	Janssens 1997	Number of currently described species is considered to be 7,213.
Protura	Imadate 1991	About 500 species have been described worldwide.
	Tuxen 1964	Only 260 were included in Tuxen's catalog. No estimate is available of the possible worldwide total.
Diplura	Conde and Pages 1991	Estimated that this group has about 800 species worldwide.
	Arnett 1993	Arnett estimated 659 described species.
Archaeognatha	Watson and Smith 1991	Stated that this order includes about 350 species.
	Arnett 1993	Arnett estimated about 250 species but indicated that many more probably remain to be discovered.
Thysanura	Smith and Watson 1991	About 370 species are known; this figure presumably does not include any estimate of undescribed species.
Ephemeroptera	Arnett 1993	Approximately 2,000 species have been described; no estimate of potential fauna worldwide is available.
Odonata	Arnett 1993	This order has 4,870 known species. Because of the intensity with which they have been collected, the group is not likely to be much larger.
Plecoptera	Arnett 1993	Estimated the number of described species to be about 1,550.
	Theischinger 1991	Theischinger estimates the number to be slightly more than 2,000. No estimate of the potential extent worldwide is available.
Blattodea	Roth 1991	About 4,000 species of cockroaches are known worldwide.
Isoptera	Arnett 1993	About 1,900 species of termites are cataloged.
	Watson and Gay 1991	Estimated 2,300 worldwide; this presumably includes an estimate of undescribed species known then.
Mantodea	Arnett 1993	Estimated 1,500 known mantid species.
	Balderson 1991	Estimated 1,800 known species.
Grylloblattodea	Storozhenko 1986	About 20 known species in this curious Northern-hemispheric group.
Dermaptera	Arnett 1993	Estimated about 1,100 known species of earwigs.
	Rentz 1991	Said that 1,800 species had been described.
Orthoptera	Rentz 1991	More than 20,000 known species in this major group.
	Arnett 1993	Estimated 12,500 species.
Phasmatodea	Key 1991	Estimate of 2,500 species is possibly an underestimate of worldwide fauna.

continues

TABLE 2 Continued

Order	Reference	Comments
Embioptera	Ross 1991 Ross 1995	Fewer than 200 species have been described, but estimates are that as many as 2,000 species exist. Provides a worldwide list of described species on a web site.
Zoraptera	Smithers 1991	About 30 species have been described.
Psocoptera	Smithers 1996	About 3,000 species worldwide have been described.
Phthiraptera	Palma and Barker 1996	More than 3,000 species have been described, but a considerable number of species probably remain undescribed.
Heteroptera	C.W. Schaefer (pers. comm.).	Estimated about 37,000 described and 24,500 undescribed species of Hemiptera-Heteroptera exist.
Homoptera	Hodkinson and Casson 1991	Estimated grand total of 48,660 described species. Extrapolated estimate of total worldwide fauna in Hemiptera and Homoptera combined is about 190,000 species.
Thysanoptera	Mound manuscript catalog	About 6,880 species are named, but this is reduced by known and suspected synonymy to about 5,000 valid species. Total fauna worldwide is possibly twice this, but sampling in tropical areas of high diversity remains inadequate for any serious estimate.
Megaloptera	Theischinger 1991	About 300 species have been described worldwide.
Raphidioptera	Aspöck and Aspöck 1991	Number of species does not exceed 200, most of which have been described.
Neuroptera	New 1991	Includes slightly more than 5,000 described species.
Coleoptera	Arnett 1993 Lawrence and Britton 1994 Lawrence 1991	Stated that approximately 290,000 species of beetles had been described. Estimated 350,000 named species. Stated that Australian fauna includes 20,000 described species of beetles; another 10,000 species likely exist. If this ratio between named and total known species is extrapolated worldwide, this would include more than 500,000 species.
	Hammond 1992, Hammond 1994	Hammond (1992) estimated about 400,000 described species and a total fauna of 2.3 million and 866,667 species (Hammond 1994). We consider 850,000 species to be a reasonable estimate, because we believe that the proportion of described species in Australia is possibly higher than it is in parts of the wet tropics.

continues

TABLE 2   Continued

Order	Reference	Comments
Strepsiptera	Kathirithamby 1991	532 species have been described.
Mecoptera	Byers 1991	About 500 species have been described.
	Penny 1995	Maintains a worldwide list on the web.
Siphonaptera	Dunnet and Mardon 1991	By 1985, a total of 2,380 species and subspecies of fleas had been described.
Diptera	Arnett 1993	98,500 species of flies have been described.
	Colless and McAlpine 1991	Indicated that if undescribed species are included, this order is likely to include at least 150,000 species.
Trichoptera	Neboiss 1991	More than 7,000 species have been described but since tropical faunas generally have been sampled poorly, the worldwide total is likely to be considerably larger. Maintains a searchable list of worldwide species on a web site.
	Morse 1997	
Lepidoptera	Heppner 1991, Hammond 1992	Concluded that 146,277 species have been named; this is close to Hammond's estimate of 150,000 species.
	N.P. Kristensen (1992, unpubl. ms., "Lepidoptera of the World. Status and Perspectives on the Inventory of a Major Insect Order."	Estimated that the total fauna ranges from more than 250,000 to fewer than 1 million species. A number of published estimates fall within the range of 360,000 to 500,000 species.
Hymenoptera	Gaston 1993	Estimated 115,000 described species after personal contact with many of the world's most experienced specialists. The number of living species is unlikely to be less than 2 times this number and, given the relatively low effort in taxonomy in tropical countries, could be considerably more than 2 times this number.

ACKNOWLEDGMENTS

We are grateful to our many colleagues in Canberra, at the Australian National Insect Collection and Australian Biological Resources Study, for their frequent advice, help, and criticism. C.W. Schaefer, of the University of Connecticut, Storrs, kindly gave us his opinion on numbers of Heteroptera species.

REFERENCES

Arnett RH. 1993. American Insects. A handbook of the insects of America north of Mexico. Gainesville: Sandhill Crane Pr. 850 p.

Aspöck H. Aspöck U. 1991. Raphidioptera (Snake-flies, camelneck-flies). In: Naumann ID, Carne PB, Lawrence JF, Nielsen ES, Spradbery JP, Taylor RW, Whitten MJ, Littlejohn MJ. 1991. (eds). The insects of Australia. 2nd ed. Carlton: Melbourne Univ Pr.

Balderson J. 1991. Mantodea (praying mantids). In: Naumann ID, Carne PB, Lawrence JF, Nielsen ES, Spradbery JP, Taylor RW, Whitten MJ, Littlejohn MJ. 1991. (eds). The insects of Australia. 2nd ed. Carlton: Melbourne Univ Pr.

Bayers GW. 1991. Mecoptera (scorpion-flies, hanging-flies). In: Naumann ID, Carne, PB, Lawrence, JF, Nielsen, ES, Spradbery, JP, Taylor, RW, Whitten, MJ, Littlehohn, MJ.1991. (eds.). The Insects of Australia. 2nd ed. Carlton: Melbourne University Press.

Colless DH, McAlpine DK, 1991. Diptera (flies). In: Naumann ID, Carne PB, Lawrence JF, Nielsen ES, Spradbery JP, Taylor RW, Whitten MJ, Littlejohn MJ. 1991. (eds). The insects of Australia. 2nd ed. Carlton: Melbourne Univ Pr.

Condé B, Pagés J. 1991. Diplura. In: Naumann ID, Carne PB, Lawrence JF, Nielsen ES, Spradbery JP, Taylor RW, Whitten MJ, Littlejohn MJ. 1991. (eds). The insects of Australia. 2nd ed. Carlton: Melbourne Univ Pr.

Erwin TL. 1982. Tropical forests: their richness in Coleoptera and other arthropod species. Coleopt Bull 36:74–5.

Evenhuis NL (ed) 1989. Catalog of the Diptera of the Australian and Oceanian regions. Honolulu: Bishop Museum.

Dunnet GM, Mardon DK. 1991. Siphonaptera (fleas). In: Naumann ID, Carne PB, Lawrence JF, Nielsen ES, Spradbery JP, Taylor RW, Whitten MJ, Littlejohn MJ. 1991. (eds). The insects of Australia. 2nd ed. Carlton: Melbourne Univ Pr.

Gaston KJ. 1991a. The magnitude of global insect species richness. Cons Biol 5:283–96.

Gaston KJ. 1991b. Estimates of the near-imponderable: a reply to Erwin. Cons Biol 5:564–6.

Gaston KJ. 1993. Spatial patterns in the description and richness of the Hymenoptera. In: LaSalle J, Gauld ID (eds). Hymenoptera and biodiversity. Wallingford UK:CAB International. p 277–93.

Gaston KJ. 1994. Spatial patterns of species description: how is our knowledge of the global insect fauna growing? Biolog Conserv 67:37–40.

Hammond P. 1992. Species inventory. In: Groombridge B (ed). Global biodiversity, status of the earth's living resources. New York NY: Chapman & Hall. p 17–39.

Hammond P. 1994. Practical approaches to the estimation of the extent of biodiversity in species groups. Philos Trans Roy Soc London B 345:119–36.

Hawksworth DL, Kalin-Arroyo MT. 1995. Magnitude and distribution of biodiversity. In: Heywood VH (ed). Global biodiversity assessment. Cambridge UK: Cambridge Univ Pr. p 107–91.

Heppner JB. 1991. Faunal regions and their diversity of Lepidoptera. Tropical Lepidoptera 2, Supplement 1.

Hodkinson ID, Casson D. 1991. A lesser predilection for bugs: Hemiptera (Insecta) diversity in tropical rain forests. Bio J Linnean Soc 43:101–9.

Hopkins SP. 1996. Biology of the springtails. Insects: Collembola. Cambridge UK: Cambridge Univ Pr.

Imadaté G. 1991. Protura. In: Naumann ID, Carne PB, Lawrence JF, Nielsen ES, Spradbery JP, Taylor RW, Whitten MJ, Littlejohn MJ. 1991. (eds). The insects of Australia. 2nd edition. Carlton: Melbourne Univ Pr.

Janssens Fl. 1997. Web site address: http://www.geocities.com/CapeCanaveral/Lab/1300/2

Kathirithamby J. 1991. Strepsiptera. In: Naumann ID, Carne PB, Lawrence JF, Nielsen ES, Spradbery JP, Taylor RW, Whitten MJ, Littlejohn MJ. 1991. (eds). The insects of Australia. 2nd ed. Carlton: Melbourne Univ Pr.

Key KHL. 1991. Phasmatodea (stick-insects). In: Naumann ID, Carne PB, Lawrence JF, Nielsen ES, Spradbery JP, Taylor RW, Whitten MJ, Littlejohn MJ. 1991. (eds). The insects of Australia. 2nd ed. Carlton: Melbourne Univ Pr.

Kitching R. 1990. The science show. Australian Broadcasting Corporation (28 July 1990).

Kristensen NP. unpublished. Lepidoptera of the world. Status and perspectives on the inventory of a major insect order.

Lawrence JF, Britton EB. 1991. Coleoptera (beetles). In: Naumann ID, Carne PB, Lawrence JF, Nielsen ES, Spradbery JP, Taylor RW, Whitten MJ, Littlejohn MJ. 1991. (eds). The insects of Australia. 2nd ed. Carlton: Melbourne Univ Pr.

Lawrence JF, Britton EB. 1994. Australian beetles. Melbourne Univ Pr.

Longino JT. 1994. How to measure arthropod diversity in a tropical rainforest. Bio Intl 28:3–13.

May RM. 1990. How many species? Philos Trans Roy Soc Lond (B) 330:293–304.

Morse JC. 1997. http://biowww.clemson.edu/ento/databases/trichoptera/trichintro.html.

Mound LA. 1998. Insect taxonomy in species-rich countries: the way forward? An Soci Entomol Brasil 27:1–8.

Naumann ID, Carne PB, Lawrence JF, Nielsen ES, Spradbery JP, Taylor RW, Whitten MJ, Littlejohn MJ. 1991. (eds). The insects of Australia. 2nd ed. Carlton: Melbourne Univ Pr.

New TR. 1991. Neuroptera (lacewings). In: Naumann ID, Carne PB, Lawrence JF, Nielsen ES, Spradbery JP, Taylor RW, Whitten MJ, Littlejohn MJ. 1991. (eds). The insects of Australia. 2nd ed. Carlton: Melbourne Univ Pr.

Nielsen ES, Edwards ED, Rangsi TV (eds). 1996. Checklist of the Lepidoptera of Australia. Monogr Austral Lepidop 4:1–529.

Nielsen EB, West JG. 1994. Biodiversity research and biological collections: transfer of information. Syste Asso Spec Vol 50:101–21.

Palma RL, Barker SC. 1996. Phthiraptera. In: Wells A (ed). Zoological catalogue of Australia. Vol 26, Psocoptera, Phthiraptera, Thysanoptera. Melbourne: CSIRO Publishing. p 81–247, 333–61 (App. I–IV), 373–96 (Index).

Penny ND. 1995. gopher://CAS.calacademy.org:70/00/depts/ent/mecolist

Recher HF, Majer J. 1996. One humble gum tree; home to 1000 species. Geo Austral 18:20–9.

Rentz DCF. 1991. Orthoptera (grasshoppers, locusts, katydids, crickets). In: Naumann ID, Carne PB, Lawrence JF, Nielsen ES, Spradbery JP, Taylor RW, Whitten MJ, Littlejohn MJ. 1991. (eds). The insects of Australia. 2nd ed. Carlton: Melbourne Univ Pr.

Rentz DCF, Kevan DKM. 1991. Dermaptera (earwigs). In: Naumann ID, Carne PB, Lawrence JF, Nielsen ES, Spradbery JP, Taylor RW, Whitten MJ, Littlejohn MJ. 1991. (eds). The insects of Australia. 2nd ed. Carlton: Melbourne Univ Pr.

Robinson GS, Nielsen ES. 1989. Tineid genera of Australia. Monogr on Austral Lepidop 2:1–344.

Ross ES. 1991. Embiotera, Embiidina (Embiids, web-spinners, foot-spinners). In: Naumann ID, Carne PB, Lawrence JF, Nielsen ES, Spradbery JP, Taylor RW, Whitten MJ, Littlejohn MJ. 1991. (eds). The insects of Australia. 2nd ed. Carlton: Melbourne Univ Pr.

Ross ES. 1995. gopher://CAS.calacademy.org:70/00/depts/ent/embilist

Roth LM. 1991. Blattodea blattaria (cockroaches). In: Naumann ID, Carne PB, Lawrence JF, Nielsen ES, Spradbery JP, Taylor RW, Whitten MJ, Littlejohn MJ. 1991. (eds). The insects of Australia. 2nd ed. Carlton: Melbourne Univ Pr.

Scoble MJ (ed). 1999. Geometrid moths of the world: a catalogue (Lepidoptera, Geometridae). Melbourne: CSIRO Publishing.

Smith GB, Watson JAL. 1991. Thysanura Zygentoma (Silverfish). In: Naumann ID, Carne PB, Lawrence JF, Nielsen ES, Spradbery JP, Taylor RW, Whitten MJ, Littlejohn MJ. 1991. (eds). The insects of Australia. 2nd ed. Carlton: Melbourne Univ Pr.

Smithers CN. 1991. Zoraptera. In: Naumann ID, Carne PB, Lawrence JF, Nielsen ES, Spradbery JP, Taylor RW, Whitten MJ, Littlejohn MJ. 1991. (eds). The insects of Australia. 2nd edition. Carlton: Melbourne Univ Pr.

Smithers CN. 1996. Psocoptera. In: Wells A (ed). Zoological catalogue of Australia. Vol 26, Psocoptera, Phthiraptera, Thysanoptera. Melbourne: CSIRO Publishing, Australia. p 1–79, 363–72 (index).

Stork NE. 1988. Insect diversity: fact, fiction and speculation. Bio J Linnaean Soc 35:321–37.

Stork NE. 1993. How many species are there? Biod Cons 2:215–32.

Storozhenko S. 1986. The annotated catalogue of living Grylloblattida (Insecta). Articulata 2:279–92.

Theischinger G. 1991a. Plectopetra (stoneflies). In: Naumann ID, Carne PB, Lawrence JF, Nielsen ES, Spradbery JP, Taylor RW, Whitten MJ, Littlejohn MJ. 1991. (eds). The insects of Australia. 2nd ed. Carlton: Melbourne Univ Pr.

Theischinger G. 1991b. Megaloptera (alderflies, dobsonflies). In: Naumann ID, Carne PB, Lawrence JF, Nielsen ES, Spradbery JP, Taylor RW, Whitten MJ, Littlejohn MJ. 1991. (eds). The insects of Australia. 2nd ed. Carlton: Melbourne Univ Pr.

Tuxen SL. 1964. The Protura. A revision of the species of the world, with keys for determination. Paris: Hermann. 360 p.

Vane-Wright RI. 1992. Species concepts. In: Groombridge B (ed). Global biodiversity: status of the Earth's living resources. New York NY: Chapman & Hall. p.13–6

Watson JAL, Gay FJ. 1991. Isoptera (termites). In: Naumann ID, Carne PB, Lawrence JF, Nielsen ES, Spradbery JP, Taylor RW, Whitten MJ, Littlejohn MJ. 1991. (eds). The insects of Australia. 2nd ed. Carlton: Melbourne Univ Pr.

Watson JAL, Smith GB. 1991. Archaeognatha microcoryphia (bristletails). In: Naumann ID, Carne PB, Lawrence JF, Nielsen ES, Spradbery JP, Taylor RW, Whitten MJ, Littlejohn MJ. 1991. (eds). The insects of Australia. 2nd ed. Carlton: Melbourne Univ Pr.

Wilson EO. 1985. The biological diversity crisis: a challenge to science. Iss Sci Tech 2:20–9.

White MJD. 1982. Karyotypes and meiosis of the morabine grasshoppers. IV. The genus Gecomima. Aust J Zool 30:1027–34.

THE ROLE OF

THE GROUP

IN BIODIVERSITY

THE WORLD BENEATH OUR FEET:
SOIL BIODIVERSITY AND ECOSYSTEM FUNCTIONING

DIANA H. WALL

Natural Resource Ecology Laboratory, College of Natural Resources,
Colorado State University, Fort Collins, Colorado 80523

ROSS A. VIRGINIA

Environmental Studies Program, Dartmouth College,
Hanover, NH 03755

ECOLOGICAL SERVICES PROVIDED BY SOIL BIODIVERSITY

THE IMPORTANCE of soil fertility as a national resource was aptly noted by Franklin D. Roosevelt: "The nation that destroys its soils destroys itself" (Roosevelt 1937). Since then, the importance of soils and the organisms within them for many vital ecosystem processes has been identified, for example, cleansing of water, detoxification of wastes, and decay of organic matter. Indeed, it is now recognized that the functioning of soils, the dark material beneath our feet, is critical for the survival of life on the planet in its present form. Almost every phylum known above the ground exists below the surface of the ground (Brussaard and others 1997). Soil biota include the microorganisms (bacteria, algae, and fungi), protozoa (single-celled animals), microscopic invertebrates that are less than 1 mm long (such as rotifers, copepods, tardigrades, nematodes, and mites), larger invertebrates up to several centimeters long such as those easily seen by the naked eye—ants, snails, earthworms, spiders, termites and so on, and vertebrates. One cubic meter of soil can harbor millions of species of microorganisms and microscopic invertebrates—organisms whose identities and contributions to sustaining our biosphere are largely undiscovered.

Life in soil is recognized as an important part of Earth's overall biodiversity, yet few studies measure the taxonomic diversity of soil or the relationship of soil biodiversity to ecosystem structure and function (Pimentel and others 1997; Swift and Anderson 1994). Understanding of the relationship between biodiversity and ecosystem function in soils is critically needed if we are to manage and predict the impacts of human activity on ecosystems effectively and ensure soil sustainability.

Species in soils perform ecological services that directly control the sustainability of human life. Soil microorganisms and invertebrates (such as fungi, bacteria, nematodes, and earthworms) provide for the purification of air and water, for the decay and recycling of organic matter and hazardous wastes, and for soil fertility. Soil organisms mediate critical ecosystem processes, particularly those in biogeochemical cycling (Swift and Anderson 1994; Matson and others 1987). Soils store vast amounts of carbon, and it is the biota in soils that most influences local and global processes involving the cycling of carbon and nitrogen, including several greenhouse gases (Coleman and Crossley 1996; Huston 1993). The organisms in soil—through their direct, indirect, and modifying effects on these ecosystem processes (Lavelle and others 1995)—provide humans with numerous services (table 1). Pimentel and others (1997) valued the function of soil biodiversity at $25 billion per year on the basis of the contributions of soil biodiversity to topsoil formation in agricultural lands; this value would increase considerably if natural terrestrial systems were included.

A single ecosystem service, such as the generation and renewal of soil and soil fertility (table 1), involves many ecosystem processes and countless organisms representing diverse phyla. These range from large vertebrates to invertebrates and smaller macrofauna such as earthworms and ants that channel through the soil, algae living on the soil surface, and microorganisms involved in the decay of organic matter (Pankhurst and Lynch 1994). The decay of a small animal (such as a piglet) in the soil requires many phyla and can involve 100–500 species of Arthropoda (Richards and Goff 1997). Knowledge of the succession of species participating in the decay of humans is used in forensic medicine to determine the time of death (Goff 1991). Information on the number and types of soil species and phyla required to decompose plant material or invertebrates might be avail-

TABLE 1   Some Ecosystem Services Provided by Soil Biota

Biota Ecosystem Services
Regulation of major elemental cycles
Retention and delivery of nutrients to plants
Generation and renewal of soil, and soil fertility
Detoxification and decomposition of wastes
Modification of the hydrological cycle
Mitigation of floods and droughts
Translocation of nutrients, particles, and gases
Regulation of atmospheric trace gases (production and consumption)
Regulation of animal and plant populations
Control of potential agricultural pests
Foundation of life from which humanity has derived elements of its agricultural, medicinal, and industrial enterprises

Source: Modified from Daily (1997).

able, but the data from many isolated field studies and from taxonomic work have not been synthesized.

Soil organisms contribute to the detoxification of pollutants on a global and a local scale—for instance, detoxifying the pollutants in our yards, farms, golf courses, and parks. These organisms, through their metabolism, are critical to detoxifying and purifying many pollutants before they are leached into ground-water and reach aquatic ecosystems (Abrams and Mitchell 1980; Sayler 1991). Finding environmentally sound ways to use organisms to renew polluted soils and decompose the garbage in our landfills is a growing industry (bioremediation) that depends on the ecosystem services of soil organisms (Sayler 1991).

Ecosystem services such as the mitigation of floods and droughts through prevention of soil erosion, the buffering and modification of the hydrological cycle and the translocation of nutrients, particles, and gases are a reult of many species accomplishing different, but linked, tasks. For example, soils are a temporary habitat for predominantly aboveground organisms (such as vertebrates, lizards, rabbits, gophers, and birds) (Anderson 1987) and invertebrates (such as ants, spiders, and beetles) that move through the soil acting as cultivators or bioturbators, some species ingesting soil and others burrowing in and moving it. Those activities affect soil porosity, the retention of soil water and its movement vertically and horizontally, the transfer of materials throughout the soil profile, and the hydrological cycle. Soil bioturbators, while changing the physical and chemical environment of the soil, also transfer other, smaller organisms and soil particles within the soil, constantly creating new soil aggregates and new surfaces as habitats for microorganisms and facilitating topsoil formation. In this way, the soil biota "plows" the soil, mixing organic matter and nutrients essential for life throughout the soil profile.

Soil organisms have long been recognized as essential for agricultural food production. Nitrogen-fixing bacteria, mycorrhizal fungi, and rhizobacteria that have beneficial relationships with plants, in consort with the decomposers, supply elements essential for plant growth. In addition, through predator and prey interactions and parasitism, soil organisms control vast numbers of agricultural pests (insects, microorganisms, and fungi) (Kerry 1987). For example, the Steinernematid and Heterorhabditid nematodes that parasitize insects above and in the ground are used as a biological control of armyworms, carpenter worms, flea beetles, crown borers, cutworms, cockroaches, leaf miners, mole crickets, root weevils, stem borers, and white grubs (Kaya 1993). Many invertebrate species yet to be discovered are expected to have enormous potential as biological control agents.

SOIL BIODIVERSITY ASSESSMENT

Despite the essential nature of services provided by the soil biota, the systematics of the majority of these organisms has not been determined. Information is lacking on how species' abundance, distribution, and interactions influence ecosystem functioning and whether there are key taxa essential for ecosystem processes. Our ecological knowledge is insufficient to make needed inferences about factors controlling the distribution and activity of the species of soil biota over broad geographic ranges and whether removal or introduction of species alters ecosystem processes. The identification of individual soil organisms to the species level is severely hampered because

- the sheer abundance of soil biota is overwhelming to describe—$1m^2$ of a pasture can contain 10 million nematodes, 45,000 oligochaetes, and 48,000 mites and collembola (Overgaard-Nielsen 1955);
- few scientists have soil taxonomic or soil ecological expertise; estimates are that only 3% of the world's scientists study microscopic and invertebrate organisms in soils;
- in situ identification of most soil organisms is difficult, so sampling and extraction techniques must be used to remove the organisms from soil, and these techniques should not affect the features used to identify and describe the individuals;
- organisms range in size from microscopic to macroscopic;
- organisms can have many different structures during their life cycle;
- methods of sampling and identification must vary with the size of the taxonomic group, for example, earthworms and bacteria (Hall 1996; Oliver and Beattie 1996); and
- promising molecular techniques for most soil organisms are still in their infancy (Blair and others 1996; Hall 1996).

Together, those factors often make the identification and enumeration of soil biota seemingly an insurmountable obstacle for soil research. Perhaps the most important part of this problem is that the decline in human resources in taxonomy overall as a result of diminished institutional support for systematic research, particularly by agricultural and natural resource agencies, has been especially severe for soil taxa (Brussaard and others 1997; Freckman 1994).

There is a poor understanding of the ecological roles played by soil species. Factors contributing to the dearth of knowledge are many and include the following:

- The diversity of soil organisms spans many phyla (from microorganisms to arthropods to vertebrates), and this makes interactions and ecological roles difficult to assess.
- The temporal (seasonal and annual population changes) and spatial scale of the soil habitat that is relevant for an organism (from soil aggregate to landscape) varies among groups.
- Soil species can live at considerable depths (Freckman and Virginia 1989; Silva and others 1989), or can be restricted to microhabitats such as near the surface of roots (rhizosphere).
- The specific taxa participating in soil food webs can change with the soil physiochemical environment, the quality of organic matter, plant species diversity, landscape characteristics and climate. All these make it difficult to compare the ecological roles of soil taxa in different ecosystems.

Other than for earthworms, termites, and other larger soil invertebrates, the use of species composition in ecosystem studies is not yet widespread, because the taxonomy of nearly all groups is incomplete, and for most species only the adult stage is described. As a result, the approach to studying the link between organisms and ecosystem processes has been to place soil organisms in functional groups at a gross level—for example, considering all oribatid mites and springtails that feed on fungi to be fungivores, all mesostigmatid mites to be predators of other microfauna, and so on. The taxonomic and ecological limitations of this approach have been emphasized (Moore and others 1996; Walter and others 1988). We lack knowledge of the feeding strategies of more than 90% of the soil biota.

There is only baseline knowledge of the soil biodiversity in a few ecosystem types, mainly those with high economic value—agricultural, grazing, and forestry (Daily 1997; Pimentel and others 1997). The soil biodiversity estimates of those types generally exclude aboveground organisms that have only one phase of their life cycle in soil or that use the soil as a habitat. Groups of invertebrates, such as wasps and bees, or vertebrates (Anderson 1987; Ingham and Detling 1984; Naiman and Rogers 1997) are studied primarily by "aboveground scientists", and the interchange of information about the functions of such organisms between these scientists and soil ecologists is rare. Some vertebrates, a group generally thought of as living predominantly aboveground, live entirely in the ground. For example, Caecilians (Wake 1983), one species of which was found living at a depth of 30 ft. We note that the recent summaries of taxonomic progress in the major soil biotic groups, which we have outlined below, do not include these predominantly aboveground organisms (Brussaard and others 1997; Groombridge 1992; Hawksworth and Ritchie 1993; O'Donnell and others 1994; Systematics Agenda 2000 1994). A brief assessment of the summaries and methods for studying soil organisms follows. We discuss the groups of soil organisms in order of increasing body size.

Viruses, Bacteria, and Fungi

There have been dramatic advances in the methods for assessing bacterial and fungal biodiversity, although no method can give the "best" quantitative estimate of diversity, because for these taxa and such invertebrates as the nematodes, the reproductive biology of the groups does not permit the application of a "species concept" (de Leij and others 2000; Zak and Visser 1996). For those groups, characteristics to define species are genetic, ecological, chemotaxonomic, and physiological (Snelgrove and others 1997). Molecular methods and chemosynthetic approaches are expanding our knowledge of bacterial and fungal diversity and of trophic relationships with soil invertebrates (Anderson 1975; Hawksworth 1991). Fungi and other microorganisms can be combined into functional groups on the basis of differences in the enzymes required to use particular carbon compounds (for example, cellulose, lignin, and sugar) (Zak and Visser 1996). The specificity of analysis has increased, allowing functional groups to be separated at finer levels of resolution, and enabling the types and numbers of microorganisms and their rates of use of primary and secondary compounds to be analyzed (Zak and Visser 1996). Another method, relying on biochemical markers of diversity termed

FAME (fatty acid, methyl esterase) profiles the relative abundance and diversity of broad groups of microorganisms (Stahl and Klug 1996). Viruses are rarely considered even though they can be potential biological control agents for soilborne pathogens of plants or plant pests. The importance of these microorganisms to ecosystem function is detailed in Lynch and others (this volume).

Microfaunal and Mesofaunal Invertebrate Groups

These groups are Protozoa, Rotifera (wheel animals), Tardigrada (water bears), Nematoda (roundworms), Acari (mites), and Collembola (springtails). No method extracts all taxa, and methods vary widely in their ability to extract these organisms from soil quantitatively and qualitatively. Many groups can be classified to the species level on the basis of morphological characteristics. More molecular methods are becoming available for classification and for assessing interspecific and intraspecific variation on a geographic basis (Avanzati and others 1994; Courtright and others in press; Oliver and Beattie 1996)

Macrofaunal Invertebrate Groups

These include Arachnida, Chilopoda, Diplopoda, Insecta, Annelida (segmented worms), and Mollusca (see figure 1). They can be more easily classified to the species level, and their ecological roles are known in general (Brown and Gange 1990). In temperate regions, their ecological roles include direct processing of organic matter, predatory regulation of population size, modification of soil structure, and production and consumption of atmospheric gases, such as methane. Organisms that cannot be readily identified to the species level include enchytraeid worms, many of the larger mites, some spiders, larval beetles and larval flies. Knowledge of these soil taxa varies dramatically between different locales, and only a few locations have well-described invertebrate macrofauna. Stable-isotope techniques have been used successfully to study trophic relationships and interactions in freshwater habitats and have great promise in soils (Barios and Lavelle 1986; Boutton and others 1983), particularly if extended to the microfauna.

CURRENT ESTIMATES OF SOIL BIODIVERSITY

Almost all aboveground phyla have representatives in soils, but there are no global assessments of the biodiversity in soils and only a few global estimates of individual taxonomic groups (Brussaard and others 1997). In figure 1, we present our estimate of soil biodiversity described to date on the basis of the literature or " best guesses" by specialists working on particular groups. Caution should be exercised when considering these numbers, inasmuch as the size of the soil samples used varies greatly with the size of the organism. Earthworm diversity might have been assessed from 1-m^2 samples to a depth of 40 cm, whereas the protozoan diversity might have been described from a 5-g sample scraped from the soil surface. There are estimates of the total number of species that exist in these groups. For some of the taxa, all or the vast majority of projected species are soil-dwellers; estimates of total species in these groups include

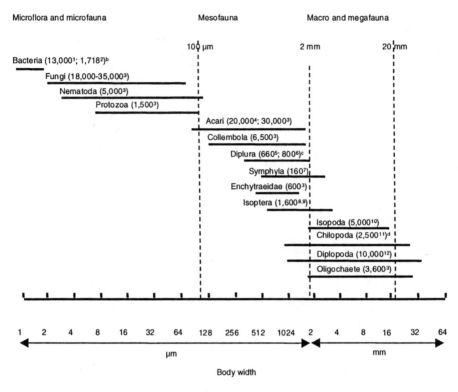

FIGURE 1   Size classification of organisms in the decomposer food web of soils[a], by body width (based on Swift, Heal and Anderson 1979), with number of species described to date for each group.

(a) Species in litter and decaying logs are included in the estimate of soil-dwelling species. (b) Torsvik and others (1994) measured 4,000 independent bacterium-sized genomes in 30 g of forest soil, using DNA analysis. They calculated that this equated to 13,000 species. In a review of Torsvik's study, Dykhuizen (1998) suggested that there could be as many as 500,000 species in the soil sample. Those numbers of species based on DNA analysis are much higher than those described from traditional bacterial isolation and culturing techniques (for example, the 1,718 according to Akimov and Hattori 1996) because culturable bacteria might represent only 0.1–1% of the species in a population. (c) Maddison (1995) gives the number of total described Diplura species (800). Because most Diplura are found in soil, we assumed this number to equal the number of soil-dwelling species. (d) Hoffman (1982a) gives the number of total Chilopoda species described (2,500). Because Chilopods are found only in soil, leaf litter, rotting wood, and caves, we assumed this number to equal the number of soil-dwelling species.

[1]Torsvik and others 1994.
[2]Akimov and Hattori 1996.
[3]Brussaard and others 1997.
[4]Walters, personal communication.
[5]Ravlin 1996a.
[6]Maddison 1995.
[7]Scheller 1982.
[8]Bignell, personal communication.
[9]Bignell and Eggleton 1998.
[10]Brusca 1997.
[11]Hoffman 1982a.
[12]Hoffman 1990.

Collembolla, 20,000–24,000 (Ravlin 1996b); Diplura, 1,647 (Ravlin 1996a); Enchytraeidae, 1,200 (Behan-Pelletier, personal communication); and Isoptera, 3,000 (Bignell and Eggleton 1998). In other groups, some of the species live outside the soil; estimates of total species in these groups include bacteria, 1,000,000; fungi 1,500,000; algae, 400,000; Nematoda, 1,000,000; protozoa, 200,000 (Hammond and others 1995); acari, 348,500–900,000 (Walter, personal communication; Walter and others 1998); and Diplopoda, 50,000–60,000 (Hoffman 1982b, 1990). These estimates of total existing species could be low; the soil component of biodiversity has traditionally been underappreciated and poorly described relative to aboveground species because of soil organisms' abundance and microscopic size and the dearth of soil taxonomists. The gulf between the numbers of species already described in soils and the projections of total numbers of species confirms what has already been heralded by others (Andre and others 1994; Wilson 2000; May 2000): that microscopic groups—such as bacteria, fungi, nematodes (Baldwin and others 2000; Bernard 1992), acari (Behan-Pelletier and Bisset 1993), Symphyla, and Enchytraeids (Healy 1980) are desperately in need of taxonomic and ecological attention.

The broad groups listed in figure 1 probably all have a global distribution. Information has not been synthesized to make general statements about diversity and geographic distribution (Brussaard and others 1997; Dighton and Jones 1994). For example, Lavelle and others (1995) analyzed earthworm community assemblages containing 8–11 species across 53 global climatic locations and found that neither species richness nor species diversity increased with decreasing distance from the tropics. However, they noted changes in the proportion of species feeding on soil and litter with decreasing latitude. That might not be unusual; Hammond (1995) noted that, for seasonally dry ecosystems, the increase in floristic diversity is not necessarily paralleled by an increase in the terrestrial diversity of invertebrates, fungi, or microorganisms.

There are more biogeographic assessments for individual groups listed in figure 1 on the political and regional scale than on a global scale (Brussaard and others 1997; Folgarait 1996; Pearce and Waite 1994). Species distribution patterns are influenced by chemical and physical factors—such as soil texture, organic matter, and soil moisture—as well as by climate and vegetation (Anderson 1975; Swift and others 1979; Wright and Coleman 1993). For example, some species of earthworms and Enchytraeids in the Oligochaetes rarely occur in deserts (Dash 1990; James 1995); and in the United States, where there are six indigenous genera of earthworms, the species distribution has been limited geographically (to such areas as those not affected by the Wisconsin glaciation, forests, mud flats, and riparian areas) (James 1995; Reynolds 1995).

On the basis of political boundaries, Coomans (1989) listed 228 terrestrial nematode species in Belgium, and Behan-Pelletier and Bissett (1993) estimated North American soil arthropods as follows: isopods, 92 described, 62 undescribed; chilopods, 850 described, 400 undescribed; diplopods, 850 described, 400 undescribed; pauropods, 70 described, 47 undescribed; symphyla, 33 described, 22 undescribed; spiders, 1,700 described, 250 undescribed; Opilionids, 235 described, 250 undescribed; acari, 2,500 described, 14,500 undescribed; and

Protura, 26 described, 104 undescribed. Lindquist and Behan-Pelletier (personal communication) estimate that the totality of described and undescribed species represents 10% of the world's total soil arthropods.[*] Lawton and others (1996) in a Cameroon tropical-forest study, identified 115 species of termites and 432 nematode morphospecies in 185 genera. More than 90% of the nematodes remain unidentified because costs for further descriptions by taxonomists are expensive and time-consuming.

There is no compilation of endemic species in soils, but there are summaries of the endemic species in some groups listed in figure 1, such as earthworm species in the United States (James 1995). It is likely that in ecosystems that have endemic species and low species diversity, such as in soils of the Antarctic dry valleys, the species are more vulnerable to loss by disturbance, but whether their loss would affect ecosystem function must be confirmed through experimentation (Freckman and Virginia 1998).

EFFECT OF DISTURBANCE ON SOIL BIODIVERSITY

Whether the presence or absence of a single soil species can affect ecosystem structure or function (for example, such biogeochemical processes as rates of decomposition and plant production) is largely unresolved (Beare and others 1995) except for agroecosystems, in which a single species becomes noted as a crop pest, affecting transfer of nutrients and plant production (Brussaard and others 1997; Swift and Anderson 1994). The effects of introductions of soil species on ecosystem functioning constitutes a new field of study by aboveground ecologists, but it is a well-established field among agriculturalists. An example of the introduction of an alien species from Asia into a natural ecosystem in the United States is the earthworm species *Amynthas hawayanus*, which reduced New York forest-floor organic matter and increased water runoff and soil erosion (Burtelow and others 1998). The effects of introductions of alien species into fields have been studied by agriculturalists worldwide because of the potential devastating economic loss to crops (Cotten and Riel 1993; Swift 1997). Of growing importance is research on the introductions of soil organisms that might enhance plant productivity while protecting plants from pests (Cook 1993).

Better known, however, are the quarantines to restrict movement of soil biota and plant pests and thereby prevent the spread of exotic or established soil-pest species. The European Community lists nine plant-parasitic nematode species that are targeted in the hope of prohibiting introduction into Europe (EEC Council Directive 77/93/EEC, see Cotten and Riel 1993); California has 25 nematode species listed in its quarantine regulations. Despite such efforts, even the tightest regulations have been only partially successful. The US Department of Agriculture, with millions of dollars allocated to prevent the spread of *Globodera rostochiensis* (the potato cyst nematode) from Long Island, NY, where it became established in 1941, was unsuccessful in preventing its spread to other North American areas (figure 2). A second species of the potato cyst nematode, *Globodera pallida*, has not established itself in the United States but since 1972

FIGURE 2 Global dispersal of the potato cyst nematodes *Globodera rostochiensis* (diagonal lines) and *G. pallida* (stippling) from their centers of origin in the Andes of South America (based on Evans and Trudgill 1992).

has spread from its place of origin in the Andes of South America to New Zealand and globally (Wood and others 1983). It appears that the quarantine measures slow movement of the pest but in the long run is unable to prevent its introduction.

Although there has been no synthesis of the effects of the loss of individual soil species, the loss of entire functional groups has been shown to influence ecosystem function. For example, loss of taxa that are symbiotic with plants, such as the mycorrhizal fungi, can affect plant community structure. Re-establishment of plants on mine spoils or young volcanic soils can be slowed if mycorrhizal fungi are not present (Allen 1991). What is difficult to assess is the importance of the interactions of functional groups or individual species with other soil biota in the food web and how these perhaps less noticeable interactive effects influence ecosystem processes.

Soil biodiversity is adversely affected by human-induced disturbances that can be classified as physical (plowing, desertification, and landfill), chemical (air pollutants, fertilizers, chemical spills, and pesticides), changes in plant diversity (introductions of plants, changing from natural to managed systems, and changing plant species in agroecosystems), and environmental changes, such as climate alteration. Most observed changes in the diversity of soil biota after disturbance have been recorded at the level of the functional group (Brussaard and others 1997; Coleman and Crossley 1996) or at taxonomic levels, such as family (Andren and others 1995). On the basis of such results, several groups are recognized as useful indicators of soil disturbance; nematodes might be the most widely studied of such indicators (Blair and others 1996; Bongers 1990; Niles and Freckman

1998), although Lawton and others (1998) found that few organism groups, whether aboveground or in the ground, were effective indicators of habitat fragmentation.

The physical changes from disruptions of the soil environment include plowing, erosion, and desertification, which induce changes in the soil profile, soil texture, amounts and types of soil organic matter, soil chemistry, and microclimate (temperature, moisture, aeration, and release of carbon, nitrogen, and other greenhouse gases). Like aboveground organisms, soil-dwelling species have habitat preferences, and disruption of their soil habitat changes the community composition (Freckman and Ettema 1993; Freckman and Virginia 1989). The number of species of soil-dwelling termites and nematodes found in the Cameroon discussed earlier (Lawton and others 1998) declined with soil disturbance, even though deforestation can also increase the amount of woody debris for termite consumption. The termite species produce methane, but how their responses to disturbance affected methane production locally or globally is unknown. Erosion of the top inch of soil can also remove many microorganisms and smaller invertebrates. Chemical disturbances—such as caused by fertilizers, oil spills, heavy metals, and pesticides—have adverse effects on biotic function and species richness in soils (Brussaard and others 1997; Korthals 1997; Niles and Freckman 1998). Changes in plant diversity in cropping systems have been used for centuries to reduce the population densities of host-specific soil parasites.

The effects of global change on soils have been summarized (Ingram and Wall-Freckman 1998: see reviewer comment 26); they can alter soil biodiversity and ecosystem processes directly and indirectly. Norby (1997) suggested that effects of global change on the soil biota can result in feedbacks that could increase cycling rates of nutrients, resulting in increased emissions of CO_2 to the atmosphere. Direct effects of temperature could influence the duration of the life cycle of organisms and affect quantities and biomass of prey, food sources, and predators.

RESEARCH PRIORITIES AND CONCLUSIONS

Among the most pressing research needs is increased training of ecologists and taxonomists to work on the natural history and identification of soil-dwelling organisms. Only with this knowledge can we determine how to manage soils sustainably around the globe. Programs such as the Tropical Soil Biology Fertility Program (Swift 1997b), which has been going on for 11 years, are using international experiments that incorporate available information on soil biota and ecosystem processes to develop sustainable soil management in tropical agriculture. Increased knowledge of the relationship of biodiversity in soils to plant species diversity and the physical and chemical environment might allow us to predict hot spots of soil diversity by using the tools of remote sensing and geographic information systems and thus contribute to better predictions for land management. Information about soil biotas alone and their interactions with plants will contribute to testing and developing the ecological theory needed to predict the response of natural systems to disturbance (Ohtonen and others 1997). Knowledge of the

TABLE 2   Scientific Disciplines Necessary for Soil-Ecology Research

Biology and Taxonomy	Soil Science
Vertebrate zoology	Pedology
Invertebrate zoology	Soil chemistry
Entomology	Soil physics
Nematology	Soil ecology
Microbiology	Geology
Bacteriology	Hydrology
Mycology	Micrometeorology
Virology	Ecosystem science
Plant sciences	Landscape ecology
Agronomy and botany	Biogeochemistry
Plant pathology	
Plant physiology	
Ecology	
Taxonomy	
Molecular biology	
Informatics	

natural history of a soil biota, such as types of physiological changes in plant hosts caused by pest species, could help to reduce crop damage on a regional scale. For example, differences in transpiration flows between winter-wheat fields infected and not infected by the soil nematode *Heterodera avenae* can be noted with thermal infrared radiometry (Rivoal and Cook 1993). The spread of the pest, which is the most important in wheat fields of Australia and causes damage to 50% of the wheat fields in Europe, could be more effectively controlled with techniques based on knowledge of the natural history of the species. Today, our knowledge of this relatively well-studied species has enabled us to use its obligate fungal parasites for effective biological, rather than chemical, control in some regions (Rivoal and Cook 1993). Without increased resources for training in soil biodiversity and ecology, we will never be able to discover (or realize) the full extent of the benefits that the life in our soils can offer.

Recently, soil ecologists have assessed the priorities for research (Brussaard and others 1997; Ingram and Wall-Freckman 1998; Klopatek and others 1992; Wall-Freckman and others 1997), some of which are as follows:

• development of new techniques, improvement of current techniques, standardization of techniques for sampling and analysis, and informatics to enhance the database on soil biodiversity;
• development of interdisciplinary (table 2) and international cross-site experiments with predictive models to quantify the relationship of soil biodiversity to critical ecosystem processes on various spatial scales (see Freckman 1994 for examples of experiments, http://www.nrel.colostate.edu/soil/lifeinthesoil.html);   a

basis for this work is being led by the UK (http://mwnta.nmw.ac.uk/soilbio/) and is being considered by other nations; and

• development of syntheses of the distributional patterns of soil biodiversity globally and regionally to understand the effects of global change on endemic and introduced species and to predict better how soils should be maintained for sustainable use; included is a proposed synthesis by a subcommittee of the Scientific Committee on Problems of the Environment that will examine the interrelationship of species biodiversity and functional groups aboveground to the belowsurface realm, soils, and freshwater and marine sediments (Wall-Freckman and others 1997).

ACKNOWLEDGMENTS

The authors are grateful to Gina Adams for her research and critical comments. DHW particularly acknowledges the continued enthusiastic encouragement of Wren Wirth, Hal Mooney, D.C. Coleman, and D.A. Crossley Jr. The support of National Science Foundation grants OPP 91-20123, DEB 96-26813, and OPP 92-11773 is appreciated.

REFERENCES

Abrams BJ, Mitchell MJ. 1980. Role of nematode-bacterial interactions in heterotrophic systems with emphasis on sewage sludge decomposition. OIKOS 35:404–10.

Akimov V. Hattori T. 1996. Towards cataloguing soil bacteria: preliminary note. Microbes Environ 11:57–60.

Allen MF. 1991. The ecology of Mycorrhizae. Cambridge UK: Cambridge Univ Pr.

Anderson DC. 1987. Below-ground herbivory in natural communities: a review emphasizing fossorial animals. Quar Rev Biol 62(3):261–86.

Anderson JM. 1975. Succession, diversity, and trophic relationships of some soil animals in decomposing leaf litter. J Anim Ecol 44:475–94.

Andre HM, Noti M-I, Lebrun P. 1994. The soil fauna: the other last biotic frontier. Biod Cons 3:45–56.

Andren O, Bengtsson J, Clarholm M. 1995. Biodiversity and species redundancy among litter decomposers. In: Collins HP, Robertson GP, Klug MJ (eds). The significance and regulation of soil biodiversity. Plant and Soil 170(1) Dordrect The Netherlands: Kluwer Acad Publ. p.141–51.

Avanzati AM, Baratti M, Bernini F. 1994. Molecular and morphological differentiation between steganacarid mites (Acari: Oribatida) from the Canary Islands. Biol J Linn Soc 52:325–40.

Baldwin JG, Nadler SA, Wall DH. 2000. Nematodes: pervading the earth and linking all life. In: Raven PH, Williams T (eds). Nature and human society: the quest for a sustainable world. Washington, DC: National Academy Press. p 176–91.

Barios I, Lavelle P. 1986. Changes in respiration rate and some physiochemical properties of a tropical soil during transit through Pontoscoloex corethurus. Soil Biol Biochem 18(5):539–41.

Beare MH, Coleman DC Jr, Hendrix PF, Odum EP. 1995. A heirarchical approach to evaluating the significance of soil biodiversity to biogeochemical cycling. In: Collins HP, Robertson GP, Klug MJ (eds). The significance and regulation of soil biodiversity. Plant and Soil 170(1) Dordrect The Netherlands: Kluwer Acad Publ. p 5–22.

Behan-Pelletier VM, Bisset, B. 1993. Biodiversity of Nearctic soil arthropods. Canad Biod 2:5–14.

Bernard EC. 1992. Soil nematode biodiversity. Biol Fertil Soils 14:99–103.

Bignell DE, Eggleton P. 1998. Termites. In: Calow P (ed). Encycolpedia of ecology and environmental management. Oxford UK: Blackwell Scientific. p 742–4.

Blair JM, Bohlen PJ, Freckman DW. 1996. Soil invertebrates as indicators of soil quality. In: Doran J, Jones A (eds). Methods for assessing soil quality. Madison WI: Soil Science Society of America. p 273–91.

Bongers T. 1990. The maturity index: an ecological measure of environmental disturbance based on nematode species composition. Oecologia 83:14–19.

Boutton TW, Arshad MA, Tieszen LL. 1983. Stable isotope analysis of termite food habits in East African grasslands. Oecologia 59:1–6.

Brown VK, Gange AC. 1990. Insect herbivory belowground. Adv Ecol Res 20:1–58.

Brusca R. 1997. Isopoda. Grice Marine Biological Laboratory, University of Charleston. Available at http://phylogeny.arizona.edu/tree/eukaryotes/animals/arthropoda/crustacea/isopoda/isopoda.html. Last accessed October 26, 1998.

Brussaard L, Behan-Pelletier VM, Bignell DE, Brown VK, Didden W, Folgarait P, Fragaso C, Wall,- Freckman D, Gupta VVSR, Hattori T, Hawksworth DL, Klopatek C, Lavelle P, Malloch DW, Rusek J. 1997. Biodiversity and ecosystem functioning in soil. Ambio 26 (8):563–70.

Burtelow A, Bohlen PJ, Groffman PM. 1998. Influence of exotic earthworm invasion on soil organic matter, microbial biomass and denitrification potential in forest soils of the northeastern US. Appl Soil Ecol 9(1–3):197–202.

Coleman DC, Crossley Jr DA. 1996. Fundamentals of soil ecology. San Diego CA: Academic Press. p 205.

Cook RJ. 1993. Making greater use of introduced microorganisms for biological control of plant pathogens. Ann Rev Plant Path 31:53–80.

Coomans A. 1989. Oversicht van vrijlevenda nematofauna van Belgie. In: Wouters K, Baert L (eds). Invertebrates of Belgium symposium papers. Brussels Belgium: Insitut Royal des Sciences Naturalles de Belgique. p 43–56.

Cotten J, Riel HV. 1993. Quarantine: problems and solutions. In: Evans K, Trudgill DL, Webster JM (eds). Plant parasitic nematodes in temperate agriculture. Wallingford UK: CABI. p 566–93.

Courtright EM, Freckman DW, Virginia RA, Frisse LM, Vida TT, Thomas WK. In press. Nuclear and mitochondrial DNA sequence diversity in the Antarctic Scottnema lindsayae. Nematology.

Daily GC. 1997. Natures services: societal dependence on natural ecosystems. Washington DC: Island Pr. p 392.

Dash MC. 1990. Oligochaeta: Enchytraeidae. In: Dindal DL (ed). Soil biology guide. New York NY: J Wiley. p 311–40.

de Leij FAAM, Hay DB, Lynch JM. 2000. Investment in diversity: the role of biological communities in soil. In: Raven PH, Williams T (eds). Nature and human society: the quest for a sustainable world. Washington, DC: National Academy Press. p 242–51.

Dighton J, Jones HE. 1994. A review of soil biodiversity. London UK: Royal Commission on Environmental Pollution.

Dykhuizen DE. 1998. Santa Rosa revisited. Why are there so many species of bacteria? Antonie van Leeuwenhoek 73:25–33.

Eisenbeis G, Wichard W. 1987. Atlas on the biology of soil arthropods. New York NY: Springer-Verlag.

Evans K, Trudgill DL. 1992. Pest aspects of potato production. Part 1—Nematode pests of potatoes. In: Harris PM (ed). The potato crop: the scientific basis for improvement. London UK: Chapman & Hall. p 445.

Folgarait PJ. 1996. Latitudinal variation in myrmecophytic Cercropia. Bull Ecol Soc Amer 77:143.

Freckman DW. 1994. Life in the soil. Soil biodiversity: its importance to ecosystem processes. Fort Collins CO: NREL, Colorado State University.

Freckman DW, Ettema CE. 1993. Assessing nematode communities in agroecosystems of varying human intervention. Agric Ecosys Environ 45:239–61.

Freckman DW, Virginia RA. 1989. Plant feeding nematodes in deep-rooting desert ecosystems. Ecology 70:1665–78.

Freckman DW, Virginia RA. 1998. Soil biodiversity and community structure in the McMurdo Dry Valleys, Antarctica. In: Priscu J. Ecosystem dynamics in a polar desert. The McMurdo Dry Valleys, Antarctica. Washington DC: American Geophysical Union. p 323–36.

Goff ML. 1991. Comparison of insect species associated with decomposing remains recovered inside dwellings and outdoors on the island of Oahu, Hawaii. J Foren Sci 36(3):748–53.

Groombridge B (ed). 1992. Global biodiversity: status of the Earth's living resources. World Conservation Monitoring Center. London UK: Chapman & Hall.

Hall GS. 1996. Methods for the examination of organismal diversity in soils and sediments. Wallingford UK: CABI. p 307.

Hammond PM. 1995. Described and estimated species numbers: an objective assessment of current knowledge. In: Allsopp D, Colwell RR, Hawksworth DL (eds). Microbial diversity and ecosystem function. Wallingford UK: CABI. p 29–71.

Hammond PM, Hawksworth DL, Kalin-Arroyo MT. 1995. Chapter 3. Magnitude and distribution of biodiversity: 3.1. The current magnitude of biodiversity. In: Heywood VH (ed). Global biodiversity assessment. Cambridge UK: Cambridge Univ Pr. p 113–38.

Hawksworth DL. 1991. The fungal dimension of biodiversity: magnitude, significance, and conservation. Mycolog Res 95:641–55.

Hawksworth DL, Ritchie JM. 1993. Biodiversity and biosystematic priorities: microorganisms and invertebrates. Wallingford UK: CABI.

Healy B. 1980. Records of Enchytraeidae (Oligochaeta) from western France and Pyrenees. Bull Mus Nation Hist Nat Paris Ser (2) A:421–43.

Hoffman RL. 1982a. Chilopoda. In: Parker SP (ed). Synopsis and classification of living organisms. New York NY: McGraw-Hill. p 681–8.

Hoffman RL. 1982b. Diplopoda. In: Parker SP (ed). Synopsis and classification of living organisms. New York NY: McGraw-Hill. p 689–724.

Hoffman RL. 1990. Diplopoda. In: Dindal DL (ed). Soil biology guide. New York NY: J Wiley. p 835–60.

Huston M. 1993. Biological diversity, soils, and economics. Science 262:1676–80.

Ingham RE, Detling JK. 1984. Plant-Herbivore interactions in a North American mixed-grass prairie. Oecologia 63:307–13.

Ingram J, Wall-Freckman D. 1998. Soil biota and global change. Special issue, preface. Global Change Biology 4(7):699–701.

James SW. 1995. Systematics, biogeography, and ecology of Nearctic earthworms from eastern, central, southern and southwestern United States. In: Hendrix PF (ed). Earthworm ecology and biogeography. Boca Raton FL: Lewis Publ. p 29–52.

Kaya HK. 1993. Entomogenous and entomopathogenic nematodes in biological control. In: Evans K, Trudgill DL, JM Webster (eds). 1993. Plant parasitic nematodes in temperate agriculture. Wallingford UK: CABI. p 565–91.

Kerry BR. 1987. Biological control. In: Brown RH, Kerry BR. Principles and practices of nematode control in crops. Sydney Australia: Acad Pr. p 233–63.

Klopatek CC. 1992. The sustainable biosphere initiative: a commentary from the US Soil Ecology Society. Bull Ecol Soc Amer 73(4):223–7.

Lavelle P, Lattaud C, Trigo D, Barois I. 1995. Mutualism and biodiversity in soils. In: Collins HP, Robertson GP, Klug MJ (eds). The Significance and regulation of soil biodiversity. Plant and Soil 170(1). Dodrecht The Netherlands: Kluwer Acad Publ. p 23–33.

Korthals GW. 1997. Pollutant-induced changes in terrestrial nematode communities. PhD thesis. Landbouwuniversiteit te Wageningen.

Lawton JH, Bignell DE, Bloemers GF, Eggleton P, Hodda ME. 1996. Carbon flux and diversity of nematodes and termites in Cameroon forest soils. Biod Cons 5:261–73.

Lawton JH, Bignell DE, Botton B, Bloemers GF, Eggleton P, Hammond PM, Hodda M, Holt RD, Larsen TB, Mawdsey NA, Stork NE, Srivastava DS, Watt AD. 1998. Biodiversity inventories, indicator taxa and effects of habitat modification in a tropical forest. Nature 391:72–6.

Maddison, David. 1995. Diplura. Department of Entomology, University of Arizona. Available at http://phylogeny.arizona.edu/tree/eukayotes/animals/arthropoda/hexapoda/diplura/diplura.html. Last accessed 26 October 1998.

Matson PA, Vitousek PM, Ewel JJ, Mazzarino MJ, Robertson GP. 1987. Nitrogen transformations following tropical forest felling and burning on a volcanic soil. Ecology 68(3):491–502.

May RM. 2000. The dimensions of life on earth. In: Raven PH, Williams T (eds). Nature and human society: the quest for a sustainable world. Washington, DC: National Academy Press. p 30–45.

Moore JC, Coleman DC, DeRuiter PC, Freckman DW, Hunt HW. 1996. Microcosms and soil ecology: critical linkages between field studies and modelling food webs. Ecology 77:694–705.

Naiman RJ, Rogers KH. 1997. Large animals and system-level characteristics in river corridors: implications for river management. BioScience 47(8):521–9.

Niles RK, Freckman DW. 1998. From the ground up: nematode ecology in bioassessment and ecosystem health. In: Barker KR, Pederson GA, Widham GL (eds). Plant-nematode interactions. Agronomy Monograph. Madison WI: Amer Society of Agronomy, Crop Science Society of America, and Soil Science Society of America. p 65–85.

Norby R. 1997. Inside the black box. Nature 388:522–3.

O'Donnell AG, Goodfellow M, Hawksworth DL. 1994. Theoretical and practical aspects of the quantification of biodiversity among microorganisms. Philos Trans Roy Soc London 345:65–73.

Ohtonen R, Aikio S, Vare H. 1997. Ecological theories in soil biology. Biochemistry 29:1613–9.

Oliver I, Beattie AJ. 1996. Designing a cost-effective invertebrate survey: a test of methods for rapid assessment of biodiversity. Ecol Appl 6(2):594–607.

Overgaard-Nielsen C. 1955. Studies on Enchytraeidae 2: field studies. Natura Jutlandica 4:5–58.

Pankhurst CE, Lynch JM. 1994. The role of the soil biota in sustainable agriculture. In: Pankhurst CE, Doube BM, Gupta VVSR, Grace PR (eds). Soil biota management in sustainable farming systems. Melbourne Australia: CSIRO. p 3–9.

Pearce MJ, Waite B. 1994. A list of termite genera with comments on taxonomic changes and regional distribution. Sociobiology 23:247–62.

Pimentel D, Wilson C, McCullum C, Huang R, Dwen P, Flack J, Tran Q, Saltman T, Cliff B. 1997. Economic and environmental benefits of biodiversity. BioScience 47(11):747–57.

Ravlin FW. 1996a. Diplura (Diplurans). Department of Entomology, Virginia Polytechnic Institute and State University. Available at http://www.gypsymoth.ento.vt.edu/~ravlin/insect_orders/diplura.html. Last accessed 26 October 1998.

Ravlin FW. 1996b. Collembola (Springtails). Department of Entomology, Virginia Polytechnic Institute and State University. Available at http://www.gypsymoth.ento.vt.edu/~ravlin/insect_orders/collembola.html. Last accessed 26 October 1998.

Reynolds JW. 1995. Status of exotic earthworm systematics and biogeography in north America. In: Hendrix PF (ed). Earthworm ecology and biogeography. Boca Raton FL: Lewis Publ. p 1–28.

Richards EN, Goff ML. 1997. Arthropod succession on exposed carrion in three contrasting tropical habitats on Hawaii island. J Med Entomol 34(3):328–39.

Rivoal R, Cook R. 1993. Nematode pests of cereals. In: Evans K, Trudgill DL, Webster JM (eds). Plant parasitic nematodes in temperate agriculture. Wallingford UK: CABI 210–59.

Roosevelt FD. 1937. Speech as United States President. Letter to all State Governors, requesting implementation of Standard State Soil Conservation District Laws.

Sayler G. 1991. Contribution of molecular biology to bioremediation. J Hazard Mat 28:13–28.

Scheller U. 1982. Symphyla. In: Parker SP (ed). Synopsis and classification of living organisms. New York: McGraw-Hill. p 96–102.

Silva S, Whitford WG, Jarrell WM, Virginia RA. 1989. The microarthropod fauna associated with a deep rooted legume, Prosopis glandulosa, in the Chihuahuan desert. Biol Fertil Soils 7:330–5.

Snelgrove PVR, Blackburn TH, Hutchings PA, Alongi DM, Grassle JF, Hummel H, King G, Koike I, Lambshead PJD, Ramsing NB, Solis-Weiss V. 1997. The importance of marine sediment biodiversity in ecosystem processes. Ambio 26(8):578–83.

Stahl PD, Klug MJ. 1996. Characterization and differentiation of filamentous fungi based on fatty acid composition. Appl Env Microbiol 62(11):4136–46.

Swift MJ. 1997. Soil biodiversity, agricultural intensification and agroecosystem function in the tropics. Appl Soil Ecol 6:1–108.

Swift MJ. 1997. Ten years of soil fertility research—where next? In: Bergstrom L, Kirchman H (eds). Carbon and nutrient dynamics in natural and agricultural tropical ecosystems. Wallingford UK: CABI. p 303–12.

Swift MJ, Anderson JM. 1994. Biodiversity and ecosystem function in agricultural ecosystems. New York NY: Springer-Verlag. p 25.

Swift MJ, Heal OW, Anderson JM. 1979. Decomposition in terrestrial ecosystems. Oxford UK: Blackwell Scientific Pr.Systematics Agenda 2000. 1994. Charting the biosphere. A consortium of the American Society of Plant Taxonomists, Society of Systematic Biologists, and the Willi Hennig Society, in cooperation with the Association of Systematists Collections. Available from: Systematics Agenda 2000, Dept. of Ornithology, American Museum of Natural History.

Thematic Issue: soil biota and global change. 1998. Global change biology.

Torsvik V, Goksoyr J, Daae FL, Sorheim R, Michalsen J, Salte K. 1994. Chapter 4. Use of DNA analysis to determine the diversity of microbial communities. In: Ritz K, Dighten KE, Giller (eds). Beyond the biomass. Chichester UK: Wiley. p 39–48.

Wake MH. 1983. Gymnopis multiplicata, Dermophis mexicanus, and Dermophis parviceps (Soldas, Suelda con Suelda, Dos Cabezas, Caecilians) 400–1. In: Janzen DH (ed). 1983. Costa Rican natural history. Chicago IL: Univ Chicago Pr. p 816.

Wall-Freckman D, Blackburn TH, Brussard L, Hutchings PA, Palmer M, Snelgrove PVR. 1997. Linking biodiversity and ecosystem functioning of soils and sediments. Ambio 26(8):556–62.

Walter DE, Hunt HW, Elliott ET. 1988. Guilds or functional groups? An analysis of predatory arthropods from shortgrass steppe soil. Pedobiologia 31:247–60.

Walter, DE, Krantz J, Lindquist E. 1998. http://phylogeny.arizona.edu/tree/eukaryotes/animals/athropoda/arachnida/acari/acari.html.

Wilson EO. 2000. The creation of biodiversity. In: Raven PH, Williams T (eds). Nature and human society: the quest for a sustainable world. Washington, DC: National Academy Press. p 22–9.

Wood FH, Foot MA, Dale PS, Barber CJ. 1983. Relative efficacy of plant sampling and soil sampling in detecting the presence of low potato cyst nematode infestations. New Zeal J Exp Agric 11:271–3.

Wright DH, Coleman DC. 1993. Patterns of survival and extinction of nematodes in isolated soil. OIKOS 67:563–72.

Zak J, Visser S. 1996. An appraisal of soil fungal biodiversity: the crossroads between taxonomic and functional biodiversity. Biod Cons 5:169–83.

NATURAL INVESTMENT IN DIVERSITY:
THE ROLE OF BIOLOGICAL COMMUNITIES IN SOIL

FRANS A.A.M. DE LEIJ
DAVID B. HAY
JAMES M. LYNCH
School of Biological Sciences, University of Surrey,
Guildford, Surrey GU2 5XH, UK

INTRODUCTION

THERE IS A persistent view that the commercial application of "environmental biotechnology" relies largely on the exploitation of single species, well-defined biochemical pathways, or the expression of novel gene sequences by genetically modified organisms. Consequently, the role of species assemblages and in particular the importance of interactions between the members of natural communities have been largely neglected in a commercial or industrial context. This oversight is important because most ecosystem processes appear to be governed by the activity of species at the guild and community level and not by species that function in isolation. Indeed, most well-studied bioremediation experiments and many examples of effective biological control show that natural, or "intrinsic," processes are at least as important in achieving economic ends as specific and targeted biotechnological interventions.

This paper suggests that exploitation of microbial communities is a potentially rewarding alternative to the "classical" or "single-species" biotechnological approach. We emphasize the need for ecosystem study in the context of biotechnology development. The development of sampling and monitoring techniques has high priority for research. Monitoring is an important technology for tracking "intrinsic" beneficial processes, and sampling will provide data to improve our fundamental understanding of ecological processes in an applied context. In this paper, the commercial uses of microorganisms and microbial communities are examined in the context of bioremediation and biological control, but they could equally well be scrutinized in relation to such processes as soil formation and bio-degradation.

BIOLOGICAL CONTROL

Development of Single Isolates as Biological Control Agents

The commercial development of *Bacillus thuringiensis* is one of the most out-standing examples of successful technology based on a specific biological resource. *B. thuringiensis* has taken the dominant share of the biopesticide market: a little over 1% of total pesticide sales. That figure could rise to 10% during the next decade (Hokkanen and Lynch 1995). Although *B. thuringiensis* is often referred to as a biological control agent, its application and action have much more in common with chemical pesticides. The toxin crystal produced by the bacterium is the active ingredient of commercial products, and the ecology of the organism itself is largely irrelevant in pest control—in stark contrast with familiar examples of biological control in which the dynamics of host-parasite or host-predator in-teractions determine product efficacy.

Where biological solutions have been sought to combat pests in well-defined (and controlled) conditions, commercial success has sometimes been based on the exploitation of single species or isolates. For example, *Peniophora gigantea* is used to control *Heterobasidion annosum* in pine forests, and *Agrobacterium radiobacter* is commercially valuable for the control of *A. tumifaciens* (crown gall) in tree nurs-eries (Deacon 1991). Nevertheless, it is the specific match between the environ-mental requirements of the biocontrol agent and the conditions in which pest populations thrive that results in high product efficacy. Such matches appear to be the exception rather than the rule.

Involvement of Many Organisms in "Natural" Biological Control

Soil communities comprise a large variety of microbial species. It is estimated that a gram of natural soil might contain as many as 4,000 or 5,000 "species" with DNA-sequence similarities of less than 70% (Tørsvik 1990). This level of diver-sity is comparable in a variety of soil habitats, but microorganisms from different localities generally show markedly different patterns of species composition. The role of these microbial "species" and of species diversity in ecosystem function is largely unknown, but it is well established that several components of the normal soil flora serve to regulate the activities of pathogens. Plant-parasitic nematode populations in soil are regulated by a large variety of egg parasites, female para-sites, nematode-trapping fungi, bacteria, and possibly viruses (Jatala 1986). Speci-ficity is common, and in suppressive soils it is the combined activities of a con-sortia of antagonists that achieve control. Nematode-trapping fungi—which produce adhesive knobs, adhesive rings, or adhesive hyphal networks—are adapted to "catch" the free-living nematodes. Likewise, the nonmotile spores of *Pasteuria penetrans* (a bacterial parasite of root-knot and cyst nematodes) attach to juvenile nematodes. *Fusarium oxysporum, Catenaria auxiliaris,* and *Nematophtora gynophila* parasitize young females before egg-laying commences, and *Peacilomyces lilacinus, Cylindrocarpon destructans,* and *Verticillium chlamydosporium* are classified as specialized egg parasites. Many other organisms inhibit soil nematode populations in a nonspecific way through toxins, competition, and predation (Jatala 1986). The combined action of the whole community of specialized and nonspecialized organisms is responsible for keeping nematode populations below the economic threshold in nematode-suppressive soils.

There is also considerable species diversity within "functional groupings." Kerry (1988) reported that as many as 150 species of fungi were isolated from eight cyst-nematode species, parasitizing 97% of adult female nematodes in suppressive soils; and Jatala (1986) reported that there are at least 100 species of nematode-trapping fungi. At any time, consortia of antagonists provide nematode suppression. The exact composition or activity of these consortia, however, is determined by spatial and temporal characteristics of the environment (Crump 1987). Different parts of the diverse community of organisms involved in nematode suppression are necessary to provide effective nematode control at any given time during the crop cycle.

Despite many studies that document the combined activities of several antagonists in the control of plant-parasitic nematodes, the commercial drive for biopesticide products has been in the development of solutions based on single microbial species or even single isolates. That approach is based on little understanding of host-parasite population dynamics, and the selection of agents for development is complicated by issues of ease of culture and product shelf-life. Biological control of cyst nematodes provides a good example; *Nematophtora gynophila* and *Pasteuria penetrans* are thought to be important agents of nematode control in natural soil (Davies and others 1992; Kerry and others 1982), but *Verticillium chlamydosporium* is the only organism chosen for commercial development, mainly because it is easy to mass-produce it in vitro.

Beneficial Use of Simple Consortia of Species If Vitro Culture Is Possible

Experimental tests have shown that the fungus *Verticillium chlamydosporium* can reduce plant-parasitic nematode populations in soil by as much as 90% (De Leij and others 1992b) but that this can be achieved only under specific environmental conditions: temperatures must be close to 20°C, appropriate host plants must be available for fungal colonization, and nematode population densities must be low (De Leij and others 1992a,b,c). Thus, *V. chlamydosporium* has a specific "window of opportunity", and its utility as a commercial biopesticide in the field is not large. Furthermore, the high multiplication rate of cyst and root-knot nematodes means that parasitism rates as high as 90% (common in laboratory tests) are insufficient to prevent nematode population increases and economic damage to crops.

Higher levels of nematode control can be achieved with combined application of *V. chlamydosporium* and *Pasteuria penetrans* (an obligate bacterial parasite of nematodes). This approach provides control that is comparable with the use of nematicides and is much more efficacious than the use of either organism in isolation (De Leij and others 1992a). *V. chlamydosporium* is unable to penetrate the plant root and parasitizes only egg masses on the root surface (De Leij and others 1992a); *P. penetrans* also parasitizes nematodes that develop deep inside the root. Even though this simple consortium approach shows promise, it is commercially unattractive because of production constraints. Furthermore, the combination of nematode host specificity of *P. penetrans* and the relative small environmental window of opportunity of *V. chlamydosporium* is commercially unattractive.

Management of the System as a Means of Inducing Biological Control

Agricultural management provides an alternative to biological intervention. Simple practices, such as crop rotation, prevent the buildup of pests and diseases to economically damaging levels. In continuous cropping, initial accumulation of pests and diseases is tolerated to allow populations of natural antagonists to reach levels that provide long-term control. Augmentation of soil with organic manure is widely used to increase nonspecific biological activities that suppress pest and disease populations. Hams and Wilkin (1961), for example, reported that augmentation of soils with farmyard manure or green manure reduced plant damage attributed to plant-parasitic nematodes; the introduction of organic substrates probably promotes general microbial activity that is antagonistic to nematode populations. Similarly, amending soil with organic residues that have relatively high carbon-to-nitrogen ratios can control fusarium root rot; the free nitrogen required by the microbial biomass to degrade these organic amendments leads to insufficient nitrogen availability for pathogen growth (Snyder and others 1959). Others (for example, Park and others 1988) have suggested that induction of fusarium-suppressive soils is a more specific process whereby nonpathogenic *Fusarium oxysporum* isolates interact with siderophore-producing fluorescent pseudomonads to provide conditions that are nonconducive for the pathogen. Lemanceau and co-workers (1992, 1993) showed that pathogenic *F. oxysporum* isolates were more sensitive than nonpathogenic isolates to the iron deficiency induced by *Pseudomonas putida*, and this difference resulted in effective biological control.

In general, disease suppression in soils is attributed to biological processes. Experiments have shown that suppression can be transferred to nonsuppressive soils by adding small quantities of suppressive soil to soils that are conducive to disease (for example, Stirling and Kerry 1983). However, attempts to attribute disease suppression to specific components of the natural microbial community have largely met with failure. Processes and species interactions at the community level, rather than the specific ecosystem services or functions of individual species, are likely to be responsible for disease control and pest control in suppressive soils. Research on and economic exploitation of processes at the community level are therefore potentially rewarding. It is also an environmentally sound, sustainable, and in many situations realistic approach. Suppressive soils need not be only "hunting grounds" for potential biological control products; an understanding of suppressive-soil community ecology is likely to lead to augmentative and manipulative management practices that are of considerable economic benefit.

BIOREMEDIATION

Intrinsic and Augmentative Approaches

The degradative enzymatic capabilities of microorganisms and microbial species assemblages play an important role in the remediation of polluted environments (see Crawford and Crawford 1996 for a review of principles and applications of microorganism exploitations and Lynch and Wiseman 1998 for details of microbial and microbial-product use in ecotoxicant monitoring). Natural communities show considerable potential to recover from small- and even medium-scale pollution effects, and with time biotic and abiotic factors interact to reduce contaminants to nondetectable levels. This is "intrinsic bioremediation", remediation that relies solely on natural processes with little or no intervention (see Ellis and Gorder 1997 for review).

As a commercial technology, the "do nothing" (but monitor) approach does not require investment in physical removal or discharge technology and is easily integrated with other pollution-control and remediation technologies. In many situations, this is a realistic and economical approach to pollution abatement. As a consequence, biodegradation by naturally occurring populations of microorganisms is a major mechanism, for example, in the removal of petroleum from coastal waters (well documented in the Prince William Sound, Alaska, after the Exxon *Valdez* oil spill). Rapid acclimation of the resident microbial population is common after hydrocarbon contamination (for example, Braddock and others 1995) and is evidence of the capacity of the community to respond to pollution. Detailed research has also shown that a 10-fold increase in population size of hydrocarbon degraders can follow substantial petroleum contamination of coastal waters (Atlas 1995a). Furthermore, petroleum degradation rates can be increased (by a factor of 3–5) by enrichment with inorganic fertilizers (Atlas, 1995b: Coffin and others 1997; Pritchard and others 1995). Similarly, in terrestrial environments, there is considerable evidence that natural microbial species assemblages respond to pollution in ways that ameliorate or remove contaminants and that this activity can be enhanced by manipulation of the physiochemical conditions to augment remediation (Liu and Suflita 1993).

Intrinsic cleanup does not require extensive knowledge of the abiotic and biotic processes and interactions by which remediation is achieved. Nevertheless, understanding of "the system" is beneficial where it leads to the ability to enhance decontamination rates by manipulating and controlling environmental conditions or augmentation of biological processes for human benefit.

Complex Interactions, Biological Diversity, and the Exploitation of Intrinsic Bioremediation Processes

Molecular studies have shown that diverse microbial species assemblages (and genes) are involved in the complete catabolism of complex substrates (for example, Vallaeys and others 1995). Metabolic capabilities are often widely dispersed among distinct taxonomic groups and environments (Mueller and others 1994), and the metabolic capabilities of microbial communities as a whole are characterized by considerable functional overlap (Pritchard and others 1995). This could be indicative of "functional redundancy," but it is more likely that patterns of susceptibility and resistance to pollutants interact with metabolic and cometabolic activities to create a mosaic of functions underpinning microbial community integrity in the context of environmental heterogeneity. Furthermore, studies of mixed cultures (enrichments of xenobiotic-degrading microorganisms in liquid culture, for example) have demonstrated the importance of secondary use of substrates in microbial populations. Complex interactions—including the provision of specific cofactors, removal of toxic products, modification of growth rates, cometabolism, and gene transfer—have now been implicated in microbial communities and are important in degradation (Weightman and Slater 1979). Much can be learned from degradation studies of pesticides: the herbicide dalapon, for example, is not readily degraded by any organism, but enzyme products in small and well-structured microbial communities can bring about complete substrate metabolism (Senior and others 1976).

Diversity of metabolic function is undoubtedly important in the ability of microbial communities to achieve bioremediation, but it is also important because of indirect and "cascade" interactions that enable complete degradation. Genetic diversity also underpins community-level responses to environmental change, species compensation, and complementarity (see Frost and others 1995); and genetic diversity is likely to be particularly important in the degradation of substrates in changeable, fluctuating, and perturbed environmental conditions (which are often characteristic of polluted environments).

Difficulty of High-Technology Bioremediation Solutions in Natural Environments

Recombinant-gene technology appears to offer appealing prospects for the design of microorganisms for use in bioremediation. Genetic manipulation has been used to expand the array of substrates that can be used by wild-type microorganisms and to restructure existing metabolic pathways (thereby avoiding the production of deleterious metabolites; see Lui and Suflita 1993). Nevertheless, the "inundative approach"—in which single species, strains, or isolates of bacteria (recombinant or wild-type) are cultured in vitro and released into the environment—has proved difficult for achieving viable populations of pollution degraders in situ. Experimental tests often show that recombinants are likely to be outcompeted by wild-type parental strains (Fleming 1994; Recorbet and others 1992; Vahjen 1997), and even indigenous species often fail to establish in field trials, because of abiotic or biotic factors (de Leij and others 1992c; Kerry and others 1993). Thus, the commercial use of recombinants and wild-type "superstrains" is not likely to be great in the context of bioremediation. As Hamer (1993) has stated, the utility of genetically engineered microorganisms in bioremediation processes is likely to be restricted to specific in situ and ex situ applications because recombinants fail to match the degradative abilities of natural microbial species assemblages (despite the addition of metabolic capabilities) and fastidiousness is likely to preclude their use in all but the most highly protected and controlled environments.

Wilson and Lindow (1993) refer to some 27 experimental releases of recombinant microorganisms in field trials. Much of the work has been done to assess the potential risks posed by releasing genetically modified organisms into the environment, but it is now clear that it remains a major technological challenge to achieve the establishment and persistence of recombinants in natural environments. The genetic and physiological "programming" of microorganisms to achieve viable and controlled phenotypic expression under variable (and largely unpredictable) physiochemical and biotic conditions is no small task (Delorenzo 1994).

High Priority of Ecological Research for Commercial Exploitation of Microorganisms in Bioremediation

Exploiting intrinsic bioremediative processes does not require a full understanding of the biochemical, physiological, and ecological interactions by which pollutant removal (or transformation) is achieved. But it is clear that the better these processes are understood, the easier it will be to improve intrinsic remediation efficiencies. A considerable body of evidence suggests that bioremediation rates in situ are determined by soil, sediment, and substrate chemistry. However, recent studies of pollutant mineralization (of hexadecane, phenanthrene, and naphthalene, for example) show that environmental factors (especially temperature, disturbance, and mixing) are at least as important as purely chemical interactions in determining rates of biodegradation by microorganisms (Sugai and others 1997), emphasizing the need for ecosystem-level study of system interactions in bioremediation processes.

Similarly, the study of interactions between microorganisms and other fauna has high priority. In terrestrial habitats, competition and grazing by microbial predators are thought to be important determinants of soil biodegradation rates (Travis and Rosenberg 1997), but models of microorganism-substrate interactions have yet to include a robust analysis of distribution and dispersal of such interactions in field situations (Dighton and others 1997). As a whole, there is a need for fundamental research to improve understanding of the complex interactions that determine the removal of pollutants by natural communities. Such research might lack the glamour of manipulative genetic technologies (to produce recombinants with "designer" functions), and the approach does not have the same value as "intellectual property," but it is likely to be much more profitable. As Price (1997) concludes in a recent review of bioremediation of marine oil spills, "understanding fundamental microbial ecology is the priority for commercial clean-up technology."

DISCUSSION

Diverse communities are likely to comprise commercially valuable individual species and strains. This is well documented and often cited to support "species conservation" (Wilson 1992). However, diverse microbial species assemblages can act in concert (via complex ecosystem-level interactions) to achieve ecologically and economically valuable processes. Furthermore, it is the innate capacity of microbial assemblages to respond to change that makes the community as a whole valuable as a resource in combating pollutants, plant pests, and diseases. The gene frequency of degradative biochemical pathways appears to be characterized by high levels of "functional overlap". Similarly, shared preferences for specific hosts are well documented among potential biological control agents. This could suggest "redundancy" in the normal community state but have considerable economic value in "variable" natural systems. Natural microbial communities comprise large numbers of unidentified or unculturable genotypes (Ward and others 1990) that can belie ecosystem integrity and contribute "function" after community disturbance. In fact, the nature and scale of microbial ecosystem processes have led some authors to conclude that such concepts as redundancy and species value have little meaning in the context of microbial community ecology (for example, Finlay and others 1997).

We conclude that intrinsic processes and ecosystem interactions can have important value for human society. Often, it is the community function as a whole (not a particular or "valuable" species) that is important. The maintenance of biodiversity itself is important for benefits derived from microbial communities. Microbial diversity needs to be conserved not only for the benefit of individual species and genotypes that function within the community as a whole, but also because microbial "ecosystem services" are often carried out at the community level. Besides development of appropriate environmental-management strategies aimed at preserving and stimulating the activity of naturally occurring communities, monitoring techniques have high priority for research and technology development. Intrinsic cleanup and biocontrol processes must be tracked so that further steps can be taken when natural abatement fails, and sampling data are also likely to improve our fundamental understanding of ecosystem processes.

REFERENCES

Atlas RM. 1995a. Bioremediation of petroleum pollutants. Intl Biodet Bioremed 35:317–27.

Atlas RM. 1995b. Petroleum biodegredation and oil-spill bioremediation. Mar Poll Bull 31:178–82.

Braddock JF, Lindstrom JE, Brown EJ. 1995. Distribution of hydrocarbon-degrading microorganisms in sediments from Prince William Sound, Alaska, following the Exxon Valdez oil spill. Mar Poll Bull 30:125–32.

Chakrabarty AM, Friello DA, Bopp LH. 1978. Transposition of plasmid DNA segments specifying hydrocarbon degradation and their expression in various microorganisms. Proc Nat Acad Sci USA 75:3109–12.

Crawford RL, Crawford DL (eds). 1996. Bioremediation: principles and applications. Cambridge UK: Cambridge Univ Pr. 400p.

Coffin RB, Cifuentes LA, Pritchard RB. 1997. Assimilation of oil-derived carbon and remedial nitrogen applications by intertidal food chains on a contaminated beach in the Prince William Sound. Mar Environ Res 44:27–39.

Crump DH. 1987. Effect of time sampling, method of isolation and age of nematode on the species of fungi isolated from females of Heterodera schachtii and H. avenae. Rev Nématol 10:369–73.

Davies KG, Flynn CA, Laird V, Kerry BR. 1990. The life-cycle, population dynamics, and host specificity of a parasite of Heterodera avenae, similar to Pasteuria penetran. Rev Nématol 13:303–9.

Deacon JW. 1991. Significance of ecology in the development of biological control agents against soil-borne plant pathogens. Biocont Sci Tech 1:5–20.

de Leij FAAM, Davies KG, Kerry BR. 1992a. The use of Verticillium chlamydosporium Goddard and *Pasteuria penetrans* (Thorne) Sayre and Starr alone and in combination to control *Meloidogyne incognita* on tomato plants. Fund Appl Nematol 15:235–42.

de Leij FAAM, Dennehy JA, Kerry BR. 1992b. The effect of temperature and nematode species on interactions between the nematophagous fungus *Verticillium chlamydosporium* and root-knot nematodes (*Meloidogyne* spp). Nematologica 38:65–79.

de Leij FAAM, Kerry BR, Dennehy JA. 1992c. The effectiveness of *Verticillium chlamydosporium* as a biological control agent for *Meloidogyne incognita* in pot tests in three soils and for control of M. *hapla* in a micro-plot test. Nematologica 39:115–26.

Delorenzo V. 1994. Designing microbial systems for gene-expression in the field. TIBTECH. 12:365–71.

Dighton J, Jones HE, Robinson CH, Beckett J. 1997. The role of abiotic factors, cultivation practices, and soil fauna in the dispersal of genetically modified microorganisms in soils. Appl Soil Ecol 5:109–31.

Ellis B, Gorder K. 1997. Intrinsic bioremediation: an economic option for cleaning up contaminated land. Chem Ind 3:95–9.

Finlay BJ, Maberly SC, Cooper JI. 1997. Microbial diversity and ecosystem function. Oikos 80:209–13.

Fleming CL, Leung, KT, Lee H, Trevors JT, Greer CW. 1994. Survival of *Lux-Lac* marked biosurfactant producing *Pseudomonas aeruginosa* UG2L in soil monitored by nonselective plating and PCR. Appl Environ Microbiol 60:1606–13.

Frost TM, Carpenter SR, Ives AR and Kratz TK. 1995. Species compensation and complementarity in ecosystem function. In: Jones CG, Lawton JH (eds). Linking Species and Ecosystems. London UK: Chapman & Hall. p 224–39.

Hamer G. 1993. Bioremediation: a response to gross environmental abuse. TIBTECH 11:317–9.

Hams AF, Wilkin GD. 1961. Observations on the use of predatious fungi for the control of *Heterodera* spp. Ann Appl Biol 49:515–23.

Hokkanen HMT, JM Lynch (eds). 1995. Biological control: benefits and risks. Cambridge UK: Cambridge Univ Pr. 300p.

Jatala P. 1986. Biological control of plant-parasitic nematodes. Ann Rev Phytopath 24:453–89.

Kerry BR. 1988. Fungal parasites of cyst nematodes. Agric Ecosys Environ 24:293–305.

Kerry BR, Crump DH, Mullen LA. 1982. Studies of the cereal cyst nematode *Heterodera avenae*, under continuous cereals, 1975–1978. II. Fungal parasitism of nematode females and eggs. Ann Appl Biol 100:489–9.

Kerry BR, Kirkwood IA, de Leij FAAM, Barba J, Leijdens MB, Brookes PC. 1993. Growth and survival of *Verticillium chlamydosporium* Goddard, a parasite of nematodes, in soil. Biocont Sci Tech 3:355–65.

Lemanceau P, Bakker PAHM, De Kogel WJ, Alabouvette C, Schippers B. 1992. Effect of pseudobactin 358 production by *Pseudomonas putida* WCS358 on suppression of Fusarium wilt of carnations by nonpathogenic *Fusarium oxysporum* Fo47. Appl Environ Microbiol 58:2978–82.

Lemanceau P, Bakker PAHM, De Kogel WJ, Alabouvette C, Schippers B. 1993. Antagonistic effect on nonpathogenic *Fusarium oxysporum* strain Fo47 and pseudobactin 358 upon pathogenic *Fusarium oxysporum* f. sp. *dianthi*. Appl Environ Microbiol 59:74–82.

Liu S, Suflita JM. 1993. Ecology and evolution of microbial populations for bioremediation. TIBTECH 11:344–52.

Lynch JM, Wiseman A (eds). 1988. Environmental biomonitoring: the ecotoxicology biotechnology interface. Cambridge UK: Cambridge Univ Pr. 299 p.

Price RC. 1997. Bioremediation of marine oil spills. TIBTECH 15:158–9.

Mueller JG, Lantz SE, Devereux R, Berg JD, Pritchard 1994. Studies on the microbial ecology of polycyclic aromatic hydrocarbon degradation. In: Hinchel RE, Leeson A, Semprini L, Ong SK (eds). Bioremediation of chlorinated polycyclic aromatic hydrocarbons. Boca Raton FL: Lewis Pr. p 218–30.

Park CS, Paulitz TC, Baker. 1988. Biocontrol of Fusarium wilt of cucumber resulting from interactions between *Pseudomonas putida* and nonpathogenic isolates of *Fusarium oxysporum*. Phytopathology 78:190–4.

Pritchard PH, Mueller JG, Lantz SE, Santavy DL. 1995. The potential importance of biodiversity in environmental biotechnology applications: bioremediation of PAH-contaminated soils and sediments. In: Allsopp D, Colwell RR, Hawksworth DL (eds). Microbial diversity and ecosystem function. Wallingford UK: CABI. 161–82 p.

DE LEIJ, HAY, and LYNCH / 251

Recorbet G, Steinberg C, Faurie G. 1992. Survival in soil of a genetically engineered *Escherichia coli* as related to inoculum density, predation, and competition. FEMS Microbiol Ecol 101:251–60.

Senior E, Bull AT, Slater JH. 1976. Enzyme evolution in a microbial community growing on the herbicide Dalapon. Nature 263:476–9.

Snyder WC, Schroth MN, Christou T. 1959. Effect of plant residues on root rot of bean. Phytopathology 49:755–6.

Stirling GR, Kerry BR. 1983. Antagonists of the cereal cyst-nematode *Heterodera avanae* Woll. in Australian soils. Aus J Agric Anim Husb 23:318–24.

Sugai SF, Lindstrom JE, Braddock JF. 1997. Environmental influences on the microbial degradation rate of Exxon Valdez oil on the shorelines of Prince William Sound, Alaska. Env Sci Tech 31:1564–72.

Tørsvik V, Gøksoyr J, Daae FL. 1990. High diversity of NNA of soil bacteria. Appl Environ Microbiol 56:782–7.

Travis BJ, Rosenberg ND. 1997. Modelling *in situ* bioremediation of TCE at Savannah River: effects of product toxicity on microbial interactions on TCE degradation. Env Sci Tech 31:3093–102.

Vahjen W, Munch JC, Tebbe CC. 1997. Fate of three genetically engineered, biotechnologically important microorganism species in soil: impact of soil properties and intraspecies competition with nonengineered strains. Can J Microbiol 45:827–34.

Vallaeys T, Persello Cartieaux F, Rouard N, Lors C, Laguerre G, Soulas G. 1995. PCR-RFLP analysis of 16S rRNA and tfdB genes reveal a diversity of 2,4-D degraders in soil aggregates. FEMS Microbiol Ecol 24:269–78.

Weightman AJ, Slater JH. 1979. The problem of xenobiotics and recalcitrance. In: Lynch JM, Hobbie JE (eds). Microorganisms in action: concepts and applications in microbial ecology. Oxford UK: Blackwell Scientific. p 322–47.

Wilson, EO. 1992. The diversity of life. London UK: Penguin Pr. 406 p.

Wilson M, Lindow SE. 1993. Release of recombinant microorganisms. Ann Rev Microbiol 47:913–44.

MEANS TO
MEASURE
BIODIVERSITY

CONSERVATION BIOLOGY AND
THE PRESERVATION OF BIODIVERSITY:
AN ASSESSMENT

GARY K. MEFFE

Department of Wildlife Ecology and Conservation, Newins Ziegler 303,
Box 110430, University of Florida, Gainesville, Florida 32611-0430

THE FIELD OF conservation biology has been formally recognized for 10, 15, or 20 or more years, depending on how one identifies its beginning (Ehrenfeld 1970; Soulé and Wilcox 1980; Soulé 1986). Regardless of which date we accept, the field has existed for a short time, yet it has had profound and far-reaching effects on the science and management of biodiversity, effects that are well out of proportion to its youthful existence. These influences, some of which I will discuss here, imply that the development of the field of conservation biology was nearly inevitable and perhaps overdue. It brought together and motivated large numbers of scientists of varied description and inclination to address, in a highly pluralistic manner, the problem of the greatest loss of biological diversity in 65 million years. It continues to do so, with some degree of success, although history must be the final judge of its efficacy. I will discuss the field of conservation biology and its contributions to the preservation of biodiversity, identify its areas of weakness, and suggest directions in which the field should go. Much of this material is opinion— my personal assessment of the field—and little more. It should not be misconstrued as a comprehensive attempt to critically assess the field; that task remains for future analysts.

WHAT IS CONSERVATION BIOLOGY?

I begin with a general (and admittedly superficial) description of the field; more-detailed treatments are available in many other sources. I offer the definition of conservation biology I have used before (Meffe and Carroll 1997):

> An integrative approach to the protection and management of biodiversity that uses appropriate principles and experiences from basic biological fields such as genetics and ecology; from natural resource management fields, such as fisheries and wildlife; and from social sciences, such as anthropology, sociology, philosophy, and economics.

An important aspect of this definition is that conservation biology borrows and synthesizes from many disciplines. It is an amalgamation of the perspectives, data, techniques, and pursuits of many natural and social sciences, all focused on the problem of the loss and protection of biodiversity.

Conservation biology differs from more traditional conservation endeavors, such as fisheries, wildlife biology, forestry, or soil conservation, in at least three ways. First, its origin was strongly academic and theoretical. The field of conservation biology was developed largely by academicians, especially population geneticists and ecologists, who applied their genetic and ecological models to the growing problem of loss of biodiversity. It subsequently was enriched by many other disciplines, in both the natural and social sciences.

Second, the field is rooted in a philosophy of stewardship rather than one of utilitarianism or consumption. The latter has been the basis of traditional resource conservation, that is, conserving resources solely for their economic use and human consumption. This change is reflected in the adoption of very different "guiding lights" in traditional resource management and modern conservation biology: Gifford Pinchot's resource conservation ethic versus Aldo Leopold's evolutionary-ecological land ethic (Callicott 1990).

Third, conservation biology includes significant contributions from nonbiologists in the various social sciences, political sciences, and economics, who join with those in the natural sciences to address our complex problems and develop perspectives and methods. Thus, conservation biology is a broad synthesis of many academic fields, and its purpose is to address the loss and stewardship of biodiversity.

Another important feature of conservation biology is its basis in and recognition of three broad underlying principles (Meffe and Carroll 1997): the inevitability of evolutionary change, the recognition that ecology is dynamic, and the need to take into account the human presence. Conservation biologists recognize that because natural systems are the result of long-term evolutionary change, they will continue to evolve. To protect the status quo, as in a museum, would be a mistake, because systems must continue to evolve. Another mistake is to not understand the evolutionary processes that led to the characteristics of a species when we are attempting to protect or recover it. Likewise, natural systems are dynamic on shorter, ecological time scales, and conservation biologists recognize that natural disturbances are critical to the integrity of ecological systems. The "balance-of-nature" paradigm has been usurped by a "flux-of-nature" viewpoint (Pickett and others 1992).

Finally, conservation biologists recognize that it would be hopelessly naive to ignore humans in the conservation equation or to focus our attention solely on highly natural or pristine systems and lock them away from humanity. In fact, the growing human population is the primary motivation for and the reason we need the field of conservation biology, and it must be considered at all times. The goal of conservation biology, then, is to understand and meld all three of these foci to help establish an ecologically sustainable world.

WHAT HAS CONSERVATION BIOLOGY CONTRIBUTED TO THE PROTECTION OF BIODIVERSITY?

I begin this section with a caveat: Although I will claim a great many advances by the field of conservation biology, I do not mean to imply that they all are the result of this field exclusively, that the field has any unique or singular claim to them, or that they would not or could not have developed otherwise. However, I do believe that conservation biology has played an important role in each of these advances.

The major contributions of conservation biology to the protection of biodiversity that I will discuss are of three kinds: new ideas and syntheses, galvanization and reform of natural-resource management, and inspiration for new and related disciplines and current natural-resource practitioners.

New Ideas and Syntheses

First and foremost, conservation biology has provided a formal, global recognition of biodiversity—what it is and what we are losing (Myers 1992; Wilson and Peter 1988). The world's attention to this crisis and our subsequent modes of dealing with it have been guided largely by this field. In defining biodiversity, we have argued with various degrees of success that biodiversity is much more than richness of species, that it ranges from genes to landscapes and includes the various processes that occur as a function of that diversity. The field has helped to define what we have, what we are losing, and how we deal with it. It is best organized around a triumvirate of *composition* (what is there), *structure* (how it is distributed in space and time), and *function* (what it does) rather than merely around counts of species (Noss 1990). We have learned that if we are to preserve biodiversity successfully, we must deal with natural complexities at multiple levels and configurations.

Next, conservation biology has acted to coalesce many scientific issues under one roof as a metadiscipline. Such issues include genetics, biogeography (including the practical application of island-biogeography theory to rates of loss of biodiversity), population ecology and dynamics, community and ecosystem ecology, evolutionary biology, landscape ecology, and numerous social-science and human dimensions. We consistently draw on these and other disciplines to address the complex interdisciplinary issues that confront us. Conservation biology also recognizes the critical importance of habitat fragmentation and edge effects in losses of biodiversity. It promotes the concept that the quality and spatial configuration of habitat is at least as important to the protection of biodiversity as the total amount of habitat available. Work on metapopulations, spatially explicit models, and the highly practical tool of population viability analysis, also developed by conservation biologists, are related to the spatial considerations of habitat fragmentation. All these address the various problems of persistence that face real populations on real landscapes.

Conservation biology has advanced considerably the serious recognition of the potentially disastrous effects of exotic species on native species and ecosystems. The influence of nonindigenous plants and animals has become a major focus in the protection of biodiversity as we have learned how such invaders not only can affect the richness of species, but also can change ecosystem functions.

Finally, conservation biology has incorporated values and ethics into its science. It clearly is a value-laden science, with a strong value base that freely recognizes that protection of biodiversity is good and necessary, not only for the benefit of humanity, but also for its own inherent good.

Galvanization and Reform of Natural Resource-Management

The second major contribution of conservation biology is that it has acted (whether intentionally or not) to galvanize and reform natural-resource management in two important ways. First, it caused some staid and conservative disciplines—such as traditional fisheries, wildlife, and forestry—to take notice of ideas, controversies, and approaches that had been simmering under the surface for some time. In fact, the clinging to tradition by these fields may have helped to spawn conservation biology, because individuals who were unhappy with the status quo searched for and developed a new discipline that offered an alternative to traditional, consumption-oriented approaches to natural-resource management. As a result, these other disciplines also seem to be moving forward as they embrace the concepts of conservation biology and make them work for natural-resource management. One need only scan the recent pages of such journals as *Fisheries* and *The Wildlife Society Bulletin* to see the influence of the last decade of conservation biology.

Conservation biology also has acted in the opposite direction, of bringing ecology and evolution out of the "pure" realm of the ivory tower and applying them to the problems of the day. Many "pure" researchers, who formerly would not dirty their hands with applied problems, now are applying what they know to real landscapes and real issues, thus enriching those endeavors. This new interplay between pure and applied research, with the breakdown of barriers between them, is possibly one of the healthiest and most positive benefits of the development of conservation biology.

Second, conservation biology has changed tangibly management practices as they actually occur on the landscape. In retrospect, the old practices were too scattered and, in many cases, had insufficient scientific justification to have lasted much longer, and their failure may have contributed to the development of conservation biology as a field. As the many pathologies (*sensu* Holling 1995, Holling and Meffe 1996) of natural-resource management became apparent, new approaches were needed and developed. This has been manifested in several ways:

• challenging and changing natural resource management practices by federal and state agencies to incorporate and accommodate the various principles promulgated by conservation biology, which now is influencing, through its science and philosophy, how we treat our resources;

• moving away from simple command-and-control approaches to management, which repeatedly have been shown to fail ultimately, toward understanding how nature operates and working within those "rules" (Knight and Meffe 1997);

• developing a greater appreciation for and understanding of uncertainty in environmental management and policy and incorporating that uncertainty into management practices. Recognition of the many natural and human sources of uncertainty has led to multiple calls for adaptive management (Gunderson and others 1995; Holling 1978; Walters 1986), which management agencies are starting to heed and embrace; and

• incorporating natural patterns of variation, such as disturbance regimes, into management. This includes such activities as reinstituting fire in appropriate ecosystems, leaving storm debris on forest floors, and mimicking a natural flood in the Grand Canyon, all designed to incorporate natural processes into management.

In addition to these changes, we are seeing environmental activists working with scientists (or with science) in their calls for policy reform. Numerous activist organizations now routinely incorporate conservation biology into their activities, a step that represents a convergence of science and activism toward the common goal of science-based policy. In sum, natural-resource managers the world over now are relying increasingly on the findings and principles of conservation biology for direction. In the United States, federal and state agencies alike are retooling, using conservation biology as a guide.

Inspiration for New Ideas, Disciplines, and Organizations

New ideas, disciplines, and organizations have been inspired by conservation biology, and a new generation of practitioners is undergoing intellectual development and professional training in this new environment. For example, the idea that cross-boundary issues are critical is now a common point of discussion among natural-resource management agencies and private landowners, whereas 10 years ago, political boundaries on a map seemed real and impermeable. Stepping back to view the larger landscape and cooperate with other land tenants, rather than hiding behind the seemingly comfortable and protective boundaries set by legal documents, is becoming a way of life rather than an unusual behavior. In general, such notions of cooperation for a common good rather than of confrontation or competition, are becoming prevalent.

New disciplines have been defined or developed further as a result of progress within conservation biology. For example, restoration ecology, landscape ecology, environmental ethics, and ecological economics all have begun to flourish as important components of conservation biology. Surely they existed beforehand, and they may have developed independently, but conservation biology seems to have been and continues to be the overarching catalyst that supported and promoted their advancement. The metadiscipline of conservation biology is the glue that binds these and other disciplines into a coherent and focused package.

An obvious but extraordinarily important catalyst for the field was development of a major international society, the Society for Conservation Biology, and its journal, *Conservation Biology,* as the focal points for intellectual activities in the field. The effects and influences of this society and journal are virtually incalculable within the academic and applied communities of conservation scientists. They help to identify and define the field and offer an intellectual home to its practitioners.

Closely related to all these factors, and ultimately feeding the further development of conservation biology, are the many courses and degree programs in conservation biology that are developing in colleges and universities around North America and the rest of the world, as well as several college textbooks that are designed specifically for use in these courses. For the first time in history, the field has reached the point at which we are formally educating a new generation of students as conservation biologists, in contrast with the founding generation, who came to the field from various specialized disciplines. These students have been inspired by the challenges and opportunities involved in the protection of biodiversity, which seems to have given greater meaning to basic programs in ecology.

Finally, training courses have developed in various natural-resource management agencies to bring the practitioners up to speed on such topics as general tenets of conservation biology, ecosystem management, and various human dimensions. My experiences as a trainer in some of these courses tells me that as a result of this reorientation natural-resource management in the United States will never be the same.

GAPS AND PROBLEMS

Although conservation biology has been considered a rousing success by most measures, it has its problems, it has experienced growing pains, and it still has some way to go to be considered a mature discipline. A useful analogy is human ontogeny. Conservation biology was born rapidly, with typical pains and shakiness; it grew quickly, feeling its way along, learning first how to walk, then to run; it became an awkward adolescent; and now it is emerging into confident maturity as a young adult. It has not reached its full potential yet, and it has a great deal to learn before it has its full effect on the world, but its future looks bright and exciting. However, hurdles must be overcome, and I present several of them here.

First, I believe the field's main problem is that it means very little globally, compared with many other human endeavors; conservation biology certainly is not yet a household term that most people can identify. Society at large does not realize what conservation biologists have to offer or the relevance of conservation biology to their lives, other than in a vague connection to a general concern for the environment. Conservation biologists have not done a good job of positioning the field to be a globally effective agent of social change.

Part of the problem is that society has not defined its environmental problems broadly enough to address them adequately. Many individuals seem to associate environmental problems with the need to recycle, with possible global climatic change, with harm to individual animals, with the problem of toxins in the air, water, and soil, or with other issues related to human health. As important as these problems are, others—such as major losses of biodiversity and their ramifications, collapse of ecosystem services, and destruction and fragmentation of habitat (apart from tropical deforestation, which much of the public recognizes)—do not seem to resonate as major environmental issues or issues that hold much threat for or relevance to humanity. Many people do not seem to make connections between the development of strip malls or golf courses, growth of population, loss of soils, withdrawal of water, and related activities and their influences on biodiversity, sustainability, human health, or social structure. In essence, I do not believe that society at large appreciates what really supports human populations or why desertification, logging of old-growth forests, and mass extinction of species are critically important to all peoples.

Second, the field of conservation biology developed with a largely terrestrial bias, which it retains. Consequently, it has lagged in addressing problems in freshwater aquatic systems and, especially, marine systems (Irish and Norse 1996). Recent attention to the marine realm, including major marine symposia at the meeting of the Society for Conservation Biology in 1997, seems to be addressing that problem.

Third, in my opinion, conservation biology is still too academic: it clings to its roots in academe and seems fearful of venturing too far into unknown territories. I believe that conservation biologists need to be more pragmatic and more practical, and the field needs more relevance to immediate problems of the day if we are to have a greater influence on the protection and recovery of biodiversity. To do this, we must dare to leave the comfort of the academic womb and take greater risks in the real world of conservation action.

Fourth, the nature of the university system itself, at least in the United States, has done little to foster risk-taking and creativity and much to promote conservatism and the status quo. With its high disciplinary walls (Meffe 1998), adherence to tradition, and rewards for conformity, academe not only frustrates progress in a new discipline, such as conservation biology, but does little to address the major environmental and social problems of the day (Orr 1994). Rather than playing a leadership role in cutting-edge ideas, universities often seem to lag behind, restricting such activities and rewarding those which bring in large sums of money for low-risk work. Much of the activity in conservation biology is taking place outside universities, in resource-management agencies, advocacy groups, and even resource-extraction industries.

WHERE DOES THE FIELD NEED TO GO?

I think the field should move in several directions and be strengthened in some areas so that conservation biology can develop further as a discipline and, more importantly, be able to influence society with more scientifically based decision-making.

• Conservation biology needs greater synthesis with other disciplines or sub-disciplines—such as restoration ecology, design (broadly defined to include all human-made products and endeavors), and ecological economics—and with various human dimensions such as sociology, psychology, and anthropology. Conservation biology has something to contribute to all these, and vice versa. Greater communication across fields, some where conversations possibly have never occurred, can help to promote problem-solving in the broadest sense.

• Conservation biologists need to do a better job of teaching about the connection between the overall ecological condition and individual human-health or social conditions. Many times, the arguments we muster for the protection of biodiversity, although compelling to scientists, do not resonate with average citizens who are just trying to make a living. In addition to the various moral arguments we typically use to justify our concerns, we need to do a better job in making it clear that functional natural ecosystems are necessary for workable human social systems and the health and vigor of all humankind. Conservation biology is concerned not just with nature, but very much with humanity as well.

• We also need to take the lead in modifying educational curricula—from kindergarten through graduate level—to reflect better the central importance of an ecological perspective in society. The primary task will be to break down the artificial disciplinary boundaries that have haunted education for centuries, to overcome departmental territorialities, and to cease the extreme specialization that so often results in narrow technical training rather than a broad education that can lead one to understand the interrelationships in complex problems and begin to address them. We need to stop teaching as though mathemathics, sociology, biology, engineering, history, and literature are unrelated. We need to do a better job of teaching the full diversity of the human experience and of centering it on functioning ecosystems that make the planet livable for all species, including humans.

• Conservation biology has a golden opportunity to join with many and varied religious interests that focus on environmental awareness and protection of life on Earth. For example, so-called "green evangelicals" fervently recognize and understand the importance of protecting biodiversity, although the term they use is different ("God's creations"). Seeing all life as the result of a single event of creation and interpreting Biblical writings on dominion as a responsibility for stewardship rather than a license for domination and control of nature, this perspective can be valuable beyond measure, reaching large numbers of people who otherwise might not identify with "biodiversity" or care much about it from a scientific perspective. Harnessing the energy of religious perspectives concerned with guardianship of creation can be a powerful boost to protection of biodiversity.

• Most important, I think, we need to do a better job of incorporating what we know into effective public policy. We need to make our science *work*; we need to put it to daily use. It is time for conservation biology to move to a new plateau in society, to make our presence known, our science relevant, and our views sought and respected. Ideally, the public should listen to what conservation biologists have to say with as much anticipation, concern, and enthusiasm as they have for daily stock-market reports, economic forecasts, or news about medical advances.

CONCLUSIONS

I believe the science of conservation biology is in an extremely active, turbulent, and exciting period of development right now. The last 15 years have seen dramatic changes in conservation priorities, techniques, philosophies, and approaches. Now is when the science is being molded, when the approaches to the enormous challenges to humanity are being mapped out, and when the future of biodiversity and humanity largely are being determined. This is a thrilling, frightening, and wonderful time to be practicing conservation science, one that I hope we can look back on with pride and satisfaction. Conservation biology has taken huge strides in the effort to protect biodiversity, but these still are only the initial, cautious steps of a long and never-ending journey; we have much yet to learn and accomplish.

REFERENCES

Callicott JB. 1990. Whither conservation ethics? Cons Biol 4:15–20.

Ehrenfeld DW. 1970. Biological conservation. New York NY: Holt, Rinehart and Winston.

Gunderson LH, Holling CS, Light SS (eds). 1995. Barriers and bridges to the renewal of ecosystems and institutions. New York NY:Columbia Univ Pr.

Holling CS. 1978. Adaptive environmental assessment and management. New York NY: J Wiley.

Holling CS. 1995. What barriers? What bridges? In: Gunderson LH, Holling CS, Light SS (eds). Barriers and bridges to the renewal of ecosystems and institutions. New York NY: Columbia Univ Pr. p 3–34

Holling CS, Meffe GK. 1996. Command and control and the pathology of natural resource management. Cons Biol 10:328–37.

Irish KE, Norse EA. 1996. Scant emphasis on marine biodiversity. Cons Biol 10:680.

Knight RL, Meffe GK. 1997. Ecosystem management: agency liberation from command and control. Wildl Soc Bull 25:676–8.

Meffe GK. 1998. Softening the boundaries (editorial). Cons Biol 12:259–60.

Meffe GK, Carroll CR. 1997. Principles of conservation biology, second edition. Sunderland MA: Sinauer.

Myers N. 1992. The primary source. New York NY: WW Norton.

Noss RF. 1990. Indicators for monitoring biodiversity: a hierarchical approach. Cons Biol 4:355–64.

Orr DW. 1994. Earth in mind: on education, environment, and the human prospect. Washington DC: Island Pr.

Pickett STA, Parker VT, Fiedler PL. 1992. The new paradigm in ecology: implications for conservation biology above the species level. In: Fiedler PL, Jain SK (eds). Conservation biology: the theory and practice of nature, conservation, preservation, and management. New York NY: Chapman & Hall. p 65–88.

Soulé ME, Wilcox BA. 1980. Conservation biology: an evolutionary-ecological approach. Sunderland MA: Sinauer.

Soulé ME (ed). 1986. Conservation biology: the science of scarcity and diversity. Sunderland MA: Sinauer.

Walters CJ. 1986. Adaptive management of renewable resources. New York NY: McGraw-Hill.

Wilson EO, Peter FM (eds). 1988. Biodiversity. Washington DC: National Acad Pr.

CONSERVATION GENETICS:
APPLYING MOLECULAR METHODS
TO MAXIMIZE THE CONSERVATION OF
TAXONOMIC AND GENETIC DIVERSITY

DON J. MELNICK*
JUAN CARLOS MORALES
Center for Environmental Research and Conservation, Columbia University,
1200 Amsterdam Avenue, New York, NY 10027-5557
*Departments of Anthropology and Biological Sciences, Columbia University, New York, NY 10027
RODNEY L. HONEYCUTT
Department of Wildlife and Fisheries Sciences,
Texas A&M University, College Station, TX 77843

CONSERVATION BIOLOGY is an applied science that involves direct human intervention into the management of diminishing natural resources. However, unlike traditional resource management, the focus of conservation biology is not necessarily driven by direct economic incentive or the desire to manage a resource for the sake of harvesting it. Instead, the primary goal of this field is to stop the downward spiral of loss of biological diversity by mitigating factors that erode the biological integrity of intact ecosystems and the long-term evolutionary viability of populations, species, and communities of organisms. In this sense, conservation biologists attempt to manage biodiversity on two time scales, the ecological (present) and the geological (future), but management itself is inevitable.

By definition, conservation biology must be multidisciplinary, requiring an integration of many areas of biology, including biogeography, systematics, plant and animal ecology, reproductive biology and physiology, range and wildlife management, environmental toxicology, population biology, genetics, and molecular biology. Moreover, the coordination or planning of any conservation effort also involves issues outside the realm of biology, because most environmental-conservation solutions are compromises between the biological requirements of a natural system and the socioeconomic and political realities of the human populations that are associated with that system. Therefore, the conservation of biodiversity must strike a balance between the needs of a growing human population and the viability of biological systems in the face of a rapidly changing environment.

The act of preserving our natural or biological resources, as in so many action-oriented fields, can be distilled into five simple questions: why, what, where, how, and who? These questions are simple, but their answers have proved vexing and indeed have generated some intense debates. The answer to the first and perhaps most important societal issue, *why*, has been framed in ways that range from the economics of ecosystem services to the psychological and cultural value of intact ecosystems and species to our moral and ethical obligations to pass on to our children the natural world in roughly the same condition in which we found it (Costanza and others 1997; Kellert and Wilson 1993; Pimm 1997). A detailed discussion of these issues is outside the scope of this paper; we will assume that the reader will find justification elsewhere for *why* we should conserve our natural resources.

Assuming we should develop a rational means to describe and conserve the world's biological diversity, we must rely on some scientific systems of measurement and theory to address the four remaining critical questions. Over the last 10 years, conservation genetics has emerged as a subfield of biological conservation that offers an objective approach to three of these questions: what, where, and how. Conservation genetics is more a focus than a field of study, but it has at its root the application of molecular and quantitative genetic methods to the preservation of genetic, species, and ecosystem diversity. Genetics can be applied to these issues in numerous ways, but the term conservation genetics usually refers specifically to molecular genetic techniques that help to

- identify evolutionarily distinct groups of organisms (for example, populations or species) that are worthy of separate conservation efforts (that is, conservation units) (Avise 1996; Moritz 1994, 1995; O'Brien 1996);
- define specific geographic regions that harbor genetically distinct populations, and/or species (that is, regions of genetic endemism) (Avise 1996; Riseberg and Swensen 1996; Templeton and Georgladis 1996; Vane-Wright and others 1994; Williams and Humphries 1994; Witting and others 1994); and
- estimate the distribution of genetic diversity within and among conservation and management units to develop plans that will conserve the greatest amount of that diversity and the evolutionary potential it offers (Burgman and others 1993; Caughley 1994; Frankham 1995; O'Brien 1994).

Researchers have used genetic approaches to address a variety of conservation problems in plants and animals found on many continents. This genetic research has involved diverse laboratory procedures and approaches to data analysis; the results have provided critical information for wildlife managers and environmental policy-makers. In the following sections, we present a brief overview of both the methods used in conservation genetics and some empirical studies that underscore the value of genetic analysis in conservation management.

MOLECULAR SYSTEMATICS—
WHAT ARE THE CONSERVATION UNITS?

A critical first step in designing appropriate conservation measures is properly defining and identifying the group one wishes to conserve, the so-called conservation unit (CU). In other words, we need to define *what* we wish to conserve before we can take measures to conserve it. Although this may seem trivial, the evolutionary process is often murky enough to lead to great difficulty in defining CUs, particularly when we are dealing with closely related species that have been thrown together recently by human-induced changes in the environment or those which were never isolated fully from one another reproductively (that is, hybridization has occurred). In these cases, careful examination of the biological characteristics of an organism that are most likely to carry evolutionary historical information (a field known as biological systematics)—such as features of anatomy, behavior, and genetics—often yields the clues necessary to place a series of populations and species on a synthetic family tree (also known as a phylogenetic tree). These clues even can be used to determine whether a group of organisms can be defined as a single evolutionarily significant unit (ESU), which may be but is not necessarily synonymous with what we usually refer to as a species.

Molecular-genetic approaches to biological systematics have emerged as one of the most exciting new areas of biological research (Hillis and others 1996). A wide array of technical and analytical methods has been used to address issues of evolution and conservation at all levels of organization, ranging from genes within populations through the process of speciation to the reconstruction of the tree of life itself (Avise 1994). Of particular importance here are the contributions of molecular techniques to the identification and phylogenetic placement of rare and endangered species. Knowledge about diversity at the molecular level can be used to reconstruct the evolutionary history of an endangered organism (Avise and Hamrick 1996) and to identify the ESUs on which to focus our attention for conservation (Moritz 1994, 1995). Because much of conservation planning depends on taxonomic or species assignments (Taberlet 1996), identifying systematically based CUs aids considerably in developing management plans and in evaluating priorities for conservation (Smith and Wayne 1996).

Case Studies

Molecular-systematic studies can help clarify taxonomic issues at three different levels. First, we can identify cases of "oversplitting", that is, when distinct morphological forms are considered different evolutionary entities but are in fact genetically indistinguishable. This implies that gene flow still may be occurring between the different forms and that they therefore should not be considered evolutionarily distinct. For example, the now-extinct dusky seaside sparrow (*Ammodramus maritimus nigrescens*) of Florida, originally described as a distinct species, was redefined later as a subspecies when it was shown to be genetically indistinguishable from other populations of seaside sparrows (Avise and Nelson 1989). Moreover, it was shown that all populations of seaside sparrows on the Atlantic coast (including the dusky seaside sparrow) are genetically more similar to each other than they are to populations of seaside sparrows that are found along the Gulf of Mexico. Clearly, molecular data supported the inclusion of the dusky seaside sparrow in the seaside-sparrow species and suggested that its loss, although regrettable, had little or no effect on the long-term evolutionary course of the entire species.

Second, molecular systematics can help us distinguish between forms that are morphologically similar but are in fact ancient, unrelated lineages with little or no gene flow between them. One example is Darwin's fox (*Dusicyon fulvipes*) on Chiloé Island in Chile. Some scientists had considered Darwin's fox a small race of the common South American grey-fox species *Dusicyon griseus*, on the basis of morphological similarities. Chiloé island is only 5 km off the coast and likely was connected to the mainland during the last glaciation (about 13,000 years ago), which would have created opportunities for gene flow between Darwin's and grey foxes. However, genetic analyses of Darwin's fox and other South American fox species suggest that Darwin's fox is at least as divergent from the grey fox as the grey fox is from another well-recognized fox species, the culpeo fox (*Dusicyon culpaeus*), and that Darwin's fox probably evolved from the first immigrant foxes into South America 2–3 million years ago. Recently, a small population of Darwin's foxes was found on mainland Chile, and they were shown to be quite genetically divergent from the grey fox but closely related to the population on Chiloé island. This suggests little or no present or historical gene flow between Darwin's and grey foxes, and it supports the distinctiveness of Darwin's fox as a separate species (Wayne 1996).

Third, systematic analyses of genetic characters can provide an objective means of identifying evolutionarily distinct lineages among closely related groups. The Iberian lynx, *Lynx pardinus*, is considered to be the most vulnerable cat in the world. Its remaining populations are highly fragmented and of limited size. The species status of the Iberian lynx is complicated: Some consider it to be a geographic variant of the Eurasian lynx, *Lynx lynx*, and others consider it to be a distinct species. Because the taxonomic status of the Iberian lynx is important to the establishment of an effective management plan for lynxes in general, a molecular-systematic study was conducted recently (Beltran and others 1996). The results of this study revealed a close relationship between the Canadian lynx (*Lynx canadensis*) and the Eurasian lynx, but the Iberian lynx is evolutionarily more distinct. Thus, these molecular data give validity to the concept that the Iberian lynx is a phylogetically distinct species that deserves separate consideration for conservation.

MOLECULAR PHYLOGEOGRAPHY—
WHERE DO THOSE CONSERVATION UNITS RESIDE?

Once we decide what groupings of organisms are distinct and worthy of separate efforts at conservation (that is, we identify our CUs), it becomes critical to determine the geographic location of important subsets of individuals within each CU. In other words, *where* will we focus our conservation efforts to preserve a CU or species?

The use of molecular systematics in a geographic context can contribute to answers to this question in two ways. First, detailed studies of intraspecific (within-species) variation can identify the geographic limits of either a CU or what Moritz (1994) calls a management unit (MU). Second, patterns of intraspecific phylogenies of unrelated groups of organisms may assist in identifying geographic regions whose populations and species have had a shared, unique evolutionary history, thus allowing for the conservation of communities of organisms that have high levels of genetic endemism or uniqueness.

Intraspecific Phylogeography

The term *intraspecific phylogeography* denotes the connection between biological systematics, population genetics, and biogeography, the study of the distribution of organisms in geographic space and the factors that led to that distribution (Avise and others 1987). In principle, any biological characteristic can be used for this purpose, but *intraspecific phylogeography* now mostly is associated with the study of molecular markers, especially mitochondrial DNA (in animals) or chloroplast DNA (in plants). By determining the detailed genetic and evolutionary relationships of populations within a species (or CU), and superimposing that intraspecific molecular phylogeny on a geographic map, one can infer the processes that historically determined the current distribution of organisms. One also can use this approach to identify the geographic location of genetically distinct populations (that is, populations that substantially differ from one another by the frequency of genetic traits rather than by the presence or absence of those traits) or MUs, which might deserve special attention if specific conservation measures become necessary to preserve a given species. The identification of specific MUs and their geographic location currently has one of the highest priorities in most efforts that use molecular markers for conservation purposes, and intraspecific phylogeography provides a theoretical framework to accomplish this.

Case Study. Among the animal species currently listed by the International Union for the Conservation of Nature (IUCN), the Convention on International Trade in Endangered Species (CITES), and the US Department of the Interior as endangered, the Sumatran rhinoceros (*Dicerorhinus sumatrensis*) is one in greatest need of special attention and immediate wild-population (or in situ) management. Historically, this species inhabited most of the Indochinese peninsula, from Burma (Myanmar) to Vietnam, and south to Malaysia and the islands of Sumatra and Borneo. Destruction of habitat and hunting have led to a rapid decline of this species over the last 2 decades. Only a few confirmed populations remain on peninsular Malaysia, Borneo, and Sumatra. Because of the dire situation of this species, translocation programs have been proposed that would move individuals that are scattered among fragments of unsustainable forest and concentrate them in protected zones of natural habitat (Foose and van Strien 1995). However, it is important to remember that the objectives of any conservation effort should be not only to maintain a collection of organisms, but also to preserve the maximal amount of existing genetic variability within a species and to maintain the evolutionary historical integrity of its wild populations.

Geographic mapping of the distribution of mitochondrial-DNA (mtDNA) variants among Sumatran rhinoceros populations (Morales and others 1997), using both molecular-systematic and population-genetic methods, reveals two phylogeographic features that are important to the conservation of the Sumatran rhinoceros. First, a phylogenetic tree of mtDNA haplotypes, overlaid on the distribution of the Sumatran rhinoceros, suggests that the population in Borneo possesses a unique mtDNA variant that is not shared with other Sumatran rhinoceroses, indicating a long history of isolation from the remaining Sumatran and peninsular Malaysian populations. Therefore, the population in Borneo should be considered a separate ESU. Second, a geographically referenced population-genetic analysis suggests that the populations outside Borneo can be divided into two groupings or MUs—west Sumatra, and east Sumatra and Malaya—on the basis of significant differences in the frequency of mtDNA variants and the restriction of gene flow that they imply. Thus, translocation and other conservation efforts should take these three distinct units (ESUs or MUs) into consideration and try to maintain the evolutionary and genetic integrity of each unit.

Regional Phylogeography

More recently, and closely related to the common use of intraspecific phylogeography, efforts have been made to map intraspecific phylogeographic patterns simultaneously among a wide variety of species that occupy overlapping geographic ranges. This has been done to identify regions that harbor populations or species that are consistently genetically distinct from other populations within their species (so-called conspecifics) or other closely related (sister) species within their genus (so-called congenerics). These regions of genetic uniqueness or genetic endemism can be used to design reserves and other mechanisms of conservation (Avise and Hamrick 1996; Templeton and Georgladis 1996; Williams and Humphries 1994; Witting and others 1994). They also provide an effective shortcut to making decisions about conservation at the levels of species and community because it would be impossible to conduct individual genetic surveys of the hundreds of thousands of species in a particular region. Thus, a consistent pattern of regional genetic uniqueness across a diverse but logistically feasible number of species (including fungi, plants, invertebrates, and vertebrates) would allow one to assume reasonably that most populations or species within that particular region, the large majority of which will not have been analyzed, are genetically unique.

Case Studies. One compilation of several studies of invertebrate and vertebrate animals of the southeastern United States found major patterns of molecular phylogeographic congruence among populations of coastally distributed species (Avise 1996). These patterns were shared among varied groups of organisms, including horseshoe crabs, American oysters, diamondback terrapins, ribbed mussels, seaside sparrows, toadfish, black sea bass, and tiger beetles. Even species that have a greater ability to disperse, like white-tailed deer, showed a similar pattern, suggesting population differentiation in this region in response to a persistent set of historical biogeographic processes (Ellsworth and others 1994). The pattern revealed by most species indicates major molecular phylogenetic discontinuities between the Gulf of Mexico and the Atlantic coastline of the southeastern United States, whereas some patterns, like that seen for deer, indicate the uniqueness of the populations in southern Florida. Together, these findings suggest that maritime and other species in this region may have been subjected to the same biogeographic influences and thus share a common biogeographic history. Although some exceptions to the common molecular-phylogeographic pattern exist, the evidence is strong that at least among animal species, populations on either side of southern Florida are likely to be genetically distinct from one another. The conservation of any one of those species must incorporate efforts on both sides of this important biogeographic divide.

POPULATION GENETICS—HOW WILL WE MANAGE GENETIC DIVERSITY WITHIN EACH CONSERVATION UNIT?

Once we have determined *what* groupings of organisms (CUs) are distinct and worthy of separate efforts at conservation, and *where* the genetically unique populations of those units (MUs) or larger regions of general genetic uniqueness (regions of genetic endemism) are, it is critical for us to devise the means to conserve those species individually or regionally. In other words, *how* will we conserve each species's populations and their underlying genetic diversity well into the future?

Population genetics offers a key perspective on this issue because most critical evolutionary events occur at the level of the population. The potential rate of evolution depends on the amount of genetic diversity in a population; processes that erode levels of genetic diversity or increase the occurrence of deleterious combinations of genes (such as inbreeding) within populations limit the rate and scope of potential evolutionary changes in those populations to meet environmental challenges (Templeton and others 1990). Furthermore, biologists agree that levels of genetic diversity within individuals may confer important advantages of fitness on those individuals (Allendorf and Leary 1986). Thus, a fundamental concern of conservation biologists is to preserve genetic diversity in populations and species and the resulting evolutionary potential. The field of population genetics plays a critical role in determining how that diversity is distributed and how best to preserve it.

Population Genetic Structure

A species's genetic diversity can be distributed in various ways, depending on historical ecological, geological, and human-induced events, as well as on the current patterns of geographic distribution, individual dispersal, social organization, ecological adaptation, demographic transition (births, deaths, and generation length and overlap), and genetic migration (the flow of genes across a landscape). How one configures a conservation-management strategy to encompass the individuals and populations that are necessary to capture the greatest amount of a species's genetic diversity will be derived largely from knowledge of the existing distribution of that diversity across the species's range, otherwise known as the population genetic structure of a species. Information of this sort is essential for conservation planning but is often difficult to obtain.

Case Study. Among Asian macaque monkeys are species that have extensive geographic distributions, either contiguous through the mainland, like the rhesus monkey (*Macaca mulatta*), or fragmented, like the long-tailed macaque (*Macaca fascicularis*), whose distribution includes part of mainland Asia and many islands in the Malay Archipelago and the Philippines. Melnick (1988) and Melnick and Hoelzer (1992) have shown that 91% of the nuclear genetic variation in the rhesus monkey can be attributed to variation within a geographic region; in the more fragmented long-tailed macaque, this figure is reduced to 67%. In other words, nearly 4 times as much of a species's genetic variation can be attributed to differences between regions in the more fragmented long-tailed macaque as in the more contiguously distributed rhesus monkey. This pattern also holds true for species that have a much more restricted distribution, like the toque macaque (*Macaca sinica*) of Sri Lanka and the Japanese macaque (*Macaca fuscata*). In the toque macaque, which exists on only one island, only 3% of the species variation can be attributed to differences between regions, whereas in the Japanese macaque, which exists on a number of islands in Japan, that figure increases to 24%.

What does all this mean in terms of conservation? Very simply, the greater the percentage of overall genetic variation in a species that can be attributed to differences between populations, the greater the number of populations that must be included in a conservation-management plan that seeks to preserve some maximal level (for example, 90%) of existing genetic diversity. The Japanese macaque is considered an endangered species by the IUCN; thus, given its current population genetic structure, efforts to conserve this species must include a broad geographic representation of different island populations to maximize the genetic diversity to preserve. If the toque macaque ever shares a similar fate in Sri Lanka, reserves that harbor only a small number of sufficiently large populations likely will capture most of the species's existing variation.

Metapopulation Management

As human populations continue to grow, landscapes that are fragmented by human activities are becoming the predominant arena within which demographic and evolutionary processes in terrestrial plants and animals occur. Nevertheless, the effects of human-induced changes in the landscape on the distribution of genetic variation in wild populations remain largely unknown. It is important to examine the long-term genetic consequences of fragmentation of habitat so we can develop appropriate strategies for maintaining viable populations in remnants of habitat over hundreds to thousands of years. Such studies are only beginning to be conducted, but the new area of metapopulation analysis and management has emerged as a result of these issues (Hanski and Gilpin 1997). A metapopulation is characterized as a network of populations that have limited gene flow between them and have population extinction and recolonization in specific localities (Levins 1969). In the context of management, we define a situation as extinction if a population either has died out or has been removed for the purpose of translocation.

Metapopulation management brings together the fields of demography, population genetics, and resource management. The primary goal is to "fool" the evolutionary process into "believing" it is acting on one large contiguous population, with all the attendant complexities of births, deaths, dispersal, and local group extinctions, when in fact the members of the species are distributed in many small patches that effectively are isolated from one another by intervening, unsuitable habitat through which they cannot cross. In general, a metapopulation-management plan involves in-depth study of the behavior, demography, and genetics of a species to determine when, how many, and where individuals should be moved among the existing patches of suitable habitat so as to mimic one large panmictic (free-mixing) population or species. The critical long-term goal of such a strategy is to provide the largest number of breeding individuals, or effective population size, to maintain most of the population's or species's genetic diversity over the course of centuries (Wade and McCauley 1988). One can demonstrate both mathematically and experimentally that the larger the effective population size, the less likely that genetic variation will be lost to random processes that generally remove genetic variants from a population (so-called genetic drift). Hence, the general goal is to maintain as large an effective population as possible, thus buffering the forces that otherwise would inevitably erode genetic diversity.

Case Studies. Populations of the ocelot (*Leopardus pardalis*) in southern Texas provide an example of a habitat specialist that has been fragmented into many small populations after 50 years of converting land to agricultural uses. A recent examination of genetic variation has revealed a lack of gene flow between the populations in southern Texas and the historical source population in northern Mexico (Walker 1997). Furthermore, genetic variation within populations in southern Texas has eroded. Assuming that a generation in the ocelot is about 2 years, this means that the fragmentation of the ocelots' range into small, relatively isolated populations has resulted in a major loss of genetic variation in only 25 generations. If genetic diversity within populations of ocelots in southern Texas is to be restored and maintained, any future conservation plan must involve exchanging cats between these isolated populations and those in northern Mexico.

Black lion tamarins (*Leontopithecus chrysopygus*), endemic to the state of São Paulo, Brazil, exist in only seven forest fragments (Coimbra-Filho 1976; Valladares-Padua 1993). Researchers from the Instituto Pesquisas Ecológicas (IPÊ) in São Paulo and from the Center for Environmental Research and Conservation (CERC) at Columbia University in New York are undertaking a project to devise a program of metapopulation (translocation) and management for these animals. The immediate goal of this effort is to translocate individuals from one forest fragment to another. The ultimate goals are to ensure proper assimilation of introduced individuals into other populations or into unoccupied but suitable patches of habitat and to conserve a "natural" amount of genetic diversity in the combined forest fragments, including the empty ones that will be colonized by translocated lion tamarins. One way to ensure proper assimilation of introduced lion tamarins is to mimic their natural dispersal patterns and their current population genetic structure. The genetic data from this study will contribute immeasurably to what is known about the social organization, dispersal patterns, and population genetics of the black lion tamarin and thus will enhance the chances of successful translocation, demographic stability, genetic management, and long-term survival of this highly endangered species.

CONSERVATION GENETICS TRAINING—WHO WILL PERFORM GENETIC ANALYSIS FOR CONSERVATION?

Conservation-management recommendations that come from outside the nation in which they are to be implemented rarely are followed. Indeed, innumerable counterexamples teach us that conservation is done best when it is done at home. For this reason, scientists from each country in which the work is to be done should be trained as conservation geneticists. Thus, when asking *who* should be performing genetic analysis for conservation, the logical answer would be the countries' scientists who would use the resulting genetic information to establish and revise their conservation programs and policies.

This means that, in addition to training our own students and future conservation geneticists, the universities and other research institutions in developed countries should be providing opportunities for in-depth technical and analytical training to young scientists who have the best chance of establishing this type of research in their own countries. One program that is doing just that is the Conservation Genetics Training Program for Southeast-Asian Scientists, which is based at CERC (see Melnick and Pearl 2000) and is funded by the MacArthur Foundation. This year-long program provides training in research design, laboratory techniques, and data analysis. As a followup to this training, CERC staff help the trainees establish research programs in their home countries. This assistance ranges from technical guidance to the actual purchase and outfitting of small laboratories to do the work. This program has trained researchers from Indonesia, Malaysia, Thailand, Vietnam, and China, and the CERC training staff now includes a postdoctoral scientist from Sri Lanka. Out of this program is a rapidly growing regional cadre of researchers who publish in peer-reviewed international journals (Wang and others 1997). This group is likely to have a major effect on future decisions about conservation within Southeast Asia.

CONCLUSION

In this chapter, we have highlighted the important uses of genetic analyses to define the units of conservation and the units of management, the geographic locations of those units, and the ways in which genetic variation is distributed within and among the populations that make up each unit. This discussion and the examples we have offered are meant to provide a brief introduction to the nonspecialist reader and to highlight the value of these approaches for wildlife managers, other conservation practitioners, and environmental policy-makers. It is important, however, to point out that other biological disciplines (for example, morphological systematics and behavioral ecology) also contribute significantly to the definition of evolutionary distinctiveness and that many other considerations—such as overall evolutionary uniqueness, current vulnerability, and socio-cultural value—must be considered when we are developing protective measures for a particular population or species. Ultimately, we must apply as much information as possible to decisions about designing and launching conservation efforts. We hope it is clear that genetic analysis is a powerful and timely mechanism for generating a great deal of valuable information for the purpose of conservation.

REFERENCES

Allendorf FW, Leary RF. 1986. Heterozygosity and fitness in natural populations of animals. In: Soulé ME (ed). Conservation biology: the science of scarcity and diversity. Sunderland MA: Sinauer. p 57–76.

Avise JC, Hamrick JL (eds). 1996. Conservation genetics: case histories from nature. New York NY: Chapman & Hall. 512 p.

Avise JC, Arnold J, Ball RM, Bermingham E, Lamb T, Neigel JE, Reeb CA, Saunders NC. 1987. Intraspecific phylogeography: the mitochondrial DNA bridge between population genetics and systematics. Ann Rev Ecol Syst 18:489–522.

Avise JC, Nelson WS. 1989. Molecular genetic relationships of the extinct dusky seaside sparrow. Science 243:646–8.

Avise JC. 1994. Molecular markers, natural history, and evolution. New York NY: Chapman & Hall. 511 p.

Avise JC. 1996. Toward a regional conservation genetics perspective: phylogeography of faunas in the southeastern United States. In: Avise JC, Hamrick JL (eds). Conservation genetics: case histories from nature. New York NY: Chapman & Hall. p 431–70.

Beltran JF, Rice JE, Honeycutt RL. 1996. Taxonomy of the Iberian lynx. Nature 379:407–8.

Burgman MA, Ferson S, Akqakaya HR. 1993. Risk assessment in conservation biology. New York NY: Chapman & Hall. 314 p.

Caughley G. 1994. Directions in conservation biology. J Anim Ecol 63:215–44.

Coimbra-Filho AF. 1976. Os saguis do genero *Leontopithecus* Lesson, 1840 (Callithicidae-Primates). Unpublished Master's thesis, Universidade Federal do Rio de Janeiro.

Costanza R, d'Arge R, de Groot R, Farber S, Grasso M, Hannon B, Limburg K, Naeem S, O'Neill RV, Paruelo J and others. 1997. The value of the world's ecosystem services and natural capital. Nature 387:253–60.

Ellsworth DL, Honeycutt RL, Silvy NL, Bickham JW, Klimstra WD. 1994. Historical biogeography and contemporary patterns of mitochondrial DNA variation in white-tailed deer from the southeastern United States. Evolution 48:122–36.

Foose TJ, van Strien NJ (eds). 1995. Asian rhinos. Newsl IUCN SSC Asian Rhino Specialist Group: No 1.

Frankham R. 1995. Conservation genetics. Ann Rev Gen 29:305–27.

Hanski IA, Gilpin ME (eds). 1997. Metapopulation biology. ecology, genetics, and evolution. San Diego CA: Academic Pr. 512 p.

Hillis DM, Moritz C, Mable BK. 1996. Molecular systematics, 2nd ed. Sunderland MA: Sinauer. 655 p.

Levins R. 1969. Some demographic and genetic consequences of environmental heterogeneity for biological control. Bull Entomol Soc Amer 15:421–31.

Kellert SR, Wilson EO (eds). 1993. The biophilia hypothesis. Washington DC: Island Pr. 439 p.

Melnick DJ. 1988. The genetic structure of a primate species: Rhesus macaques and other cercopithecine monkeys. Intl J Primat 9:195–231.

Melnick DJ, Hoelzer GA. 1992. Differences in male and female macaque dispersal lead to contrasting distributions of nuclear and mitochondrial DNA variation. Intl J Primat 13:379–93.

Melnick DJ, Pearl MC. 2000. Center for Environmental Research and Conservation (CERC): a new multi-institutional partnership to prepare the next generation of environmental leaders. In: Raven PH, Williams T (eds). Nature and human society: the quest for a sustainable world. Washington, DC: National Academy Press. p. 462–70.

Morales JC, Andau PM, Supriatna J, Zainuddin ZZ, Melnick DJ. 1997. Mitochondrial DNA variability and conservation genetics of the Sumatran rhinoceros. Cons Biol 11:539–43.

Moritz C. 1994. Defining evolutionarily significant units for conservation. Trends Ecol Evol 9:373–5.

Moritz C. 1995. Uses of molecular phylogenies for conservation. Phil Trans R Soc London Series B 249:113–8.

O'Brien S. 1994. A role for molecular genetics in biological conservation. Proc Natl Acad Sci USA 91:5748–55.

O'Brien S. 1996. Conservation genetics of the Felidae. In: Avise JC, Hamrick JL (eds). Conservation genetics: case histories from nature. New York NY: Chapman & Hall. p 50–74.

Pimm SL. 1997. The value of everything. Nature 387:231–2.

Rieseberg LH, Swensen SM. 1996. Conservation genetics of endangered island plants. In: Avise JC, Hamrick JL (eds). Conservation genetics: case histories from nature. New York NY: Chapman & Hall. p 305–34.

Smith TB, Wayne RK (eds). 1996. Molecular genetic approaches in conservation. New York, NY: Oxford Univ Pr. 483 p.

Taberlet P. 1996. The use of mitochondrial DNA Control region sequencing in conservation genetics. In: Smith TB, Wayne RK (eds.). Molecular genetic approaches in conservation. New York NY: Oxford Univ Pr. p 125–42.

Templeton AR, Shaw K, Routman E, Davis SK. 1990. The genetic consequences of habitat fragmentation. Ann Missouri Botan Garden 77:13–27.

Templeton AR, Georgladis NJ. 1996. A landscape approach in conservation genetics: conserving evolutionary processes in the African Bovidae. In: Avise JC, Hamrick JL (eds). Conservation genetics: case histories from nature. New York NY: Chapman & Hall p 398–430.

Valladares-Padua C. 1993. The ecology, behavior and conservation of the BLTs (*Leontopithecus chrysopygus* Mikan, 1823). Unpublished PhD dissertation. Gainesville FL: Univ Florida.

Vane-Wright RI, Smith CR, Kitching IJ. 1994. Systematic assessment of taxic diversity by summation. In: Forey PL, Humphries CJ, Vane-Wright RI (eds). Systematics and conservation evaluation, systematics association special volume 50. Oxford UK: Clarendon Pr. p 309–26.

Wade MJ, McCauley DE. 1988. Extinction and recolonization: their effects on the genetic differentiation of local populations. Evolution 42:995–1005.

Walker CW. 1997. Patterns of genetic variation in ocelot (*Leopardus pardalis*) populations for south Texas and northern Mexico. Unpublished PhD Thesis. College Station TX: Texas A&M Univ.

Wang W, Forstner MRJ, Zhang Y, Liu Z, Wei Y, Hu H, Xie Y, Wu D, Melnick DJ. 1997. A phylogeny of Chinese leaf monkeys using mitochondrial ND3-ND4 gene sequences. Intl J Primat 18:305–20.

Wayne RK. 1996. Conservation genetics in the Canidae. In: Avise JC, Hanuick JL (eds). Conservation genetics: case histories from nature. New York NY: Chapman & Hall. p 75–118.

Williams PH, Humphries CJ. 1994. Biodiversity, taxonomic relatedness, and endemism in conservation. In: Forey PL, Humphries CJ, Vane-Wright RI (eds). Systematics and conservation evaluation, systematics association special volume 50. Oxford UK: Clarendon Pr. p 269–87.

Witting L, McCarthy MA, Loeschcke V. 1994. Multi-species risk analysis, species evaluation and biodiversity conservation. In: Loeschcke V, Tomiuk J, Jain SK (eds). Conservation genetics. Basel Germany: Birkhduser Verlag. p 239–49.

APPLICATION OF GEOSPATIAL INFORMATION
FOR IDENTIFYING PRIORITY AREAS
FOR BIODIVERSITY CONSERVATION

ASHBINDU SINGH

Division of Environmental Information, Assessment & Early Warning—North America,
United Nations Environment Programme,
EROS Data Center, Sioux Falls, SD 57198

INTRODUCTION

BIOLOGICAL DIVERSITY, the variety and variability among living organisms and the environment in which they occur, is important to maintain life-sustaining systems of the biosphere, but it is threatened by many human activities. Recently, the United Nations Environment Programme (UNEP) Global Biodiversity Assessment concluded that "the adverse effects of human impacts on biodiversity are increasing dramatically and threatening the very foundation of sustainable development" (UNEP 1995). The total number of species that inhabit the planet is unknown, and most extinctions occur before the species have been named and described. It is estimated that 85–90% of all species can be protected by setting aside areas of high biodiversity before they are further degraded, without the need to inventory species individually. It is generally assumed that most terrestrial species are in the tropics. Realistically, only a relatively small portion of the total tropical land area is likely to be devoted to biodiversity conservation, so it is critical to identify areas rich in species diversity and endemism (the characteristics of species that are native or confined to a particular area) as a first step toward protection of remaining natural habitat before the areas are destroyed.

In the past, protected areas were often set aside without regard to the biodiversity within their boundaries. As a result, many protected areas now have little importance with respect to biodiversity; conversely, many areas of habitat with important biodiversity lack protection. The study discussed here seeks to identify relationships between land cover, human population, and protected areas

276

through the analysis of comprehensive and consistent spatial datasets at 1-km resolution to answer the following two questions: Are African ecoregions with a high degree of biodiversity adequately protected? Is biodiversity within Africa threatened by human population pressure and land use?

THE STUDY AREA

The present study dealt with two areas: the continental area consisted of the African continent, including Madagascar; and the regional area consisted of the African Great Lakes Region, including Burundi, Democratic Republic of Congo, Kenya, Malawi, Mozambique, Rwanda, Tanzania, Uganda, Zambia, and Zimbabwe.

DATA SOURCES

The analysis was carried out with geographic information systems, remote-sensing technologies, and the most comprehensive and consistent 1-km spatial datasets. The land-cover dataset was derived from the International Geosphere-Biosphere Program land-cover classification, which was based on the National Oceanic and Atmospheric Administration's 1-km Advanced Very High Resolution Radiometer satellite data spanning a 12-month period (April 1992–March 1993). The land-cover characteristic database was produced at the US Geological Survey EROS Data Center. Political boundary data were from the Digital Chart of the World. The protected-areas database from the World Conservation Monitoring Centre and World Resources Africa Data Sampler and the population-density database for Africa from the UNEP Global Resource Information Database were used in the analysis.

Some of the smaller protected areas might not have been accounted for, because of the coarse resolution of analysis. The protected-area database is not current for all countries. The land-cover and population datasets were the best available ones covering all of Africa. Considerable errors are known to exist in the mapped distribution of croplands. The population dataset is generated with a model incorporating many variables, including the location of protected areas, so the areas of intersection between population and protected areas are compromised. However, that does not invalidate conclusions drawn from the analysis regarding the proximity of the protected areas to the areas of high population. None of the datasets has been rigorously validated, so local relationships and distributions should be viewed with caution. Availability of high-quality current data remains a stubborn barrier in such analytical analysis, and this highlights the need to support development and updating of databases.

RESULTS

The Continental Area

Protected areas in Africa occupy slightly over 2 million square kilometers or 7% of the continent's 30 million square kilometers (figure 1). Among various ecoregions, barren and sparsely vegetated lands make up about 9.6 million square

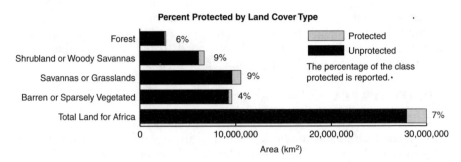

FIGURE 1   Africa: histogram comparing protected and unprotected areas. Percentages of class protected are reported.

kilometers, whereas biodiversity-rich, tropical evergreen broadleaf forests make up about 3 million square kilometers. Of the barren and sparsely vegetated lands, about 4% of discrete pieces of land are protected, whereas less than 6% of the tropical evergreen broadleaf forests are protected. Closed shrublands, which are estimated to be over 700,000 square kilometers in extent, have the largest proportion of protected area, namely 14%. About 2 million square kilometers, or about 8%, of croplands and a mosaic of croplands mixed with natural vegetation are under protected status.

The Regional Area

The 10-country African Great Lakes Region contains a wide range of habitats, including deserts, savannas, and dry and humid tropical forests. In this region of 6 million square kilometers, 12% of the area is protected. Biodiversity-rich, tropical evergreen broadleaf forests cover about 1.4 million square kilometers of the region, and about 100,000 square kilometers, or slightly less than 7%, is protected, leaving the bulk of the tropical evergreen broadleaf forest unprotected. In contrast, protected areas make up about 9–15% of the areas in the category of woody savannas, savannas, grasslands, croplands, and croplands-natural vegetation mosaic.

Furthermore, the degree to which the forests listed as protected are actually protected varies. In the African Great Lakes Region, for example, about 125,000 square kilometers of croplands and croplands interspersed with natural vegetation mosaic is found in protected areas. This apparent encroachment of agriculture highlights the lack of enforcement of protection of the natural flora and fauna in designated protected areas in the region.

The highest human population densities are found in Rwanda, Burundi, and Uganda around Lake Victoria and in scattered areas in Malawi, Zambia, and Kenya. Areas of low population density coincide with many protected areas, and smaller areas of medium and high population density are found in and adjacent to protected areas.

SUMMARY FOR POLICY-MAKERS

The geographic analysis of relationships between protected areas, distribution of land-cover types, and population density clearly revealed the following:

• Lack of protection status and effective implementation of protection measures in the designated protected areas seems to pose a serious threat to forest biodiversity in Africa.

• As estimated with a geographic information system, about 7% of the total land area of Africa is protected; this is much higher than the estimate of about 5%, compiled from official statistics, usually cited in international sources. Thus, there are substantial differences between protected-area statistics derived from actual planar area on the ground, as estimated by calculations of a geographic information system, and estimates based on official statistics. The differences, reflecting data of different sources, highlight the need to provide more resources to improve the environmental-information infrastructure in countries so that accurate and up-to-date environmental data can be generated and maintained for planning and policy purposes.

• About 6% of the area covered by biodiversity-rich, tropical evergreen broadleaf forests in Africa is protected. Most of these valuable ecoregions, rich in biodiversity and endemic species, are concentrated in countries like the Democratic Republic of Congo and Madagascar, and seem to lack adequate protection. Practical action programs that include accelerated establishment of a network of protected areas are needed urgently.

• In Africa, drier ecoregions are generally better protected than tropical evergreen broadleaf forests. That is contrary to the widely held belief that moist habitat, such as tropical rain forests, is generally better protected than drier zones, such as dry forests and grasslands.

• The presence of croplands in protected areas indicates that legal designation of areas as protected is not sufficient for the protection of biodiversity in the face of human competition for the same land. Protected status must be accompanied by effective enforcement measures over the long term to ensure protection of biodiversity and endemic and endangered species. Additional resources should be applied to understand socioeconomic factors associated with protection of biodiversity, and local stakeholders should be included by giving them a role and economic incentives to conserve biodiversity.

• In contrast with many other regions, low human population densities in many areas of Africa provide an opportunity to protect such areas for conservation purposes.

• A shift in national and international policy formulation and planning processes based on targeting biodiversity-rich areas is needed to protect biodiversity in Africa more effectively. Geographic targeting and programmatic focus are needed to conserve species ecoregions rich in biodiversity and endemism and to address the socioeconomic causes of encroachment and loss of biodiversity.

DISCLAIMERS

The views expressed in this text do not necessarily reflect those of the agencies cooperating in this project. The designations used and material presented above do not imply the expression of any opinion whatsoever on the part of the cooperating agencies concerning the legal status of any country, territory, city, or area or of its authorities or of the delineation of its frontiers or boundaries. The work was jointly funded by the United Nations Environment Programme, the US National Aeronautics and Space Administration, and the US Geological Survey.

ACKNOWLEDGMENTS

My sincere thanks to a number of scientists, including Bhaska Ramachandran, Gene Fosnight, Tom Crawford, Grey Tappan, Brad Reed, Eric Wood, Jim Rowland, Steve Howard of Hughes STX, and Anna Stabrawa of UNEP, who contributed to this work and made valuable suggestions.

REFERENCE

UNEP [United Nations Environment Programme]. 1995. Heywood VH (ed.). Global biodiversity assessment. New York NY: Cambridge Univ Pr. 1140 p.

HAWAII BIOLOGICAL SURVEY:
MUSEUM RESOURCES IN SUPPORT OF CONSERVATION

ALLEN ALLISON
Hawaii Biological Survey, Bishop Museum,
1525 Bernice Street, Honolulu, HI 96817

SCOTT E. MILLER
International Center of Insect Physiology and Ecology,
Box 30772, Nairobi, Kenya

INTRODUCTION

HAWAII—because of its geographic isolation, rich volcanic soils, and enormous topographic and climatic diversity—has produced a biota with a very high percentage of endemism among multicellular terrestrial organisms. The native biota includes about 18,000 species (Eldredge and Miller 1998) (table 1). The 8,500 terrestrial and aquatic plants and animals might have evolved from as few as 1,000 original colonists (Gagné 1988; see also Sakai, and others 1995) in the absence of many biotic influences that are present on larger land masses (such as grazing herbivores), and they have proved vulnerable to extreme population reduction and even extinction owing to introduced predators, competitors, and diseases. Although Hawaii accounts for only about 0.2% of the land area of the United States, it has 31% of the nation's endangered species and 42% of its endangered birds. Of the 1,023 species of native flowering plants 73 are down to about 20 or fewer individuals in the wild, and nine are down to one (US Fish and Wildlife Service 1999). Almost 75% of the historically documented extinctions of plants and animals in the United States have occurred in Hawaii.

About 15 years ago, as the dimensions of this extinction crisis were beginning to become clear, a wide array of state, federal, and private organizations, catalyzed by The Nature Conservancy and the Hawaii Audubon Society, redoubled their efforts to develop effective mitigative measures. More recently, a formal consortium of agencies developed the Hawaii Conservation Biology Secretariat, which has raised the profile of these important issues and helped to coordinate responses.

TABLE 1 Numbers of Species Known from Hawaii and Surrounding Waters[a]

	Total Species	Endemic Species	Alien Species	Species at Risk	Extinct Species
Algae and other protists	1,939	4	5	0	0
Fungi and Lichens	2,080	240	?	0	0
Flowering plants	2,074	908	1051	546	91
Other plants	763	241	44	0	
Mollusks	1,650	956	86	115	500?
Insects	7,998	5,245	2,589	308	
Other arthropods	2,109	324	577	2	
Other invertebrates	2,281	824	71	1	
Fish	1,197	139	73	1	0
Amphibians	5	0	5	0	0
Reptiles	27	0	23	3	0
Birds	294	63	46	39	50+
Mammals	44	1	19	2	1
Totals	22,462	8,864	4,598	1,017	642+

[a] Endemic species are restricted to Hawaii; nonindigenous alien, (includes introduced) do not naturally occur in Hawaii. Total includes endemic, alien, and indigenous (occur naturally in Hawaii but not endemic) species and species of unknown status. "Species at risk" include federally endangered, threatened, and candidate species, plus "species of concern." "Extinct" includes pre-Captain Cook extinctions.
Source: Based on Eldredge and Miller 1998.

Those efforts have been seriously hampered by lack of fundamental information. The basic taxonomy of many groups has not been fully worked out, and information on the ranges or identities of many species was until recently available only from scattered research publications or museum collections. Although a substantial amount of information has been assembled on endangered plants, vertebrates, and a few invertebrate taxa, successful efforts to manage Hawaiian ecosystems requires information about *all* species, native and alien. In fact, the greatest threat to Hawaiian organisms and to the integrity of Hawaiian ecosystems is posed by alien species. To address the information need, the Hawaii legislature in 1992 designated the Bishop Museum, which houses the world's largest natural-history collections from Hawaii (nearly 4 million specimens) as the Hawaii Biological Survey (HBS) and charged it with the task of compiling comprehensive information on the entire biota of the state (Allison and others 1995).

The Bishop Museum developed a six-stage process to implement the biological survey. Briefly, this involves, for each major group of organisms,

- developing a computerized database of the literature;
- preparing a species checklist based on the literature, collections, and consultation with experts;
- developing a database of the collections, including coding localities to facilitate geographic information system (GIS) analysis and presentation;

- developing a database of information from other collections or from other organizations that are conducting biological surveys (or establishing computer linkage to such information);
- directing research efforts to high-priority needs; and
- filling gaps in information through additional field surveys.

In practice, many of these are concurrent activities. The literature databases
and species checklists developed by HBS scientists and collaborators provide a
firm foundation for the computerization of specimen-based data from collections.
When specimen data are computerized and incorporated into an environmental
information system, one can easily determine the range of a species, document
how it has changed, identify broad multispecies patterns of distribution and diversity (ecosystem characteristics), and evaluate how these features are related to
various environmental factors (such as climate and soils) and have been or are
likely to be affected by resource-management and land-use strategies. It is important to emphasize that specimen collections constitute the most accessible and
cost-effective source of data for the development of comprehensive environmental-information systems (Allison 1991; Nielsen and West 1994). Those information systems, involving GIS and other spatial-analysis and database technology,
are crucial to the efficient management of Hawaii's fragile ecosystems and are in
use by all the state's natural-resource management and land-use agencies.

In its role as HBS, the museum is providing a service to the scientific and local
communities as an information clearinghouse. It gathers, processes, synthesizes,
and distributes to a variety of partners information related to the biological resources of Hawaii. Information from the collections is crucial to provide authority files, data points for distribution maps, additional ecological information, and
a historical perspective on the biota of Hawaii. Inasmuch as completeness is necessary for functionality, HBS also plays a crucial role in centralizing and facilitating distribution of information from partner organizations. The overall strategy
is to streamline the process of developing information products while continuing
the development of longer-term projects and continuously improving and refining all products.

In this paper, we discuss the overall strategy of HBS and its accomplishments
to date. Although our efforts arose out of an urgent need to address critical conservation issues in a relatively small geographic area, we feel that they can serve
as an effective model for the role that museums can play in understanding and
managing biodiversity. Our overall theme is that museum collections and associated databases are crucial information resources for understanding and managing
biological diversity. With more than 400 million specimens in US museums alone,
and perhaps 2 billion museum specimens worldwide (Duckworth and others
1993), the implications are enormous.

INFORMATION MANAGEMENT

The information-management strategy developed for HBS is represented sche-
matically in figure 1. Information sources for HBS include those listed on the left

INFORMATION MANAGEMENT

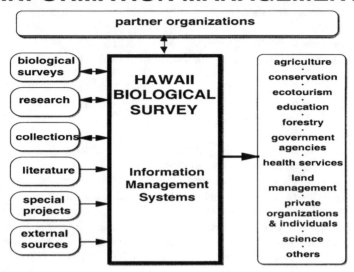

FIGURE 1   Hawaii Biological Survey obtains data from a variety of sources and processes them into information useful to diverse stakeholders.

side, including biological surveys, research projects, existing collections, existing literature, and special projects (such as syntheses undertaken with collaborators). In many cases, the flow of information is reciprocal; this is especially true for the collections, where there is constant interaction between scientists producing reports based on the collections, which result in improved quality of identifications and localization. HBS activities are undertaken in collaboration with an array of partner organizations. The collaboration in some cases is formalized at an institutional level, and an informal network of collaboration by scientific staff extends internationally, especially in systematics research. HBS information-based products are used by government, commercial, and private clients for a variety of purposes, including agriculture, conservation, education, fisheries, forestry, health services, land management, quarantine and regulatory services, and other research, as shown on the right side of figure 1.

Some of the primary partners of HBS in recent years have been state and federal natural-resource management agencies (Hawaii Department of Agriculture, Hawaii Department of Land and Natural Resources, US Department of Agriculture, and US Department of the Interior, especially the Fish and Wildlife Service and the former National Biological Survey, now part of the US Geological Survey), conservation organizations (Center for Plant Conservation, Ducks Unlimited, Hawaii Conservation Biology Forum, and The Nature Conservancy), educational organizations (Hawaii Department of Education and University of Hawaii), and other biodiversity research organizations (including Cornell University,

National Tropical Botanic Garden, New York Botanic Garden, and Smithsonian Institution). Many of these organizations maintain specialized databases related to specific applications in conservation or agriculture or to specific taxonomic groups. Rather than duplicate these efforts, we seek to link with them through the development of authority files, data standards, and information models (http://www.bishop.hawaii.org/asc-cnc/).

FIVE-YEAR ACCOMPLISHMENTS

During the last 5 years, HBS has developed comprehensive bibliographies and species checklists of all major groups of plants and animals and some fungi, protists, and algae in Hawaii—terrestrial, freshwater, and marine. Hawaii is the only state in the United States other than Illinois (Post 1991) and the only large tropical area in the world in which the total number of described species is accurately known (Eldredge and Miller 1995, 1997, 1998; Miller and Eldredge 1996; http://www.bishop.hawaii.org/bishop/HBS/hispp.html).

HBS provides a venue for disseminating work of individual scientists to a variety of users. Most individual researchers do not have at their disposal the contacts, time, or technology needed to deliver their products to all potential users, especially land managers. A researcher might be the world's expert on a particular taxon that occurs in Hawaii but have neither the time nor the means to circulate research results widely within the state. HBS provides an efficient and cost-effective means of disseminating varied research products and extending the useful life of datasets beyond the funding of a particular project or the career of an individual investigator (for example, Helly and others, 1996; US National Committee for CODATA 1995; http://www.sdsc.edu/compeco_workshop/report/helly_publication.html).

The products of HBS take various forms to meet our diverse user community, as shown in figure 2. We see our primary product as information on our World Wide Web (WWW) server. The WWW server makes large amounts of information available worldwide 24 hours a day, and we can update or post information immediately at low cost. Information on the WWW should be our most recent version and should end confusion about versions of information distributed in other media or the use of outdated information that might have been gathered from our collections years ago. The "self-serve" approach also lowers our personnel costs in handling frequently asked questions. Other products beyond the WWW include information services provided directly by staff, enhancements of collections (for example, returning improved identifications of specimens), such technical publications as checklists (Cowie and others 1995; Nishida 1997) and systematic monographs (Gagné 1997), popular publications like our nascent series of user-friendly identification handbooks (Polhemus and Asquith 1996), contributions to formal and informal education, exhibits and internships, and products developed from various partnerships.

One of the products of HBS is the annual publication of a compilation of changes in our understanding of the status and distribution of the Hawaiian biota

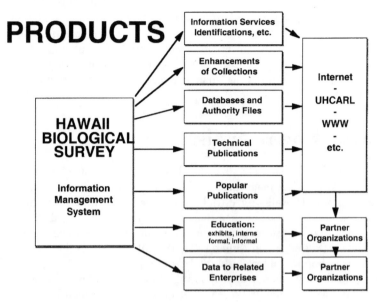

FIGURE 2   Data managed by Hawaii Biological Survey are made available to others and synthesized into various information products. Data and products are generally accessible through the World Wide Web.

titled *Records of the Hawaii Biological Survey*. *Records* is published annually in the journal *Bishop Museum Occasional Papers*. It has been especially effective at providing a publishing vehicle for short papers to document distribution or taxonomic changes that are important in the Hawaii context but might not have an appropriate venue elsewhere in the scientific literature. A number of agencies use the information from HBS to support their own products. One, the Hawaii Ecosystems at Risk project, a consortium led by the US Geological Survey Biological Resources Division, depends on *Records* as its primary source of documentation of new records of weeds and of taxonomic validation of these records.

We have largely completed the first two of the three levels of databases that provide the foundation for HBS. The first is literature databases. These focus on the taxonomic and distributional literature but include any other publications and reports that come to our attention. The second is taxonomic authority files or species checklists. These databases, compiled largely from the literature with extensive consultation with specialists, provide an index to and synthesis of what has been learned in over 100 years of biological research on Hawaii; without them, much historical information would remain unrecognized or inaccessible. The third is databases of Hawaiian specimens in the Bishop Museum's extensive collections (table 2). Progress in each category of database for each taxon depends on the level of knowledge of the taxon, the expertise available to help, funding priorities, and the curatorial condition of our collections.

DISCUSSION

Biological surveys are fundamental to the documentation of the plants and animals of the earth (Blackmore and others 1997) and are one of the major reasons for the founding of the world's great natural-history museums (Cotterill 1997; Lane 1996; Raven and others 1993). Early biological surveys were closely associated with exploration of the earth during the last three centuries and had as their purpose documentation of the general biota of scientifically unexplored areas (see, for example, Viola and Margolis 1985). As major biological features of Earth became known, museums' scientific interest shifted more toward detailed taxonomic studies of plant and animal groups. Government agencies were formed to manage natural resources, and they have conducted much of the biological survey work during the last century; for example, in 1939, the Bureau of Biological Survey, an agency in the US Department of Agriculture that developed in close association with the Smithsonian Institution, was, with the Bureau of Fisheries, transferred to the Department of the Interior and later became the Fish and Wildlife Service). With rising human populations and increasing demand for land and natural resources, public and private agencies are now facing tremendous challenges in their efforts to obtain sufficient information to manage and preserve the world's biodiversity.

With the advent of modern database technology, the information in museum collections can be made available for a wide range of uses. This has led to the development of new and strengthened partnerships between museums and resource-management agencies, for example, creation of the National Biological Survey in 1993. These partnerships have focused mostly on the need for detailed information on the distribution of plants and animals to support management

TABLE 2 Estimated Numbers of Hawaiian Collection Records (Specimens or Specimen Lots) in Bishop Museum Databases As of March 1999

Organisms	Units in Databases[a]	Total Units[a]	Percent Complete
Birds and mammals (recent)[b]	7,000	7,000	100
Fossil birds	100	10,000	1
Reptiles and amphibians	400	900	44
Fish (mostly marine)	5,000	5,000	100
Mollusks (terrestrial and marine)	68,000	140,000	49
Insects and mites	40,000	500,000	8
Other invertebrates (mostly marine)	25,000	25,000	100
Algae (mostly marine)	25,000	25,000	100
Fungi and lower plants	3,000	6,000	50
Vascular plants	45,000	145,000	31
TOTALS	218,500	863,900	25

[a] Units are specimens, except for fish, invertebrates, and mollusks, which are in lots (one or more conspecific specimens with identical data).
[b] The Bishop Museum also maintains a database of some 55,000 sighting records of Hawaiian birds.

efforts. Museums are the primary repositories of such information. For example, although the systematics of vascular plants of the United States is reasonably well known, precise distributional details on many species are not readily available, and many of the data reside in museum collections; it is therefore urgent to mobilize information from museum collections into databases and to link the databases into information systems.

The major strength of HBS is its comprehensive approach and the fact that its activities are undertaken in close partnership with management agencies. This helps to ensure that HBS products and services meet user needs. In addition, working with partners helps to ensure that collections are built in a purposeful way (see Hawksworth 1991) and have maximal utility. We have emphasized conservation applications in this paper, but biological surveys also have important applications in agriculture, medicine, and recreation (Klassen 1986; Roberts 1992).

The approach of HBS is unique in attempting to provide at least basic information on all organisms while focusing more detailed surveys or products on taxa of concern to specific users. The All Taxa Biodiversity Inventory (ATBI) approach is similar in covering all organisms (Miller 1993; Yoon 1993), but the approaches differ in that HBS synthesizes the literature first and then undertakes surveys to update data and fill gaps, whereas the ATBI emphasizes intensive surveys in smaller areas.

Although many conservation agencies are moving away from efforts to protect individual species and are instead highlighting the need to protect entire ecosystems (Kirlin and others 1994), the classification of ecosystems tends to be rather arbitrary. In a promising alternative approach that has been recently developed (Kiester and others 1996; White and others 1997), species occurrence data (presence or absence) are assembled into map layers and grouped into classes. This method, which can readily use museum-specimen data, involves a high level of objectivity and therefore has many advantages, particularly in public-policy debates, over the use of classed data, such as on vegetation. A particular strength of this approach is that it facilitates analysis and modeling of the risk to biodiversity, including individual species and populations, posed by different land-use strategies.

The scientific importance of museum collections has been well documented (Nudds and Pettitt 1997), but this value is poorly reflected in public policy. Indeed, most museums initially began computerizing their collections to gain internal management efficiency and have been slow to develop scientific products and services outside the traditional research enterprise. The systematics community has also been slow in providing authority files in readily accessible forms, although the recent production of a checklist of almost 100,000 species of North American insects shows what can be done (Poole and Gentili 1997). We agree with Lane (1996) that computerization of collections is central to an expanded role for museums in serving science and society, and nowhere is that more urgent than in the conservation of biodiversity. New organizations throughout the world—such as INBio, ERIN, and CONABIO (Anonymous 1994; Gámez 1991; Soberón and others 1996)—and long-established organizations, such as the Illinois Natural

History Survey (Anonymous 1996), have proved the importance of museum collections for understanding and managing biodiversity. The recent formation of the US Organization for Biodiversity Information (USOBI) signifies a trend to unite individual institutional efforts into a federation to achieve economies of scale and develop standards and common gateways to highly dispersed data (NRC 1993:94–5).

ACKNOWLEDGMENTS

We thank our many collaborating individuals and institutions, especially the John D. and Catherine T. MacArthur Foundation and the National Science Foundation for major funding. Gordon Nishida prepared the figures.

REFERENCES

Allison A. 1991. The role of museums and zoos in conserving biological diversity in Papua New Guinea. In: Pearl M, Beehler BM, Allison A, and Taylor M (eds), Conservation and environment in Papua New Guinea: establishing research priorities. Washington DC: The Government of Papua New Guinea and Wildlife Conservation International. p 59–63.

Allison A, Miller SE, and Nishida GM. 1995. Hawaii Biological Survey—a model for the Pacific Region. In: Maragos JE, Peterson MNA, Eldredge LG, Bardach JE, and Takeuchi HF (eds). Marine and coastal biodiversity in the tropical island Pacific region. Vol 1, p 349–55. Honolulu HI: East-West Center.

Anonymous. 1994. Collaboration with biodiversity agencies. Erinyes 20:1–8.

Anonymous. 1996. Illinois Natural History Survey Annual Report 1995–1996. Champaign IL: Illinois Natural History Survey, 55 p.

Blackmore S, Donlon N, and Watson E. 1997. Calculating the financial value of systematic biology collections. In: Nudds JR, and Pettitt CW (eds). The value and valuation of natural science collections. Proceedings of the International Conference, 1995, Manchester UK. London UK: Geological Society. p 17–21.

Cotterill FPD. 1997. The second Alexandrian tragedy and the fundamental relationship between biological collections and scientific knowledge. In Nudds, JR & Pettit, CW(eds), Proceedings of the International Conference on the Values and Valuation of Natural Science Collections. p. 227–241. Manchester: Manchester Museum.

Cowie RH, Evenhuis NL, Christensen CC. 1995. Catalog of the native land and freshwater molluscs of Hawaii. Leiden Netherlands: Backhuys Publ. 248 p.

Duckworth WD, Genoways HH, Rose CL. 1993. Preserving natural science collections: chronicle of our environmental heritage. Washington DC: National Institute for the Preservation of Cultural Property Inc. 140 p.

Eldredge LG, Miller SE. 1995. Records of the Hawaii Biological Survey for 1994. How many species are there in Hawaii? Bishop Mus Occas Pap 41:1–18.

Eldredge LG, Miller SE. 1997. Numbers of Hawaiian species: supplement 2, including a review of freshwater invertebrates. Bishop Mus Occas Pap 48:3–22.

Eldredge LG, Miller SE. 1998. Numbers of Hawaiian species: supplement 3, with notes on fossil species. Bishop Mus Occas Pap 55:3–15.

Gagné WC. 1988. Conservation priorities in Hawaiian natural systems. BioScience 38:264–71.

Gagné WC. 1997. Insular evolution, speciation, and revision of the Hawaiian genus Nesiomiris (Hemiptera: Miridae). Bishop Mus Bull Entomol 7: i–x, p 1–226.

Gámez R. 1991. Biodiversity conservation through facilitation of its sustainable use: Costa Rica's National Biodiversity Institute. Trends Ecol Evol 6(12):377–8.

Hawksworth DL (ed). 1991. Improving the stability of names: needs and options. Regnum Vegetabile No. 123. Königstein : Koeltz Scientific Bk.

Helly JT, Case T, Davis F, Levin S, Michener W(eds.). 1996. The state of computational ecology. San Diego CA: San Diego Supercomputer Center. 20pp. [also at http://www.sdsc.edu/compeco_workshop/report/helly_publication.html].

Kiester AR, Scott JM, Csuti B, Noss RF, Butterfield B, Sahr K, and White D. 1996. Conservation priorization using GAP data. Cons Biol 10(5):1332–42.

Kirlin JJ, Asmus P, Thompson R. 1994. Species conservation through ecosystem management. California Policy Choices 9:143–171.

Klassen W. 1986. Agricultural research: the importance of a national biological survey to food production. In: Kim KC Knutson L (eds). Foundations for a national biological survey. Lawrence KS: Assoc of Systematics Collections. p 65–76.

Lane MA. 1996. Roles of natural history collections. Ann Missouri Botan Gard 83:536–45.

Miller SE. 1993. All Taxa Biological Inventory workshop. Assoc Syst Coll News 21(4): 41, 46–7.

Miller SE, Eldredge LG. 1996. Number of Hawaiian species: supplement 1. Bishop Mus Occais Pap 45:8–17.

Mlot C. 1995. In Hawaii, taking inventory of a biological hot spot. Science 269:322–3.

NRC [National Research Council]. 1993. A biological survey for the nation. Washington DC: NatlAcad Pr. 205 p.

Nielsen ES, West JG. 1994. Biodiversity research and biological collections: transfer of information. In: Forey PL, Humphries CJ, Vane-Wright RI (eds). Systematics and conservation evaluation. Oxford UK: Clarendon Pr. p 101–21.

Nishida GM. 1997. Hawaiian terrestrial arthropod checklist. Third edition. Bishop Mus Tech Rep 12.

Nudds JR, Pettitt CW (eds). 1997. The value and valuation of natural science collections. Proceedings of the International Conference, Manchester, 1995. Manchester UK: Manchester Museum. p xii + 276.

Polhemus D, Asquith A. 1996. Hawaiian damselflies: a field identification guide. Hawaii Biol Surv Handbook. Honolulu HI: Bishop Museum Pr. 122 p.

Poole RW, Gentili P (eds). 1997. Nomina Insecta Nearctica: a check list of the insects of North America. Rockville MD: Entomological Information Services. CD-ROM. Also published as check list in four paper volumes. 1996–1997.

Post SL. 1991. Native Illinois species and related bibliography. Illinois Nat Hist Surv Bull 34:463–75.

Roberts L. 1992. Chemical prospecting: hope for vanishing ecosystems. Science 256:1142–3.

Sakai AK, Wagner WL, Ferguson DM, Herbst DR. 1995. Origins of dioecy in the Hawaiian flora. Ecology 76:2517–29.

Soberón J, Llorente J, Benítez H. 1996. An international view of national biological surveys. Ann Missouri Botan Gard 83:562–73.

US Fish and Wildlife Service. 1999. US Fish and Wildlife Service Species List, March 23, 1999. Honolulu: unpubl.

US National Committee for CODATA, Committee for Pilot Study on Database Interfaces. 1995. Finding the forest in the trees: the challenge of combining diverse environmental data: selected case studies. Washington DC: Natl Acad Pr. 129 p.

Viola HJ, Margolis C. 1985. Magnificent voyagers: the US Exploring Expedition, 1838–1842. Washington DC: Smithsonian Inst Pr. 303p.

White D, Minotti PG, Barczak MJ, Sifneos JC, Freemark KE, Santelmann MV, Steinitz CF, Kiester AR, Preston EM. 1997. Assessing risks to biodiversity from future landscape change. Cons Biol 11(2):249–360.

Yoon CK. 1993. Counting creatures great and small. Science 260:620–2.

BUILDING THE NEXT-GENERATION
BIOLOGICAL-INFORMATION INFRASTRUCTURE

JOHN L. SCHNASE

Center for Botanical Informatics, LLC. St. Louis, MO
(Current address: NASA Goddard Space Flight Center, Greenbelt, MD 20771)

MEREDITH A. LANE

The Academy of Natural Sciences, 1900 Benjamin Franklin Parkway, Philadelphia, PA 19103

GEOFFREY C. BOWKER

SUSAN LEIGH STAR

Graduate School of Library and Information Sciences, University of Illinois at Urbana-Champaign,
Champaign, IL
(Current address: Department of Communication, University of California, San Diego,
9500 Gilman Drive, La Jolla, CA 92093-0503)

ABRAHAM SILBERSCHATZ

Information Sciences Research Center, Lucent Technologies-Bell Laboratories,
600 Mountain Avenue, Murray Hill, NJ 07974

A GRAND challenge for the 21st century is to harness the accumulating knowledge of Earth's biodiversity and the ecosystems that support it. To accomplish that, we must mobilize biological information—assemble it, organize it, and deliver it with dramatically increased capacity. We must elevate the global biological-information infrastructure to a new level of capability—a "next generation"—that will allow people to share on a worldwide basis the knowledge created by biodiversity and ecosystems research.

Recognizing the urgency of the task, the President's Committee of Advisors on Science and Technology, through its Panel on Biodiversity and Ecosystems, recently coordinated a review of the US National Biological Information Infrastructure (NBII) (PCAST 1998). Over a 6-month period in 1997, people from a broad cross section of the public and private sectors contributed their insights, experiences, concerns, and hopes. What emerged was a renewed understanding of the importance of biological information to all aspects of human society. It became clear that much remains to be done to ensure that this information is complete and usable. Although the purpose of the review was to develop recommendations to build capacity in the United States, many of the panel's findings address global concerns of relevance to biodiversity research wherever it occurs. In this paper, we provide a summary of the panel's report, a view of what a next-generation biological-information infrastructure might encompass, and suggestions about how it might be achieved.

BACKGROUND

In the United States, NBII is the primary mechanism whereby biodiversity and ecosystem information is made available to all sectors of society. It is the biological component of the National Information Infrastructure and is the framework that connects US activities to the global biodiversity and ecosystem research enterprise. Its meaning is expansive and intended to convey the idea that an information infrastructure comprises not only computers, networks, and the like, but also the information, policies, standards, and people who use it. Initiation of the NBII was one of the primary recommendations made by the 1993 National Research Council report A *Biological Survey for the Nation* (NRC 1993).

Because our fate and economic prosperity are so completely linked to the natural world, information about biodiversity and ecosystems—as well as the infrastructure that supports it—is vital to a wide range of scientific, educational, commercial, and government uses. Most of this information now exists in forms that are not easily accessed or used. From traditional paper-based libraries to scattered databases and physical specimens preserved in natural-history collections throughout the world, our record of biodiversity and ecosystem resources is uncoordinated, and large parts of it are isolated from general use. It is not being used effectively by scientists, resource mangers, policy-makers, or other potential client communities (National Performance Review 1997; NRC 1997).

Research activities are being conducted around the world that could improve our ability to manage biological information. In the United States, the Human Genome Project is producing new medical therapies and developments in computer and information science. Geographic information systems (GISs) are expanding the ability of federal agencies to conduct data-gathering and data-synthesis activities more responsibly and creating opportunities for commercial partnerships that can lead to new software tools. The National Spatial Data Infrastructure (http://nsdi.usgs.gov) is improving the management of geographic, geological, and satellite datasets; the Digital Libraries (http://www.cise.nsf.gov/iis/dli_home.html) projects are beginning to produce useful results for some information domains; and the High-Performance Computing and Communications Initiative (http://www.hpcc.gov) has enhanced some computation-intensive engineering and science fields.

But little attention has been paid to computer and information science and technology research in the biodiversity and ecosystem domain. We must produce mechanisms that can efficiently search through terabytes of Mission to Planet Earth satellite data and other biodiversity and ecosystem datasets, make correlations among data from disparate sources, compile those data in new ways, analyze and synthesize them, and present the results in an understandable and usable manner. Despite encouraging advances in computation and communication performance in recent years, we are able to perform these activities on only a very small scale. We can, however, make rapid progress if the computer and information science and technology research community becomes focused on the needs of the biodiversity and ecosystem research community (Robbins 1996).

MANAGING COMPLEXITY

Knowledge about biodiversity and ecosystems is vast and complex. The complexity arises from two sources. The first is the underlying biological complexity of the organisms themselves. There are millions of species, each of which is highly variable across individual organisms, populations, and time. Species have complex chemistries, physiologies, developmental cycles, and behaviors resulting from more than 3 billion years of evolution. There are hundreds, if not thousands, of ecosystems, each comprising complex interaction among large numbers of species and between those species and multiple abiotic factors.

The second source of complexity in biodiversity and ecosystem information is sociologically generated. The sociological complexity includes problems of communication and coordination—among agencies, among divergent interests, and among groups of people from different regions and different backgrounds (academe, industry, and government) and with different views and requirements. The kinds of data that humans have collected about organisms and their relationships vary in precision, in accuracy, and in numerous other ways. Biodiversity data types include text and numerical measurements, images, sound, and video. The range of other databases with which biodiversity datasets must interact is also broad, including geographic, meteorological, geological, chemical, and physical databases. The mechanisms used to collect and store biological data are almost as varied as the natural world that they document. In addition, biological data can be politically and commercially sensitive and can entail conflicts of interest. Users' skill levels are highly variable, and training in this field is not well developed.

Because of those complexities, humans still play a crucial role in the processing of biological data. Biological information is not as amenable to automatic correlation, analysis, synthesis, and presentation as many other types of information, such as that in radioastronomy, where there is more coherent global organization and the problems being studied are often conducive to automatic analysis. In biodiversity research, people act as sophisticated filters and query processors—locating resources on the Internet, downloading datasets, reformatting and organizing data for input to analysis tools, then reformatting again to visualize results. This process of creating higher-order understanding from dispersed datasets is a fundamental intellectual process, but it breaks down quickly as the volume and dimensionality of the data increase. Who could be expected to "understand" millions of cases, each having hundreds of attributes? Yet problems on this scale are common in biodiversity and ecosystem research (Schnase and others 1997).

For a biological-information infrastructure to be effective, it must provide the means to manage complexity. It must allow scientists to extract new knowledge from the aggregate mass of information generated by the data-gathering and synthesis activities of other scientists. It must use the power of computers to facilitate the queries, correlations, and processing that are impossible for humans to perform alone. And it must deliver this functionality within a physically and intellectually accessible framework. That means developing ways of delivering information to a wide array of users with differing skills, ages, and investment in the material.

We are only beginning to develop a vocabulary to describe these large-scale, synthetic, information-processing activities. Some sociologists use the term *distributed cognitive system* to emphasize the role of humans in a synergistic information-processing network (Hutchins 1995). *Data mining* is often used by the database community. Whatever the name, the activities form only a part of a process of knowledge discovery that includes the large-scale, interactive storage of information (known by the unintentionally uninspiring term *data warehousing*); the cataloging, cleaning, preprocessing, transformation, verification, and reduction of data; and the generation and use of models, evaluation and interpretation, personal communication, the evolution of sophisticated user interfaces, and finally consolidation and use of the newly extracted knowledge. Those processes will become increasingly important if we are to use what we know and expand our knowledge in useful directions.

At present, the NBII provides little support for these activities. At best, it can be used to access information in databases held by federal agencies and other institutions around the country. Once the information is accessed, however, the task of organizing, integrating, and interpreting it remains, for the most part, a laborious, manual process. The development of computational tools for the biodiversity and ecosystem enterprise lags behind other sciences. Important classes of information are missing (information on fewer than 1% of the specimens in our natural-history collections has been entered in databases!), and databases are uneven in the types of information that they hold. It is difficult for individual scientists to publish their data electronically in useful ways. Standards for information exchange have not been widely adopted. We have no mechanism for archiving data over generations of use and generations of technologies. And the power of communication networks to build communities remains largely untapped. In summary, the NBII is neither a system nor an infrastructure: it is a cumbersome and brittle patchwork that presents as many obstacles to scientific work as it does opportunities. It clearly is time to transform it into a coherent and empowering capability.

THE NEXT GENERATION—NBII-2

We envision a "next generation" National Biological Information Infrastructure, NBII-2, that would address many of the concerns described above. The overarching goal of NBII-2 would be to become a fully accessible, distributed, interactive digital library. NBII-2 would provide an organizing framework from which scientists could extract useful information—new knowledge—from the aggregate mass of information generated by various data-gathering activities. That would be accomplished by using the power of computers and communication networks to augment the processing activities that now require a human mind. It would make analysis and synthesis of vast amounts of data from multiple datasets easier and more accessible to a variety of users. It would also serve management and policy decision-making, education, recreation, and industry by presenting data to each user in a manner tailored to that user's needs and skill level.

We envision NBII-2 as a distributed facility that would be considerably different from a "data center," considerably more functional than a traditional library, considerably more encompassing than a typical research institute. Unlike a data center, NBII-2 would have the objective of automatic discovery, indexing, and linking of datasets rather than collection of all datasets on a given topic into one facility. Following the best practice of traditional libraries, this special library would update the form of storage and upgrade information content as technologies evolve. Unlike a typical research institute, it would provide services to research going on elsewhere, and its own staff would conduct biodiversity and ecosystem research and research in biological informatics. The facility would offer "library" storage and access to diverse constituencies.

The core of NBII-2 would be a "research library system" that would comprise at least five regional nodes sited at appropriate institutions (national laboratories, universities, museums, and so on) and connected to each other and to the nearest telecommunication providers by the highest-bandwidth network available. In addition, NBII-2 would seamlessly integrate all computers—laptops, workstations, fileservers, and supercomputers—capable of storing and serving biodiversity and ecosystem data via the Internet. The providers of information would have complete control over their own data but have the opportunity to benefit from (and the right to refuse) the data-indexing, cleansing, and long-term storage services of the system as a whole.

NBII-2 would be

- the framework to support knowledge discovery for the nation's biodiversity and ecosystem enterprise and to involve many client and potential-client groups;
- a common focus for independent research efforts and a global context for sharing information among those efforts;
- an accrete-only, no-delete facility from which all information would be available on line—24 hours a day, 7 days a week—in a variety of formats;
- a facility that would serve the needs of (and eventually be supported by partnership with) government, the private sector, education, and individuals;
- an organized framework for collaboration among federal, regional, state, and local organizations in the public and private sectors that would provide improved programmatic efficiencies and economies of scale through better coordination of efforts;
- a commodity-based infrastructure that uses readily available, off-the-shelf hardware and software and the products of digital-library research wherever possible;
- an electronic facility where scientists and others could "publish" biodiversity and ecosystem information for cataloging, automatic indexing, access, analysis, and dissemination;
- a place where intensive work on how people use large information systems would be conducted, including studies of human-computer interaction, the

sociology of scientific practice, computer-supported cooperative work, and user-interface design;

• a place for developing the organizational and educational infrastructure that will support sharing, use, and coordination of massive datasets;

• a facility that would provide content storage resources, registration of datasets, and "curation" of datasets (including migration, cleansing, and indexing);

• an applied biodiversity and ecosystem informatics research facility that would develop new technologies and offer training in informatics; and

• a facility that would provide high-end computation and communication to researchers and institutions throughout the country.

The facility would not be a purely technical and technological construct but, would also encompass sociological, legal, and economic issues within its research purview. These would include intellectual-property rights management, public access to the scholarly record, and the characteristics of evolving systems in the networked information environment. The human dimensions of the interaction with computers, networks, and information will be particularly important subjects of research as systems are designed for the greatest flexibility and usefulness to people.

The research nodes of NBII-2 must address many needs, including

• new statistical pattern-recognition and modeling techniques that can work with high-dimensional, large-volume data;

• workable data-cleaning methods that automatically correct input and other types of errors in databases;

• strategies for sampling and selecting data;

• algorithms for classification, clustering, dependency analysis, and change and deviation detection that scale to large databases;

• visualization techniques that scale to large and multiple databases;

• metadata encoding routines that will make data mining meaningful when multiple distributed sources are searched;

• methods for improving connectivity of databases, integrating data-mining tools, and developing better synthetic technologies;

• methods for improving large-scale project coordination and scientific collaborations;

• continuing, formative evaluation, detailed user studies, and quick feedback between domain experts, users, developers, and researchers;

• methods for facilitating data entry and the digitization of large amounts of irregularly structured information; and

• ways of engaging society in the pursuit of global information-sharing.

None of those problems is peculiar to biodiversity research. However, there is an urgent need to address them in the biodiversity domain because research has demonstrated that there can be no domain-independent solutions. We cannot "borrow" discoveries wholesale from other disciplines; we must work through these problems ourselves (Star and Ruhleder 1996). To comprehend and use our biodiversity and ecosystem resources, we must learn how to exploit massive datasets, learn how to store and access them for analytical purposes, and develop methods to cope with growth and change in data. NBII-2 as envisioned here can be the enabling framework that unlocks the knowledge and economic power lying dormant in the masses of biodiversity and ecosystem data that we have on hand now and will accumulate in the future.

INFRASTRUCTURE REQUIREMENTS

The total volume of biodiversity and ecosystem information is almost impossible to measure. We do know that whatever the total, only a fraction has been captured in digital form. Our natural-history museums, for example, contain at least 750 million specimens, the vast majority of which have not been recorded in databases. The same holds for the published record, where most biodiversity and ecosystem information still resides in paper-based journals, books, field notes, and the like. Clearly, one of the most important infrastructure issues is to move the biodiversity and ecosystem enterprise into a digital world—to create the content for the NBII-2 digital library—by digitizing the existing corpus of scholarly work on a large scale.

The NBII-2 digital library will place challenging demands on network hardware services and on software services related to authentication, integrity, and security. Needed are both a fuller implementation of current technologies, such as digital signatures and a public-key infrastructure for managing cryptographic key distribution, and consideration of tools and services in a broader context related to library use. For example, the library system might have to identify whether a user is a member of an organization that has some set of access rights to an information resource. As a national and international enterprise that serves a large range of users, the library must be designed to detect and adapt to various degrees of accessibility of resources connected to the Internet.

A fully digital, interactive library system, such as NBII-2, will require substantial computational resources, although little is known now about the precise scope of the necessary resources. In many aspects that are critical to digital libraries, such as knowledge representation and resource description or summarization and navigation, even the basic algorithms and approaches are not yet well defined, so it is difficult to project computational requirements. Many information-retrieval techniques are intensive in their computational and input-output demands as they evaluate, structure, and compare large databases in a distributed environment. Distributed-database searching, resource discovery, automatic classification and summarization, visualization, and presentation are also computationally intensive activities that are likely to be common in the NBII-2 digital library.

Finally, NBII-2 will require enormous storage capacity. Even though the library system we are proposing would not set out to accrue datasets to become the repository of all biodiversity data—many other federal agencies have their own storage facilities, and various data-providers will want to retain control over their own data—large amounts of storage on disk, tape, optical media, and other future storage forms will still be required. As research is conducted to produce new ways to manipulate large datasets, these will have to be sought out, copied from their original sources, and stored for use in the research. And in serving its long-term curation function, NBII-2 will accumulate substantial amounts of data for which it will be responsible, including redundant datasets that will have to be maintained in case of loss.

RESEARCH AGENDA

New approaches to managing information must be developed in the context of NBII-2. Massive datasets can lead to the collapse of traditional approaches in database management, statistics, pattern recognition, personal-information management, and visualization. For example, a statistical-analysis package assumes that all the data to be analyzed can be loaded into memory and then manipulated. What happens when the dataset does not fit into main memory? What happens if the database is on a remote server and will never permit a naive scan of the data? What happens if queries for stratified samples cannot be accepted, because data fields in the database being accessed are not indexed and the appropriate data therefore cannot be located? What if the database is structured with only sparse relations among tables or if the dataset can be accessed only through a hierarchical set of fields?

Furthermore, challenges often are not restricted to issues of scalability of storage or access. For example, what if a user of a large data repository does not know how to specify the desired query? It is not clear that a structured query language (SQL) statement—or even a program—can be written to retrieve the information needed to answer a query like, "Show me the list of gene sequences for which voucher specimens exist in natural-history collections and for which we also know the physiology and ecological associates of those species." Many of the interesting questions that users of biodiversity and ecosystem information would like to ask are of this type: they are "fuzzy," the data needed to answer them must come from multiple sources that will be inherently different in structure and conceptually incompatible, and the answers might be approximate.

Major advances are needed in methods for knowledge representation and interchange, database management and federation, navigation, modeling, and data-driven simulation; in approaches to describing large, complex networked information resources; and in techniques to support networked information discovery and retrieval in extremely large-scale distributed systems. In addition to near-term operational solutions, new approaches are needed to longer-term issues, such as the preservation of digital information across generations of storage, processing, and representation technology. Traditional information-science skills, such as thesaurus construction and indexing, must be elaborated on and scaled to accommodate large information sources. We need to preserve and support the knowledge of library-science and information-science researchers and help to scale up the skills of knowledge organization and information retrieval.

Also much needed are software applications that provide more-natural interfaces between humans and databases than are now available. For example, a valuable data-cleansing activity might be to "show the data related to all specimens in our natural-history collections whose likelihood of being mislabeled exceeds 0.75." Assuming that some cases in the database can be identified as "labeled correctly" and others "known to be mislabeled," a training sample for a data-mining algorithm could be constructed. The algorithm would build a predictive model and retrieve records matching that model rather than a structured query that a person might write. This is an example of a much needed and much more natural interface between humans and databases than is currently available. In this case, it eliminates the requirement that the user adapt to the machine's needs rather than the other way around. We must refine and augment the interactions between people and machines, expand the role of agentry in information systems, and discover more-powerful and more-natural ways of navigating the scientific record.

In return, research in computer and information science and technology in the biodiversity and ecosystem domain is likely to yield discoveries of value to other fields (Spasser 1998). Nowhere do we find the problems of heterogeneous database federation more challenging than in the life sciences. A fully implemented digital library for biology would include everything from ideas to physical objects and enormous amounts of information in every medium type imaginable. Research on global climate change, habitat destruction, and the discovery of species is among the most distributed of our scientific activities and creates extraordinary opportunities to learn about computer-mediated project coordination and communication. At almost every turn, scale, complexity, and urgency conspire to create a particularly wicked set of problems. Working on these problems will undoubtedly advance our understanding and use of information technologies, perhaps more than in any other circumstance.

ACTION PLAN

We have laid out the case for building a fully digital, interactive, research-library system for biodiversity and ecosystem information and the basic requirements of and goals for the library and its research and service. But how much will it cost, and how long will it take to build?

We estimate that each of the regional nodes that will form the core of NBII-2 will require an annual operating budget of at least $8 million—probably more. Minimally, supporting five such nodes would require at least $40 million per year, an amount that is a small fraction of the funds spent nationwide each year to collect data (conservatively estimated at $500 million for federal government projects alone). As with the Internet itself, the federal government should provide the "jump start" for this new infrastructure by investing heavily in its formative stages. Part of the investment should be devoted to developing incentives for the participation of private-sector partners. Gradually, support and operation of the infrastructure should be shared by nongovernment participants, as has happened with the Internet.

The planning and request-for-proposals process should be conducted within a year. Merit review and selection of sites should be complete within the following six months. The staffing of the sites and initial coordination of research and outreach activities should take no more than a year after initial funding is provided. The "lifetime" of each facility should not be guaranteed for more than 5 years, but the system must be considered a long-term activity so that data access is guaranteed in perpetuity. Evaluation of the sites and of the system should be regular and rigorous, although the milestones whereby success can be measured will be the incremental improvements in ease of use of the system by students, policy-makers, scientists, and others. In addition, an increasing number of public-private partnerships that fund the research and other operations will indicate the usefulness of accessible, integrated information to commercial and government interests.

CONCLUSION

In the 21st century, work will depend increasingly on rapid, coordinated access to shared information. Through the shared digital library of NBII-2, scientists and policy-makers will be able to collaborate with colleagues who are geographically and temporally distant. They will use the library to catalog and organize information, perform analyses, test hypotheses, make decisions, and discover new ideas. Educators will use its systems to read, write, teach, and learn. In traditional fashion, intellectual work will be shared with others through the medium of the library—but these contributions and interactions will be elements of a global and universally accessible library that can be used by many different people and many different communities. By increasing the effectiveness of information, NBII-2 is likely to lead to scientific discoveries, advance existing fields of study, promote disciplinary fusions, and enable new research traditions. And most important, it could help us to protect and manage our natural capital so as to provide a stable and prosperous future.

REFERENCES

Hutchins E. 1995. Cognition in the wild. Cambridge MA: MIT. 381 p.

National Performance Review. 1997. Access America: reengineering through information technology. Report of the National Performance Review and the Government Information Technology Services Board. Washington DC: GPO. 97 p.

NRC [National Research Council]. 1993. A biological survey for the nation. Washington DC: National Acad Pr. 205 p.

NRC [National Research Council]. 1997. Bits of power: issues in global access to scientific data. Washington DC: National Acad Pr. 235 p.

PCAST [President's Committee of Advisors on Science and Technology]. 1998. Teaming with life: investing in science to understand and use America's living capital. Report to the President of the United States from the PCAST Panel on Biodiversity and Ecosystems. Washington DC: GPO.

Robbins RJ. 1996. Bioinformatics: essential infrastructure for global biology. J Comp Biol 3(4):465–78.

Schnase JL, Kama DL, Tomlinson KL, Sánchez JA, Cunnius EL, Morin NR. 1997. The flora of North America digital library: a case study in biodiversity database publishing. J Network Comp Applica 20:87–103.

Spasser MA. 1998. Articulating collaborative activity: design-in-use of collaborative publishing services in the Flora of North America Project. Proceedings of ISCRAT '98 (Århus, Denmark, June 1998).

Star SL, Ruhleder K. 1996. Steps toward an ecology of infrastructure: design and access for large information spaces. Info Syst Res 7(1):111–34.

PART

5

THREATS TO

SUSTAINABILITY

NATURE DISPLACED:
HUMAN POPULATION TRENDS AND PROJECTIONS
AND THEIR MEANINGS

RICHARD P. CINCOTTA
ROBERT ENGELMAN
Population Action International,
1120 19th Street, NW Ste. 550, Washington, DC 20036-3678

UNLIKE THE GREAT species extinctions of Earth's past, the one occurring today is less an episode than a process, whose full results will not be known for hundreds of years. Between the linked human-induced phenomena of global climate change and biodiversity loss, the planet could be passing into the equivalent of an entirely new geological epoch in just a few human generations. Or it could be that biodiversity loss will amount to little more than a manageable depletion, incurring regrettable scientific and economic losses but leaving the basic services provided by most major ecosystems largely intact.

The size and distribution of human population over the near and distant future will surely be a dominant factor in determining whether the loss of biodiversity that the world faces turns into merely a source of wistful regret for future generations, a planetary catastrophe, or something in between. Population growth enlarges the scale and extent of the human enterprise and hence inflates the likelihood that human activities will push native nonhuman populations and biotic communities past critical thresholds of tolerance and renewal.

Demands for housing (Mason 1996), food energy and arable land (Bongaarts 1994; Engelman and LeRoy 1995; Smil 1994), freshwater (Engelman and LeRoy 1993; Falkenmark and Widstrand 1992), and industrially fixed nitrogen (Howarth and others 1996; Smil 1991; Vitousek and others 1997) appear more sensitive to the growth of human population than to the growth of per capita income or even to recent changes in technological efficiency. Habitat conversion, historically the greatest threat to biodiversity, has been driven by these very demands—by housing needs, pressures to expand and intensify agriculture, and the quest to harness additional freshwater supplies. Climate change, the demise of commercial fish populations and coastal reefs, widespread soil degradation, and the re-emergence of infectious disease also reflect the strong influence of population dynamics and take a growing toll on biodiversity. These global changes threaten ecosystem function and raise the risk of future extinction. It thus makes sense to consider the prospects for human population growth.

In this article, we consider those prospects by examining the United Nations (UN) population projections—both *how* and *what* they project. The methods and meaning of UN projections are poorly understood by scientists outside the field of demography. And the recent misuses of the projections in the press have confounded the public.

Despite widespread perceptions to the contrary, there is nothing inevitable about most future human population growth. Our species now numbers 6 billion and is growing at a pace of just over 80 million per year. More than 95% of this growth is occurring in countries of the developing world. Most demographers expect human population at least to approach 8 billion in the next half-century. Beyond that expectation, however, no one can be certain that world population will ever rise to greater levels. There is equal uncertainty that population will stop growing at any particular time in the not too distant future.

We can be certain, however, that today women in most developing countries desire fewer children than their mothers or even their older sisters sought or had (Westoff 1991). Over the last 30 years, that trend, when supplemented with access to modern contraception and the information needed to use it safely and effectively, clearly has resulted in lower rates of childbearing in countries with traditionally high fertility (Robey and others 1994). In the future, changes could occur even more rapidly. Decisions made today will have an enormous influence on the demographic future. These decisions are likely to be among the most important that we can make to conserve as much as possible of the planet's remaining biodiversity.

HUMANITY'S PLACE IN NATURE

Few scientists outside the field of ecology are aware of how ecologically unprecedented is the scale of human numbers—not just *present* numbers, but also those of the last several millennia. No other mammal of comparable body weight has ever attained anywhere near such abundance. By manipulating the qualities and quantities of other species through agriculture, *Homo sapiens* broke through the energy and nutrient constraints that limited it as a hunter-gatherer.

Statistical models relating the adult body weight of mammals to their observed abundance (Peters 1983, p 166–7) predict that the equilibrium density of mammalian species in their home ranges will vary according to the following relationships: $D_C = 15 \, W^{-1.16}$ for carnivores and $D_H = 103 \, W^{-0.93}$ for herbivores (grazers and browsers), where D is animal density expressed in individuals per square kilometer, and W is the adult body weight in kilograms. For a carnivorous mammal or herbivore the size of *Homo sapiens* (roughly 65 kg), these relationships predict 0.12 individual/km^2, and 2.1 individuals/km^2, respectively. The natural availability of preagriculture human diets, however, fell between carnivorous and herbivorous diets. In fact, we are still largely grain-, fruit-, and tuber-eating with a predilection for meat. A liberal estimate of the average density that our species would likely have attained *without agriculture* is around 1 individual/km²—similar to the density at which hunter-gatherers and nomadic pastoralists lived until relatively recently.

If preagriculture humans at that density were to exploit every square kilometer of Earth's habitable terrestrial surface, about 130 million square kilometers (Hannah and others 1994), the world would support roughly 130 million people. According to one estimate, world population surpassed that number during the early years of the Roman Empire (Biraben 1979 reprinted in Livi-Bacci 1992; Cohen 1995). The United States alone surpassed it just before World War II (US Bureau of the Census 1995).

DEMOGRAPHICS THEN AND NOW

We know with reasonable certainty that *Homo sapiens* has expanded in numbers from at most a few tens of millions in prehistory to nearly 6 billion at the close of the 20th century. Most of these billions arrived in the 20th century, as the march of technology (especially in sanitation, immunization, and agriculture) allowed, for the first time, the vast majority of babies born to survive to become parents themselves. Some of the most rapid population growth during the 19th century occurred in the United States, where annual increases, roughly 2.5–3%, were as high then as in sub-Saharan Africa today. The consistently high 19th-century growth rates are a major reason that the United States is today the third most populous country in the world.

The result of the victory over infant and child death is evident in every region and major city. The planet sustains nearly half its humans in urban areas. Roughly three of every five people live in Asia. Each of the other major world regions is home to several hundred million people, but the populations of the continents are growing at markedly different paces: Europe, with about 730 million people, at a mere 0.2% per year (UN 1996a); North America (mostly the United States and Canada), with 300 million, at about 1.0% annually; Asia, with 3.5 billion, at about 1.5% per year; and the Latin American and Caribbean region, with about 485 million, at about 1.7% per year. Standing apart from the rest of the world demographically is Africa, with 708 million, where population growth has continued for decades at nearly 3% per year, falling slightly now to 2.7%. The average of all uneven rates of growth worldwide is equivalent to that of Asia, or about 1.5% per year.

Despite the ever-larger population base, world population growth is gradually slowing. The annual rate peaked at 2.1% in the late 1960s and has drifted down since. When a growth *rate* decreases, however, *growth* itself continues until the rate reaches zero. And substantial growth continues decades after fertility descends to replacement levels (slightly more than two children per woman) or even dips below. That effect, known as population momentum, is due to the long lag time between birth and reproductive maturity that characterizes our species. The lag allows past growth to continue to augment the absolute size of the reproductive segment of the population (women roughly 15–49 years old), thus supporting high numbers of births despite a decrease in fertility to replacement level.

As world population increases, more modest rates of growth can add larger annual increments to the population base. That has occurred although the highest rates of global population growth, estimated to have occurred around 1970, saw only about 72 million people added to world population each year. Current lower rates of growth are adding more than 80 million people per year. The global annual growth increment itself has declined since 1988 and could continue to decline—although by how much and for how long is unknown. A previous temporary decline during the middle 1970s, reflecting devastating effects of famine and political upheaval on the age structure of China's huge population (NRC 1984), illustrates how uncertain demographic projections can be.

During the 1970s and 1980s, human fertility in industrialized countries, which was already near replacement levels, declined once more. Nearly all the European countries fell below the roughly two-children-per-couple average that, in the absence of immigration, is necessary to replace each generation with the one that follows. The meaning of that trend for future population is potentially enormous.

Throughout the developing world, couples desire smaller families and later childbirths and they increasingly have the means to achieve the family size they seek. Several good examples can be gleaned from East Asia and Southeast Asia. During the middle 1960s, South Korea, Taiwan, Singapore, Thailand, and the former Hong Kong Territory began effective programs to lower infant mortality, establish easy access to family-planning services (ADB 1997; Tsui 1996), and increase primary-school enrollments and educational attainment (ADB 1997; Birdsall and Sabot 1993; Birdsall and others 1996; UNDP 1996; World Bank 1993). Thirty years later, average fertility in each of these Asian states is below two children per woman (the US average).

Other developing countries—including Mexico (3.1 children per woman), Brazil (2.4), Indonesia (2.9), Tunisia (3.3), and Sri Lanka (2.2)—are also experiencing downward trends in fertility (UN 1996b). A recent analysis of regional patterns of demographic change (Bongaarts and Watkins 1996) suggests that the first country in each developing region to begin its transition to lower fertility was endowed with relatively high indicators of social and economic progress, as measured by the UN's Human Development Index (UNDP 1996). In each case, however, fertility decline spread to nearby countries—probably via transfers of expertise, experience, and information at the government and local levels— despite the neighbors' lower scores for economic and social development.

In most developed countries, where there is access to affordable, effective contraception and safe abortion, women are more likely to have the number of children they want than are women in developing countries. Where these circumstances prevail and where childbearing and rearing are expensive or constrain economic mobility, total fertility consistently remains below the replacement level of slightly more than two children per woman (Potts 1996).

The other great trend shaping world population, aside from changes in fertility, is rapid mortality decline—or rapid increase in average life expectancy. Life expectancy began its climb in middle-18th-century Europe (figure 1). By the late 20th century, people in all corners of the world had longer life expectancies. The dominant influences are at both ends of the age spectrum: smaller proportions of children are dying in the first few years of life, and larger proportions of adults are surviving to old age.

Demographers assume that the mortality decline will continue, placing some further upward pressure on the pace of population growth. Falling mortality, however, could moderate worldwide as additional improvements in health care and nutrition become more difficult to achieve. In eastern Europe, mortality has actually risen in recent years; and in sub-Saharan Africa, the AIDS pandemic is reversing recent progress in infant mortality (US Bureau of the Census 1994). Both trends and the growing specter of emerging infectious diseases (Olshansky and others 1997) raise questions about the strength of the UN's assumption of continued mortality decline well into the 21st century.

THE PROJECT OF PROJECTING

In projecting an image of the future, the challenge for demographers is to understand the complex and uneven trends in fertility, mortality, and migration and to consider to what extent they are likely to continue and—perhaps most critical—at what levels they might end. Given the hodgepodge of modern demographic trends, all that can be said with certainty about future trends and end points is that *we cannot be certain*. The UN Population Division, which produces the most widely cited tables of international population information, has addressed such uncertainty by computing every 2 years a three-piece set of population projections. The most recent series, published in 1996, projects populations for each of the UN's 185 member countries to 2050 (see UN 1996b).

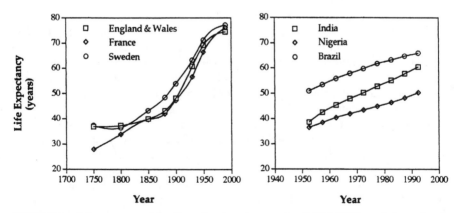

FIGURE 1   Life expectancy in three developed countries (1750–2000) and three developing countries (1950–2000). European life expectancies for years before 1950 from tables compiled by Livi-Bacci (1992), citing various authors who have analyzed historical records. Data from 1950 and beyond from current UN tables (UN 1996b).

But scientists and journalists should take note: the UN projections are *not* statistically predictive. They are not estimates calculated from models of underlying behavioral relationships, nor are they the extrapolated curves with which biologists are most familiar. For that reason, the projections tend to be poorly understood and commonly misused.

The three pieces making up the UN's set of projections are its low, medium, and high variants generated for each country. Each variant differs from the other two in just a couple of key assumptions—its fertility end point and the path of fertility to that end point (see country examples in figure 2). When plugged into a model that generates births and eliminates the dead from each age group (and adds immigrants and subtracts emigrants where necessary), each variant traces a different population trajectory through the future.

To generate the medium variant's fertility curve, UN demographers use assessments of each country's situation and progress to make an educated guess of when each country will achieve replacement-level fertility. In each case, for projection purposes, this date is assumed to fall before midcentury. A fertility trajectory is then created that allows national fertility to fall—or, where needed, to rise—smoothly to its replacement-level end point. Once fertility arrives at this point, it is assumed to stay there indefinitely. By completing this exercise for fertility of each country (using standard mortality assumptions) and migration, adding all national populations for each year computed, the UN arrives at a continuous medium variant trajectory for world population.

The 1996 UN medium variant projects a global population of about 9.4 billion people around the middle of the 21st century, compared with the known 2.5 billion in 1950 and the 5.9 billion in 1998. If extended beyond 2050, as the UN does in its long-range projections (UN 1992; also see Haub and Yinger 1992; McNicoll 1992), population then grows fairly slowly, stabilizing at around 12 billion early in the 22nd century. The medium variant, however, is only one element of the projections.

To generate the other elements, the low and high variants, the UN adheres to the same model used to generate the medium variant but adjusts the fertility end point and the path of fertility to that point. In the low variant—a lower bound for plausible scenarios—each country's fertility end point is reset to achieve 1.6 children per woman before 2050 and held constant thereafter. To fix an upper bound of plausibility, the high variant applies the same schedule to settle at 2.6 children per woman.

Those are not error limits. Instead, the low and high variants are distinct, but extreme, scenarios of demographic change applied to every country. The low variant mimics the behavior of many European and several East Asian populations that over the last 2 decades have dipped below replacement fertility (1.2–1.9 children per woman). The high variant mimics a number of Central American and South American countries that have momentarily stabilized at levels somewhat higher than replacement (Haub and Yinger 1992). For example, total fertility of both Uruguay and Argentina has fluctuated erratically below 3.5 children per woman for at least 50 years without ever having reached replacement levels. In

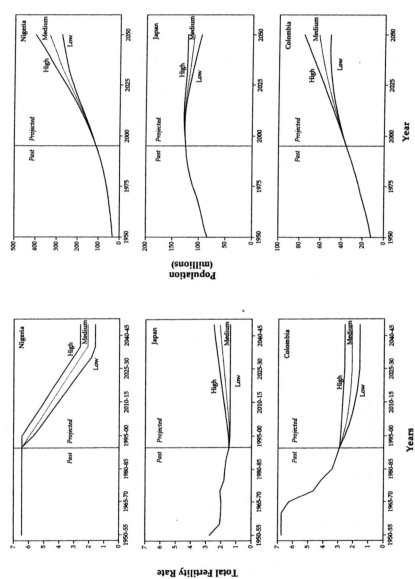

FIGURE 2 Fertility curves for three countries (UN 1996b)—past data and projected high-, medium-, and low-variant scenarios and corresponding national population projections (UN 1996c).

Costa Rica and Chile, fertility declined rapidly during the 1970s but stalled at similar levels.

By generating low and high variants, the UN projections present an envelope of plausibility, suggesting that a range of population futures is possible. Those two scenarios project a 2050 world population between 7.9 billion and 11.9 billion (figure 3).

A SEPARATE DEMOGRAPHIC REALITY

The demographic experience of the world suggests that total fertility is dynamic and highly responsive to the circumstances of women and couples. The UN series of projections, however, must necessarily remain mechanical and thus reproducible every 2 years. Perhaps the most mechanical feature is the UN's assumption of a stable fertility end point for each variant. But even the paths drawn to those end points often appear inconsistent with past data.

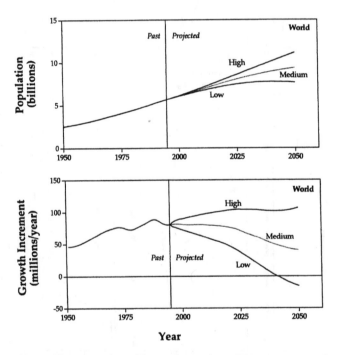

FIGURE 3   Past and projected world population from UN estimates and projections (1996c) and annual increment of world population growth (annual change in growth) derived from these data and projections. The trough in the world-population growth increment that began in the middle 1970s was caused by irregularities in China's growth increment. China's irregular growth was due to an age structure shaped by high mortality (NRC 1984) and low fertility (Coale and Li 1987) that occurred in the wake of famine and political upheaval during the Great Leap Forward from 1958 to 1960.

Three examples illustrate these points (see figure 2). In the case of Nigeria, where there is still little evidence of substantial demographic change, all three fertility variants seem highly speculative and contrived. For Japan's projected fertility, at least two of the variants seem difficult to reconcile with past trends. In the case of Colombia, however, high-, medium-, and low-variant fertility curves all seem similarly consistent with past data.

Although there are good reasons to expect fertility decline to continue where families are typically large, there is no particular reason to assume that fertility rates will settle between 2.0 and 2.1 or at 2.6 or 1.6 children per woman. In fact, Sweden, Luxembourg, the United Kingdom, and France—each below replacement-level fertility today—have been there before (UN 1996b; Livi-Bacci 1992), bobbing back up above replacement-level during national baby booms and moving downward during political and economic turmoil.

Recently, the UN low-variant population projection has been used by several analysts and journalists (Buchanan 1997; Eberstadt 1997; Wattenberg 1997) as evidence that UN demographers are predicting alarming declines in global population beginning as "soon" as 2040. That is either a gross misunderstanding or a misuse of the projections. The low variant does, in fact, trace a downward path after that date. But eventually it must, by its very nature. With each national population in the world fixed forever at 1.6 children per woman—about a half-child below the replacement level—there is ultimately nowhere for the calculated population to go but down. The high variant, just as artificially, forces the trajectory upward, and the medium variant ultimately forces stability. Clearly, it makes little sense to use one variant without reference to the others.

There is no good reason to assume that the below-replacement-level fertility experienced in some industrialized countries today will be sustained long enough to lead to a substantial net population decline in the long run. Fertility rates might well rise again if the direct costs or opportunity costs of childrearing decline or if larger families regain social approval. Nor does it make sense to assume that below-replacement-level fertility returns to and stabilizes exactly at replacement-level fertility. Realistically speaking, we do not know.

In practice, most journalists and analysts take the UN's "medium variant," or middle trajectory, to be the most probable one, whether for national, regional, or global population figures. It is often expressed inaccurately as the "expected" population future. That hardly makes the medium projection the "most likely" scenario within the wide range of plausible paths described by the high and low variants. True, neither the high nor the low extreme could be properly considered as "likely"—they are extremes, after all—but neither is there any special center of gravity midway between them. In fact, the medium projection uses a reasonable, repeatable method that cuts a path through a future without surprises, one in which demographic change is gradual and limited.

A relative absence of demographic surprise, however, has not always been the rule. Until the 1950s, demographers most often underestimated population growth. The largest cause of failure among early population projections was that their authors missed the fact that mortality was falling at an increasing rate in the developing world. The country-by-country triumphs of sanitation, clean water

312 / NATURE AND HUMAN SOCIETY

supply, antibiotics, and vaccinations were the big surprise. Of less impact, but still important, were increases in fertility in the industrialized countries after World War II—the so-called baby boom—that were difficult to foretell.

Strictly speaking, no population growth—not even tomorrow's—is really *certain*. Obviously, in the unlikely event that a nuclear war breaks out or a comet hits, all demographic bets are off. But leaving aside those unlikely possibilities, it is sobering to remind ourselves that infectious disease, war, and economic disruption still strongly influence population dynamics of individual nations and could do so more forcefully in the future. Clearly, words like *inevitable* and *certain* overstate the case.

More important, such language lulls observers into a conviction that no action in the present can influence the near demographic future. With much of the developing world exposed to television and computers in the span of a mere decade or two, a revolution in fertility patterns cannot be ruled out either. The likelihood of such changes is discounted in projections, perhaps reasonably, but such assumptions receive no discussion when the projection results are released to the public.

NATURE'S PLACE IN HUMANITY

Humanity is pushing our planet across a series of important environmental thresholds at a time when our institutions—even in democratic societies—seem disinclined to take such threats seriously. This is the case whether the need is to secure the future or to help those whose well-being is most threatened today (Cincotta and Engelman 1997). What nonhuman genetic endowment shall we strive to preserve for future generations? Although the question is largely ethical and biological, we can be sure that demographics and economics will ultimately provide history with much or most of the answer (Morowitz 1991).

Among the complex factors that drive these changes in our ecology, human population growth is arguably the most easily addressed. Ten years ago, that statement would have seemed absurd. Five years ago, it might have been considered bravado. Now demographers tell us that the cessation of human population growth is within reach during the 21st century. We stand on that century's doorstep.

How, then, should scientists view and represent the prospects for world population? Certainly not in terms of any inevitable figure. In peering into the future, it is useful to consider the UN population projections—the entire range described by the variants, not just medium variants—as a reasonably sound basis for describing a demographic future without substantial surprises.

We must loosen the grip that the medium projections have on the limited attention of policy-makers and the public. We need at least to bring attention to the range of growth suggested by the low and high projections for the next century and beyond. And, despite its necessarily artificial quality, we should hold forward the realistic hope offered by the low variant. Population growth might well slow further within the next few decades. Population size might peak before adding more than 2 billion people to our current numbers. And the world might someday experience a degree of population decrease before attaining relative long-term stability.

If the scenario becomes real, it will not be the product of "population control" or coercive government family-size targets. Rather, such a world will grow out of consistently pursued development initiatives that focus largely on the capacity of women to manage their own lives, especially their reproductive options. Such initiatives slow population growth while serving more immediate human needs. And in slowing and eventually easing to a halt the growth of human population, such a strategy can help to ensure that nature and its myriad ecosystems and species do not recede forever from their rightful place on the planet.

REFERENCES

ADB [Asian Development Bank]. 1997. Emerging Asia. Manila Phillipines: Asian Development Bank.

Biraben J-N. 1979. Essai sur l'Evolution du Nombre des Hommes. Population (Paris) 34(1):13–25.

Birdsall N, Bruns B, Sabot RH. 1996. Education in Brazil: playing a bad hand badly. In: Birdsall N, Sabot RH (ed). Opportunity foregone. Washington DC: Inter-American Development Bank. p 7–47.

Birdsall N, Sabot RH. 1993. Virtuous circles: human capital growth and equity in East Asia. World Bank Working Paper, Policy Research Department. Washington DC: World Bank.

Bongaarts J. 1994. Can the growing human population feed itself? Sci Amer 270(3):35–42.

Bongaarts J, Watkins SC. 1996. Social interactions and contemporary fertility transitions. Pop Devel Rev 22(4):639–82.

Buchanan P. 1997. Demographic decline of west. New York Times, 17 Nov. p A23.

Cincotta RP, Engelman R. 1997. Economics and rapid change: the influence of population growth. PAI Occas Pap No 3. Washington DC: PAI.

Coale AJ, Li CS. 1987. Basic data on fertility in the provinces of China, 1940–1982. Pap East-West Pop Inst, No 104. Honolulu HI: East-West Center Program on Population.

Cohen JE. 1995. How many people can the earth support? New York NY: WW Norton.

Eberstadt N. 1997. The population implosion. The Wall Street Journal, 16 Oct. p A22.

Engelman R, LeRoy P. 1993. Sustaining water: population and the future of renewable water supplies. Washington DC: PAI.

Engelman R, LeRoy P. 1995. Conserving land: population and sustainable food production. Washington DC: PAI.

Falkenmark M, Widstrand C. 1992. Population and water resources: a delicate balance. Washington DC: Population Reference Bur.

Hannah L, Lohse D, Hutchinson C, Carr JL, Lankerani A. 1994. A preliminary inventory of human disturbance of world ecosystems. Ambio 23(4–5):246–50.

Haub C, Yinger N. 1992. The UN long-range population projections: what they tell us. Washington DC: Population Reference Bur.

Howarth R W, Billen G, Swaney D, Townsend A, Jaworski N, Lajtha K, Downing JA, Elmgren R, Caraco N, Jordan T, Berendse F, Freney J, Kudeyarov V, Murdoch P, Zhu Z-L. 1996. Regional nitrogen budgets and riverine N & P fluxes for the drainages to the North Atlantic Ocean: natural and human influences. Biogeochemistry 35:181–226.

Livi-Bacci M. 1992. A concise history of world population. Cambridge MA: Blackwell.

Mason A. 1996. Population, housing, and the economy. In: Ahlburg DA, Kelley AC, Mason KO (eds). The impact of population growth on well-being in developing countries. Berlin Germany: Springer. p 175–360.

McNicoll G. 1992. The United Nations' long-range population projections. Pop Devel Rev 18(2):333–40.

Morowitz HJ. 1991. Balancing species preservation and economic considerations. Science 253:752–4.

NRC [National Research Council]. 1984. Rapid population change in China, 1952–1982. Report No 17. Washington DC: National Acad Pr.

Olshansky SJ, Carnes B, Rogers RG, Smith L. 1997. Infectious diseases—new and ancient threats to world health. Washington DC: Population Reference Bur.

Peter RH. 1983. The ecological implications of body size. Cambridge UK: Cambridge Univ Pr.

Potts M. 1997. Sex and the birth rate: human biology, demographic change, and access to fertility regulation methods. Pop Devel Rev 23(1):1–39.

Robey B, Rutstein SO, Morris L. 1993. Fertility decline in developing countries. Sci Amer. 269:60–7.

Smil V. 1991. Population growth and nitrogen: an exploration of a critical existential link. Pop Dev Rev 17(4):569–601.

Smil V. 1994. How many people can the Earth feed? Pop and Dev Rev 20(2): 255–92.

Tsui AO. 1996. Family planning programs in Asia: approaching a half-century of effort. Asia Pop Res Rep No 8. Honolulu HI: East-West Center Program on Population.

UN [United Nations]. 1992. Long-range world population projections: two centuries of population growth, 1950–2150. New York NY: UN Department of International Economic and Social Affairs.

UN [United Nations]. 1996a. World population 1996, wall chart. New York NY: UN Department for Economic and Social Information and Policy Analysis, Population Division.

UN [United Nations]. 1996b. World population prospects: the 1996 revision. New York NY: UN.

UN [United Nations]. 1996c. World population prospects: the 1996 revision. computer diskettes. New York NY: UN. (Note: the diskettes feature annual population data, rather than in 5-year intervals as appears in the printed publication).

UNDP [United Nations Development Programme]. 1996. Human development report, 1996. New York NY: UN.

US Bureau of the Census. 1995. Census of population and housing. CPH-2-1, Population and Housing Unit Counts, Table 16. Washington DC: US Dept of Commerce. (http://www.census.gov/population/www/ censusdata/ pop-hc.html). (last accessed : January 5, 1998)

US Bureau of the Census. 1994. The Impact of HIV/AIDS on world population. Washington DC: US Government Printing Office.

Vitousek PM, Aber JD, Howarth RW, Likens GE, Matson PA, Schindler DW, Schlesinger WH, Tilman D. 1997. Human alteration of the global nitrogen cycle: sources and consequences. Ecol Appl 7(3):737–50.

Wattenberg BJ. 1997. The population explosion is over. New York Times Magazine, 23 Nov. p 60–3.

Westoff CF. 1991. Reproductive preferences: a comparative view. Demographic and Health Surveys Comparative Studies No 3. Columbia MD: Institute for Resource Development/Macro Systems.

World Bank. 1993. The East Asian miracle: economic growth and public policy. Oxford UK: Oxford Univ Pr.

POPULATION GROWTH, SUSTAINABLE DEVELOPMENT, AND THE ENVIRONMENT

SERGEY KAPITZA

Population Action International,
Washington, DC
(Current address: Institute for Physical Problems, Russian Academy of Sciences,
2 Kosygina St., Moscow 117334, Russia)

THE LONG-TERM state of the biosphere—the conservation of species and of bio-diversity—will depend to a great extent on the population growth of the world and the demographic pressure on the environment. At present, the population of the world is 5.9 billion, and it is growing by 1.5% a year; 250,000 inhabitants are added every day. Practically all population growth occurs in the developing world. At the same time, much of the industrial development also happens there. On the other hand, in the foreseeable future, because of transition in the population, global population is expected to level off at 12–14 billion. It is in these terms that one should consider the effect of humankind on the future of our planet, taking into account the trends in development seen in the larger context of the dynamics of the world-population system.

The links of population growth and development were the subject of a seminal statement of the Royal Society of London and the US National Academy of Sciences titled *Population Growth, Resource Consumption, and a Sustainable World* and signed by the presidents of the two societies (Atiyah and Press 1993). In that statement, probably for the first time, two great academies voiced their opinion on this all-important and sensitive subject. They came to the following conclusions:

> The applications of science and technology to global problems are a key component of providing a decent standard of living for a majority of the human race. Science and technology have an especially important role to play in developing countries in helping them to manage their resources effectively and to participate fully in worldwide initiatives for common benefit. Capabilities in science

and technology must be strengthened in LDCs [less-developed countries] as a matter of urgency through joint initiatives from the developed and developing worlds. But science and technology alone are not enough. Global policies are urgently needed to promote more rapid economic development throughout the world, more environmentally benign patterns of human activity, and more rapid stabilization of world population. The future of our planet is in the balance. Sustainable development can be achieved, but only if irreversible degradation of the environment can be halted in time. The next 30 years may be crucial.

The issues raised in that statement have grown in importance, and a major international debate has followed. The problem of global population growth has been reviewed thoroughly, and it is probably best summed up in the book *The Future Population of the World: What Can We Assume Today?*, edited by Lutz of the International Institute for Applied Systems Analysis, IIASA (Lutz 1994). An extensive study by Cohen (1995), *How Many People Can the Earth Support?*, reviews a vast amount of data and many ideas and misconceptions. Perhaps the complexity and difficulty of these subjects, which are essential to discussions of sustainable development, can be seen in the fact that the titles of both books are questions.

The book by Cohen and chapter 10 of the book by Lutz, "How Many People Can Be Fed on Earth?", show that the idea of carrying capacity is counterproductive, if not misleading or even wrong. That point is well illustrated by the "limits" suggested by various writers since 1600, when fewer than a half-billion people lived. Until 1900, the limits indicated are rather similar and mostly reasonable, around 10–15 billion; these figures compare well with modern assessments. The huge discrepancies—from 1 billion to 1 trillion—seen in projections from the last 50–100 years more likely indicate the ups and downs of the subjective mood, private and public, that in its own way expresses the turbulent history and transitory nature of 20th century, rather than a trend toward a greater understanding of human destiny.

PROJECTIONS OF POPULATION GROWTH BY DEMOGRAPHIC METHODS

Those projections of the world's population were based on assumptions of the availability of resources, mainly land. Most modern estimates, however, are the result of extensive studies of population growth that look at population dynamics rather than resources. Results can be obtained with standard demographic methods and are valid for 1 or 1.5 generations. For periods of longer than about 30–50 years, demographic calculations become computationally unstable. All extrapolations farther into the future come from plausible hypotheses regarding the future development of humankind, which set guidelines for the calculations made. For the year 2100, the most probable projections by a team at IIASA indicate a population of 11.5 ± 1 billion. That estimate indicates that in the next 100 years, the population of our planet will barely double (Lutz and others 1994).

Of critical importance is that the future population of the world will be determined not by the incessant growth that has marked development until the present but by a complex transition to a stabilized world population. This demographic transition, as it has become known, is the crucial feature of modern human population growth. The phenomenon, first recognized by Notenstein in 1950, now is determining the lack of population growth in the developed countries; in the next 50 years, it undoubtedly will lead to a decrease in the rate of growth and the final leveling off of the world's population. For the more-distant future, demographers can make only plausible guesses, estimating a stabilized world population of some 12 billion.

The demographic transition will be accompanied by a marked change in the age structure of the population. Currently, people younger than 15 years make up 32% of the world population and people older than 65 years, 7%; after the transition, the numbers will be 18% and 22%, respectively. At the same time, a huge number of people will move to towns in a global pattern of urban development. Ultimately, the demographic transition will lead to major changes in the lifestyle and values of billions of people. It is certainly the most significant event in human history, seen on a large scale. That is why it is well worth the effort to study the demographic transition with methods other than those provided by modern demography.

MODELING THE GROWTH OF HUMANKIND

I have developed an alternative approach (Kapitza 1994, 1996a,b), which interprets the global population as an interactive dynamic system. The entire population of the world is the object of study. The global-population system is considered to be an entity, coupled with interactions that determine its long-term growth and development, rather than as a mere sum of countries and regions, each following its own pattern of growth. This is the next step in generalizing population growth. In fact, when describing the demographics of China or India, we already are summing up, in a single measure of growth, a vast country that has many regions and cultures of great ethnic diversity and that encompasses 17–20% of all humanity.

In treating the whole of humankind in such a general way, it is also possible to expand the time scale considerably. One must break away from the unit of a generation that is used customarily in demography and even go beyond the millennia of history to the millions of years that provide the scale for human development as seen in anthropology. For shorter intervals, this way of describing the growth of human population must merge with and rely on the methods and concepts of demography. Thus, the two ways of describing our growth and development complement each other and provide mutual support and justification of their results.

If we consider the long-term pattern of the growth rate of the human population, we see that the rate is proportional to the square of the total number of people. This nonlinear growth corresponds to a hyperbolic growth curve and is well known for describing explosive systemic development. For humankind, quadratic growth is valid for more than 1 million years into the past, right from the appearance of *Homo habilis*, the primeval tool-maker. The quadratic law of growth is applicable only to the total number of people; it cannot be applied to describe regional growth. But every region and country participates in and is influenced by it because we are dealing with a nonlinear law that implies a global interaction in the complex population system of the entire world.

However, the quadratic law leads to a divergence in a finite time. The divergence has begun already; if the growth rate that has been valid for so long persists, the runaway into infinity will happen in 2025. The pattern of quadratic growth can describe the first stage of the population transition—the population explosion. This is a new way of looking at the demographic transition as a phase transition, describing it in terms and methods that come from nonlinear physics of systems.

For explosive divergences, it is well known that a cutoff must be brought in, taking into account factors of less importance during the main period of growth. As we approach the singularity, the concepts of demography become significant. They indicate that if we take into account the finite human life span and reproductive time, we can expect the whole pattern of growth to change. The development of this phenomenon, following well-established methods of systems analysis and physics, envisages an asymptotic transition to a stabilized world population of 12–14 billion, with 12 billion reached in 2100. A time constant of 42 years, characterizing the human life span, has to be incorporated into the calculations.

This model has developed into a theory that describes the gross features of the growth of humankind. It provides an estimate of the beginning of human development, some 4–5 million years ago, and an estimate of the number of people who ever lived, of 100 billion. As we approach the critical year of 2007 (it shifts from 2025 when the 42-year cutoff is taken into account), the dynamics of growth indicate a logarithmic compression of the time scale of history. In 2007, the maximal annual growth of 85–90 million is expected, but the relative growth rate already reached its peak, 17% per year, in 1989.

My approach now reconciles well with the methods of demography and can be seen as complementary in treating the same problem on a larger scale: the way human population has led to the concept of the population imperative. This means that growth has not been limited globally by resources, but is governed by the inherent nature of the systemic interactive dynamics of the global population system. On a large scale, growth has been systemically stable, although local and temporal variations have occurred.

FACTORS THAT LIMIT POPULATION GROWTH

At this point, it is appropriate to mention the fundamental differences between my model and the Malthusian population principle. According to Malthus and those who developed the same ideas later into the "limits-to-growth" concept, growth is limited by resources (Meadows and others 1972). In the case of Malthus, the lack of food led to hunger, which decreased the birth rate. But this way of treating the origin of growth and the factors that control it does not consider that humankind is a highly interactive system. In the case of my model, the rise in the number of people, observed over the ages, is the outcome of all factors of a biological, technological, economic, social, and cultural nature relevant to growth in society and expressed by the quadratic-growth law. In other words, we must treat humankind as an entity, an open system, and integrate everything that is going on inside it. That is the meaning of systemic development, as opposed to the reductionist approach that is pursued in most global models, in which all relevant processes and resources are purportedly taken into account separately. With great expertise and effort, this is done in demography and is the reason for its limited temporal horizon in describing growth, although in this case we get insight into the details of growth and its distribution in age groups and space.

It should be noted that humans are not only qualitatively different but also quantitatively different from any mammals of comparable size and position in the food chain; humans are 100,000 times more numerous. Only domestic animals that accompany humans outnumber by far their relatives that live in comparative equilibrium in the wild. Humankind has broken away from the rest of nature and has developed a habitat of its own. On the other hand, in the last million years, humans have scarcely evolved biologically at all since the appearance of *Homo sapiens*. These are the basic reasons for considering the development of humankind as a separate entity, a system of its own, in which, in the process of sapientation, social, technological, and economic development have determined our incessant growth.

In the global interaction of all people, the exchange of information and the transfer of knowledge are instrumental in how growth becomes the outcome of all processes in the complex nonlinear global system. The interaction described by the quadratic growth rate can be seen as a collective phenomenon, an expression of consciousness, peculiar to humans and making them fundamentally different from all other animals. By speech and language, information is transferred vertically from the past and into the future. At the same time, information is spread horizontally, synchronizing human development globally. The nonlinear systemic theory of the growth of humankind indicates the synchronous development of the large-scale features of history and prehistory that are well substantiated by observations of historians and anthropologists.

Finally, this theory indicates that our development over the vast period of growth can be seen best on a logarithmic scale. This has been intuitively done by anthropologists, who otherwise could not accommodate on a single chart the million years of the lower Paleolithic age with the 10,000 years of the Neolithic age. Calculations show that the time of the development of all humankind should be shown logarithmically, reckoning time from the year 2007—the peak of the demographic transition—so that the whole human story can be shown in the same table (see figure 1). A table like this also offers an explanation of the nonuniform way in which time has passed during the course of our development. This change in the relative duration of events is a direct kinematic result of the accelerated growth of humankind proportional to the population of the world. As we approach the singularity of the demographic transition, the transformation and compression of time in history are striking. In the theory of growth, the large-scale features appear as epochs and periods of population growth. The first epoch, A, which lasted 2.8 million years, corresponds to the time it took for *Homo*

Epoch	Period	Date (Yr)	Number of People	Cultural Period	$\Delta T(Yr)$	History, Culture, Technology
		2175 A.D.	13×10^9	Stabilized Population	125	Changing Age Distribution
C		2050 A.D.	10.5×10^9	World	42	
	T_1	2007 A.D.	7×10^9	Demographic	42	Urbanization
	11	1965 A.D.		Transition	42	NOW ←
	10	1840 A.D.	10^9	Recent	125	Computers World Wars Electric Power
	9	1500 A.D.		Modern	340	Industrial Revolution Printing
	8	500 A.D.	10^8	Middle Ages	1000	Geographic Discoveries Fall of Rome
	7	2000 B.C.		Ancient World	2500	Greek Civilization Buddha
B	6	9000 B.C.		Neolithic	7,000	Writing Domestication
	5	29,000 B.C.	10^7	Mesolithic	20,000	Ceramics, Bronze Microliths America Populated
	4	80,000 B.C.		Moustier	51,000	Homo Sapiens
	3	0.22 Ma	10^6	Acheulean	1.4×10^5	Speech, Fire Europe and Asia Populated
	2	0.60 Ma		Chelles	3.8×10^5	Hand Axe Choppers
	1	1.6 Ma	10^5	Olduvai	1.0×10^6	Homo Habilis
A	0 T_0	4.4 Ma	(1)		2.8×10^6	Hominida Separate from Hominoids

FIGURE 1   The Development of Humankind on a Logarithmic Scale.

habilis to emerge during the evolution of early hominids. Epoch B, the time of quadratic growth, began 1.6 million years ago and culminated in 1965 in the advent of the demographic transition. This led to epoch C, the transition to a stabilized population of the world. The periods traditionally identified by anthropology and history subdivide epochs A and B into 12 intervals of ΔT years. These intervals become shorter and shorter as we approach 2007, the critical date of the transition.

This novel way of presenting human development is the result of a consistent and straightforward mathematical model for interpreting the general features of the development of humankind. It comes from applying methods of sciences that arrogantly call themselves exact to problems of the humanities—an effort that is far from easy, because the sides need to learn to understand each other but have long been separated in our culture. It seems that this can be done only through an interdisciplinary endeavor, and global problems likely are best suited for this purpose. Population growth now practically has reached the peak of the transition to a stabilized world population for the foreseeable future, and the period 1965–2050 is the time of this transition. The transition is remarkably short if we compare it with the million years of our development, but 10% of all the people who ever lived will experience this period of rapid change. The pace and width of the transition are the result of interactions in the global population and the outcome of the complex behavior of a highly nonlinear dynamic system. During this eventful period of 85 years, the population of the world will become 3 times larger and much older. It is the most critical and singular period ever to be experienced by humankind. All through the ages, humankind has followed a stable, persistent pattern of growth; this pattern is changing rapidly now to a stabilized global population. In fact, it cannot change faster (barring an all-out nuclear war or extraterrestrial intervention!), and it is this rapid change—from blowup to saturation—that must be kept in mind in any attempt to understand the global problems, that now face the world.

SUSTAINABLE DEVELOPMENT

Since the Conference on Development and the Environment was held in Rio de Janeiro in 1992, the concept of sustainable development has emerged as an important landmark in the international debate on world affairs. In the summer of 1997, a review conference in New York showed the difficulties—even a split in the attitudes toward development and the environment—between the developed and developing nations. The consensus reached in Rio de Janeiro is being questioned now, and the origin of the differences in attitudes needs to be investigated, taking into account the population transition.

Because the transition first happened in the so-called developed world and now is proceeding to the developing world, it would be better to speak in terms of the countries where the population has already stabilized and the greater part of the world, which now is passing through the demographic transition. To see the magnitude of these events, the population undergoing the transition now is 15 times larger than, and the rate of change is twice as great as, in the developed world. Today's annual growth rate in China is 1.1% in a population of 1.3 billion; in India, the annual growth rate is 1.9% in a population of 930 million. The respective economic growth rates of these countries are 12% and 6–7%. What should be of greatest concern for the world community is stability during this remarkable time of development (Kapitza 1996c). On the other hand, the differences in the stages of the demographic transition provide the demographic and economic backdrop against which the concept of sustainable development must be examined.

In the developed world, the demographic transition already has led to a stabilized and rapidly aging population of predominantly senior citizens. The process of urban development is slowing down. Indications of such a stable, affluent, and highly developed society can be seen in many ways. Extensive service sectors of the economy—of health, education, and social security—are developing. The change in values of the fundamental paradigm of development—from growth on all counts in terms of children, cars, or soldiers to that of limited growth and concern for the environment—is important. We, hopefully, can see a trend to abate consumption, making it a subject of public awareness. As an expression of a new consciousness, global responsibility for the environment is gaining ground. It is from these premises that the idea of sustainable development has sprung (Kapitza 1997).

The other side, that is, the developing world, has quite different circumstances. The younger generations are predominant. There is a vast migration from villages to towns, leading to rapid urbanization. The young migrants are the new working class, who are active and unsettled and who can man armies or leave the country or, in the case of unemployment, become a source of unrest. The possible scenarios are well known. In the developed world, we need to look back only 100 or 150 years to find a similar situation, but we must keep in mind that growth and everything that accompanies it occur twice as fast today as they did then.

In assessing changes and development in the world system, one also needs to think in terms not of averages, but of distributions—distributions by sex and age in populations, in wealth and income, in education and health, and in the very nonuniform distribution of people in towns and villages. Without studying the evolution of these distributions, it is practically impossible to describe the changes that are happening. Because of a lack of understanding of the statistical origins and social relevance of these distributions, ideas have evolved on drastic cuts in world population, of a "golden billion", and of extrapolating the southern California lifestyle worldwide.

All distributions of land, food, energy, and wealth show that the world population system is far from equilibrium. The origin of these distributions is most important; it indicates rapid growth, which increases as a country approaches the demographic transition. On the other hand, the evolution of these distributions shows that, in processes of growth, the world population system was dynamically sustainable—otherwise it could not have evolved consistently for 1 million years as it has. In this context, Vishnevsky made an interesting observation in interpreting the dynamic model. He remarked that the history of humankind preceding the demographic transition can be seen as a rapid passage, a nonequilibrium, a transitory state of growth and self-organization toward a stabilized world population, which will be the long-term asymptotic and stable state of humankind. This point is important to remember when we are addressing global problems of the present age.

We must look into the meaning of sustainability in a world of zero or very low population growth. We should assume that the world population is moving rapidly toward stabilization and that, in promoting and propagating the idea of a sustainable world, this must be taken into account. But will we run out of global resources at the expected levels of consumption? That is what matters and what led to the split at the 1997 New York conference.

The point is often made that we are living in a common world and that we must consider the common heritage that we all share—be it the atmosphere, the oceans, or the complexity of the biosphere. That is certainly true, but where are the limits of demanding a common policy on these issues? In this new, stabilized world, the population by the end of the next century will be twice as large as it is today. The energy produced—the best way of estimating the use of resources— will be 4–6 times as large as it is now (Holden 1991). Can our planet carry this load without collapsing? Probably so, but great changes will take place. It is best to remember that the environment in every populated part of the world—from Europe to China, India, and much of North America—has a highly transformed natural habitat. There are still large sparsely populated spaces that escape our attention. A comparison of Argentina and India is instructive. India's area is about 40% larger than Argentina's, but its population is 30 times greater. India is one of the oldest civilizations, if not *the* oldest, whereas Argentina, as a nation, is only 200 years old. But Argentina reportedly could feed the entire world.

As long as such large discrepancies exist, it can be assumed that the global population system is open and has enough resources to support its development in the foreseeable future. The first indication of a global shortage will be a more uniform pattern of the use of resources. On this scale of events, the next century will be crucial for humankind to negotiate the last stage of adaptation to the stabilized state of its future, when, we hope, we can carry out a pattern of sustainable development. At that stage, all progress will need to be reckoned with by means that do not involve numerical growth, the stereotype of development that has dominated humankind for 1 million years and tens of thousands of generations. History and our present experience show that our "software"—our ideas and values—evolves much more slowly than our "hardware", which for ages was geared for maximal growth and productivity. Under the pressure of rapid development, these long-entrenched attitudes will have to change. Of all factors, this probably is central to resolving the issue of sustainability.

SUSTAINING BIODIVERSITY

These ideas provide the historical context for considering the sustainability of biodiversity. As recent environmental research has shown, we can expect to lose biodiversity mainly during the period of rapid growth, as happened in the developed world two or three generations ago, during the first stage of the demographic transition—the stage of rapid growth. Today, many see the very fast growth of the developing world as the primary menace to the global environment, with biodiversity in first place over the short term, compared with long-term environmental issues. The sheer rate of growth and the rapid transition to a stabilized new world are competing factors that will determine the outcome and the state of the world in the foreseeable future. What can and will resolve these issues to some extent is a change of values that will determine our patterns of social behavior. At the peak rate of the present stage of development, material growth by far outstrips the development of humankind's "software".

The differences in our values, ideas, and material development are influenced to a great extent by the processes of globalization. If the spread of technology, money, and industrial know-how is accelerating development, the appropriate diffusion of ideas and values is lagging. The sheer complexity of global society is complicating matters, for it takes much time for our social habits and customs to be established and even longer for international institutions to evolve. The time scales involved can be traced to the fact that it takes only 9 months to produce a human's "hardware" but at least 20 years to program a human's "software". These are the fundamental biological and human constants that finally determine both our personal development and the fate of humankind. Ultimately, it is the interplay and balance of matter and mind that will resolve our predicament.

REFERENCES

Atiyah M, Press F. 1993. Population growth, resource consumption, and a sustainable world. Statement of the Royal Society of London and US National Academy of Sciences. Washington DC: National Academy Press.

Cohen J. 1995. How many people can the world support? New York: Norton.

Holdren J. 1991. Population and the energy problem. Population and environment. J Interdisc Stud 12(3):231–55.

Kapitza SP. 1995. Population dynamics and the future of the world. In: Towards a war-free world. Proceedings of the 44th Pugwash conference on science and world affairs. Singapore: World Scientific.

Kapitza SP. 1996a. The phenomenological theory of world population growth. Physics—Uspekhi 39(1):57–72

Kapitza SP. 1994. The impact of the demographic transition. In: Schwab K (ed). Overcoming indifference: ten key challenges in today's world. New York NY: New York Univ Pr.

Kapitza SP. 1996b. Population: past and future. A mathematical model of the world population system. Science Spectra 2(4).

Kapitza SP. 1996c. Population dynamics and the West-East Development. Annals of the 7th Engelberg Forum.

Kapitza SP. 1997. Population growth and sustainable development. The 47th Pugwash Conference on Science and World Affairs. Lillehammer, Norway: World Scientific.

Lutz W (ed). 1994. The future population of the world: what can we assume today? London UK: IIASA and Earthscan Press.

Lutz W, Sanderson W, Scherbov S. 1997. Doubling of world population unlikely. Nature (387):803–5.

Meadows D and others. 1972. Limits to growth. New York NY: Universe Bk.

NONINDIGENOUS SPECIES—A GLOBAL THREAT TO BIODIVERSITY AND STABILITY

DANIEL SIMBERLOFF

Department of Ecology and Evolutionary Biology, University of Tennessee, 569 Dabney Hall, Knoxville, TN 37996

THE WORLD'S biota is being rapidly homogenized. This global change constitutes a major threat to biodiversity and to our ability to extract resources sustainably from many ecosystems. The threat was first recognized 50 years ago, but its extent is only now being realized as burgeoning tourism and unfettered international trade expand the opportunity for species to get from one region to another. In the past, a desired immigrant species or one furtively hitching a ride often had to survive a sea voyage of months. Now, over 280 million passengers use commercial airliners each year worldwide, as do millions of tons of cargo. The brown tree snake (*Boiga irregularis*) occasionally arrives in Honolulu in wheel wells and cargo bays of planes from Guam, where it has devastated forest birds after introduction from the Admiralty Islands (Rodda and others 1992). Similarly, mosquitoes arrive in Great Britain from Africa in airliner passenger cabins (Bright 1996), and the giant African snail (*Achatina fulica*), which has ravaged agriculture on many Pacific islands, was carried by a boy from Hawaii to Florida as a gift to his grandmother (Simberloff 1997a).

Of course, on every continent, many of the most venerated plants and animals *were* introduced intentionally. In many parts of the world, the major crop plants are almost all introduced, as are livestock. For example, of nine crop plants in the United States classified as "major" (USDA 1997), one (corn) is native and five were introduced from the Old World, one from the Andes, and two from Central America. Pets and ornamental plants are also usually of exotic origin. So what is the threat, exactly?

EFFECTS OF NONINDIGENOUS SPECIES

The biggest threat posed by introduced species is the disruption of ecosystems, often by invasive plant species that replace the native species. The Australian tree *Melaleuca quinquenervia*, until recently increasing its range in southern Florida by more than 20 ha/day, replaces cypress and other native plants. It now covers about 200,000 ha, provides poor habitat for many native animals, affects the fire regime, and causes water loss (Schmitz and others 1997). South American water hyacinth (*Eichhornia crassipes*) now blankets many near-shore areas of Africa's Lake Victoria, blocking light and killing plants at the bottom of the food chain. The death and decay of plants that make up the water hyacinth mat remove still more oxygen from the water, and the major fisheries are in drastic decline. In addition to the ecological damage, water hyacinths are an economic nightmare, fouling engines and propellers of cargo ships and ferries, preventing docking, and clogging power-plant pipes and so causing numerous blackouts (McKinley 1996).

Introduced plants can also change an ecosystem without smothering the native plants. For example, on the island of Hawaii, the eastern Atlantic island shrub *Myrica faya* has invaded nitrogen-poor lava flows and ash deposits. A nitrogen-fixer, it favors other introduced species over the native plants adapted to low nitrogen (Vitousek 1986). In much of the American West and in Hawaii, Old World grasses, such as cheatgrass (*Bromus tectorum*), increase the frequency and intensity of fires to the great detriment of native plants and the animals that use them (D'Antonio and Vitousek 1992; Macdonald and others 1989).

Entire marine ecosystems can be radically changed by the invasion of a single plant species. The Pacific seaweed *Caulerpa taxifolia*, released from the Oceanographic Museum of Monaco into the Mediterranean about 15 years ago, now covers over 4,000 ha and has locally smothered native seagrass beds that harbor many native animals (Boudouresque and others 1994; Simons 1997). Introduced red mangrove (*Rhizophora mangle*) trees from Florida on the coasts of the Hawaiian islands and Australian "pine" trees (*Casuarina* spp.) on the Florida coast have come to dominate their new homes, displacing native plants and animals (Schmitz and others 1997; Walsh 1967).

Just as an introduced plant can modify an ecosystem, a species that eliminates a plant can have a drastic effect. The Asian chestnut blight fungus (*Cryphonectria parasitica*), which arrived in New York City on nursery stock in the late 19th century, spread over about 100 million ha of the eastern United States in less than 50 years, destroying almost all chestnut trees (von Broembsen 1989). Chestnut had been the most common tree in many forests, making up one-fourth or more of the canopy trees, so the cascading ecosystem effects of this invasion were substantial. For example, several insect species that were host-specific to chestnuts were extinguished (Opler 1979); that chestnut leaves decompose faster than leaves of the oaks that largely replaced them suggests that the invasion greatly affected nutrient cycling (K. Cromack, Oregon State University, pers. comm.), although systematic data were not gathered. The North American pine wood nematode (*Bursaphelenchus xylophilus*) reached Japan in timber and spread among the islands, killing more than 10 million pine trees and affecting 25% of Japan's pine forests (von Broembsen 1989). The effects on other forest species must have been dramatic.

In addition to ecosystem effects, nonindigenous species have myriad effects on particular native species or groups of them. They can eat them, for example. The Nile perch (*Lates niloticus*), after introduction into Lake Victoria, eliminated many species of endemic cichlid fishes, which had undergone perhaps the greatest evolutionary radiation that scientists have studied (Goldschmidt 1996). Introduced rats (*Rattus* spp.) on many islands have destroyed at least 37 species and subspecies of island birds (Atkinson 1985; King 1985). The impact of the brown tree snake on the Guam avifauna is noted above. Introduced herbivores can similarly drive species to extinction, especially on islands where plants are less likely to have a refuge, an area that herbivores cannot reach. For example, goats introduced to St. Helena in 1513 almost certainly eliminated over 50 endemic plant species, although only seven were scientifically described before they disappeared (Groombridge 1992).

Introduced pathogens, often carried by introduced plants and animals, can also devastate native species. The chestnut blight was noted above. As another example, in the Hawaiian islands, the extensive introduction of Asian songbirds has brought avian pox and avian malaria, which have contributed to the decline and extinction of numerous native forest-bird species (van Riper and others 1986). The introduction into Africa of the virus rinderpest, native to India, in cattle in the 1890s led to the infection of many native ungulate species; mortality in some species reached 90%, and the distribution of some species is still affected by the virus (Dobson 1995).

Nonindigenous species can compete with native ones, although competition for resources is often difficult to demonstrate. Some well-studied examples provide good evidence. The house gecko (*Hemidactylus frenatus*) has invaded many Pacific islands; this has led to drastic declines in the population of some native gecko species. Experiments suggest that at least one of the natives, *Lepidodactylus lugubris*, avoids the larger house gecko, thereby suffering food shortage (Petren and others 1993), and that the invader depletes the insect food base sufficiently to reduce the food available for the native (Petren and Case 1996). The continuing replacement in the United Kingdom of the native red squirrel (*Sciurus vulgaris*) with the introduced American gray squirrel (*S. carolinensis*) is now attributed largely to the greater foraging efficiency of the invader and concomitant lowering of food available to the native (Williamson 1996).

Many instances are known in which introduced species affect native ones by interfering with them directly rather than indirectly through resource depletion. The South American fire ant (*Solenopsis invicta*), which has spread throughout the southeastern United States, attacks individuals of native ant species and is replacing the latter in many habitats (Tschinkel 1993). In a plant analogue of aggression, the African crystalline ice plant (*Mesembryanthemum crystallinum*) accumulates salt, which remains in the soil when the plant decomposes. In California, this plant thus excludes native plants that are intolerant of such salty soil (Vivrette and Muller 1977).

Nonindigenous species also eliminate native species by mating with them; this threat is especially strong if the native species is much less numerous than the introduced one. For example, the New Zealand gray duck (*Anas superciliosa superciliosa*) and the Hawaiian duck (*A. wyvilliana*) are threatened with a sort of genetic extinction because of rampant hybridization and introgression with the introduced North American mallard (*A. platyrhynchos*) (Rhymer and Simberloff 1996). Likewise, Europe's rarest duck, the white-headed duck (*Oxyura leucocephala*), is threatened in its last redoubt in Spain by hybridization and introgression with North American ruddy ducks (*O. jamaicensis*), which were introduced into England as an amenity, escaped, and made their way to Spain (Rhymer and Simberloff 1996). This sort of threat is far more common in regions that exchange closely related species (such as Europe and North America) than in those whose species are so distantly related that they are unlikely to be able to mate and exchange genes (such as Australia and either Europe or North America). A native species can be threatened by hybridization with an introduced one even if no genes are exchanged, simply by the reproductive reduction effected by fruitless matings. Females of the endangered European mink (*Mustela lutreola*) mate with male introduced American mink (*M. vison*); although the embryos are aborted, the loss of reproduction by the European mink exacerbates their population decline (Rozhnov 1993).

SLOWING THE FLOW

The first line of defense against nonindigenous species is to keep them from being introduced. There are both practical and legal impediments to doing so. The sheer volume of tourism and trade dictates that inspection is destined to miss many inadvertent immigrants. Agricultural pests insinuate themselves into foodstuffs, woodboring beetles into timber, rodents into cargo containers—virtually any product shipped in bulk can carry many hitchhikers. Routine purging of ship's ballast water has released hundreds of nonindigenous species in waters throughout the world (Carlton and Geller 1993). Tourists can easily import species inadvertently in baggage, even if they heed warnings about which items are the most likely carriers of immigrants. In 1990, about 333 million nonindigenous plants were imported into the United States through Miami International Airport alone (OTA 1993). Economic resources are insufficient to examine everything that crosses a nation's borders.

Furthermore, liberalization of trade through such treaties as the General Agreement on Tariffs and Trade (GATT) and the North American Free Trade Agreement (NAFTA) is bound to increase the flow of nonindigenous species, and not only as a result of the increased volume. Under GATT and NAFTA, restrictions claimed as environmental measures can be challenged on the grounds that they are protectionist. The relevant regulatory authority must then adjudicate the dispute. Aside from the overwhelming appeal of free trade, both GATT and NAFTA require that species exclusions be based on risk assessments. However, risk-assessment procedures for introduced species are in their infancy and do not appear to be scientific, often resting on undefended judgments by experts and on arbitrary algorithms for combining risks (Simberloff and Alexander 1998). Furthermore, these risk assessments are expensive; one conducted by the US Department of Agriculture (USDA) on risks associated with importing larch from Siberia into the United States (USDA 1991) cost $500,000 (Jenkins 1996). It is difficult to imagine finding funding sources sufficient to mount risk assessments for all the challenges that might appear even to an educated layperson to be justified on prima facie grounds.

Virtually every specialist in invasion biology who has examined the matter concludes that aspects of the ecological impact of a nonindigenous species are inherently unpredictable (for example, Hobbs and Humphries 1995), and many scientists argue that every species should be considered a potential threat to biodiversity and sustainability if it were to be introduced (for example, Ruesink and others 1995). That implies that every species proposed for deliberate introduction, whether or not it appears superficially to be innocuous, necessitates some formal risk assessment. The cost would be staggering if the USDA process (USDA 1991) were the model.

In addition, many parties introduce species not inadvertently, but deliberately. These range from the boy smuggling giant African snails to his grandmother, who released them in her yard in Miami (Simberloff 1997a), to such large industries as the pet and ornamental-pet trades, which lobby vigorously against many restrictions. In the United States, recommendations that all species proposed for introduction must be on "white lists"—lists of species whose invasive potential has been assessed and has been approved for introduction—have been systematically attacked by those interest groups. Rather, the major laws that restrict entry of species use "black lists"—lists of species that have already been shown to be damaging or are strongly suspected of being dangerous; a species is prohibited only if it is on such a list (Schmitz and Simberloff 1997). Rarely is blacklisting forward-looking.

Thus, there will always be a flow of nonindigenous species. However, the flow can be lessened. Undoubtedly, increased public education as to the risks would lead to fewer deliberate and inadvertent introductions as people strive to be good environmental citizens. The Convention on Biological Diversity mandates that its signatories "as far as possible and as appropriate...prevent the introduction of...those alien species which threaten ecosystems, habitats, or species." Bean (1996) suggests that this statement reflects widespread recognition that nations are obliged to attempt to prevent introductions, and he cites as an example New Zealand's 1993 Biosecurity Act, which subjects all incoming persons and goods to rigorous inspection and prevents the importation of any species not already cleared by government authorities for inclusion on a white list. He also notes the increasing international and national regulation of purging of ballast water and points out that the considerable legal framework and effort that many nations use to attempt to prevent agriculturally harmful introductions could be adapted and expanded to prevent ecologically harmful ones. The problem is educating the public sufficiently that they demand regulation of nonindigenous species.

MANAGING NONINDIGENOUS SPECIES

Once a species enters a new region, there are several options for managing it. The most obvious one is to attempt to eradicate it. This approach is often feasible if the invasion is recognized and targeted early enough (Simberloff 1997a), but several factors militate against its success. Perhaps foremost, almost no countries have an early-warning system in place that is charged with determining when an invasion has occurred, much less a procedure to generate a rapid, coordinated response while the invasion is still restricted geographically. The reaction is usually only after an invasion has existed for so long that it has become noticeable, and by then eradication is often impossible (Schmitz and Simberloff 1997). Second, for species deliberately introduced, the same forces that conspired to allow introduction in the first place act to prevent eradication. In addition, many invasions appear innocuous for long periods (Crooks and Soulé 1996; Williamson 1996); by the time they are recognized as ecologically or economically damaging, they are so widespread that they cannot be eradicated.

Minimizing ecological and economic damage if eradication proves impossible is usually attempted by one or more of three routes (Simberloff 1996): chemical, mechanical, and biological control. The environmental and human health effects of broad-spectrum pesticides are legendary. Although some newer chemicals have far fewer side effects, their high cost and the necessity of repeated application and the frequent evolution of resistance by the target pest have led to great interest in alternative methods. Also, if pesticides were used to prevent damage by introduced species both to vast areas of natural habitats and to agriculture, all the above problems would be exacerbated.

Mechanical methods, either alone or in concert with pesticides, are sometimes feasible. For example, water hyacinth has been successfully controlled in Florida for over 20 years by a combination of mechanical harvesting and treatment with the herbicide 2,4-D (Schardt 1997). However, mechanical devices are often expensive and would be less likely to work on widespread invasions.

Biological control—the introduction of a natural enemy of the pest—has seemed an extremely attractive alternative to chemical and mechanical control on both ecological and economic grounds. Many biological-control projects have provided continuing suppression of a pest to acceptably low levels with the sole costs being those of the initial exploration to find natural enemies and the testing for efficacy and safety. Odour (1996) cites the control of water hyacinth in Sudan by three South American insects, of prickly pear cactus (*Opuntia inermis* and *O. stricta*) in Australia by the moth *Cactoblastis cactorum* from Argentina, and of the South American cassava mealybug (*Phenacoccus manihoti*) in Africa by a South American encyrtid wasp. In each instance, the natural enemies maintain populations in perpetuity without further human intercession.

More recently, biological control has been subjected to critical scrutiny on the grounds that nontarget species, some of conservation concern, have been attacked and even driven to extinction (Howarth 1991; Simberloff 1992). Early biological control projects using vertebrates, such as the small Indian mongoose or the cane toad, and the widespread dissemination of the New World predatory snail

Euglandina rosea to control the giant African snail were disastrous, and biological-control professionals now eschew the use of vertebrates, except for fishes. However, insects tested for host specificity have also attacked nontarget species. For example, the Eurasian weevil *Rhinocyllus conicus*, introduced into North America to control musk thistle (*Carduus nutans*), is now attacking native nonpest thistles, including narrowly restricted endemic species in nature reserves (Louda and others 1997). Although the extent of such problems is controversial, the fact that biological control agents can both disperse and evolve, just as any other introduced species can, suggests great caution in their use and extensive preliminary testing before their release.

ACTION NEEDED NOW

Burgeoning international interest in invasive nonindigenous species has led to several international meetings (for example, Sandlund and others 1996), new monographs (for example, Williamson 1996; Simberloff and others 1997), increased news coverage (for example, McKinley 1996; Simons 1997), and widespread appeals for action (for example, Glowka and de Klemm 1996). Nevertheless, there is no evidence that the flow of exotics is decelerating under the pressures of increased trade and tourism described above. What else must be done?

Glowka and de Klemm (1996) feel that inclusion of nonindigenous species as a priority item for the Conference of Parties to the Convention on Biological Diversity, which has been ratified by 172 nations, is necessary to prevent a fragmented approach to the problem. Schmitz and Simberloff (1997) see the effort in the United States as also bedeviled by fragmentation. In short, as long as one program deals with aquatic plants, another aquatic animals, another agricultural weeds, and yet another bird introductions, the effort is bound to be frustrated if only because species often interact synergistically to generate an environmental or economic problem (Simberloff 1997b). Furthermore, because nonindigenous species do not recognize political boundaries, both regulatory and management responsibility must also cross for them to be effective. Thus, the Convention on Biological Diversity, an international instrument, is highly appropriate as one locus of action. It is important to observe, however, that, even if no species were henceforth able to cross a national border, introduced species would still be a major problem. In the United States, for example, interstate movement of introduced species has the same effect as importing such species from other countries: ecosystems are subjected to invasion and disruption by species that have evolved elsewhere. And, within-country transport can threaten invasion of neighboring countries.

A major current lacuna is a comprehensive database on introduced species that is associated with an early-warning system and a rapid-response team. For most taxa in most countries, someone who finds a species suspected to be nonindigenous and potentially invasive has nowhere to turn to examine this possibility. There is no emergency telephone number to use to determine whether it is a newly recorded species or a species that is spreading after introduction. Even if there were an organization charged with receiving such queries, there is no list of species to which it could turn to give an answer. For most species, there is no systematic effort to record where they have been introduced or their suspected effects. And there is rarely a procedure in place to respond rapidly to a newly recorded invasion, partly because of the fragmentation of authority described above.

The white-list approach advocated by Ruesink and others (1995) and Wade (1995) and discussed above needs to be adopted in some form both nationally and internationally. Black lists have never worked well, and the inherent unpredictability and idiosyncrasy of introductions dictate that all potential introductions be subjected to scrutiny—with no blanket exceptions. That requirement, of course, would mean that funding would be needed to process applications and to give them the necessary attention. Whether the costs of white listing are borne by the party wishing to import a species or by society as a whole will have to be addressed. For that matter, so will the costs of an unforeseen disaster if a white-listed species turned out not to be innocuous. Should an applicant be required to post a bond? Should an applicant be able to be indemnified by purchasing disaster insurance? Should society as a whole bear the cost? These matters have barely been broached.

How a species proposed for introduction should be assessed is yet another crucial issue that has been at best cursorily considered. As noted above, standard risk-assessment procedures for chemical and physical stressors do not appear to work well for biological introductions, for which the probabilities of such events as evolution and long-distance dispersal are so difficult to evaluate as to be mere guesses (Simberloff and Alexander 1998). The concatenation of guesses and arbitrary assignment of risk categories that pervades the current USDA risk-assessment procedure (see, for example, USDA 1991) hardly seems scientific, but no general alternative has been widely considered (O'Brien 1994). Having agreed that risk assessment will be the appropriate procedure to adjudicate disputes, we must determine how to do risk assessments en masse for nonindigenous species.

REFERENCES

Atkinson IAE. 1985. The spread of commensal species of Rattus to oceanic islands and their effects on island avifaunas. In: Moors PJ (ed). Conservation of island birds. Cambridge UK: International Council for Bird Preservation. p 35–81.

Bean MJ. 1996. Legal authorities for controlling alien species: a survey of tools and their effectiveness. In: Sandlund OT, Schei PJ, Viken A (eds). Proceedings of the Norway/UN conference on alien species. Trondheim Norway: Dir for Nature Management and Norwegian Inst for Nature Research. p 204–10.

Boudouresque CF, Meinesz A, Gravez V (eds). 1994. First international workshop on Caulerpa taxifolia. Marseille France: GIS Posidonie.

Bright C. 1996. Understanding the threat of bioinvasions. In: Worldwatch Institute, State of the World 1996. New York NY: WW Norton. p 95–113.

Carlton JT, Geller J. 1993. Ecological roulette: the global transport and invasion of nonindigenous marine organisms. Science 261:78–82.

Crooks J, Soulé ME. 1996. Lag times in population explosions of invasive species: causes and implications. In: Sandlund OT, Schei PJ, Viken A (eds). Proceedings of the Norway/UN Conference on

Alien Species. Trondheim Norway: Dir for Nature Management and Norwegian Inst for Nature Research. p 39–46.

D'Antonio CM, Vitousek PM. 1992. Biological invasions by exotic grasses, the grass/fire cycle, and global change. Ann Rev Ecol Syst 23:63–87.

Dobson A. 1995. The ecology and epidemiology of rinderpest virus in Serengeti and Ngorongoro Conservation Area. In: Sinclair ARE, Arcese P (eds). Serengeti 2. Chicago IL: Univ Chicago Pr. p 485–505.

Glowka L, de Klemm C. 1996. International instruments, processes, organizations, and nonindigenous species introductions: is a protocol to the convention on biological diversity necessary? In: Sandlund OT, Schei PJ, Viken A (eds). Proceedings of the Norway/UN Conference on Alien Species. Trondheim Norway: Dir for Nature Management and Norwegian Inst for Nature Research. p 211–8.

Goldschmidt T. 1996. Darwin's dreampond, drama in Lake Victoria. Cambridge MA: MIT Pr.

Groombridge B (ed). 1992. Global biodiversity: status of the earth's living resources. London UK: Chapman & Hall.

Hobbs RJ, Humphries SE. 1995. An integrated approach to the ecology and management of plant invasions. Cons Biol 9:761–70.

Howarth FG. 1991. Environmental impacts of classical biological control. Ann Rev Entom 36:485–509.

Jenkins P. 1996. Free trade and exotic species introductions. In: Sandlund OT, Schei PJ, Viken A (eds). Proceedings of the Norway/UN Conference on Alien Species. Trondheim Norway: Dir for Nature Management and Norwegian Inst for Nature Research. p 145–7.

King WB. 1985. Island birds: will the future repeat the past? In: Conservation of island birds. Cambridge UK: International Council for Bird Preservation. p 3–15.

Louda SM, Kendall D, Connor J, Simberloff D. 1997. Ecological effects of an insect introduced for the biological control of weeds. Science 277:1088–90.

Macdonald IAW, Loope LL, Usher MB, Hamann O. 1989. Wildlife conservation and the invasion of nature reserves by introduced species: a global perspective. In: Drake JA, Mooney HA, di Castri F, Groves RH, Kruger FJ, Rejmanek M, Williamson M (eds). Biological invasions: a global perspective. Chichester UK: Wiley 215–55.

McKinley JC. 1996. An Amazon weed clogs an African lake. New York Times, 5 Aug 1996. p A5.

O'Brien MH. 1994. The scientific imperative to move society beyond "just not quite fatal." Environ Prof 16:356–65.

Odour G. 1996. Biological pest control and invasives. In: Sandlund OT, Schei PJ, Viken A (eds). Proceedings of the Norway/UN Conference on Alien Species. Trondheim Norway: Dir for Nature Management and Norwegian Inst for Nature Research. p 116–22.

Opler PA. 1979. Insects of the American chestnut: possible importance and conservation concern. In: McDonald W (ed). The American chestnut symposium. Morgantown WV: Univ West Virginia Pr. p 83–5.

Petren K, Bolger DT, Case TJ. 1993. Mechanisms in the competitive success of an invading sexual gecko over an asexual native. Science 259:354–8.

Petren K, Case TJ. 1996. An experimental demonstration of exploitation competition in an ongoing invasion. Ecology 77:118–32.

Rhymer J, Simberloff D. 1996. Extinction by hybridization and introgression. Ann Rev Ecol Syst 27:83–109.

Rodda GH, Fritts TH, Conry PJ. 1992. Origin and population growth of the brown tree snake, Boiga irregularis, on Guam. Pacif Sci 46:46–57.

Rozhnov VV. 1993. Extinction of the European mink: ecological catastrophe or natural process? Lutreola 1:10–16.

Ruesink JL, Parker IM, Groom MJ, Kareiva PK. 1995. Reducing the risks of nonindigenous species introductions. BioScience 45:465–77.

Sandlund OT, Schei PJ, Viken A (eds). 1996. Proceedings of the Norway/UN Conference on Alien Species. Trondheim Norway: Dir for Nature Management and Norwegian Inst for Nature Research.

Schardt JD. 1997. Maintenance control. In: Simberloff D, Schmitz DC, Brown TC (eds). Strangers in paradise. impact and management of nonindigenous species in Florida. Washington DC: Island Pr. p 229–43.

Schmitz DC, Simberloff D. 1997. Biological invasions: a growing threat. Iss Sci Technol 13(4):33–40.

Schmitz DC, Simberloff D, Hofstetter RH, Haller W, Sutton D. 1997. The ecological impact of nonindigenous plants. In: Simberloff D, Schmitz DC, Brown TC (eds). Strangers in paradise: impact and management of nonindigenous species in Florida. Washington DC: Island Pr. p 39–61.

Simberloff D. 1992. Conservation of pristine habitats and unintended effects of biological control. In: Kauffman WC, Nechols JE (eds). Selection criteria and ecological consequences of importing natural enemies. Lanham MD: Entomological Soc America. p 103–17.

Simberloff D. 1996. Impacts of introduced species in the United States. Consequences 2(2):13–23.

Simberloff D. 1997a. Eradication. In: Simberloff D, Schmitz DC, Brown TC (eds). Strangers in paradise: impact and management of nonindigenous species in Florida. Washington DC: Island Pr. p 221–8.

Simberloff D. 1997b. The biology of invasions. In: Simberloff D, Schmitz DC, Brown TC (eds). Strangers in paradise: impact and management of nonindigenous species in Florida. Washington DC: Island Pr. p 3–17.

Simberloff D, Alexander M. 1998. Assessing risks from biological introductions (excluding GMOs) for ecological systems. In: Calow P (ed.). Handbook of environmental risk assessment and management. Oxford UK: Blackwell. p 147–76.

Simberloff D, Schmitz DC, Brown TC (eds). 1997. Strangers in paradise: impact and management of nonindigenous species in Florida. Washington DC: Island Pr.

Simons M. 1997. A delicate Pacific seaweed is now a monster of the deep. New York Times 16 Aug 1997. p A1, A4.

Tschinkel WR. 1993. The fire ant (Solenopsis invicta): still unvanquished. In: McKnight BN (ed). Biological pollution: the control and impact of invasive exotic species. Indianapolis IN: Indiana Acad Sci. p 121–36.

OTA [US Congress, Office of Technology Assessment]. 1993. Harmful nonindigenous species in the United States. Washington DC: US GPO.

USDA [US Department of Agriculture]. 1991. Pest risk assessment of the importation of larch from Siberia and the Soviet Far East, miscellaneous publication no 1495. Washington DC: USDA Forest Service.

USDA [US Department of Agriculture]. 1997. Agricultural statistics 1997. US Department of Agriculture, National Agricultural Statistics Service. Washington DC: US GPO.

van Riper C, van Riper SG, Goff ML, Laird M. 1986. The epizootiology and ecological significance of malaria in Hawaiian land birds. Ecol Monogr 56:327–44.

Vitousek P. 1986. Biological invasions and ecosystem properties: can species make a difference? In: Mooney HA, Drake JA (eds). Ecology of biological invasions of North America and Hawaii. New York NY: Springer-Verlag. p 163–76.

Vivrette NJ, Muller CH. 1977. Mechanism of invasion and dominance of coastal grassland by Mesembryanthemum crystallinum. Ecol Monogr 47:301–18.

von Broembsen SL. 1989. Invasions of natural ecosystems by plant pathogens. In: Drake JA, Mooney HA, di Castri F, Groves RH, Kruger FJ, Rejmanek M, Williamson M (eds). Biological invasions: a global perspective, Chichester UK: Wiley. p 77–83.

Wade SA. 1995. Stemming the tide: a plea for new exotic species legislation. J Land Use Environ Law 10:343–70.

Walsh GE. 1967. An ecological study of a Hawaiian mangrove swamp. In: Lauff GH (ed.). Estuaries. Washington DC: AAAS. p 420–31.

Williamson M. 1996. Biological invasions. London UK: Chapman & Hall.

P A R T

6

INFRASTRUCTURE
FOR SUSTAINING
BIODIVERSITY—
SCIENCE

SCIENCE AND THE PUBLIC TRUST IN A FULL WORLD: FUNCTION AND DYSFUNCTION IN SCIENCE AND THE BIOSPHERE

GEORGE M. WOODWELL
Woods Hole Research Center,
13 Church Street, P.O. Box 296, Woods Hole, MA 02543

The ecological symptoms of unsustainability include shrinking forests, thinning soils, falling aquifers, collapsing fisheries, expanding deserts, and rising global temperatures. The economic symptoms include economic decline, falling incomes, rising unemployment, price instability and loss of investor confidence. The political and social symptoms include hunger and malnutrition, and, in extreme cases, mass starvation; environmental and economic refugees; social conflicts along ethnic, tribal, and religious lines; and riots and insurgencies. As stresses build on political systems, governments weaken, losing their capacity to govern and to provide basic services, such as police protection. At this point the nation-state disintegrates, replaced by a feudal social structure governed by local warlords as in Somalia, now a nation-state in name only.

Lester R. Brown, 1995

THE TRANSITION FROM EMPTY TO FULL

THAT GRIM PROSPECT from Lester Brown summarizes lucidly the course of the current civilization in the eyes of pragmatic ecologists who deal daily with the dependence of the human undertaking on the long-sustained biotic functions of the earth. It has little to do with "biotic diversity" and much to do with the erosion of the capacity of the biotic systems of the earth to continue to support a vigorous, successful, and continuing civilization. The phenomenal technological and economic successes of the current moment mask the elementary fact of the

337

dependence of all life on a habitat of diminishing dimensions. It is the current diminution of the biosphere that is the subject of this forum and this paper.

Herman Daly, the economic philosopher, has observed that the world has made a transition from "empty" to "full" and that the rules for success in management of human affairs have changed (Daly 1993). No longer are resources large in proportion to demands; the easy compromises available among competing interests in an empty world are of the past. The transition is recent, the product of the decades since 1960 as the human population has doubled once again and technology has offered an even more comprehensive capacity for turning the earth to human succor. The intensification of use of the whole earth comes to focus in a series of problems with biotic resources, although the immediate issues might appear to be energy, such as oil in the coastal zone, or the disposal of wastes, or the commitment of land to roads or to shopping malls or to industrial uses. The critical issue in each instance is a threat to one or more biotic resources, including food, human health, and the normally biotically controlled function of the biosphere.

Science has a special role in defining what will work in a biophysical sense in this new world, in which intensification of use will continue but in which each use must be held within dimensions of resources that are in fact diminished by the current use. The sum of these local activities is the world as a whole, the biosphere. Suddenly, in recent decades, within this century, incremental local disruptions of normal biotic functions are accumulating as global disruptions. The transition presents a major political challenge to governmental systems that were developed when resources seemed globally abundant and opens a new realm for the definition of civil rights. In a democracy, we establish government to protect each from all and all from each. What are the dimensions of protection as challenges to the human habitat become more acute and effects of local actions accumulate as global disruptions or impoverishment? The issue of how the world works and how it can be kept working in the largest interest of the public becomes central. The question is biophysical first and only secondarily economic and political, but success in the evolution of all three realms is essential. Science in general and ecology in particular have responsibility in joining in the definition of human rights in this new world—rights to clean air, clean water, food that is free of poison, a wholesome habitat that is not drifting into biotic impoverishment, and a world that is not itself being steadily impoverished biotically. What is clean air? Clean water? A stable and healthful habitat? What are essential human rights in a full world? What is it that we form governments to do for us all? And who will define that task and hold governments to it?

THE EVIDENCE THAT THE EARTH IS FULL IS GLOBAL BIOTIC INSTABILITY

The most powerful evidence of the transition to a world that is full, as opposed to empty, is the series of global transitions under way now. The most important are the warming of the earth and the progressive reductions in the capacity of the earth for supporting life: biotic impoverishment. The two are mutually reinforcing. The accumulation of heat-trapping gases in the atmosphere is the cause of the warming. The accumulation is due in part to the destruction of forests. A rapid warming has the potential for speeding the destruction of forests and accelerating the warming (Houghton and others 1998; Woodwell 1995; Woodwell and others 1998). The two processes are also open-ended, actively developing, directly threatening to human welfare, and, at the moment, not addressed effectively by any government or society despite various agreements to act. We have squandered trillions of dollars in the second half of this century on the mere possibility that the mismanagement of international affairs might lead to a nuclear war that could reduce the earth to a cinder in a few hours. We are currently engaged in vicious arguments over whether it is worth any effort to deflect the global changes that are in fact bringing increments of global impoverishment that move the world toward the same end, only more slowly. The difficulty is in part that the increments of change are small to the point of being inconspicuous to ordinary people; they are obscure for the moment but have the potential intrinsic in exponential growth for emerging suddenly as overwhelming problems that might, at that moment, have surged beyond control. The difficulty is also that action requires a reduction in the use of fossil fuels, a step that is unpopular with politically powerful commercial interests around the world.

The fact is that all interests, commercial and public, will suffer in a world afflicted by the chronic and rapid climatic disruptions already inevitable as a result of past accumulations of heat-trapping gases in the atmosphere. The changes entail cumulative and progressive increments of biotic impoverishment. Although the increments might be obscure minute by minute and are further obscured generation by generation as each generation starts with a baseline that is already eroded, the effects ultimately become conspicuous as erosion of the human habitat.

The rate of the warming offers one criterion for appraising the global rate of disruption. The warming has proceeded at a global average over recent decades of 0.1–0.2°C per decade. It is expected to proceed at that rate or higher throughout at least the next century. It has proceeded and will continue to proceed at 2–3 times that average rate in the higher latitudes, according to both experience and the most widely accepted projections (Houghton and others 1996). While the global warming was about 0.5°C between 1895 and 1990, the average for Canada as a whole was about 1.0°C and, for the Mackenzie District of northwestern Canada, about 1.7°C (Gullet and Skinner 1992). We might inquire as to the historical rates over recent millennia to establish a basis for judgment of how the biosphere was functioning before massive intrusions by humans. Even during the glacial periods, the rates of temperature change globally appear to have been closer to 0.1°C per century than per decade. Such a rate is consistent with the time required for the regeneration of forests and fish populations that must establish themselves in new habitats and consistent with adjustments in migratory patterns of animals.

The greatest hazard associated with the warming may be the systematic and rapid impoverishment of forests and tundra of higher latitudes of the Northern Hemisphere in response to the speed of the warming with the release of large additional quantities of carbon dioxide and methane into the atmosphere (Woodwell and others 1995). Insurance against such an event—a disaster in any appraisal—would argue for intensive efforts now to stabilize, or even to reduce, the current burden of heat-trapping gases in the atmosphere. The effects go far beyond forests to involve virtually every use of land, including agriculture, aggravating well-known problems there by introducing continuous changes in patterns of precipitation and temperature globally.

If there is doubt as to the details of the effects, examples of the extremes of impoverishment are abundant. Locally, they appear as the salinized playas of agricultural India that support no agriculture or higher plants or as land eroded to rock by the effects of the combination of intensive agriculture, intensive grazing, and erosion under monsoonal rains, a baking sun, and winds. Government experts in India a few years ago acknowledged that one-third of the land area had been removed from agriculture into impoverishment by those processes and other human uses. Such land has little or no value and is not normally incorporated into national statistics or economic appraisals, but the transition from forest through various forms of agriculture to impoverishment is probably the greatest current land-use transition (Houghton 1997). It is already affecting human food supplies, as summarized so brilliantly over recent decades by Lester Brown (1997). Irrigation from the earliest times, including the civilizations of the Tigris and Euphrates Rivers, has resulted in salinization and the destruction of agricultural productivity and contributed to the demise of successive waves of civilization (Fagan 1999). The process continues, and the effects are accumulating and are all too often irreversible.

THE STARTING POINT FOR A WORLD THAT WORKS

The causes of biotic impoverishment include virtually any chronic disturbance, from mechanical and physical to chemical and biotic (Woodwell 1990). The effects are similar in all instances. But the question of where to start the measurement of incremental change remains. It is one of the classical questions in ecology, similar to "What is undisturbed?" and "What is climax?" The analysis is useful, but a definitive answer is hardly necessary. Our interest is pragmatic, immediate: we might identify it as the "integrity of biotic function", thereby setting forth a new goal, whose identification, measurement, and defense become major challenges to science. In so doing, we acknowledge that we know more about the conditions necessary to keep biotic functions substantially intact than we know about the functions themselves. And it is possible that we will know how to tell in a simple, comprehensive way the extent to which we are successful in protecting details of the human habitat. Most of all we need a simple, quantitative basis for appraising increments of impoverishment.

MEASUREMENTS OF IMPOVERISHMENT

The most systematic approach to definition, where the degree of disturbance could be measured directly and objectively, has come from experimental studies of systematic disturbance. One of the most revealing studies involved the effects of chronic exposure to ionizing radiation on a late successional oak-pine forest in central Long Island, New York (Woodwell and Houghton 1990). In that instance, perhaps surprisingly, a virtually perfect physiognomic gradient in size and structural complexity was produced in both the residual community and the successional community that developed later. The most sensitive species was the pine *Pinus rigida*, which was removed from the intact oak-pine forest at exposures that were low enough to have little or no effect on the oaks or other species. At slightly higher exposures, the oaks, with the exception of the scrub oak (*Quercus ilicifolia*), were eliminated. The scrub oak, a high shrub, was eliminated at slightly lower exposures than the shrub cover of Vacciniaceae. Within the shrubs, the taller-statured huckleberry (*Vaccinium baccata*) was more sensitive than the ground-hugging lowbush blueberry (*V. pennsylvanicum*). The pattern of greater resistance in low-growing, ground-hugging species persisted within the herbaceous plant community and extended to mosses, lichens, and soil fungi. The less the stature, the more resistant to disturbance. The response left certain mosses and lichens to the inner zones where the radiation exposures were higher and certain soil fungi to the innermost zone from which even the most resistant lichens were excluded. The gradient was spectacular and obvious, although there was no basis in earlier studies for the assumption that chronic exposure to ionizing radiation would produce anything approaching a systematic community-level response.

The results, however, were startling in their similarity to familiar gradients of structure in vegetation produced by gradients of stress elsewhere, including chronic disturbance. The immediately obvious parallel was with the transition from forest to tundra, which is compressed on mountains in New England to a few thousand feet of elevation and involves some of the same species and most of the same genera. The same pattern of structural change emerged from later studies of gradients of pollution downwind of smelters (Woodwell and Houghton 1990). Again, the list of species emerges as the most informative data on the status of the community.

If we use the experience gleaned from those gradients, we can establish a scale against which to test other transitions and on which to hang new data as they accumulate. I have pooled my own experience with the effects of ionizing radiation and other chronic disturbances, such as pollution from smelters, with F.H. Bormann's (1990) experience and observations of the effects of air pollution, including acid rain, to prepare a tabular scale showing the steps in impoverishment of forests (table 1).

Bormann came to the conclusion that most of the forests of eastern North America are being affected now by air pollution in various forms and that the effects include not only a reduction in the growth of trees, but also an increase in mortality over large areas. These transitions are in the range of stages IIB, the open-canopy stage, and IIIA-3, the herb stage of treeless savanna, in the classification of damage outlined in table 1. There is little question that the death of red spruce (*Picea rubens*) on the western slopes of the Appalachians is due to acid rain and air pollution. Succession is under way (the second sorting), and the impoverishment has not yet progressed to the cryptogam or erosion stage, but

TABLE 1   Stages of impoverishment of forests under stress

Stages of Impoverishment	Disturbance	Effect on Structure	Effect on Function
0: Intact forest	None	None	None
I: Stressed forest	Low-intermittent	Below threshold of detection	May serve as sink for pollutant or as corrective influence for other disturbance
IIA: Symptomatic stress (species)	Chronic	Effects on sensitive species conspicuous; selection of resistant genotypes	Changes in chemistry of environment detectable in plants, soil, groundwater, streams
IIB: Open-canopy stage	Intensified chronic	Sensitive species eliminated; effects on more resistant species obvious; first sorting conspicuous in thinning of tree canopy	Pollution accumulating as chemical changes in environment; evapo-transpiration affected; light reaches ground cover; warming of soil; photosynthesis and respiration affected; primary productivity reduced
IIIA: Savanna stages of impoverishment A1: High shrubs A2: Low shrubs A3: Herb stage A4: Cryptogam stage	Severe chronic	First sorting is severe with loss of tree canopy, forest is treeless savanna with high shrubs surviving, patches reduced to low shrubs and ground cover; signs of second sorting as succession of hardy, small-bodied, rapid reproducers among plants and animals proceeds	Energy budget clearly shifted to ground surface heating, evapo-transpiration affected to point where groundwater increased; runoff increased; nutrient budgets affected and water quality declines with increases in nitrogen, organic matter, and silt
IIIB: Erosion stage	Long-continued severe chronic	Landscape conspicuously dysfunctional: Haiti, Madagascar; no forests; no ground cover over much of land; erosion conspicuous	Runoff is immediate through gullies and new channels; rivers filled with sediment; water flows massive, sudden, erratic, and not restricted to well-defined courses; slopes eroding; soil temperature vulnerable to extremes; agriculture tenuous

Source: Modified from Bormann (1990).

continued chronic disturbance in those zones has the potential for producing these stages as well.

Similar effects are now accumulating in the much more diverse mixed mesophytic forests of the Appalachian plateau to the west (Little 1995). The region would be described in the scale of table 1 as now in stage IIB, the open-canopy stage.

Bormann (1990) also reported the results of research with special chambers designed to measure the growth of trees fed with ambient air and with air treated only by filtration through charcoal. The experiment was carried out in eastern New York in the Hudson Valley and showed that the filtering increased the growth of populus saplings by 15–20%. The implication is that in rural New York in a region that probably has air similar to much of the rest of eastern North America, there is an air-caused inhibition of growth of around 15–20% that does not produce conspicuous symptoms of damage to leaves or other plant parts. The implications are profound: a 15–20% reduction in the amount of energy fixed by forests over very large areas. Similar studies of agricultural crops have shown similar inhibition of growth (Heck and others 1982). The reduction in total energy available to support life in this region is prodigious. By this criterion, the forests of eastern North America, presumably over large areas, are in the stages described in table 1 as I, stressed, and IIA, symptomatic stress.

A somewhat different series of changes in Alaskan forests is being reported by Juday (1997) and Stevens (1997) in response to the warming of Alaska as permafrost melts and destroys roads and as insect pests of forest trees appear and linger in places heretofore protected by climate. The process has long been expected and can only be amplified as the warming proceeds (Univ. of Alaska 1983).

One of the greatest natural tragedies of the century occurred in the tropical moist forests of the Amazon Basin and Kalimantan, the southern two-thirds of the island of Borneo, in 1997–1998. Both regions suffered from an unprecedented drought as a result of the strongest El Niño yet experienced. The El Niño involves a warming of the surface waters of the central and eastern Pacific and global climatic changes that include the severe droughts in the normally moist regions of the southwest Pacific and central South America. Both regions have forests that are being heavily cut, opening the forests to further drying and susceptibility to fires. Both regions are also being settled by governmental programs that open the land to those displaced from industrialized agriculture elsewhere or from overpopulated urban areas. Sources of ignition are abundant, and thousands of acres burned in 1997–1998, covering both regions with smoke so dense that breathing was difficult and airports were closed for days to weeks at a time. A major airplane crash and a collision of ships were attributed to the smoke from Kalimantan, which was dense from Celebes to Singapore. The effect was the substantial destruction of the forests in both places, well within the range of stage IIIA, the savanna stage, in our scale, probably reaching IIIA3, the herb stage of treeless savanna, in extensive areas.

Coastal marine waters are subject to similar impoverishment, although the changes are less conspicuous.

THE PUBLIC INTEREST IN A FULL WORLD:
HUMAN RIGHTS REQUIRE DEFINITION BY SCIENCE

Recognition that the continuation of current trends in human use of the earth is leading to progressive biotic impoverishment raises basic questions of the role of governments and the recognition and protection of human rights. Again, a focus on the biophysical aspects helps to clarify the social, economic, and political objectives. If the biophysical objective becomes the protection of biotic functions in maintaining the global and local environment, we should have little difficulty in defining the qualities of air, water, and land required to protect those functions. The biota will run itself and perform the functions without human guidance, but the conditions under which the biota can run itself without chronic disruption and systemic impoverishment must be defined and maintained. Success requires that the public recognize an overwhelming human interest in the protection of the biosphere as the only human habitat.

The challenges to science are large: What does it take to keep the biosphere functioning with substantial stability decade by decade when human populations are increasing and human effectiveness in capturing resources for human use increases daily? How much forest does it take to defend the public's interests in a stable and wholesome landscape, in a stable global carbon budget, in water flows that support the diversity of resources that have evolved over time in each region, and in water quality that is also consistent with stability of the landscape? Such questions challenge virtually all conventional approaches to the environment and to economics and government, but they are scientific and technical issues first and political and economic issues only secondarily. They are, however, the focus of increasing interest in basic human rights in a democracy, as outlined in detail recently for forests by Ann Hooker (1994) in a discussion of the public's interests in forests.

The answers will address the need for defining how land and water are to be used in this world of intensified demands. Answers will involve zoning of land and water in a pattern already becoming clear as attempts are made to preserve coastal fisheries in the United States. The establishment of the system of "marine sanctuaries" ringing the nation offers one of the most progressive steps in acknowledging the absolute need for defining the steps required to keep biotic resources functioning and available in the long term. The program is embryonic and only feebly supported by the public and by government, but it is an essential step that requires intensive scientific support now to determine what will work in restoring the coastal zone. Much is known, but much remains to be learned, especially at the regional level in determining how to provide for both the protection of the zone and its use in the production of indigenous fisheries.

A similar challenge exists on the land starting from both the bottom and the top. The global challenge is conspicuous as climatic disruption at the moment. But the global challenge is also in restoration of normalcy in the global cycles of carbon, nitrogen, and sulfur, for example. The local challenge might be conspicuous in the need for restoring whole landscapes in Haiti; India; West Africa; Madagascar; Sudbury, Ontario; and Krasnoyarsk, Siberia. But it, too, is global in that

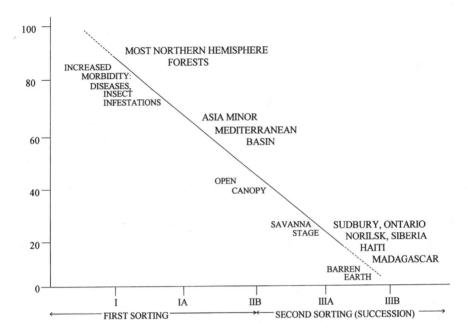

STAGES OF IMPOVERISHMENT OF FORESTS

FIGURE 1 Continuum of biotic impoverishment as appraised by systemic reduction in primary productivity. Assumption is made that continuum is linear. It might deviate from linearity in many circumstances where structure of vegetation changes discontinuously under chronic disturbance.

no corner of the earth is unaffected by human disruptions that are having biotic consequences and causing increments of erosion measurable on the scalar system of table 1, shown graphically in figure 1.

The stage is set for a rejuvenation of science in definition and defense of the broad public interest in the preservation of a habitable earth. It should come not through an endangered species act or an emphasis on an inchoate interest in biodiversity, but through emphasis on the preservation of the biotic functions locally that keep the water clean, the air clean, and the landscape intact.

REFERENCES

Bormann FH. 1990. Air pollution and temperate forests: creeping degradation. In: Woodwell GM (ed). The earth in transition: patterns and processes of biotic impoverishment. New York NY: Cambridge Univ Pr. p 25–44.

Brown LR. 1995. Nature's limits. In: State of the world 1995. Washington DC: The Worldwatch Inst. p 14.

Brown LR. 1997. The agricultural link: how environmental deterioration could disrupt economic progress. Worldwatch Paper No 136(August):73.

Daly HE. 1993. From empty-world economics to full-world economics: a historical turning point in economic development. In: Ramakrishna K, Woodwell GM (eds). World forests for the future. New Haven CT: Yale Univ Pr. p 79–91.

Fagan B. 1999. Floods, famines and emperors. Perseus NY: Basic Books. 284p.

Gullett DW, Skinner WR. 1992. The state of Canada's climate: temperature changes in Canada 1895–1991. State Envir Rep 92(2):36.

Heck WW, Taylor OC, Adams R, Bingham G, Miller J, Preston E, Weinstein L. 1982. Assessment of crop loss from ozone. J Air Pollu Contr Asso 32(4):353–61.

Hooker A. 1994. The international law of forests. Nat Res J 34:823–77.

Houghton JT, Callander BA, Harts N, Kattenberg A, Maskell K. 1996. Climate change 1995: the science of climate change. London UK, New York NY: IPCC, Cambridge Univ Pr.

Houghton RA. 1997. Forests and agriculture. Ms for The Scientific Council, World Commission on Forests.

Houghton, RA, Davidson EA, Woodwell GM. 1998. Missing sinks, feedbacks, and understanding the role of terrestrial ecosystems in the global carbon cycle. Glob Biogeochem Cyc 12(1):25–34.

Juday. 1997. *Fide.* The Boston Globe, Science and Health, 15 Sep.

Little CE. 1995. The dying of the trees: the pandemic in America's forests. New York NY: Viking.

Stevens WK. 1997. If climate changes, who is vulnerable? Panels offer projections. New York Times, 30 Sept.

University of Alaska. 1983. The potential effects of carbon dioxide-induced climatic changes in Alaska. Miscellaneous publication #83. Fairbanks AK: Univ of Alaska.

Woodwell GM, Houghton RA. 1990. The experimental impoverishment of natural communities: effects of ionizing radiation on plant communities, 1961–1976. In: Woodwell GM (ed). The earth in transition: patterns and processes of biotic impoverishment. New York NY: Cambridge Univ Pr. p 9–24.

Woodwell GM (ed). 1990. The earth in transition: patterns and processes of biotic impoverishment. New York NY: Cambridge Univ Pr. p 530.

Woodwell GM. 1995. Biotic feedbacks from the warming of the earth. In: Woodwell GM, Mackenzie FT (eds). Biotic feedbacks in the global climatic system: will the warming speed the warming? New York NY: Oxford Univ Pr.13–21.

Woodwell GM, Mackenzie FT, Houghton RA, Apps MJ, Gorham E, Davidson EA. 1995. Will the warming speed the warming? In: Woodwell GM, Mackenzie FT (eds). Biotic feedbacks in the global climatic system: will the warming speed the warming? New York NY: Oxford Univ Pr. p 393–411.

THE RESPONSE OF
THE INTERNATIONAL SCIENTIFIC COMMUNITY
TO THE CHALLENGE OF BIODIVERSITY

DAVID L. HAWKSWORTH
MycoNova, 114 Finchley Lane, Hendon, London NW4 1DG, UK

BIOLOGISTS BECAME increasingly alarmed at the loss of biodiversity during the 1970s and 1980s. The issue was treated mainly at the national level at that time and was primarily the concern of conservationists rather than scientists. The prospect of the Earth Summit in Rio de Janeiro in 1992, encompassing such issues and raising them to treaty status, found the international scientific community ill prepared.

Nevertheless, before the summit, the international scientific community had taken some action to identify the scientific issues requiring action. For example, the International Union of Biological Sciences (IUBS) and the Scientific Committee on Problems of the Environment (SCOPE) held a workshop on the ecosystem function of biological diversity in Washington, DC, in 1989 (DiCastri and Youngè 1990), which identified possible subjects for study and was adopted as the framework of an international program of biodiversity science, named DIVERSITAS in 1992. The World Conservation Union (IUCN), the United Nations Environment Programme (UNEP), and the World Resources Institute (WRI) had been most progressive in holding a series of regional consultations and workshops in developing a global biodiversity strategy (WRI and others 1992) containing a daunting 85 recommended actions. Finally, the International Council of Scientific Unions (ICSU) organized ASCEND 21 in Vienna in 1991 to develop an agenda for science and development in the next century (Doodge and others 1992). ASCEND 21 not only reviewed problems of environment and development, scientific understanding of the Earth system, and responses and strategies, and made eight recommendations for action where focused on additional research required, but also recognized the need for the scientific community to strengthen links with development agencies and other organizations charged with addressing environmental problems.

Science was full of expectations at the time. Issues of concern to scientists were on the agenda of ministers at a level that scarcely could have been dreamed of even in the middle 1980s. The anticipation was that key scientific questions and other issues related to the magnitude and description of biodiversity and its significance at the ecosystem level were to be addressed. Scientific questions transcend and do not recognize national boundaries; indeed, that is one of the beauties of science. They also can require concerted international efforts and resources for their elucidation. In this paper, I examine the extent to which the aspirations of the scientific community have been met, with particular reference to major international biodiversity initiatives.

THE CONVENTION ON BIOLOGICAL BIODIVERSITY

The Convention on Biological Diversity (UNEP 1992), now ratified by 175 governments, is concerned with the sustainable use of biodiversity and the equitable sharing of benefits arising from it. For what is essentially a political treaty, the Convention contains a surprising number of articles related to scientific issues or requiring a scientific base for their implementation. These include articles related to inventorying and monitoring, in situ and ex situ conservation, research and training, public awareness, assessment and minimization of adverse effects, technology transfer, and technical and scientific cooperation.

The implementation and work plans of the Convention are discussed by government delegations and observers at conferences of the parties (COPs), which have met almost annually since the convention came into effect in 1994. Also, a Subsidiary Body for Scientific, Technical, and Technological Advice (SBSTTA), which has met annually since 1995, considers matters referred to it after each COP. Topics discussed to date have included agrobiodiversity, a clearinghouse mechanism for information exchange, marine and coastal biodiversity, freshwater ecosystems, forest biodiversity, indicators, monitoring and assessment, and taxonomic capacity-building.

The current financial mechanism for implementing the Convention is the Global Environment Facility (GEF). Most of the funds it supplies support the incremental cost of projects within less-developed signatory countries, although some regional projects and a few enabling activities unrelated to a country have been supported. The effectiveness of the GEF is a continuing cause of concern to the parties to the Convention.

Progress in implementation has been slow, an overriding view that emerged from Earth Summit + 5, which was convened by the United Nations in New York City in June 1997. Many of the issues highlighted as continuing concerns depend on the biological sciences for progress: the preservation and sustainable use of biodiversity in freshwater, oceans and forests, and progress toward sustainable agriculture.

A major development linking the Convention to the scientific community was the signing of a memorandum of understanding with DIVERSITAS (see below) in November 1997. As a result, a meeting of experts was convened in Mexico City in March 1998 to prepare recommendations on scientific research that should be undertaken for the effective implementation of the Convention. The report was welcomed by the fourth meeting of the Conference of the Parties to the Convention in Bratislava in May 1998 and referred to the next meeting of SBSTTA, to be held in Montreal in June 1999, for further consideration. The Bratislava meeting was also important in agreeing on the need to develop a Global Taxonomy Initiative.

GLOBAL BIODIVERSITY ASSESSMENT

The first major global scientific project in support of the Convention was the Global Biodiversity Assessment (GBA) (Heywood 1995). This work aimed to provide an extensively peer-reviewed assessment of our current state of knowledge on all aspects of biodiversity. The project was initiated by UNEP. The steering group first met in Trondheim in May 1993, funding details were finally agreed on early in 1994, and the 1,140-page volume was published in November 1995. The GBA was an extensively reviewed assessment of the known, so it was not appropriate for it to make recommendations, which is not always appreciated. However, it did tackle thorny questions such as the numbers of known and estimated species, extinction rates, and the ecosystems at greatest risk.

The statistics related to the GBA are impressive: The exercise involved 16 steering-group members, 26 section coordinators, five major workshops and review meetings, five editorial-group meetings, three section workshops, 385 contributors, and 536 peer reviewers—overall, 1,003 scientists (not allowing for those acting in more than one capacity). The project was made possible through a US $3.1 million award from the GEF. Although at first this might appear excessive, the true cost has been estimated to be about 6 times that figure in a GEF-commissioned independent review of the project.

The successful realization of such a major work by the world's scientists in so short a time demonstrates unequivocally that if the resources are available, scientists are prepared to change their itineraries and commit to deliver the required product.

DIVERSITAS

The DIVERSITAS program is the major international response to the scientific challenges of the Convention. Conceived at the Workshop on Ecosystem Function and Biological Diversity held by IUBS and SCOPE in Washington in 1989 and named DIVERSITAS in 1992, it had as parents SCOPE, the UN Educational, Scientific, and Cultural Organization (UNESCO), and IUBS. The conceptual frameworks and agendas for various aspects of biodiversity research being developed under the program attracted increasing interest, and the sponsoring organizations now include ICSU, International Union of Microbiological Societies (IUMS), International Geosphere Biosphere Programme (IGBP), Global Change and Terrestrial Ecosystems (GCTE), and IUCN. ICSU is now the lead body on the scientific organizational front, and substantial financial support has been received for the secretariat from UNESCO. The steering committee is currently chaired by José Sarukhán. Through the links of the biological unions to the various international specialist scientific organizations (for example, the 83 scientific members of IUBS), DIVERSITAS has the potential to obtain input from an enormous treasure-house of expertise and to bring together biologists who rarely see or communicate with those in related disciplines, and who even might speak in different languages or "biobabble" (Lovelock 1995). One important achievement of both DIVERSITAS and the Gaia-hypothesis debates has been to bring together scientists from disparate fields and to focus them on common problems. The resulting synergism is not only stimulating and intellectually challenging, but also facilitates the holistic approaches demanded by considerations of both the conservation and the sustainable use of biodiversity and global ecology.

DIVERSITAS is divided into five core programs and five special target areas for research, or STARs (figure 1). These are all interlinked and related to a

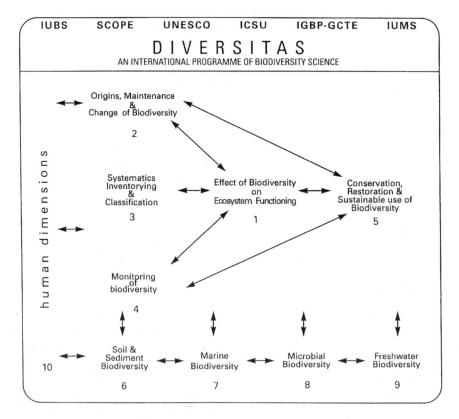

FIGURE 1   The Core programs (numbered 1–5) and special target areas (STARs, numbered 6–10) of DIVERSITAS.

consideration of the human dimension. The core programs focus on the effect of biodiversity on ecosystem functioning; the origins and maintenance of and change in biodiversity; systematics, inventorying, and classification; monitoring of biodiversity; and conservation, restoration, and sustainable use of biodiversity. The STARs, selected because they were judged to be particularly neglected topics of crucial importance to our overall understanding of biodiversity, are devoted to soil and sediment biodiversity, marine biodiversity, microbial biodiversity, and freshwater biodiversity.

For example, in the case of soil, although we know something about the functional interconnections of the groups of organisms present, we are almost totally ignorant of the numbers of species of bacteria, fungi, and nematodes, in particular, that are involved in specific ecological processes. Even the techniques that exist for examining soil biodiversity are not standardized, and a first step in this STAR was to prepare an authoritative synopsis of the methods now in use (Hall 1995).

Although DIVERSITAS is still in the process of developing action plans and seeking funding for the core programs and STARs, substantial progress has been made in several. In 1991, the IUBS-IUMS Committee on Microbial Biodiversity, the implementing body for the microbial biodiversity STAR, recommended the development of a Biodiversity Information Network. Later named BIN21 and expanded to all aspects of biodiversity, this now operates from the Fundação de Pesquisas e Tecnologia André Tosello in Campinas, Brazil, and is used extensively (Canhos and others 1994).

The importance of wild relatives of domesticated organisms was recognized from the outset as an issue that needed to be addressed. The importance of this subject has been confirmed by the development by the UN Food and Agricultural Organization (FAO) of a global plan of action for plant genetic resources (FAO 1996). As a component of its conservation core program, DIVERSITAS is contributing to plans for implementing that action plan.

SPECIES 2000, an element of the systematics, inventorying, and classification core program, originally launched by IUBS and cosponsored by the Committee on Data for Science and Technology (CODATA) and IUMS, has been supported by both UNEP and the Convention. This project aims to index the world's species by the establishment of a federation of interlinked global master species databases and a comprehensive name-finder tool (Bisby and Smith 1996; figure 2). The formation of a federation between key organizations that have data pertinent to the mission of SPECIES 2000 has been impressive. A pilot system is already operational on the World Wide Web, and now funding is needed to build authoritative databases for the groups of organisms that lack them.

Establishing a firm basis for communication necessitates both a system like that being developed for SPECIES 2000 and a more stable protocol than currently used for the scientific naming of organisms. This is clearly a responsibility of international science, and to this end, IUBS and IUMS are collaborating in the production of proposals for a unified BioCode to regulate the names of all organisms from a date to be agreed on (Greuter and others 1998; Hawksworth

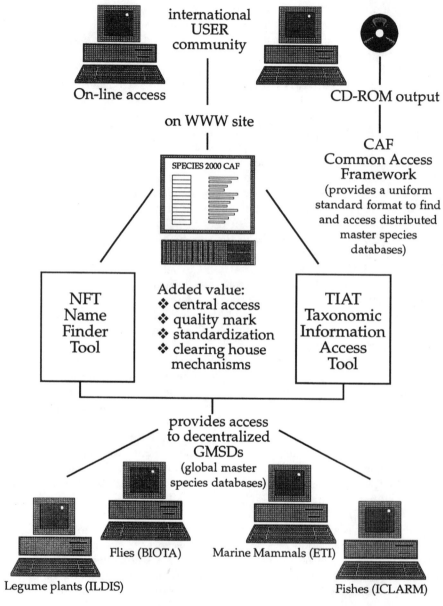

Transparent Interoperability of GMSDs.
Distributed databases of Federated Members

FIGURE 2   The Structure envisaged for SPECIES 2000.

1995). The IUBS General Assembly in Taipei in November 1997 has recommended this for consideration; the bodies concerned with the different codes were urged to also incorporate elements of it into existing codes.

A prerequisite to address our ignorance of perhaps 80% of the species with which we share the planet is a sufficiency of scientists who are able to recognize the known and describe the newly discovered. A shortage of biosystematic skills is inhibiting the ability of nations to implement the Convention, as recognized in the decision of COP3 in Buenos Aires in 1996 to support a global taxonomy initiative. The shortage of biosystematists has been a concern of biologists for more than 50 years. Data in the GBA show that only 6,989 biosystematists published new scientific names in 1992 (Heywood 1995). The Systematics Agenda 2000/International component of the systematics core program of DIVERSITAS is developing plans to address these gaps in knowledge and skills, and progress is reported by Joel Cracraft (this volume).

The need to implement a Global Taxonomy Initiative was endorsed again at COP4 in 1998 (see above), which considered the Darwin Declaration drawn up by representatives of various systematic institutions and organizations in Darwin, Australia, in February 1998. DIVERSITAS, in conjunction with Environment Australia and the GEF, then met at the Linnean Society in London in September 1998 to consider how to develop this initiative (Australian Biological Resources Study 1998). Later meetings organized by DIVERSITAS and Systematics Agenda 200/International in New York in September 1998 and by DIVERSITAS in Paris in February 1999 assessed needs and ways of defining priorities and making recommendations for consideration by SBSTTA in June 1999. This is a most welcome development involving scientists as partners in developing guidelines for actions to be taken by governments and supported by international agencies, such as the GEF.

Also pertinent to the shortage of biosystematics capability is a complementary intergovernment initiative, BioNET-INTERNATIONAL (BI). Launched in 1993 and facilitated by CAB INTERNATIONAL, BI is a strategy for enabling developing countries to establish and sustain realistic self-reliance in biosystematics (Jones 1997). The seed was sown at the Golden Jubilee of the Systematics Association in 1987 (Haskell and Morgan 1988). BI is organized into a series of seven regional locally organized and operated partnerships (LOOPs; figure 3). The LOOPs are established with the support of the governments involved and develop agendas and work programs appropriate to their needs. The focus is on the species-rich groups that are least understood, notably arthropods, fungi, and nematodes. The LOOPs are supported by networks of institutions in developed countries that commission the work. The Technical Secretariat of BI supports the establishment of LOOPs and assists in obtaining donor funding for the implementation of their programs. BI has been successful in securing funds from a wide array of donors, including the Swiss Development Corporation, the UK Department for International Development, UN Development Programme (UNDP), Deutsche Gesellschaft für Technick Zusammenarbeit (GTZ), and the Centre for Technical and Rural Co-operation.

354

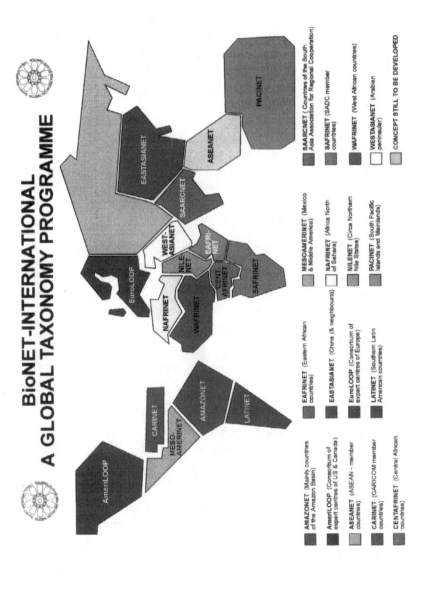

FIGURE 3 The Regional LOOPs of BioNET-INTERNATIONAL.

THE CHALLENGE

This paper has been eclectic in focusing primarily on two major international initiatives. My intention was not to denigrate in any way the remarkable work of such bodies as the IUCN, the World Conservation and Monitoring Centre, WRI, and the UN agencies. My purpose has been threefold: to show something of what can be achieved if the resources are available, to discuss the vision of scientists who are concerned with organizing a major thrust to address issues of biodiversity, and to consider the challenges that must be faced in transforming this vision into reality.

Pleading does not work with funding agencies, especially if we seem to ask for funds to do what we have always done and enjoy doing. Agendas must mesh, as they did with the GBA, and do with BioNET-INTERNATIONAL and the Global Taxonomy Initiative. Neither is it a matter of being just a salesperson. Arriving at donors' doors with our wares to sell and expecting them to open their checkbooks—even if we believe that what we have to offer is in danger of being lost—does not work.

We need new skills and approaches. Major funding is always linked to political agendas, and it is those we must influence at the formative stage. McNeely (1995) has encapsulated the requirements to be met at the political level. We also need to learn how to talk to politicians. Scientists are cautious by nature, tending to present tentative results that always seem to call for further research, but politicians want answers as quickly as possible.

We must avoid what is viewed as a "green maze" between science and politics: when there are conflicting opinions, the reaction of politicians and donors is to leave well enough alone. In August 1997, Secretary of the Interior Bruce Babbitt told the annual meeting of the Ecological Society of America that in the case of climate change, although there was a scientific consensus, there was not a public consensus (Macilwain 1997). A key virtue of the assessment approach, as seen in the GBA, is to present a scientific consensus view of what we know. This approach is being translated into a consensus of what biodiversity science should be doing in the DIVERSITAS model, but if these thrusts are to realize the required level of support, they need to take the wider public along too.

A credibility gap also has to be filled. Scientists involved in field studies of organisms have always been a source of amusement to cartoonists, and the scientist stereotypes portrayed in current television and cinema productions do not help. We need to meet this challenge of credibility by presenting ourselves as being capable of helping politicians implement their agendas. This is an issue for individual, as well as collective, action. As individual scientists, we must take time out from the pursuit of knowledge to be public-relations workers from the local to the national and international levels.

At the international level, ICSU has a key role to play in the elevation of science in the political arena. Now the primary nongovernment sponsor of DIVERSITAS, ICSU is composed of the various international scientific unions and national academies of sciences and their equivalents. It has the potential to be perceived as the voice of world science, and it merits recognition and representation at the highest political levels whenever scientific issues and priorities are under debate. ICSU is undergoing a review of its structure and role, in which raising profile and credibility must be seen as key matters for the future health of scientific endeavors.

We scientists also must be prepared to act in concert now. It is no good to say that we will get organized tomorrow. The window of opportunity to secure major international funding in some parts of biodiversity science might already be closing. A worthwhile reflection is that it took 7 years to provide DIVERSITAS with a secretariat that could begin serious consensus-building on the scientific tasks required.

In this article, I have endeavored to show by example how the international scientific community is responding to the challenge of biodiversity. We have seen what can be achieved through coordinated and adequately funded efforts in the GBA, and now we have a vision of what subjects need to be addressed through DIVERSITAS. The challenge is to put energy into working at the political and donor levels if we are ever to transform scientific potential into reality—results required by and delivered to others.

ACKNOWLEDGMENTS

I am indebted to Colleen Skule-Adam and Patricia Taylor for comments on an earlier draft.

REFERENCES

Australian Biological Resources Study. 1998. The global taxonomy initiative: shortening the distance between discovery and delivery. Canberra Australia: Australian Biological Resources Study, Environment Australia.

Bisby FA, Smith P. 1996. Species 2000: indexing the world's known species. Project Plan Version 3. Southampton: SPECIES 2000.

Canhos DAL, Canhos VP, Kirsop B (eds.). 1994. Linking mechanisms for biodiversity information. Campinas: Fundação de Pesquisas e Tecnologia André Tosello.

DiCastri F, Youngè T (eds). 1990. Ecosystem functioning and biological diversity. Biol Intl Spec Iss 22:1–20.

Dooge JCI, Goodman GT, la Riviere JWM, Marton-Leffevre J, O'Riordan T, Praderie F (eds.). 1992. An agenda of science for environment and development into the 21st Century. Cambridge UK: Cambridge Univ Pr.

FAO [United Nations Food and Agricultural Organization]. 1996. A global plan of action for plant genetic resources. Rome, Italy: FAO.

Greuter W, Hawksworth DL, McNeill J, Mayo MA, Minelli A, Sneath PHA, Tindall BJ, Trehane P, Tubbs P. 1996. Draft BioCode: the prospective international rules for the scientific names of organisms. Taxon 47:127–50.

Hall GS, ed. 1995. Methods for the examination of organismal diversity in soils and sediments. Wallingford UK: CAB INTERNATIONAL.

Haskell PT, Morgan PJ. 1988. User needs in systematics and obstacles to their fulfillment. In: Hawksworth DL (ed). Prospects in systematics. Oxford UK: Clarendon Pr. p 399–413.

Hawksworth DL. 1995. Steps along the road to a harmonized bionomenclature. Taxon 44:447–56.

Heywood VH (ed). 1995. Global biodiversity assessment. Cambridge UK: Cambridge Univ Pr.

Jones T. 1997. The BioNET-INTERNATIONAL approach. Biol Intl 35:40–6.

Lovelock J. 1995. The real value of science on the BBC. The Times 65257 (3 May 1995):23.

Macilwain C. 1997. Ecologists urged to win climate debate. Nature 388:704.

McNeeley J. 1995. Conservation with a human face. In: Bennan LA, Aman RA, Crafter SA (eds.). Conservation of biodiversity in Africa. Nairobi Kenya: National Museum of Kenya. p 383–8.

UNEP. 1992. Convention on Biological Diversity. Nairobi Kenya: UNEP.

WRI, IUCN, UNEP [World Resources Institute, The World Conservation Union, UN Environment Programme]. 1992. Global Biodiversity Strategy. Washington DC: WRI, IUCN, UNEP.

THE MILLENNIUM SEED BANK
AT THE ROYAL BOTANIC GARDENS, KEW

GHILLEAN T. PRANCE
ROGER D. SMITH
Royal Botanic Gardens, Kew, Richmond, Surrey, TW9 3AB United Kingdom

INTRODUCTION

IN RESPONSE TO the increasing rate of extinction of plant species (Ehrlich and Ehrlich 1981; Prance and Elias 1977), the Royal Botanic Gardens, Kew, decided in 1972 to begin seed-banking and research on seed conservation. Accordingly, the seed bank was set up at our second garden in the country at Wakehurst Place in Sussex, a safe location removed from a major urban area, not under the flight path of Heathrow Airport, and at an altitude of 200 m, well above sea level. Although we regard in situ conservation as the ideal, the world will certainly lose many species if we do not promote ex situ methods as well.

Seed banks are one of the most effective and economical means of conserving plant species where habitats are under threat (Miller and others 1995). In the last 2 decades, the current Kew Seed Bank has undertaken collaborative collecting expeditions in over 20 countries, and collecting activity has increased substantially over the last few years. These efforts have made the Kew Seed Bank the largest and most diverse bank that is devoted to wild plants and is run according to internationally approved standards. However, because financial resources are sparse, the bank collection still represents less than 2% of the world's flowering-plant flora. Viewed against the background of a rapidly increasing loss of biodiversity, that prompted us to investigate the possibility of increasing even more the rate of seed conservation and seed-banking undertaken by Kew.

In 1994, aided by a consultant, Sir Jeffery Bowman, we carried out a detailed study of the worldwide situation. Our survey indicated that there was very little coverage of noncrop plants in seed banks and minimal research into optimal collection, processing, and storage procedures for such species. That finding was supported by a recent review of the state of the world's plant genetic resources for food and agriculture by the Food and Agriculture Organization (FAO 1996), which concluded that there was a clear need to strengthen capacities for ex situ conservation cost-effectively and that the necessary increase in seed-banking activity and supporting research would require national, subregional, regional, and international collaboration. The report also confirmed the heavy emphasis (94%) on crop plants among the 6 million seed accessions held worldwide, the minimal coverage of truly wild species, and the slight coverage of forest, forage, ornamental, aromatic, and medicinal species and underused crops. Moreover, the FAO report highlighted the fact that only 13% of the 6 million accessions were held in secure long-term facilities—that is, where seed was stored according to internationally approved standards of temperature and moisture content, where the power supply was reliable, and where procedures for safe duplication and regeneration were in place. The current Kew Seed Bank, which was mentioned specifically in the FAO report, meets all those criteria for a secure, long-term seed bank.

On the basis of such information and with impetus added by the UK ratification of the Convention on Biological Diversity (UNEP 1992a), Kew concluded that a substantial increase in efforts to collect, conserve, and research seeds of wild species was vital. Moreover, Kew feels that it is uniquely placed to play a leading role in this process, not only because of its existing collections, its expertise in seed conservation, and its location within a geologically and politically stable country, but also because of its well-established network of collaborators and its horticultural and taxonomic expertise, which has earned it an international reputation as a center of excellence for botanical research. That expertise and a belief in collaboration will be vital to ensuring the international cooperation needed to tackle a problem of this scale.

The opportunity to achieve the great increase in Kew's seed-conservation activities was provided by the Millennium Commission, one of the distributors of national lottery proceeds in the UK, which was set up for partial funding of projects to celebrate the new millennium. In December 1995, the commission awarded Kew's Millennium Seed Bank (MSB) project a grant that would eventually total up to £30 million, which is just over one-third of the total cost of the project. With the help of the Kew Foundation, we have raised over £16 million in counterpart funding, including a grant of £9.2 million from the Wellcome Trust and a sponsorship of £2.5 million from Orange, a UK communication company. The MSB will continue to focus on wild species rather than crop species, many of which have their own seed banks or germplasm collections.

The MSB project has been presented to and discussed with representatives of national and international organizations involved in plant genetic-resources conservation, including FAO, the Consultative Group for International Agricultural Research, the International Plant Genetic Resources Institute, Botanic Gardens Conservation International, the World Conservation Union, the United Nations Environment Program (including the Secretariat to the Convention on Biological

Diversity), the United Nations Development Program, the Global Environment Facility, and the World Bank. In addition, relevant UK government departments (Department for International Development formerly ODA, Department of the Environment, and Ministry of Agriculture, Fisheries and Food) and conservation bodies have been consulted, and the proposal has been presented at several scientific conferences. All such meetings have confirmed not only that a large-scale seed-conservation project is necessary and would not duplicate any existing activity but also, inasmuch as Kew is a world leader in seed-banking for wild plants, that it is ideally placed to be the focus for such a major conservation effort.

THE AIMS OF THE MILLENNIUM SEED BANK

The MSB project will establish an international center of excellence for seed conservation at the Royal Botanic Gardens, Kew, Wakehurst Place. The project has six main aims:

- to collect and conserve seeds of most of the UK spermatophyte flora (seed-bearing plants) and a further 10% of the world's spermatophyte flora, principally from the drylands;
- to encourage plant conservation throughout the world by facilitating access to and transfer of seed-conservation technology;
- to carry out research to improve all aspects of seed conservation;
- to make seeds available for species reintroduction into the wild, for academic research, and for screening for potential new uses of plants;
- to develop the public's interest in the need for plant conservation; and
- to provide a world-class building as the focus for this activity.

THE UK SEED CONSERVATION PROGRAM

For Kew to function actively in seed conservation overseas, it is important that it make an input into plant conservation within the UK, where genetic erosion and endangerment are also high (Anon 1994; Wynne and others 1995). Common species will be included to supply material off-season or abroad, to add seed-biology information, to compare with in situ populations through time, and to guard against changing fortunes resulting from climate change (see Jackson and others 1990). No country holds a near-complete representation of its spermatophyte flora. Kew aims to enable the UK to be the first such country and hopes that the example will stimulate other countries to follow suit.

Our initial objective is to have conserved within the MSB, by the year 2000, seed from at least one population sample of every native UK plant species that produces bankable seed.

Stace (1991 and pers. comm.) has indicated that the native flora of the British Isles consists of some 1,571 species of vascular plants, of which 1,442 are spermatophytes native to the UK. The remainder are ferns or plants that occur only in Eire. Those figures do not include the microspecies of the apomictic (reproducing without meiosis or formation of gametes) genera *Rubus, Hieracium,* and *Taraxacum,* which have been grouped together into the more distinct aggregates numbering 13, 11, and 9 respectively. The Kew Seed Bank holds 552 species (plus some microspecies), so 890 need to be collected; 361 of these are restricted or rare. Within the target for collection, we estimate that six species produce seeds that cannot be banked (the recalcitrant species), and 11 rarely or never produce seed. That leaves 873, of which 163 (including 77 aquatics and 51 orchids) produce seed that will need research work before we can ascertain the likely success of banking.

Seeds for the MSB will be collected in collaboration with many individuals and conservation organizations throughout the UK, including the statutory bodies—English Nature, Scottish National Heritage, the Countryside Council for Wales, and the Department of the Environment (Northern Ireland)—and those like the Botanical Society of the British Isles and the Wildlife Trusts. A member of our collecting staff will work full-time on the project for the next $2^{1}/_{2}$ years, but many of the remaining collections will be made by partner organizations; their work will be coordinated by the MSB. Training sessions for volunteer collectors from within the partner organizations are under way.

Where appropriate, difficult species will be collected through a contract arrangement with specialist organizations, and the more common species will be collected by voluntary groups, such as the Wildlife Trusts, and will attract the payment of an honorarium. There will thus be high public involvement. If the seed-production seasons are abnormal, it might be necessary to get the volunteers to retry some species the next year. It is envisaged that the less than complete genetic representation (especially of inbreeding species) that results from stopping at the initial objective of one population sample per species will be gradually improved by donation to the collection and by further collecting later in the project.

Target-species lists have been provided to the Wildlife Trusts and several other organizations to induce offers of collection. These lists have been produced from an extensive Excel database, now substantially developed, that lists all the UK native spermatophytes by their scientific and common names broadly where they occur, the status of existing collections, their rarity, and any special problems related to them (for example, that they are aquatic). It is proposed that other useful information, such as flowering or seeding date, be added to the database in due course. By May 1999, seeds from 71% of the species that are native to the UK have been collected and booked.

THE ARID-LAND SEED-CONSERVATION PROGRAM

Since the early 1980s, the focus of the Kew Seed Bank has been on tropical drylands. Such lands, which are experiencing habitat loss because of desertification, particularly in Africa (UNEP 1994), have been identified as the ecosystem for which ex situ conservation is most appropriate, compared with the tropical rain forests (the other ecosystem undergoing extensive damage). The origin of habitat loss in the drylands—drought and the factors that exacerbate it, such as overgrazing (Binns 1985)—is less open to substantial manipulation with local political or economic tools. Consequently, other actions are necessary to underwrite the survival in situ of biological diversity in the tropical drylands.

The drylands cover one-third of Earth's land surface, including many of the world's poorest countries, and support almost one-fifth of its human population (UNEP 1992b). Rural people rely on plants in almost every aspect of their lives, and Kew's Survey of Economic Plants for Arid and Semi-Arid Lands (SEPASAL) database lists over 6,000 such plants with uses as varied as land stabilization, hedging, nitrogen fixation, contraceptives, dyes, and cooking utensils (Davis and others 1996). Products derived from plants in the drylands are also important to people in developed countries, for example, the pharmaceuticals sennoside A and B, atropine, and ephedrine and such industrial products as gums, resins, waxes, and oils (Goodin and Northington 1985). There is great scope for many more dryland plants and their products to be developed for human welfare, including those with unique morphological, physiological, and chemical adaptations induced by the particular environmental stresses of arid lands. Some of these adaptations, such as salt tolerance and the C4 and CAM mechanisms of photosynthesis, might be valuable sources of material for plant-breeding. Other characteristics that confer tolerance to drought, predation, and disease are also likely to be present in dryland plant material. In addition, many arid-land species have evolved elaborate chemical defenses that make them important potential sources of insecticides.

Some practical benefits of seed-banking follow the choice of the drylands as a target for seed collection: species are often found within discrete populations; populations often exhibit defined flowering and fruiting periods in response to climatic conditions; and vegetation is usually low (often less than 5 m high) and relatively open, allowing convenient access for seed collection. Furthermore, although the seed-storage physiology of only 2% of the world's flora has been studied, it is thought that the majority of higher plant species from drylands will exhibit "orthodox" seed-storage behavior (retain their viability after drying) and therefore be suitable for long-term conservation in seed banks.

Prospective countries have been identified for partnership on the basis of several factors: existing successful collaboration; ease of access; extent of arid, semiarid, and dry subhumid land (Goodin and Northington 1985); number of endemic plant species (WCMC 1992); and the floristic regions in which they occur (Taktahjan 1986). On that basis, 18 countries have been identified as having high priority for collaboration; of these, eight (Australia, Brazil, Kenya, Madagascar, Mexico, Morocco, South Africa, and the United States) contain the most diverse dryland floras (table 1). In addition, we will accept donations of seed from nondryland countries (or from the wetter regions of the high-priority countries), provided that collections have been made in accordance with national and international regulations.

The world's spermatophyte flora is estimated to number 242,000 (Mabberley 1990); 10% is therefore 24,200 species. Within the collaborating countries, success in collecting and conserving seeds of 10% of the world's plant species will depend heavily on input from the countries' conservation organizations themselves; the overriding consideration in any targeting of taxa for collection will be

TABLE 1   Proposed Countries for Collaboration with MSB Project

Country	% Arid and Hyperarid	% Semiarid	Taktahjan Floristic Regions	Total No. of Plant Species	No. Endemic Species
Australia	49	20	29,30,31	15,990	13,240
Brazil	0	5	25,26	56,215	?
Egypt	100	0	7	2,076	70
India	4	17	8,12,16	15,000	5,000
Kenya	49	37	10,12	6,506	265
Madagascar	2	8	15	9,505	6,000
Mexico	26	20	9,23	26,071	13,000
Morocco	67	25	6,7	3,675	625
Namibia	56	39	12,13	3,174	?
South Africa	31	22	11,12,13,28	23,420	?
Syria	50	39	6,8	3,000	330
United States	8	22	3,4,9,23	19,473	4,036
Venezuela	1	2	23,27	21,073	8,000
Zambia	0	11	12	4,747	211
Zimbabwe	0	33	12	440	95

our partners' priorities for conservation, in line with the Convention on Biological Diversity.

Nevertheless, it was recognized early in the project's development that some focusing of collecting activity is important. For the last few years, the collectors for the current Kew Seed Bank have given special interest to 30 plant families selected on the basis of an analysis of species represented in Kew's SEPASAL database (table 2).

This list of target families is now being reviewed and revised to include coverage of globally threatened species and endemics, with input from the World Conservation Monitoring Center and partner institutes in collaborating countries, such as the National Museums of Kenya, and adequate representation of "higher-level" (order and above) taxonomic diversity as a surrogate for character and evolutionary diversity (see, for example, Williams and others 1994). It is also intended to give due weight to various ecological and functional considerations, such as appropriate representation of "keystone species" in particular plant communities, as more becomes known about such species.

Obviously, some flexibility will be essential to take account of changing circumstances. We expect a continuous process of review and refinement of any lists of desiderata as the project develops.

The arid-land collecting program will build on existing links between the current Kew Seed Bank and institutes in many dryland countries established during collaborative collecting expeditions in over 20 countries in the last 2 decades. The main planning phase for the overseas conservation program will be in 1996-1999; the main collecting phase will be in 1999-2009. Initial efforts will focus on seeking collaboration with the high-priority countries, including the United States, so that seed-collecting in these countries as part of the project could start by the

TABLE 2   Top 30 Families Targeted by Utility and Biodiversity[a]

Family[a]	Class[b]	Subclass[b]	Order[b]
Agavaceae	Monocotyledonae	Liliidae	Liliales
Amaranthaceae	Dicotyledonae	Caryophyllidae	Caryophyllales
Anarcardiaceae	Dicotyledonae	Rosidae	Sapindales
Burseraceae	Dicotyledonae	Rosidae	Sapindales
Cactaceae	Dicotyledonae	Caryophyllidae	Caryophyllales
Capparidaceae	Dicotyledonae	Dilleniidae	Capparidales
Chenopodiaceae	Dicotyledonae	Caryophyllidae	Caryophyllales
Combretaceae	Dicotyledonae	Rosidae	Myrtales
Compositae	Dicotyledonae	Asteridae	Asterales
Cruciferae	Dicotyledonae	Dilleniidae	Capparidales
Cucurbitaceae	Dicotyledonae	Dilleniidae	Violales
Cupressaceae	Gymnospermae	Pinopsida	—
Ebenaceae	Dicotyledonae	Dilleniidae	Ebenales
Ephedraceae	Gymnospermae	Gnetopsida	—
Euphorbiaceae	Dicotyledonae	Rosidae	Eurphorbiales
Geraniaceae	Dicotyledonae	Rosidae	Geraniales
Gramineae	Monocotyledonae	Commelinidae	Cyperales
Leguminosae	Dicotyledonae	Rosidae	Fabales
Malvaceae	Dicotyledonae	Dilleniidae	Malvales
Meliaceae	Dicotyledonae	Rosidae	Sapindales
Moraceae	Dicotyledonae	Hamamelidae	Urticales
Myrtaceae	Dicotyledonae	Rosidae	Myrtales
Palmae	Monocotyledonae	Arecidae	Arecales
Pinaceae	Gymnospermae	Pinopsida	—
Portulacaceae	Dicotyledonae	Caryophyllidae	Caryophyllales
Rhamnaceae	Dicotyledonae	Rosidae	Rhamnales
Solanaceae	Dicotyledonae	Asteridae	Solanales
Tamaricaceae	Dicotyledonae	Dilleniidae	Violales
Tiliaceae	Dicotyledonae	Dilleniidae	Malvales
Zygophyllaceae	Dicotyledonae	Rosidae	Sapindales

[a] A ranking procedure was devised to identify families that have the greatest part of their biological diversity adapted to arid and semiarid lands and that are also of greatest human utility. The rankings were calculated in the following way for each family:

$$\frac{\text{No. genera from family in SEPASAL}}{\text{No. genera in family}} \times \frac{\text{No. species from family in SEPASAL}}{\text{No. species in family}} \times \frac{\text{No. species from family in SEPASAL}}{\text{Total No. species in SEPASAL}}$$

The resulting top 30 families are listed above alphabetically. These 30 families account for 72% of the species listed in SEPASAL.
[b] According to Mabberley's use of Cronquist's System.

year 2000. For the remaining dryland countries, contacts with those with which we are collaborating or have collaborated in the past will be maintained or renewed, and contacts will be established in the others during the next 2 years with a view to securing partnership for later in the project period.

The needs of collaborating countries will differ according to their current levels of expertise and their national priorities for biodiversity conservation, so the MSB project will aim to provide a comprehensive seed-conservation service to meet varied requirements. The key services offered to collaborators are training and technology transfer, long-term storage of seeds until facilities exist in their own countries, benefit-sharing from the use of seeds, and access to our research expertise and data.

All seed-collecting, storage, and distribution will be carried out under bilateral agreements that cover profit-sharing as a result of intellectual property rights. The agreements have been developed as part of Kew's institutional program to ensure full compliance with the Convention on Biological Diversity. We are also in consultation with Michael Gollin (attorney at Spencer & Frank, Washington, DC, who has particular experience in contracts related to pharmaceutical screening) and Neil Hamilton (director of the Agricultural Law Center, Drake University Law School, University of Iowa).

The project will increase the number of collectors from the current two to around 28 in the year 2000. Five collectors and a coordinator will be based at the MSB; the remainder will be based overseas in partner countries. It is hoped that most of the overseas-based collectors will be recruited from the collaborating countries to collect their national floras and will be funded by international donor agencies, such as the Global Environment Facility and the European Union.

The key aim is to sample the breadth of dryland plant diversity, concentrating on a wide range of species. Interspecific variation is often the initial screen for potential use. Although only one population sample per species will be collected as a key objective, substantial genetic variation is likely to be present in collections; most species will be outbreeders (see Richards 1986), and intrapopulation variation will be greater than with inbreeders (von Bothmer and Seberg 1995). The samples collected can be used to research species botany, including breeding system, and so more carefully tailor later sampling strategy. To sample genetic diversity within a species more fully, more populations would need to be collected (Brown and Marshall 1995), but that is more appropriate after successful initial trials or research. The sampling strategy for each population will be that practiced for over 20 years by the Kew Seed Bank and is similar to that recently reiterated by Brown and Marshall (1995); key factors will be to sample randomly and evenly within a population and from sufficient individuals (at least 50 where the size of the population permits). Where populations are very small (20 or fewer), collections from individual plants will be kept separate—something that is impractical for more individuals. Collectors will visit most populations once during the seeding period, so it is proposed that no more than 20% of the seed available on the harvest date should be taken from annuals, biennials, and short-lived perennials. Thus, the survival of the parent plant population should not be threatened.

In achieving the 24,000-species target, allowances have been made for duplicate and poor-quality collections. In addition, it is envisaged that 10% of the target will consist of unsolicited samples, some from within the target area and many from outside.

Voucher specimens, representing the population sampled, are always collected. At least one of these specimens is deposited in the national herbarium of the host country. The specimen returned to the MSB will be identified by reference to Kew's comprehensive herbarium collection and associated bibliography. Data recorded in the field will be as objective as possible. All populations will be located with Global Positioning Systems, either in the partner countries or in the MSB. Such systems will be used, where possible, to guide collectors to likely diversity hot spots and, in some circumstances, to predict where a particular species might be found (see Guarino 1995). They might even be able to guide collectors to areas where the greatest genetic diversity would be expected within a species range. For instance, Nevo and Beiles (1989) hypothesize that species ranging across mesic and xeric environments display the greatest levels of genetic diversity in hot deserts, where rainfall and climate unpredictability are highest.

All collections destined for the MSB will be returned from the field as quickly as possible, thereby minimizing the loss of initial seed viability, which can strongly influence longevity (see Smith 1995). Samples thereafter will be processed as they are in the current Kew Seed Bank (see Prendergast and others 1992). The procedures are adapted from those used by seed banks that store crop germplasm (see Ellis and others 1985). They involve accessioning, drying, cleaning, x-ray examination, counting, packaging, freezing, and testing of germination. The main differences between processing of wild and crop germplasm are related to the handling of "empty" or insect-damaged seed (Linington and others 1995), seed dormancy (Linington and others 1996), and identification of seed storage behavior by testing germination after drying and freezing (Smith and Linington 1996). Storage of seed will be mainly at –20°C in a variety of storage containers. Subsamples will be rechecked for germination initially every 10 years, as in the current Kew Seed Bank, but retest intervals are expected to be modified for different collections in the light of results from the seed-research program of the MSB project. It is expected that 22 processing staff will be based in the collaborating countries and a further 22 in the UK.

Seed samples collected will be shared equally between partner countries and the MSB and deposited in facilities in the country of origin, if available. The MSB will act as a backup, providing a duplicate store for an agreed proportion of the seeds. For countries where local facilities are not available, the MSB will store independently both partners' shares of each collection and provide advice and assistance on establishing a bank in the country of origin. The MSB will, in turn, back up some of its collections at the Scottish Agricultural Sciences Agency, East Craigs, Scotland, to achieve a double indemnity against loss.

SEED-BANKING IN THE UNITED STATES

The United States contains a considerable area of arid lands, and three collaborative expeditions have already taken place in the country. A meeting was held in November 1996 between one of us (R.S.) and Peggy Olwell, chairperson of the Native Plant Conservation Committee, a partnership of nine federal agencies and 54 nonfederal cooperators, to discuss the MSB project. Considerable support for the project was shown, and an invitation was extended to attend its bimonthly meetings and become a nonfederal member. Contact about the MSB has also been established with the Center for Plant Conservation (CPC), a nonfederal member of the above committee and an entity with which we have previously collaborated on technical matters in wild-species seed-banking. Michael Bennett, keeper of the Jodrell Laboratory, gave a presentation on the MSB during his visit to the Missouri Botanical Garden (home of the CPC) in May 1997. Again, an interest in collaborating was expressed. Similar enthusiasm for involvement with the MSB has been shown by representatives of the Boyce Thompson Southern Arboretum, the Desert Botanical Garden, and the University of Arizona Desert Legume Program.

A meeting with the staff of the US Department of Agriculture's National Seed Storage Laboratory at Fort Collins to discuss collaboration with the MSB took place in August 1997 to coincide with a conference on plant genetic resources.

RESEARCH, INFORMATION FLOW, TECHNOLOGY TRANSFER, AND TRAINING

For the MSB project to succeed, it must not be merely a museum of seeds. It will also be accompanied by considerable research on seed-collecting, processing, and storage and by extensive training and, where needed, technology transfer to our collaborators.

The integrated seed-banking and seed-conservation research program of the MSB offers a unique opportunity to increase knowledge on the seed biology of a considerable amount of dryland biodiversity. The value of this information will be fully realized if it is made readily available to potential end users, and this will be facilitated through technology transfer and the provision of advice and training. Thus, this part of the MSB project will have the following main objectives:

• To generate detailed primary datasets on the seed storage and germination of about 1,500 species through research.
• To construct a seed-information database on about 20,000 species using inhouse and public-domain information.
• To ensure benefit-sharing through information flow, technology transfer, and formal training.

Research

The Seed Conservation Section has been involved in research in the conservation of seeds of wild (nondomesticated) species for over 25 years. The research has already established that wild species differ from domesticated species in their seed-storage and germination behavior, but their conservation and use as seed are generally practicable.

The seed-storage behavior of only about 7,000 species has been investigated to any degree of certainty; and of the remainder of the world's spermatophyte flora, an estimated 37% of species occur in families where less than 1% of those species have had their behavior investigated (derived from Hong and others 1996).

Moreover, the dryland floras are among the most poorly known botanically of all the biomes (Frodin 1984), and details of seed characteristics of dryland species are similarly restricted to about 300 species (Gutterman 1993). Thus, with the key aim of sampling the breadth of dryland plant diversity, the MSB project will inevitably be dealing primarily with species new to seed-conservation science. Two main problems are envisaged. First, about 4% of species handled are not expected to be readily suited to conventional storage protocols, and modifications might be required of all phases of the banking activity (collecting, processing, and storage) to ensure their conservation. Second, a smaller proportion of species are likely to require detailed investigation of their germination requirements so that their genetic potential can be readily released. In addition, research on predicting seed longevity of bank collections will be required as a management tool for setting seed-viability retest intervals to improve the balance of seed consumption during the monitoring of viability and the need to maintain seed stocks. Overall, it is envisaged that the total number of species requiring research on germination and storage will be about 150 per year. There will be 21 research staff and space for 15 visiting researchers.

Research to improve collecting. The collecting phase of species conservation offers the first opportunity to identify potential seed-storage problems and apply modified handling procedures to ensure that seeds retain maximal quality before seed-banking. Thus, the current research program in this regard will be expanded and focus on improving the field diagnosis of seed-storage behavior and maximizing the harvest quality of the collections.

On the basis of information generated automatically during the processing of seed for storage in the MSB and other databases worldwide as they become available, a relational database of seed information will be constructed for over 20,000 species. It will be compatible with other databases within Kew (such as SEPASAL) and outside Kew and be used to develop a field diagnostic algorithm for potential routine seed conservation.

The vast majority (86%) of desiccation-tolerant seeds collected should remain viable for at least 200 years (Hong and others 1996) under international standard bank conditions. Historical data (see Bewley and Black 1994) and retest data from the Kew Seed Bank support that contention. However, although all collections made will be moved from the field to Wakehurst Place as quickly as possible, at least two aspects of their physiology might change before their arrival: desiccation tolerance and potential longevity (Hay and Probert 1995; Smith 1995). Developing methods to minimize and control such changes is of paramount importance if long-term storage is to be guaranteed for all collections.

Research to improve processing. The rapid and reliable distinction between the two extremes of seed-storage behavior (desiccation-tolerant and -intolerant) and subjecting of collections to appropriate processing are ever more urgent as the systematic ex situ conservation of species progresses.

Most bank collections undergo some field drying as part of the natural process of maturation of the parent plant. And it is often necessary to clean and partially dry some fleshy fruits for logistical reasons before dispatch and to reduce the opportunity for fungal infestation of the seed lot during transit to the bank. However, it has become clear recently that such postharvest practices have potentially large effects on the long-term maintenance of seed quality. For example, studies on crop seed have indicated that even a small alteration in moisture content could switch the physiological mode of the seed into or out of self-repair and hence affect seed quality. Also, the method of seed dehydration can have a profound bearing on the extent of desiccation tolerance.

To complement the first-step diagnosis of behavior in the field, more-detailed laboratory-based investigations are needed at a more mechanistic level to understand the process of desiccation intolerance. It would be an important advance, not only to seed conservation but also to seed science in general, if a universal set of markers of desiccation tolerance could be identified. Such a development would allow the screening of seed lots that had been identified by the field diagnostic as being of highest banking uncertainty to undergo rapid biochemical diagnosis.

Although studies, principally at the Kew Seed Bank, have resulted in the development of germination algorithms for many families, it is still estimated that a substantial number of previously untested species in a broad range of families will require further research to allow efficient germination. Families that do not respond to normal germination algorithms, such as Compositae (Linington and others 1996), and families for which no germination information exists require particular attention. This element of the research program will provide a unique opportunity to make detailed studies of dryland-species regeneration strategies and to model population responses to environmental cues, thereby substantially increasing our knowledge of dryland-seed biology.

Thus, our research objectives are to improve seed-drying methods, to continue the search for a biochemical diagnostic procedure for desiccation tolerance, and to develop effective germination-test regimens for dryland species further.

Research to improve storage. A large amount of seed is needed to quantify the longevity response at a single temperature, so there is also a need to provide a fundamental understanding of the mechanism of viability loss to develop a rapid and efficient system of diagnosing the storage potential of collections from a small quantity of seed.

Earlier sections of this review have clearly identified the need to quantify further how seed longevity in a species can be affected at all stages of the conservation process (collecting, processing, and storage). Of particular importance is the recent suggestion that the optimal storage conditions for orthodox seeds can differ (Vertucci and others 1994). Thus, long-term experiments, some of which should be at seed-bank temperatures, are required to establish whether there really are long-term implications of this suggestion. Our review of orthodox seeds (Hong and others 1996) has revealed that the potential longevity in storage under seed-bank conditions is known to vary considerably; predicted longevities vary over a factor of 200 in the 52 species representing 23 families for which seed-storage responses have been quantified sufficiently to allow comparison under identical conditions.

Further quantification of the variation in the rates of viability loss is needed so that appropriate retest intervals can be set and unnecessary depletion of the collections avoided. Moreover, it is appropriate to consider the cause of the intrinsic differences in dry-seed longevity. Increasing evidence of nonorthodox seed-storage behavior across species of many families demands that nonconventional storage environments be considered for the storage of some species. It is predicted that 163 species in the native UK spermatophyte flora will be difficult to collect or conserve and will therefore require researching. A brief survey of dryland species suggests that a substantial number will possess nonorthodox seeds as a result of either low desiccation tolerance or sensitivity to seed-bank temperatures. Improved short-term storage protocols are required for whole seeds to allow high-viability seed to be available as the starting material for long-term conservation. In addition, long-term storage of nonorthodox seeds under nonconventional temperatures for seed-banking, including cryopreservation, is suggested.

Seed-Information Database

The computerization of records for the seed-bank collections was started in 1981 and now includes information on nearly 11,000 accessions. Information recorded includes passport and management data. Passport data include date of collection, name and affiliation of collector (Kew or other), geographic location, type of material, number of individuals and proportion of population sampled, voucher and taxonomy, and distribution policy. For regenerated seed stocks, the following details are recorded: parent plant plus sibling data, generation, where grown, isolation conditions if any, number of seeds sown and number of plants harvested, date of harvest, voucher, and distribution policy. Management data cover x-ray record, bank location, original and current seed number, storage temperature, number and type of container, original and retest germination results and conditions, verification details and taxonomy, location of duplicate collections, interval of retest, and distribution date.

Some of the seed-bank database and summarized research datasets will be used to develop a relational seed-information database that should cover more than 20,000 species by the year 2010. Information will probably include the physical and chemical characteristics of the seed and optimal collecting, storage, and germination details. The database design is expected to ensure a high level of connectivity to other databases in and outside Kew, thus maximizing the potential use of the database as a management tool for our collections and as a means of providing advice on seed conservation to our collaborators and the public, for example, throughout data outlets in the public interpretation area (winter garden) of the MSB building.

Information Flow, Technology Transfer, and Training

Since 1992, the Seed Conservation Section has published over 90 scientific articles. In addition, the list of seeds that is used to publicize the material available for use and summarize the group's activities has been published biennially. More than 280 scientific visitors to our facilities have been accommodated since 1992;

during the same period, the research group welcomed 12 foreign visiting scientists at undergraduate to postdoctoral levels. Moreover, collaborative projects were initiated with 20 institutes in the UK and abroad, including five of the eight countries with the highest-priority for collaboration with the MSB. The group has considerable experience in organizing conferences, running training courses (over 280 students have attended our formal training courses in the last 5 years), and supervising PhD and MSc student projects.

It is envisaged that each year up to 57 trainees or researchers from collaborating countries will visit the MSB for at least a month, and there will be shorter-term visitors. The new MSB building will include accommodation for up to 28 visitors from collaborating countries at any time. That will facilitate training and technology transfer, which we see as being achieved in a number of ways: advisory visits by MSB staff, for example, to help develop seed-storage facilities; opportunities for scientists to come to the MSB to gain practical experience in specialized seed-conservation techniques, such as the identification of high-quality seeds and long-term seed-storage techniques; and opportunities to attend formal training courses. Training will be available at all levels of expertise, from technician to postdoctoral, and for various periods, from 1 month to several years.

In addition to the data produced from the routine processing of and research on seeds, the MSB will provide collaborators with general information service on many aspects of seed conservation. Visiting scientists will have opportunities to access Kew's vast resources, including the herbarium and library, and, by arrangement, other parts of Kew, such as the Jodrell Laboratory and the Center for Economic Botany. In addition to existing databases, such as SEPASAL, further databases, such as the seed-information database, will be developed throughout the project for use by collaborators. Collaborators will receive updates of important developments in seed conservation, including details of the latest key publications.

CONCLUSION

The MSB is one of the most ambitious projects ever undertaken by the Royal Botanic Gardens, Kew. However, the biodiversity crisis that the world is facing calls for such large-scale remedies to avoid disaster. We have been encouraged by the response to the MSB project both by the public and by many sources of funding as expressed in the fact that within the brief period of 2 years of planning we have been able to obtain £45 million ($73 million) for the project to add to Kew's own commitment of about £8 million ($13 million). That seeds will be stored in both the MSB at Kew and seed banks of many collaborating countries must not detract from the need to maximize the efforts of in situ conservation, which allows species to continue to interact with their environment and allows the process of evolution to continue.

ACKNOWLEDGMENTS

We thank Gillian Wechsberg for compiling much of the information presented here. We are also grateful for the help of Simon Linington, John Dickie, Hugh Pritchard, and Robin Probert.

REFERENCES

Anon. 1994. Biodiversity. The UK action plan. London: HMSO.

Bewley JD, Black M. 1994. Seeds: physiology of development and germination. New York: Plenum Press.

Binns T, editor. 1995. People and environment in Africa. Chicester, UK: John Wiley & Sons Ltd.

Brown AHD, Marshall DR. 1995. A basic sampling strategy: theory and practice. In: Guarino L, Ramantha Rao V, Reid R, editors. Collecting Plant Genetic Diversity, Technical Guidelines. CAB International.

Davis SD, Sinclair NJ, Cook FEM. 1996. The work of Kew's Center for economic botany and the survey of economic plants for arid and semi-arid lands (SEPASAL). In: West NE, editor. Rangelands in a Sustainable Biosphere. Proceedings of the Fifth International Rangeland Congress 1:111-2.

Ellis RH, Hong TD, Roberts EH. 1985. Handbook of Seed Technology for Genebanks, Vol. 1. Principles and Methodology. Rome: International Board for Plant Genetic Resources.

Ehrlich PR, Ehrlich AH. 1981. Extinction: the causes and consequences of the disappearance of species. New York: Random House.

FAO. 1996. The state of the world's plant genetic resources for food and agriculture. Rome: Food and Agriculture Organization of the United Nations.

FAO/IPGRI. 1994. Genebank standards. Rome: Food and Agriculture Organization of the United Nations, Rome and International Board for Plant Genetic Resources.

Frodin DG. 1984. Guide to standard floras of the world. Cambridge, UK: Cambridge University Press.

Goodin JR, Northington DK, editors. 1985. Plant resources of arid and semi-arid lands: a global perspective. London: Academic Press.

Guarino L. 1995. Geographic information systems and remote sensing for plant germplasm collectors. In: Guarino L, Ramantha Rao R, Reid R, editors. Collecting Plant Genetic Diversity, Technical Guidelines. p 316-28.

Gutterman Y. 1993. Seed germination in desert plants. Berlin: Springer Verlag.

Hay FR, Probert RJ. 1995. Seed maturity and the effects of different drying conditions on desiccation tolerance and seed longevity in Foxglove (*Digitalis purpurea* L.). Annals of Botany 76:639-47.

Hong TD, Linington S, Ellis RH. 1996. Seed storage behavior: a compendium. Rome: International Plant Genetic Resources Institute.

Jackson M, Ford-Lloyd BV, Parry ML, editors. 1990. Climate change and plant genetic resources. London: Belhaven Press.

Leprince O, Hendry GAF, McKersie BD. 1993. Seed Science Research 3:231-46.

Linington S, Terry J, Parsons J. 1995. X-ray analysis of empty and insect-damaged seeds in an *ex situ* wild species collection. IPGRI / FAO Plant Genetic Resources Newsletter 102:18-25.

Linington S, Mkhohta D, Pritchard HW, Terry J. 1996. A provisional germination testing scheme for seed of the Compositae. In: Hind DJN, editor. 1994. Proceedings of the International Compositae Conference, Kew. vol.2. Royal Botanic Gardens, Kew.

Mabberly DJ. 1990. The Plant-Book. Cambridge: Cambridge University Press.

Miller K, Allegretti MH, Johnson N, Jonsson B. 1995. Measures for conservation of biodiversity and sustainable use of its components. In: Hewwood, VH, Watson RT, editors. Cambridge: Global Biodiversity Assessment. Cambridge University Press.

Nevo E, Beiles A. 1989. Genetic diversity in the desert: patterns and testable hypotheses. Journal of Arid Environments 17:241-4.

Ponquett RT, Smith MT, Ross G. 1992. Lipid autoxidation and seed ageing: putative relationships between seed longevity and lipid stability. Seed Science Research 2:51-5.

Prendergast HDV, Linington S, Smith RD. 1992. The Kew Seed Bank and the collection, storage and utilization of arid and semi-arid zone grasses. In: Chapman GP, editor. Desertified Grasslands, their Biology and Management. London: Academic Press.

Prançe GT, Elias TS, editors. 1977. Extinction is forever. The New York Botanical Garden. p 437.

Richards AJ. 1986. Plant breeding systems. London: Unwin Hyman.

Smith RD. 1995. Collecting and handling seeds in the field. In: Guarino L, Ramantha Rao V, Reid R, editors. Collecting Plant Genetic Diversity, Technical Guidelines CAB International.

Smith RD, Linington, SH. 1996. Practical management of the Kew Seed Bank for the conservation of arid land and UK wild species. In: Proceedings of the Workshop on the Conservation of Wild Relatives of European Cultivated Plants. Council of Europe.

Stace C. 1991. New Flora of the British Isles. Cambridge: Cambridge University Press.

Taktahjan A. 1986. Floristic regions of the World (Translated: Cronquist A, editor). Berkeley, CA: California University Press.

UNEP. 1992a. Convention on Biological Diversity. United Nations Environment Program.

UNEP. 1992b. World Atlas of Desertification. London: Edward Arnold.

UNEP. 1994. United Nations Convention to Combat Desertification in those Countries Experiencing Drought and/or Desertification, Particularly in Africa. United Nations Environment Program.

Vertucci CW, Roos EE, Crane J. 1994. Theoretical basis of protocols for seed storage III. Optimum moisture contents for pea seeds stored at different temperatures. Annals of Botany 74:531-40.

Von Bothmer R, Seberg O. 1995. Strategies for the collecting of wild species. In: Guarino L, Ramantha Rao V, Reid R, editors. Collecting Plant Genetic Diversity, Technical Guidelines. CAB International.

Williams PH, Gaston KJ, Humphries CJ. 1994. Do conservationists and molecular biologists value differences between organisms in the same way? Biodiversity Letters 2, 67-8.

Williams RJ, Leopold AC. 1989. The glassy state in corn embryos. Plant Physiology 89, 977-81.

World Conservation Monitoring Center (WCMC). 1992. Global Biodiversity: Status of the Earth's Living Resources. London: Chapman & Hall.

Wynne G, Avery M, Campbell L, Gubbay S, Hawkswell S, Juniper T, King M, Newberry P, Smart J, Steel C, Stones T, Stubbs A, Taylor J, Tydeman C, Wynde R. 1995. Biodiversity Challenge. RSPB, Sandy.

CHARTING THE BIOSPHERE:
BUILDING GLOBAL CAPACITY
FOR SYSTEMATICS SCIENCE

JOEL L. CRACRAFT

Department of Ornithology, American Museum of Natural History,
Central Park West, 79th Street, New York NY 10024

MANAGING THE BIOSPHERE:
THE ESSENTIAL ROLE OF BIODIVERSITY SCIENCE

ABOUT 175 NATIONS have ratified the Convention on Biological Diversity and thereby signaled their intention to strive, in principle, for a sustainable world. This raises a simple question: Do we possess sufficient scientific information about the biosphere to manage it sustainably, even assuming that the political will for doing so exists? The answer to this question clearly is no.

That being the case, the pessimists among us might claim that we now live in the best of all possible worlds with respect to what we know versus what we need to know. The pessimists would therefore argue that our ignorance can only get worse as the global trends of environmental transformation accelerate, because as the world's ecosystems get more and more degraded and destroyed, it will require an increasing amount of knowledge to put things back together again and to make up for the lost goods and services provided by these biotic landscapes.

At the other extreme, the optimists among us might claim that, given a political imperative to use our biological resources sustainably, we already have a sufficiently large body of knowledge, and if only it were made available to the world's nations, resource management could become much more efficient and cost-effective and move us far in the direction of sustainability.

Contributing to the pessimists' view is the fact that the world community, sometimes including scientists who study biodiversity, often fails to recognize how much knowledge it will require to manage the biosphere to the point where it can provide meaningful and healthy lives for the world's people into the future. Obviously, scientific information is not sufficient by itself to right the world's environmental wrongs, but it is essential (Cracraft 1996). Several vignettes will emphasize this point.

First, the United States spends more money each year on environmental science than any other country, yet the evidence suggests that we are not managing our lands sustainably (PCAST 1998). Although much of the reason for this lies in a political-economic imperative to exploit our resources for short-term gains, an insufficiency of scientific knowledge has hindered proper management in many cases (NRC 1993a,b). Land managers are continually saying that they lack sufficient knowledge about the resources under their stewardship. One has only to examine how forest lands are being managed in North America to see the extent to which that insufficiency contributes to inappropriate land management (papers in Kohm and Franklin 1997; Pickett and others 1997).

Second, we are not the world. Many of us live in industrial economies that are privileged beyond belief. Much of the world, in contrast, is relatively poor and lacks even the rudiments of decent scientific infrastructure (Cracraft 1995). It is said that countries housing 80% of the world's biodiversity have only about 6% of the world's scientists. We can quibble about the numbers, but the observation is correct enough to make the point: Most of the world's nations will not have a reasonable chance of achieving a sustainable future unless knowledge about their natural resources is improved dramatically and quickly. Consider one simple example. In a recent perspective on biological research efforts in Serengeti National Park, Sinclair (1995) listed numerous gaps in basic biological knowledge of that system that impede efforts at effective resource management. Not knowing the causes of death in the wild dog (*Lycaon pictus*), for instance, hinders any informed design for its recovery program. Given that the Serengeti is probably the most thoroughly studied protected area in Africa, the obvious question is, What about the protected areas in other countries of that magnificent continent? Where will the biological knowledge to manage those ecosystems come from? If Serengeti is taken as an exemplar of the amount of knowledge that will be required to achieve effective conservation management, it is difficult to believe that inputs of researchers and financial support from developed countries will ever be sufficient to address similar needs in other parts of Africa. The only solution is to see the capacity in each country increase.

Third, because it is exceedingly difficult to comprehend the extraordinary dependence of most of the world's people on natural ecosystems, we tend to underestimate the magnitude of the problem confronting us. Around the world, people use tens of thousands of species to meet their daily needs. If these uses are to be managed in a sustainable manner, much more biological information will be required than the scientific community can deliver today. And, to emphasize the depth of the problem, that information will generally have to be gathered at, and applied to, the local level, much like the information needed for the wild dog in the Serengeti. We cannot expect to accumulate knowledge in some abstract database and not have it mean something to the people whose livelihoods and future depend on it.

Scientific knowledge of biodiversity must accumulate year after year if the biosphere is to be managed effectively. The health of the world's people, their food supply, and the ecological services provided by intact ecosystems are all threatened when knowledge of biodiversity does not advance.

One way of seeing the need is to do a simple thought experiment. Ask what might be the consequences for society if systematics knowledge had been frozen 40 years ago, with no new advances allowed. Here are some examples:

• Society would be without the benefit of all the agricultural systematics research that has mitigated the devastating effects of pests and invasive species over the last 40 years.

• Society would be without an understanding or identification of many vectors of disease that were discovered during this period.

• There would be no knowledge about many of the newly emergent diseases that have ravaged human societies, AIDS being the most pernicious.

• Medical science and biotechnology would be years behind current levels because the thermophilic bacteria that have made possible the polymerase chain reaction and all its benefits for diagnostic medicine would not have been discovered.

• None of the wild crop relatives that were discovered in the last 40 years would be available for improving our foods.

This demonstration of the importance of systematics to society could be expanded easily (*Annals of the Missouri Botanical Garden* 1996; *Biodiversity and Conservation* 1995; *BioScience* 1995; Cotterill 1995; Janzen 1993; Miller and Rossman 1997; Patrick 1997; Systematics Agenda 2000 1994a,b; Thompson 1997). The other biodiversity sciences are equally important for society, and "freezing" their knowledge at what it was 40 years ago would have similar adverse consequences. In ecology, for example, we would lack much of the basic science that has underpinned the new disciplines of landscape ecology, restoration ecology, and conservation biology. Without that knowledge, managing our biosphere would be essentially impossible.

Investment in biodiversity science—even what is often thought to be mundane, unexciting, or old-fashioned—is one of the best investments society can make for its long-term well-being. The poor old systematist toiling over the discovery, description, and identification of groups of insect pests or disease vectors potentially will contribute as much to society, in saving millions of lives and billions of dollars, as will most so-called modern research. We need to cherish and nourish all biodiversity scientists because our future depends on them (Cracraft 1996).

SYSTEMATICS-SCIENCE CAPACITY: WHAT IS IT?

Systematics is the most fundamental of the biodiversity sciences inasmuch as it is concerned with discovering, describing, and monographing Earth's species diversity. Like most sciences, systematics can be defined by its research questions and objectives. Within systematics, taxonomy is the science of discovering, describing, and classifying species and groups of species; phylogenetics is the discipline that attempts to understand the evolutionary (historical) relationships among species and groups; and classification is the means by which that understanding is translated into hierarchical (Linnaean) groupings and information systems that form the basis for effective communication about life's diversity (Systematics Agenda 2000 1994a,b).

Given that broad view of the systematics enterprise, systematics-science capacity can be taken to include all the components of infrastructure and human resources that support the systematics research effort and make its results available to those who need them. The most important infrastructure relevant to systematics is specimen-based collections housed in systematics research institutions of various kinds (Cotterill 1995, 1997). The world's collections contain over 2 billion specimens, and these constitute society's only permanent record of Earth's biodiversity. Collections take many forms, and for systematics to flourish, systematists must have access to them: natural-history museums, herbariums, frozen-tissue collections, seed banks, type-culture collections, and, for some types of studies, living material in zoos and botanical gardens. Systematics infrastructure includes the computational means to store information about collections, particularly the information associated with specimens; to analyze character-based information for phylogenetic analysis; and to facilitate communication with systematists at other institutions. Infrastructure also includes libraries through which a researcher can obtain access to prior systematics work and facilities for training of professional and paraprofessional scientists and support staff; these constitute the human resources needed for systematics research.

Systematics collections serve a much broader role than providing a basis for scientific research, and it is the broader role that is often important for many countries (Cotterill 1997). Through their exhibits and other programs, collection-based institutions, such as museums and botanical gardens, are essential in educating the public about the benefits of, and threats to, biodiversity. These institutions also are sites for formal science education of people as varied as young schoolchildren, professionals, and paraprofessionals. Little of this could take place without the scientific collections that form the foundation of educational programs.

AN AGENDA FOR SYSTEMATICS

Earth's biodiversity is poorly known. Although 1.7 million species have been recognized and described (Hammond 1995; Heywood 1995; May this volume), many specialists think that tens of millions of species are unknown to science. Our understanding of the relationships of these taxa is still in its infancy, but it is this understanding that serves as an organizing framework for information systems useful to both basic and applied biology. The world's natural-history collections house a treasury of biodiversity information associated with their specimens; for the most part, very little of this information is available digitally to the world user community (Blackmore 1996; Systematics Agenda 2000 1994a,b).

SYSTEMATICS-SCIENCE CAPACITY IN THE DEVELOPING WORLD

Countries differ greatly in their capacity to undertake research in systematics. Recent compilations in the UN Environment Program's *Global Biodiversity Assessment* (Heywood 1995) describe the global patterns of numbers and sizes of plant collections and numbers of institutions that house collections of various sorts (museums, zoos, aquariums, and botanical gardens); these patterns can be expected to reflect the general level of systematics capacity in each country and among regions. Figure 1 summarizes the numbers for six regions. Europe and North America, not unexpectedly, have the highest capacity, followed by Asia. South America, Australasia, and Africa have the least capacity. It is enormously difficult to obtain accurate numbers because such collections are defined, counted, or estimated in different ways; but the figure shows the pattern mentioned earlier: the species-rich areas of the world have the least capacity. The situation could be even worse than the figure suggests; within many of these regions, one country, such as South Africa within Africa or Australia within Australasia, dominates the statistics. Many countries lack the rudiments of capacity, and a surprising number have no botanical or zoological collections.

The numbers of natural-history collections, zoos, and other infrastructure also constitute a measure of the availability of scientists and training facilities essential

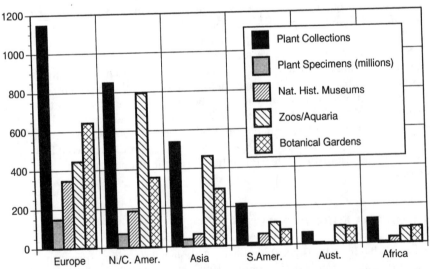

FIGURE 1   Systematics capacity can be measured by numbers of natural-history collections, here categorized into six regions: Europe, including former Soviet Union; North America and Central America; Asia, including China, southern and southeast Asia, and Japan; South America; Australasia and Oceania; and Africa. For all regions except Europe, single country dominates numbers. This means that systematics capacity in other countries in those regions is worse than implied by regional numbers alone. As measured by these collections, regions with least capacity include most island nations, Africa, Central America, eastern Europe, and countries making up former Soviet Union.

for developing human resources. The numbers indicate that many regions of the world lack adequate capabilities for professional and paraprofessional training.

As mentioned earlier, systematics capacity in the developing world is inadequate to confront the loss of biodiversity and to serve as a basis for its effective management as a resource for sustainable development. At the same time, systematists recognize that systematics capacity in the wealthy countries is incapable of filling the need. In fact, systematics capacity in the developed nations is barely adequate—many authorities would say totally inadequate—to meet those countries' own demands for systematics information (Blackmore 1996; Oliver 1988; Parnell 1993; Systematics Agenda 2000 1994a,b). We have no choice but to develop systematics capacity in all nations, particularly in the species-rich regions where the need is greatest.

AN OVERVIEW OF SYSTEMATICS AGENDA 2000 INTERNATIONAL: BUILDING A GLOBAL SCIENCE INITIATIVE

Many international organizations have called for a more thorough understanding of life's diversity through increased systematics research. It is generally estimated that we know perhaps 5% of Earth's species. Given that current knowledge and use of the known species generate trillions of dollars of economic activity—indeed, that use is the engine of the world economy—and sustains the lives of all of us, it is reasonable to expect that substantial increase in knowledge of the world's biota will add immeasurably to societal well-being in the form of new uses and benefits. The international biodiversity science program, DIVERSITAS, has recognized the need for increased research in systematics. Thus, systematic biology was recently added as a core research element of the program. Filling that role is Systematics Agenda 2000 International (SA2000I). Systematics Agenda 2000 began as a consortium of systematics societies in the United States but has expanded internationally as a program of the International Union of Biological Sciences and as a component of DIVERSITAS (Blackmore and Cutler 1996).

The activities of SA2000I are organized around three broad missions encompassing the major research fields of systematic biology: inventorying and describing of biodiversity, understanding the history of life, and using that understanding to create predictive classifications and information systems for the world user community. Since its inception, SA2000I has advanced the view that systematics knowledge of life's diversity is essential to ensure societal well being. To fulfill this societal role, systematics must solve the problem of expansion of relevant infrastructure and human resources, especially in countries that now have little or no capacity.

Inventory

Inventories are at the heart of the global discovery effort, but many countries are ill equipped to take stock of their biological heritage according to their own needs and priorities. In an effort to correct that, SA2000I held a workshop on inventories at the American Museum of Natural History, New York, in September

1998. The workshop was designed to assess country requirements for inventories, establish how priorities can be set to meet inventory needs, determine the best research strategies to satisfy country goals, and undertake an assessment of current capacity. The workshop also addressed issues and strategies for building capacity (AMNH 1999).

Phylogeny

Knowledge of phylogenetic relationships is often seen as an academic exercise of little practical importance. In fact, phylogenetic hierarchies are the foundation for creating the predictive classifications and information systems that are of immeasurable value to society. The use of biodiversity is made possible by understanding where a taxon belongs in the hierarchy and how its characteristics compare with those of close relatives. Indeed, at some level, all uses of biodiversity depend on knowledge of phylogenetic relationships and how they are translated into information systems.

Although systematists have made major strides in the last decade in understanding the interrelationships of life, corroborated hypotheses of relationships are still lacking for most groups, including some of the best studied, such as birds and mammals. That lack of understanding constitutes a critical impediment to developing efficient information systems. Phylogenetic research is global in its perspective; given the rate at which phylogenetic relationships are being resolved and the uncoordinated nature of present research, it will take many decades to achieve a satisfactory overview of the history of life. Such a delay will hinder our efforts to build bioinformatics systems that are maximally predictive—a goal that is integral to the clearinghouse mechanism of the Convention on Biological Diversity.

To address this need, SA2000I will be organizing an international research effort to produce a corroborated phylogeny of the higher taxa by the year 2010. This will be accomplished by coordinating research activities within and among working groups of investigators focusing on specific major taxa, with priority given to those of high societal importance. SA2000I's international research effort will also be concerned with incorporating new technologies, such as those associated with the Human Genome Project, and with building capacity for phylogenetic research in countries that lack it.

Systematics Bioinformatics

SA2000I has major efforts under way to improve the accessibility of systematics information. The research program on phylogenetics will contain a component on how phylogenetic information can be made widely available. A very successful effort within SA2000I and DIVERSITAS is Species 2000, an international initiative to assemble a scientifically reliable database of all the world's currently described species in a framework that links species names to other databases that house information about them. This program is of immense importance for managing what we know about biodiversity.

The systematics community recognizes that a major impediment to managing and sustainably using biodiversity is that the vast majority of the information associated with the specimens housed in the world's natural-history institutions is unavailable to users of systematics information. Many museums and herbariums are making an effort to put their collections on databases, but for the largest institutions this is a formidable and expensive task. The costs of verifying the information and maintaining it electronically are also high. Yet, the benefits of this information to the world's nations are too substantial to ignore. The industrial nations, which house 80–90% of all the biological specimens, have an obligation to repatriate the information so that it can be used for resource management and other activities. How to overcome the many challenges, particularly in terms of costs and effective information management, has not been addressed sufficiently at the international level.

BUILDING SYSTEMATICS CAPACITY: SOME EXAMPLES

What strategies should be adopted to confront current impediments to building systematics science internationally and to redress the imbalance in capacity between the developed and the developing countries? Two general things must happen. First, the wealthy countries must increase their commitment to promoting systematics. This includes not only increasing their own scientific research and capacity, but also ensuring that those programs benefit developing countries as well (for example, through training); and they must increase aid targeted to building and improving systematics capacity in developing countries. Second, the developing countries must do more to help themselves. Even if needed financial resources ultimately come from outside, developing nations must recognize the importance of systematics for their future prosperity and seek ways to increase its capacity.

Many positive things are happening, of course. The world's nations, through the Convention on Biological Diversity and its Global Taxonomy Initiative (Australian Biological Resources Study 1998; Environment Australia 1991), have acknowledged the critical role of systematics and have called for countries to increase their capacity. Funding for systematics has increased in many developed countries, and that has provided benefits to nations in developing regions as well. And many developing nations have themselves initiated programs to increase systematics capacity. A number of these efforts are worth highlighting because they provide models for other countries in their efforts to create or improve systematics capacity. The projects discussed below are by no means the only successful initiatives, and a particular example might not be the most effective or appropriate for another country, but they encompass an array of different approaches.

Costa Rica: INBio. The Instituto Nacional de Biodiversidad (INBio) of Costa Rica was established in 1989 and has gained worldwide renown for its program of national biodiversity inventory, bioprospecting, and training (Reid and others 1993). Inventory efforts not only are designed to increase knowledge of the Costa Rican biota and to incorporate it into electronic databases, but also are a major component of the country's bioprospecting efforts. Costa Rica is taking a lead role in transferring its successes to other tropical countries through training workshops. [For a more detailed description, see Gámez this volume.]

Mexico: CONABIO. In 1992, Mexico established the Comision Nacional para el Conocimiento y Uso de la Biodiversidad (CONABIO) to coordinate and promote research activities in many Mexican institutions. A major objective of CONABIO is to inventory the biota of Mexico; to accomplish this, CONABIO has begun to form databases and network its own national collections and has sent scientists to museums and herbariums around the world to create databases of Mexican specimens in these collections. The result is one of the most comprehensive geo-referenced sets of biodiversity information linked to voucher specimens found anywhere in the world. [See Soberón this volume for a description.] In addition, CONABIO has major programs designed to train professional and paraprofessional taxonomists.

Indonesia: LIPI. To meet its obligations under the Convention on Biological Diversity, Indonesia has undertaken an ambitious Global Environmental Facility project designed to increase systematics capacity and provide a framework for documenting and managing its biodiversity. Through the Research and Development Center of Biology in the Indonesian Institute of Sciences (LIPI), a nondepartmental government institution, infrastructure and human resources are being strengthened. New collections and research facilities are being built, and an intensive program of training of professional systematists at overseas institutions has begun. In addition, information associated with specimens is being put into databases, and computer facilities are being expanded to provide managerial support for the collections.

Bangladesh: National Herbarium. Another example of building systematics capacity is the construction of the new Bangladesh National Herbarium. The government of Bangladesh included a new herbarium in its aid proposal to the United Kingdom's Overseas Development Administration (ODA) in 1989. The project was accepted, and ODA (now the Department for International Development) asked systematist Vernon Heywood to act as consultant to plan the building, equip it, and set up a staff training program. The UK contribution to the project is over £1.2 million (US $2 million), and the government of Bangladesh covered the cost of site preparation. The project includes, in addition to the new herbarium building, laboratories, a library, and modern equipment and electronic communication systems. For its part, the government of Bangladesh is providing running costs, a scientific staff of 14, and technical and support personnel. The herbarium opened in 1998, and already there are plans to expand its original scope. This effort is noteworthy in that it was possible with little initial investment to create a locus for infrastructure-building and capacity-building that will extend well into the future.

Southern Africa: SABONET. A final example is the Southern Africa Botanical Network (SABONET), a consortium of the herbariums in 10 southern African nations. Supported by funds from the Global Environmental Facility (GEF) and the US Agency for International Development, SABONET is building capacity through improved and expanded infrastructure, training, inventorying, databases, and information networks. A major goal of the project is to strengthen the core group of professional and paraprofessional botanists in each of the 10 countries so that programs of inventorying and monitoring can be undertaken, collection management strengthened, and training expanded.

RECOMMENDATIONS FOR BUILDING SYSTEMATICS CAPACITY

These are encouraging times for systematics (Scoble 1997). The preceding examples show that many countries are improving their systematics capacity dramatically. But much remains to be accomplished. Most countries in the species-rich regions of the world have little capability for systematics research and training, and wealthy countries, although providing support through various national or international aid agencies, are not providing sufficient support to have a major effect in most of the poorer countries.

The various activities described above provide a framework for formulating some recommendations that have relevance for countries that wish to improve their systematics capacity, even if sometimes it must be at a relatively low level (see also AMNH 1999; Wheeler and Cracraft 1997).

Regional Cooperation

The significance of SABONET is that it shows how a regional cooperative program can synergistically improve the capacity and scientific knowledge base of many countries for less money than would be required if it were undertaken country by country. Such cooperative ventures also raise the capacity of countries that have the least capacity to the point where they might be able to pursue systematics programs independently. SABONET began in the scientific community itself and shows what can result when scientists in different countries work together. Such regional cooperation makes sense, especially because many of the countries lack sufficient capacity to undertake research or training programs on their own. No country can house all the necessary expertise, but regional cooperation and sharing of information are possible. This can be particularly effective if the countries involved share a regional, ecologically coherent biota. That would be true, for example, of the countries in East Africa, the countries of Central Africa that share the Congo Basin, countries in West Africa, the Andean countries of South America, and countries that share the Amazon Basin.

Building Capacity within Countries

The previous discussion described how different countries have found distinct ways of improving systematics capacity. Some, such as Indonesia, have undertaken major programs to improve their national systematics collections. Others, such as Mexico, have attempted to enhance cooperation and coordination among existing collections. CONABIO, moreover, has invested a relatively small sum of money to form databases of collections in other countries and has thereby substantially expanded its systematics knowledge base and its ability to manage Mexico's biological resources.

Individual countries can take the initiative to seek donor funds to improve systematics capacity. Most of the countries discussed have sought GEF funding or are in partnership with donor countries. In many poor countries, a relatively small amount of money can have a large long-term effect. The creation of the Bangladesh National Herbarium is a case in point, and such cooperative programs can lead to long-term commitments on the part of the recipient nations to maintain human resources and training.

Many aid proposals from developing countries could include a systematics research component that would establish or upgrade their capacity to preserve in natural-history institutions a permanent record of their biological diversity. Such collections would also provide key support for long-term monitoring and management programs. A particularly cost-effective approach to incorporating systematics information into biodiversity activities would be to emulate the example of CONABIO and seek funds to form databases of collections that have large holdings of specimens, which could then be used in electronic databases for management purposes.

The Role of the Wealthy Countries

Wealthy countries must do more. Small programs in wealthy countries can have large effects. The United Kingdom's contribution to building the Bangladesh National Herbarium is an important example. In the United States, the National Science Foundation initiated a short-term competition, Partnerships for Enhancing Expertise in Taxonomy (PEET), designed to improve systematics knowledge of little-known and neglected taxa and to train new students in them. A small number of funding cycles have already had a substantial impact, and continued support is certain to produce a pool of expertise that will have a long-term and worldwide influence because many of the students being trained are from developing countries.

Wealthy countries need to make a substantial contribution to building worldwide systematics capacity. Perhaps no other initiative would be as effective as providing funds for compiling databases of the largest natural-history collections and making that information available to other countries.

The Role of Systematists

Very few of the activities described above could have taken place without the leadership of the systematics community itself. Scientists must convince policymakers of the importance of systematics research and systematics infrastructure and work with them to design effective programs. Such programmatic activities as DIVERSITAS and SA2000I will be particularly helpful in providing a framework for promoting systematics within countries and establishing regional consortia.

ACKNOWLEDGMENTS

I thank Peter Raven for his inspiration and leadership in pulling the Forum on Biodiversity together and for inviting me to participate. I am grateful to Vernon

Heywood for sharing information about the Bangladesh National Herbarium. Tania Williams and the staff of the National Research Council provided considerable help, for which I am appreciative.

REFERENCES

AMNH [American Museum of Natural History]. 1999. The global taxonomy initiative: using systematic inventories to meet country and regional needs. New York NY: Center for Biodiversity and Conservation, American Museum of Natural History.

Australian Biological Resources Study. 1998. The global taxonomy initiative: shortening the distance between discovery and delivery. Australian Biological Resources Study. Canberra Australia: Environment Australia. 18p.

Annals of the Missouri Botanical Garden. 1996. The Systematics Agenda 2000 Symposium. Ann Missouri Bot Gard 83:1–66.

Biodiversity and Conservation. 1995. Special issue: Systematics Agenda 2000. Biodiv Cons 4:451–519.

BioScience. 1995. Special issue: systematics. BioScience 45: 670–714.

Blackmore S. 1996. Knowing the Earth's biodiversity: challenges for the infrastructure of systematics biology. Science 274:63–4.

Blackmore S, Cutler D. 1996. Systematics Agenda 2000: the challenge for Europe. Linn Soc Occas Publ No 1. Cardigan UK: Samara Pub Ltd.

Cotterill FPD. 1995. Systematics, biological knowledge and environmental conservation. Biodiv Cons 4:183–205.

Cotterill FPD. 1997. The second Alexandrian tragedy, and the fundamental relationship between biological collections and scientific knowledge. In: Nudds JR, Pettitt CW (eds). The values and valuation of natural science collections. London UK: Geological Soc. p 227–41.

Cracraft J. 1995. The urgency of building global capacity for biodiversity science. Biodiv Cons 4:463–75.

Cracraft J. 1996. Systematics, biodiversity science, and the conservation of the Earth's biota. Verh Dtsch Zool Ges 89(2):41–7.

Environment Australia. 1998. The Darwin declaration. Australian Biological Resources Study. Canberra Australia: Environment Australia. 14 p.

Hammond PM. 1995. Described and estimated species numbers: an objective assessment of current knowledge. In: Allsopp CD, Colwell RR, Hawksworth DL (eds). Microbial diversity and ecosystem function. Wallingford UK: CAB Intl p 29–71.

Heywood VH (ed). 1995. Global biodiversity assessment. Cambridge UK: Cambridge Univ Pr.

Janzen DH. 1993. Taxonomy: universal and essential infrastructure for development and management of tropical wildland diversity. In: Sandlund OT, Schei P (eds). Trondheim Norway. Proc Norway/UNEP Expert Conf Biodiversity. p 100–13.

Kohm KA, Franklin JF. 1997. Creating a forestry for the 21st century. Washington DC: Island Pr.

Miller DR, Rossman AY. 1997. Biodiversity and systematics: their application to agriculture. In: Reaka-Kudla ML, Wilson DE, Wilson EO (eds). Biodiversity II. Washington DC: Joseph Henry Pr. p 217–29.

NRC [National Research Council]. 1993a. Research to protect, restore, and manage the environment. Washington DC: National Acad Pr.

NRC [National Research Council]. 1993b. A biological survey for the nation. Washington DC: National Acad Pr.

Oliver JH Jr. 1988. Crisis in biosystematics of arthropods. Science 239:967.

Parnell J. 1993. Plant taxonomic research, with special reference to the tropics: problems and potential solutions. Cons Biol 7:809–14.

Patrick R. 1997. Systematics: a keystone to understanding biodiversity. In: Reaka-Kudla ML, Wilson DE, Wilson EO (eds). Biodiversity II. Washington DC: Joseph Henry Pr. p 213–6.

Pickett STA, Ostfeld RS, Shachak M, Likens GE. 1997. The ecological basis of conservation. New York NY: Chapman & Hall.

PCAST [President's Committee of Advisors on Science and Technology]. 1998. Teaming with life: investing in science to understand and use America's living capital. Washington DC: Office of Technology Assessment. 86 p.

Reid WV, Laird AL, Mayer AM, Gamez R, Sittenfeld A, Janzen DH, Gollin MA, Juma C. 1993. Biodiversity prospecting: using genetic resources for sustainable development. Washington DC: World Resources Inst.

Scoble MJ. 1997. The transformation of systematics? Trends Ecol Evol 12:465–6.

Sinclair ARE. 1995. Serengeti past and present. In: Sinclair ARE, Arcese P (eds). Serengeti II. Chicago IL: Univ Chicago Pr. p 3–30.

Systematics Agenda 2000. 1994a. Systematics Agenda 2000: charting the biosphere. New York NY: Systematics Agenda 2000, a consortium of the American Society of Plant Taxonomists, the Society of Systematics Biologists, and the Willi Hennig Society in cooperation with the Association of Systematics Collections. p 1–20.

Systematics Agenda 2000. 1994b. Systematics Agenda 2000: charting the biosphere. Technical report. New York NY: Systematics Agenda 2000, a consortium of the American Society of plant Taxonomists, the Society of Systematics Biologists, and the Willi Hennig Society, in cooperation with the Association of Systematics Collections. p 1–34.

Thompson FC. 1997. Names: the keys to biodiversity. In: Reaka-Kudla ML, Wilson DE, Wilson EO (eds). Biodiversity II. Washington DC: Joseph Henry Pr. p 199–211.

Wheeler QD, Cracraft J. 1997. Taxonomic preparedness: are we ready to meet the biodiversity challenge? In: Reaka-Kudla ML, Wilson DE, Wilson EO (eds). Biodiversity II. Washington DC: Joseph Henry Pr. p 435–46.

SCIENCE AND TECHNOLOGY IN THE
CONVENTION ON BIOLOGICAL DIVERSITY

CALESTOUS JUMA
Center for International Development, Kennedy School of Government, Harvard University,
79 John F. Kennedy Street, Cambridge, MA 02138

GUDRUN HENNE
Secretariat of the Convention on Biological Diversity, Montréal, Canada, World Trade Centre,
393 St Jacques Street, Office 300, Montréal, Québec, Canada H2Y 1N9

THE CONVENTION ON BIOLOGICAL DIVERSITY (United Nations Environment Programme 1992) made its debut at the UN Conference on Environment and Development (UNCED) in 1992, where it was presented for signature (UN Conference on Environment and Development 1992). At this historic event, 152 states and the European Community signed the convention. Since then, 175 states and one regional economic integration organization have ratified the convention (http://www.biodiv.org [1999, July 28]). With this near-universal membership, the convention rapidly has become one of the most important forums for international environmental-policy guidance.

One of the most important features of the convention is its relationship with the scientific and technological communities. The scientific community, operating through a wide network of institutions and individuals, provided the scientific basis for international action on the conservation of biological diversity. The problem was defined in terms of institutional change. The outcome not only was a diplomatic effort to consolidate existing subregimes dealing with the conservation of biological diversity under the auspices of what became the Convention on Biological Diversity, but also went beyond that consolidation and embedded existing regimes in a much broader context.

The process of regime creation took the form of the convention's medium-term program of work, which ended with the fourth meeting of the Conference of the Parties (COP), held in Bratislava, Slovakia, in May 1998. This meeting had the important task of reviewing the implementation of the convention, evaluating the effectiveness of its internal organization, and establishing a longer-term work program for the convention. One of the main issues discussed in Bratislava was the place of science and technology in the evolution of the convention. To put this issue into perspective, we place the discussion in the context of the institutional structure and functioning of the convention. The institutional history of the convention is evolving in three phases. We believe that the success of the convention will depend largely on the degree to which scientific and technological issues will be integrated into its operations during its upcoming third phase.

THE GENESIS OF THE CONVENTION

Role of Epistemic Communities

Interest in the fate and state of life forms is not new; it has been a dominant feature of intellectual inquiry and popular perception for centuries. The organization of this process into international concerns is associated with the post-World War II period, especially with the establishment of the World Conservation Union (IUCN) in 1948. This and many national institutions around the world, as well as activities in sections of the UN system, provided a basis for the emergence of an epistemic community that is devoted to a variety of concerns related to the conservation of biological diversity.

This community has made politicians and international negotiators aware of the need for international instruments on different aspects of biological diversity. Scientists—particularly those from the biological disciplines—in leading research institutions and universities all over the world, particularly those from the United States, have emphasized the need to conserve biological diversity in all its aspects and by all means.

Much of this work has been done through ad hoc scientific activities, such as those which resulted in the formulation of major biodiversity-related initiatives. Most of these efforts concentrated on the traditional field of conserving wild species and uncultivated land through the establishment of national parks. However, in the 1960s, concerns about integrating conservation with human activities started to play a prominent role in international forums. The Intergovernmental Conference of Experts on the Scientific Basis for Rational Use and Conservation of the Resources of the Biosphere, which convened in Paris in September 1968 under the auspices of the UN Scientific, Educational, and Cultural Organization (UNESCO), was a major step in this process and resulted in the establishment of the Man and Biosphere Programme, emphasizing humanity's place in the natural order of things and the importance of the ecosystem approach to conservation of nature (Di Castri and others 1981; UNESCO 1993).

Science and International Action

The UN Conference on Human Environment, held in Stockholm in 1972, gave high priority to the need to conserve natural resources, including natural ecosystems and endangered wild species and their habitats (Stockholm Declaration of the Conference on Human Environment and Action Plan 1972). The Action Plan on Programme Development and Priorities, adopted in 1973 at the first session of the Governing Council of the UN Environment Programme (UNEP), identified the conservation of nature, wildlife, and genetic resources as having high priority. Since then, conserving of biological diversity has remained one of the most important functions of UNEP. While these groups focused on conservation, the Food and Agriculture Organization (FAO) emphasized the use of genetic resources. Institutional innovations to respond to the scientific and technological aspects of conserving and using the genetic resources of plants for food and agriculture were developed within the framework of the Consultative Group for International Agricultural Research (CGIAR) (Pistorius 1997; Fowler and Mooney 1990).

In the meantime, the number of international and regional legal instruments related to biological diversity increased, all of which sought to address specific aspects of conservation and sustainable use. The Ramsar Convention on Wetlands (Convention on Wetlands of International Importance Especially as Waterfowl Habitat of 2 February 1971, 1982), the Convention for the Protection of the World Heritage (Convention for the Protection of the World Cultural and Natural heritage of 23 November 1972, 1982), the Convention on International Trade in Endangered Species of Wild Fauna and Flora (Convention on International Trade in Endangered Species of Wild Fauna and Flora of 3 March 1973, 1982), and the Berne Convention on the Conservation of European Wildlife and Natural Habitats (Berne Convention on the Conservation of European Wildlife and Natural Habitats of 19 September 1979, 1982), to name but a few, were adopted. However, no common framework existed to deal with the different levels of biological diversity, that is, genes, species, and ecosystems. Furthermore, little was done over this time to provide a global view of trends in biological diversity.

In the middle 1970s, interest grew in providing a global picture of the loss of species. Much of the statistical information was provided by a few agencies of the UN but research institutions, the US National Academy of Sciences, and especially the scientific journals started to call for a fresh look at the issue of the loss of species. One of the most important efforts to provide such a picture was the *Global 2000 Report to the President of the United States*, commissioned by President Ronald Reagan (Council on Environmental Quality and the US State Department 1980). Although this report focused primarily on tropical forests, it laid the basis for further global assessments of the status of biological diversity. It was also here that the term *biological diversity* started to get special attention. The report not only dealt with conservation, but also emphasized the economic importance of biological resources.

The historic National Forum on BioDiversity, held in Washington, DC, September 21–24, 1986, under the auspices of the National Academy of Sciences and the Smithsonian Institution, gave prominence to the term *biodiversity* (Wilson 1988). That meeting and other complementary events within the framework of the IUCN provided the scientific basis for creating an international regime for the conservation and sustainable use of biological diversity. This received much-needed political impetus from the World Commission on Environment and Development, chaired by Gro Harlem Brundtland. Its report in 1987, *Our Common Future*, called for a Species Convention, emphasizing global cooperation but also recognizing the sovereign rights of states to the natural resources under their jurisdiction (World Commission on Environment and Development 1987).

International Negotiations on Biological Diversity and Their Results

The process of creating this regime fell to UNEP, which convened the Ad Hoc Working Group of Experts on Biological Diversity in June 1987 to harmonize the existing conventions related to biological diversity. With this decision, what was originally a scientific endeavor became the subject of international diplomacy. The group agreed on the need to create a binding international instrument on biological diversity.

In May 1989, the Governing Council of UNEP established the Ad Hoc Working Group of Experts on Biological Diversity to prepare an international legal instrument for the conservation and sustainable use of biological diversity (Decision 15/34 of 25 May 1989). During its second special session in August 1990, the Governing Council of UNEP again discussed the mandate of the working group and the possible content of a convention. Decision SS II/5 asked the working group to consider the need to share costs and benefits between developed and developing countries and the ways and means to support innovation by local people (Decision SS II/5 of 3 August 1990). The ad hoc working group, which came to be known in February 1991 as the Intergovernmental Negotiating Committee (INC), held seven working sessions, which culminated in the adoption of the Nairobi Final Act of the Conference for the Adoption of the Agreed Text of the Convention on Biological Diversity. After 5 years of negotiations (Sanchez and Juma 1994; McConnell 1996), the convention was presented for signature on June 5, 1992.

The convention defines biological diversity as "the variability among living organisms from all sources including, *inter alia,* terrestrial, marine, and other aquatic ecosystems and the ecological complexes of which they are part; this includes diversity within species, between species, and of ecosystems" (article 2). The objectives of the convention are the conservation of biological diversity, the sustainable use of its components, and the fair and equitable sharing of the benefits arising from the use of genetic resources, including the appropriate access to genetic resources, appropriate transfer of relevant technologies (taking into account all rights over those resources and technologies), and appropriate funding (article 1).

The character of the convention was shaped mainly by issues that dominated the preparations for UNCED, and it is a convergence of the conservation efforts that arose from the work of such institutions as IUCN. It also has taken on a number of issues concerned with international equity. The only major feature that is peculiar to the convention is the promotion and regulation of access to genetic resources, as outlined in article 15 and other relevant provisions.

On the whole, the convention has retained its scientific and technological character, as reflected in the number of its articles that deal with technical issues. Article 7 covers identification and monitoring of biological diversity and of processes and categories of activities that have possible substantial adverse effects on the conservation and sustainable use of biological diversity. This article includes the obligation to maintain and organize relevant data. Article 8, on in situ conservation, is committed to a variety of activities that range from establishing a system of protected areas to restoring and rehabilitating of degraded ecosystems to controlling alien species and modified organisms. Article 9, on ex situ conservation, looks to conserve the complementary components of biological diversity outside their natural habitats. Article 10 stipulates obligations about the sustainable use of biological diversity, including cooperation between government authorities and the private sector in the development of methods for the sustainable use of biological resources. Article 12 strives for increased research and education in the field of biological diversity. Article 14 covers the impact assessment of effects and the minimization of adverse effects. Article 18 asks contracting parties to promote international technical and scientific cooperation in the conservation and sustainable use of biological diversity. Finally, article 26 is the control provision which strives to have nations report on the measures that they have taken to implement the convention and their effectiveness in meeting its objectives. The main challenge is how to relate the important promise of the convention to practice.

STRUCTURE AND FUNCTIONING OF THE CONVENTION

The Structure

The convention's structure is defined by a number of internal and designated organizations. The main internal organizations, within the convention are the Conference of the Parties (COP), the Secretariat, and the Subsidiary Body for Scientific, Technical, and Technological Advice (SBSTTA). The convention also has established two mechanisms: the financial mechanism and the clearinghouse mechanism. At its first meeting, the COP designated the Global Environment Facility (GEF) as the institutional structure that implements the financial mechanism on an interim basis. The clearinghouse mechanism, which is devoted to technical and scientific cooperation, is implemented through the Secretariat in Montreal.

The COP. The COP operates by consensus and deals with the internal governance of the convention. Its main function is to keep the implementation of the convention under review by considering national reports and the advice of the SBSTTA or any other advisory body or processes and by adopting protocols to the convention. The COP also reviews the implementation of the convention by contacting the executive bodies of other conventions that deal with the same matters with a view to establishing appropriate forms of cooperation with them. It makes these contacts through the Secretariat.

The COP also may establish subsidiary bodies to obtain whatever scientific and technical advice is deemed necessary for implementation of the convention. Finally, the COP may consider and undertake any additional action that may be required to achieve the purposes of the convention in the light of experience gained in its operation.

The Secretariat. The COP is supported by the Secretariat, which was established under article 24 to arrange for and provide service to meetings of the COP, to perform the functions assigned to it by any protocol, and to prepare reports on the execution of its functions under the convention and to present them to the COP. The Secretariat is also charged with the mandate of coordinating with other relevant international bodies and, in particular, entering into such administrative and contractual arrangements as may be required for the effective discharge of its functions.

SBSTTA. Article 25 of the convention established the SBSTTA to provide the COP and, as appropriate, its other subsidiary bodies with timely advice relating to the implementation of this convention. The SBSTTA was designed to be open to participation by all parties and to have a multidisciplinary approach. Its members are government representatives who are competent in relevant fields of expertise, and it reports to the COP on all aspects of its work.

The specific responsibilities of the SBSTTA are to provide scientific and technical assessments of the status of biological diversity; to prepare scientific and technical assessments of the effects of types of measures taken in accordance with the provisions of this convention; to identify innovative, efficient, state-of-the-art technologies and know-how related to the conservation and sustainable use of biological diversity and to advise on how to promote the development and transfer of such technologies; to provide advice on scientific programs and international cooperation in research and development related to conservation and sustainable use of biological diversity; and to respond to scientific, technical, technological, and methodological questions that the COP and its subsidiary bodies may ask.

The Financial Mechanism. The convention established its financial mechanism to provide financial resources to developing-country parties as grants or concessions. The mechanism functions under the authority and guidance of and is accountable to the COP. The GEF conducts the operations of the mechanism. The COP determines the policy, strategy, program priorities, and eligibility criteria for access to and use of financial resources under the mechanism. The function of the GEF as the institutional structure that implements the financial mechanism on an interim basis is governed by a memorandum of understanding signed by the COP and the Council of the GEF.

The Clearinghouse Mechanism. Article 18(3) of the convention established the clearinghouse mechanism to promote and facilitate scientific and technical cooperation. At its second meeting, the COP established a pilot phase of the clearinghouse mechanism and agreed that this phase would start by promoting the exchange of information, with emphasis on the role of emerging information and communication technologies. The clearinghouse mechanism works closely with the financial mechanism in promoting the establishment of basic communication facilities for the parties of the convention.

Functioning: Experiences Gained

Global learning: Normative and programmatic functions. Because of the nature of the convention as a legally binding instrument, meetings have focused on normative and programmatic activities, leaving operational activities to governments and international institutions. The normative activities of the convention include, first of all, providing overall guidance on policy and advice on biodiversity-related activities through decisions of the COP.

This includes the decision to take an ecosystematic approach to the objectives of the convention. Detailed programs of work have been or are being elaborated on themes of biological diversity in marine and coastal areas, agricultural areas, inland waters, and forests. The programs of work include elements of integrated management, living resources, protected areas, alien species and genotypes, and methods of production, such as mariculture or agroforestry.

Interpretive community. The COP functions as an interpretive community that seeks to clarify certain aspects of the provisions of the convention as well as those of other relevant bodies. So far, the most advanced interpretive activities of the COP have been related to such issues as the role of the clearinghouse mechanism and the transfer of technology. This interpretive function also has benefited from the advice of the SBSTTA and has drawn on the results of other international processes and meetings.

In addition, the COP has sought to clarify the interpretation of biodiversity-related activities in other forums. For example, the work of the Intergovernmental Panel on Forests (IPF) under the Commission on Sustainable Development (CSD) benefited from input from the convention. A more elaborate interpretive effort is the current process to renegotiate the International Undertaking on Plant Genetic Resources of the FAO to bring it into harmony with the convention. Another interpretive activity is the realignment of the work program of the Intergovernmental Oceanographic Commission (IOC) with the convention's Jakarta Mandate on Marine and Coastal Biological Diversity.

Guidelines for national implementation. Other normative activities include providing flexible guidelines for national implementation, for example, in the access to and benefit-sharing related to genetic resources and in the protection and promotion of, and reward for local and indigenous innovations, knowledge, and practices. The COP also has provided guidelines for preparing national reports in accordance with article 26 and has developed indicators for biological diversity to be used at the national level.

Harmonization of procedures, standards, criteria, and indicators. The biodiversity regime is setting norms and standardizing procedures, especially through continuing negotiations under the open-ended Ad Hoc Working Group on Biosafety, which has finalized a protocol for adoption by the COP at the end of 1998. This activity also is contributing to the development of international environmental laws related to the precautionary principle. Further work on identifying opportunities for harmonization has resulted from the advice of the SBSTTA on criteria and indicators for biological diversity in forests.

Scientific and technical assessments. The use of scientific and technical input in the implementation of the convention has been debated considerably, especially in the context of reviewing the operations of the SBSTTA. Scientific input has been either parallel to the convention process or on an ad hoc basis. This has been partly because SBSTTA meetings are held annually, which does not allow effective mobilization of the available scientific and technical knowledge.

For example, the Norway-UNEP Expert Conference on Biodiversity, which was convened in May 1993 in Trondheim, Norway (Sandlund and Schei 1993), played a key role in bringing the biodiversity community together. Its results were used in the preparation for the first Intergovernmental Committee on the Convention on Biological Diversity (ICCBD) meeting held in Geneva in September 1993. The Norway-United Nations Conference on Alien Species was hosted by the Norwegian Ministry of the Environment in July 1996 (Sandlund and others 1996); the proceedings provided input to the SBSTTA and to the third COP in November 1996.

Unlike other environmental treaties, such as the Convention on Climate Change and the Montreal Protocol on Substances That Deplete the Ozone Layer, the Convention on Biological Diversity has conducted no formal knowledge assessments. Instead, a Global Biodiversity Assessment (GBA) was undertaken by UNEP after the convention went into force (Heywood 1995). The GBA was an independent, peer-reviewed scientific analysis by more than 300 experts from more than 50 countries on current issues, theories, and views about the main aspects of biological diversity. Governments were invited to nominate experts to review the GBA in their personal capacity; more than 1,100 experts from more than 80 nations participated in this peer-review process. The report, however, has been used only informally in the framework of the convention, and there has been no followup by the SBSTTA, although many of the reports prepared by the secretariat have relied on the GBA as one of the most authoritative sources of information available about biological diversity.

Numerous research institutions and networks are seeking to incorporate the agenda of the convention in their programs, and some of them are becoming participants in the activities of the convention. One of these is DIVERSITAS, a scientific research program sponsored by UNESCO. Furthermore, considerable scientific work, management methods, and techniques for the conservation and sustainable use of biological diversity and its components are already available in the relevant institutions.

National reporting. One key instrument for promoting the implementation of the convention is national reporting. The first national reports were made available to the secretariat at the end of 1997. They will form the basis of a synthesized document that will be presented to the COP for consideration and further decision-making. Strengthening of national capabilities for reporting will require concerted effort by the convention in conjunction with its financial mechanism. These reports not only will provide the COP with the basis for further guidance on policy, but also represent one of the most important instruments for monitoring progress. In this regard, the work being carried out by the SBSTTA on biological-diversity indicators will be important for enhancing the normative role of the convention.

ENGAGING THE SCIENTIFIC AND TECHNOLOGICAL COMMUNITY

Mobilization of Science

It is evident that if the convention is to conduct its normative functions effectively, it will need to devise methods to mobilize the best available scientific and technical expertise. The main medium for such activities is the SBSTTA. Central to this issue is the continuing debate about the modus operandi of the SBSTTA. A number of options are open to the convention, the first of which is that the role of the SBSTTA itself needs to be reviewed in light of its operating experience. Some evidence suggests that the SBSTTA is emerging as a platform and focal point for international scientific networks. Benefiting from this opportunity will require adjustments in how the SBSTTA functions, especially in relation to its expert groups and meetings. Such groups and meetings, as well as liaison groups, can form the basis for a wide range of intersessional scientific activities.

Application of Technology

The role of technology in the implementation of the convention is now considered to be part of the thematic areas. So far, little work has been done under the convention on technological issues, although further discussion, especially on biotechnology, is expected at the next meeting of the COP. During consideration of this issue, it will be important to remember that many of the technological options available for implementing the convention are in the private sector. In this regard, the convention, possibly through the clearinghouse mechanism or other measures that the COP may wish to exact, could play a key role in encouraging the private sector to participate in the process of implementing the convention.

The Role of the United States

The United States is a leader in the scientific and technological fields that are related to biological diversity. This knowledge is generated by various stakeholders in both the private and public sectors. In the public sector, the federal government supports the generation of scientific knowledge through its various national research institutions.

Conservation and the sustainable use of biological diversity are an integral part of policy and law in the United States. Numerous task forces have been set up to formulate strategies for integrating biological diversity into sectoral activities and for developing methods of ecosystem management. Conservation is carried out jointly through national partnerships between federal, state, and nonprofit groups.

The United States has a diversified system of protected areas and biosphere reserves, including national wilderness-preservation and wildlife-refuge systems. The national-park system includes 374 areas, covering more than 83 million acres. Various programs aim at conservation and sustainable use, such as those working toward the recovery of threatened or endangered species and habitats and the restoration and enhancement of coastal zones.

Internationally, the United States has set up a variety of global programs on conservation and sustainable use of biological diversity and the fair and equitable sharing of benefits arising out of genetic resources. In accordance with its overall environmental policy, the United States is participating actively in the regime-building process of the Convention on Biological Diversity. It plays an important role in the convention process and has played a key role in seeking to maintain the scientific and technical role of the SBSTTA. The full scientific, technical, and technological contributions of the evolving convention will be enhanced further when the United States becomes a full party.

REFERENCES

Berne Convention on the Conservation of European Wildlife and Natural Habitats of 19 September 1979. 1982 In: Kiss A (ed). Selected multilateral treaties in the field of the environment. Nairobi Kenya: UNEP. p 509.

Convention on International Trade in Endangered Species of Wild Fauna and Flora of 3 March 1973. 1982. In: Kiss A (ed). Selected multilateral treaties in the field of the environment. Nairobi Kenya: UNEP. p 289.

Convention for the Protection of the World Cultural and Natural Heritage of 23 November 1972. 1982. In: Kiss A (ed). Selected multilateral treaties in the field of the environment. Nairobi Kenya: UNEP. p 276.

Convention on Wetlands of International Importance Especially as Waterfowl Habitat of 2 February 1971. 1982. In: Kiss A (ed). Selected multilateral treaties in the field of the environment. Nairobi Kenya: UNEP. p 246.

Council on Environmental Quality and the US State Department. 1980. The global 2000 report to the President. Washington DC: US GPO.

Di Castri F, Hadley M, Damlamian J. 1981. MAB: the Man and the Biosphere Programme as an evolving system. Ambio 10(2–3):52–7.

Fowler C, Mooney P. 1990. Shattering: food, politics, and the loss of genetic diversity. Tuscon AZ: Univ of Arizona Pr.

Heywood VH (ed). 1995. Global biodiversity assessment. Cambridge UK: Cambridge Univ Pr and UNEP.

McConnell F. 1996. The convention on biological diversity. A negotiation history. Amsterdam Netherlands: Kluwer Intl.

Pistorius R. 1997. Scientists, plants, and politics. A history of the plant genetic resources movement. Rome Italy: Intl Plant Genetic Res Inst.

Sanchez V. 1994. The convention on biological diversity: negotiation and content. In: Sanchez V, Juma C. Genetic resources and international relations. Nairobi Kenya: African Centre for Technology Studies Pr. p 7–18.

Sandlund OT, Schei PJ (eds). 1993. Proceedings of the Norway/UNEP expert conference on biodiversity. The Trondheim conference on biodiver, 24–28 May 1993. Trondheim Norway: UNEP

Sandlund OT, Schei PJ , Viken A (eds). 1996. Proceedings. Norway/UN conference on alien species. The Trondheim conference on biodiver, 1–5 July 1996. Trondheim Norway: UNEP.

Stockholm Declaration of the Conference on Human Environment and Action Plan. 1972. Intl Legal Materials 11:1416.

UN Conference on Environment and Development. 1992. UN Conference on Environment and Development, 5–14 June 1992, A/Conf. 151/26. Available: http://www.biodiv.org.

UNEP [United Nations Environment Programme]. 1989. Decision 15/34 of 25 May 1989, A/44/25, p 161.

UNEP [United Nations Environment Programme]. 1990. Decision SS II/5 of 3 August 1990, UNEP/GCSS.II/3, Annex I, S. 42.

UNEP [United Nations Environment Programme]. 1992. Convention on biological diversity, June 1992. Nairobi Kenya: Environ Law and Inst Prog Activity Center.

Wilson EO. 1988. Biodiversity. Washington DC: Nat Acad Pr.

WCED [World Commission on Environment and Development]. 1987. Our common future (The "Brundtland Report"). Oxford UK: Oxford Univ Pr.

ECOLOGY AND THE KNOWLEDGE REVOLUTION

GRACIELA CHICHILNISKY
UNESCO Professor of Mathematics and Economics and Director,
Program on Information Resources and Columbia Center for Risk Management,
Columbia University, 405 Low Memorial Library, New York, NY 10027

THE GOLDEN AGE OF INDUSTRIAL SOCIETY

SINCE WORLD WAR II, the world economy has expanded at a record pace, and world trade has increased at least three times faster than world production. During this period, industrialization has become an irresistible trend, made global by the dynamics of international markets and, more recently, information technology. This has been the golden age of industrial society.

The industrial society now faces the risks created by its own success. Its growth has been based on a voracious use of natural resources (Chichilnisky 1995-6), the rapid burning of fossil fuels to produce energy, and massive clearing of wooded lands and other ecosystems where most of the world's biodiversity is found. Economic activity is the fundamental driving force of the two most pressing global environmental problems: climate change and biodiversity destruction.

Only 20% of the world's population lives in industrial societies, but through global trade the success of industrialization has magnified the use of fossil fuels and other natural resources worldwide. Industrial nations consume most natural resources and originate 60% of global emissions of carbon dioxide, which can precipitate global climate change; they consume on the average 10 times as much copper, three times more roundwood, 15 times more aluminum, and 10 times more fossil fuel per capita than the developing countries. The international market mediates the relationship between industrial nations and developing countries—generally called the North and South, respectively (WRI/UNEP/UNDP 1995). The developing South specializes in resources, which account for 70% of the exports of Latin America and almost all those of Africa; the industrial North specializes in products that are intensive in capital and knowledge. The South houses most of the world's biodiversity, and the current pattern of trade is contributing to its destruction.

The trend is global. Since the end of colonialism, the Bretton Woods institutions (for example, the World Bank and the International Monetary Fund) have encouraged a pattern of resource-intensive development for the world's less advanced countries. Developing countries today play the role of resource producers, overextracting resources that are traded below their real costs and thus overconsumed in the industrial nations (WRI/UNEP/UNDP 1995). This pattern of trade and low resource prices has been explained by the historical difference in property rights between the North and the South in the context of a rapid expansion of global markets (Chichilnisky 1994a): in a world where agricultural societies trade with industrial societies, global markets magnify the extraction of natural resources and depress their prices, and as a result world exports and consumption of resources exceed what is optimal. This is at the core of the world's environmental problems; through forests' and fisheries' destruction, it leads to rapid biodiversity loss.

Today's global environmental problems are connected with the role of global markets in magnifying unsustainable patterns of consumption and resource use in industrial nations. These patterns are responsible for most of the world's ecosystem destruction. In the long run, however, the fate of the world's resources could depend on the developing world. This paper therefore concentrates on today's patterns of development in industrial nations and on future patterns of development in the rest of the world. It advances a vision of a new society in which humans could live in harmony with each other and with nature, and it describes the transition to this new society as a "knowledge revolution." That phrase refers to a swift period of change that is already under way in industrial nations, a change that requires new institutions and policies to reach a sustainable outcome. I analyze a new type of markets that will play a crucial role in tomorrow's societies—markets in knowledge and in environmental assets—and I analyze the property-rights regimes that are needed in these markets to achieve efficient, equitable, and sustainable development.

THE NEW GLOBAL MARKETS

Markets are a dominant institution in the global economy. As the century turns, however, markets themselves are evolving. Two major trends are knowledge markets and global environmental markets. Knowledge markets hold the key to the dynamics of the world economy: telecommunication and electronics, biotechnology and financial products—all involve trading products that use knowledge rather than resources as their most important input. The first global environmental market is about to emerge: following our earlier proposal to the UN Climate Convention (Chichilnisky 1993a, 1995b, 1996), the 166 nations that were parties to the Framework Convention for Climate Change (FCCC) agreed in Kyoto in December 1997 to create a framework to trade carbon-emission credits among industrial nations.

Knowledge markets and environmental markets are different from traditional markets in that they trade what I call privately produced public goods rather than private goods. Private goods—such as apples and machines—are chosen by each trader independently from each other and are "rival" in consumption. Not so with knowledge (Shulman 1999) and environmental goods: the carbon concentration in the planet's atmosphere is the same for all, and knowledge can be shared without losing it. Trading knowledge and environmental "rights to use" could lead to the most important markets of the future. The trading rights to use knowledge and environmental resources are key trends in the world economy; these trends lead the transformation that I call the knowledge revolution™ (Chichilnisky 1997a,b,c, 1998; Shulman 1999).

Focusing on those new markets, I analyze here the introduction of new institutions and the policies that can lead the transformation of industrial society into a sustainable knowledge-based society. I propose the creation of a new type of economic organization, which involves markets that trade a mixture of private and public goods to reach efficiency. The new markets require new regimes of property rights that are proposed here (Chichilnisky 1997a,b,c, 1998). They carry the seed of a human-oriented society that by its own functioning encourages the creation and diffusion of knowledge and a sustainable and equitable better use of the world's natural resources.

ECOLOGY AND THE KNOWLEDGE REVOLUTION

A major challenge is to find practical paths for sustainable development. This requires reorienting consumption patterns and the use of natural resources in ways that improve the quality of human life while living within the carrying capacity of supporting ecosystems. It will require building economic systems in which the basic needs of people are satisfied across the world, while protecting resources and ecosystems so as not to deprive the people of the future from satisfying their own needs. That is the definition of sustainability adopted by the Brundtland report, and it is anchored in the concept of development based on the satisfaction of "basic needs," a concept that was introduced and developed empirically in Chichilnisky 1997a, b. Sustainable development has also been explored in *Caring for the Earth*, a joint publication of The World Conservation Union, UN Environment Programme and the World Wildlife Fund. It requires building a future in which humans live in harmony with nature. We are far from that goal; indeed, in many ways, the world economy is moving in the opposite direction.

Just as the environmental problems generated by industrial society are becoming a threat to human welfare, industrial society is in the process of transforming itself. The rapid pace of the change has led me to call it a revolution. The change is centered in the use of knowledge, so I call it the knowledge revolution. What characterizes this revolution?

The question is best answered in a historical context, by contrasting the current situation with the agricultural and the industrial revolutions, two landmarks in social evolution. Neither of the two previous revolutions is complete. Across the world, we find today preagricultural societies populated by nomadic hunters and gatherers, and most of the developing world is still working its way through the industrial revolution. Nevertheless, in many societies, knowledge is becoming a leading indicator of change. Knowledge means the ability to choose wisely what to produce and how to produce it. That ability is becoming the most important input of production and the most important determinant of wealth and economic progress. It resides mostly in human brains rather than in physical entities, such as machines or land. It is worth pointing out that the important input is *knowledge* rather than information. That difference distinguishes between the computer industry, which is based on information technology, from other sectors—such as telecommunication, biotechnology, and financial sectors—that involve knowledge other than computers. Knowledge is key to sustainability. Indeed, the value of biodiversity resides mostly in its knowledge content, according to such ecologists as EO Wilson and Tom Lovejoy. In a nutshell, knowledge is the content, and information is the medium. The content (knowledge) is driving change, and this change is facilitated by the medium (information). Information technology is the fuel for knowledge sectors because it performs the important role of allowing the human brain to expand its limits in the production, organization, and communication of knowledge. The most important input of production today is not information technology itself; it is knowledge (Chichilnisky 1997a,b,c, 1998; Shulman 1999).

CHARACTERIZING THE KNOWLEDGE REVOLUTION

We may characterize the knowledge revolution as a period of rapid transition at the end of which knowledge itself becomes the most important input of production, the most important factor of economic progress and wealth. For example, the knowledge content of biodiversity becomes a key input for improving public health and human welfare, and, as pointed out above, it is identified as a crucial source of the economic value of biodiversity. In contrast the most important actual inputs of production in prior revolutions were land (in the agricultural revolution) and machines (in the industrial revolution), inputs that became better used because of new knowledge. ("Capital," in the sense of economic value, shows the same trend: it was associated mostly with land holdings in the agricultural society, with machinery in the industrial society, and with ideas in the knowledge society.) Knowledge differs fundamentally from land and machines in that it is not rival in consumption, so the knowledge revolution is based on a radically different type of input of production. Property rights to inputs of production matter a great deal: for example, property rights to industrial capital determine the difference between socialism and capitalism and have led to global strife in most of this century. Property rights to knowledge are now becoming equally important (Shulman 1999).

The knowledge revolution is already taking place. One indication of that is that the value of corporations in the stock exchanges of the world is increasingly measured according to their knowledge assets—such as discoveries, patents, brand names, and innovative products—rather than their capital base or physical assets. Knowledge-related assets (such as patents) are increasingly regarded as the most important source of economic progress in a corporation and of its value. At the level of the economy as a whole, knowledge of mathematics and science has become a good predictor of national economic progress across the world. In this period of change, the United States leads the pack (Chichilinsky 1997a). Today, more Americans make semiconductors than construction machinery. The telecommunication industry in the United States and Canada employs more people than the automobile and automobile-parts industries combined. The US health and medical "industry" has become larger than its defense industry and larger than its oil refining, aircraft, automobiles, automobile-parts, logging, steel, and shipping industries put together. More Americans work in biotechnology than in the machine-tools industry. Most US jobs in the last 20 years were generated in smaller, knowledge-intensive firms driven by risk capital. One-third of US growth is accounted for by the knowledge sectors; thus, knowledge is an increasingly important determinant of economic progress. The knowledge sectors of the US economy already grow about twice as fast as the rest of the economy and therefore account for most of the dynamics of economic growth (Chichilinsky 1997a). That is despite the fact that current systems of accounting undervalue the contributions of electronics, which are extraordinarily productive and therefore offer rapidly lowering costs for their products. In a nutshell, knowledge products in the United States are rapidly becoming the most important input of production, source of value, and economic progress. Development of knowledge sectors is slower in Europe than in the United States because Europe's financial markets and property-rights systems are not as flexible, well developed, and regulated and this inhibits the creation, development, and commercialization of knowledge through new risk venture corporations.

Knowledge sectors have lower consumption of resources and less ecological impact than the rest, so they could decrease environmental damage once they become dominant in the economy (Chichilinsky 1997a). That is partly because of our new knowledge about the environmental consequences (costs) of our economic behavior. The question is whether the pace and scope of this process of change will foster a sustainable society on a time scale that matters. It is important to encourage and accelerate the transition in the right direction. The economic transformation depends on, among other things, the evolution of the new markets for knowledge and for environmental assets. These require special analysis because, as already mentioned, knowledge and environmental assets are privately produced public goods and lead to new types of markets with new challenges and new opportunities for action.

A SERVICE ECONOMY

It is important to differentiate the knowledge revolution from the so-called service economy, which used to be thought of as the latest stage of the industrial society. A service economy is characterized by the production of services more than goods, and it is similar to a knowledge economy in that knowledge sectors often involve services (such as finance). The inevitable concern about the service economy is that it could lead mostly to service-oriented labor, such as the labor used in the food services or in bank processing, which requires little skill and achieves lower wages. Services now make up the largest part of advanced industrial economies, but the analogy ends there. A difference between the service economy and the knowledge society is that in the latter the typical worker is highly skilled and generally well paid. Furthermore, workers' knowledge resides mostly in their own brains and life experiences rather than in the machines that complement labor. Therefore, the knowledge economy could result, with proper institutions, in a society that is more human-oriented than the industrial or the service society. Such a society would involve more human connection and therefore would have different values, being more sensitive to others' needs and the effects of our actions on them.

KNOWLEDGE AS A PRIVATELY PRODUCED PUBLIC GOOD

As knowledge itself becomes the most important input to production, economic behavior changes because knowledge is a special type of good. It is called a public good by economists, not because it is produced by governments, but because, as already pointed out, it is not "rival" in consumption. This means that we can share knowledge without losing it; this is a physical property of knowledge, not an economic property, and it is independent of the organization of society. However, the economic rules governing the use of knowledge—for example, whether patents can be used to restrict its use—can have a major impact on human welfare and organization.

Knowledge is also different from conventional public goods of the type that economists have studied for many years, such as law and order or defense, which are supplied by governments in a centralized fashion. What is unique about knowledge among public goods is that it is typically supplied by *private* individuals who are its creators. At the level of production, therefore, knowledge is like any other private good: expensive to produce, and produced from private rival resources (human time) that often cannot be used simultaneously for other purposes. Producing knowledge requires economic incentives similar to those for producing any other private good.

A VISION OF THE KNOWLEDGE SOCIETY

Following the knowledge revolution, a new society could well develop that is centered in human creativity and diversity and that uses information technology rather than fossil fuels to power economic growth. The vision is a human-centered society that is innovative with respect to knowledge and at the same time conservative in its use of natural resources. The consumption of resources might not be as voracious as that in the industrial society and could be better distributed across societies and across the globe. The knowledge society might achieve economic progress that is harmonious with nature.

That vision is only a possibility at present. Without developing the right institutions and incentives, it might never be realized, and a historical opportunity would be lost; we need institutions to bridge the gap between a grim present and a bright and positive future. The rest of this paper addresses this issue.

THE PARADOX OF KNOWLEDGE

To produce new knowledge, creators need economic incentives. This could involve restricting the use of knowledge by others. Patents on new discoveries work in this fashion: by restricting others' use of knowledge. That creates a problem: any restriction in the sharing of knowledge is inefficient because knowledge can be shared at no cost and its sharing can make others better off. However, without some restrictions there might be no incentive to create *new* knowledge. I call this the paradox of knowledge; resolving this is at the heart of the success of the knowledge society, of its ability to bring human development for many and not only a wealthy few.

A NEW PROPERTY-RIGHTS REGIME

New regimes for property rights are needed to deal simultaneously with the need to share the use of knowledge for efficiency, and the need to preserve private incentives for production (Shulman 1999). I propose complementing patents with a system of *compulsory and negotiable licenses that are traded competitively in the market along with all other goods in the economy*, and which are offered in prederential terms to lower income groups. In this new scheme, the right to use knowledge is unrestricted, and by law everyone should have access to it. However, users must pay the creator each time they use the knowledge. Trading of the licenses competitively in markets ensures that the creators of knowledge are compensated for their labor in a way that reflects the demand for their products and therefore their usefulness for society. Furthermore, the prices paid for the use of licenses are uniform and determined by competitive markets. This new regime differs fundamentally from the current system of patents in that, in principle, patents can restrict the use of knowledge—licenses related to patents can be negotiated, but they do not have to be. Today owners of patents are legally entitled not to negotiate licenses, and thus in effect to create a monopoly during the patents' life (Shulman 1999). Furthermore, even if they are traded, there is no requirement that the market for patents be competitive. By contrast, no restriction in the use of knowledge is allowed in the system I propose (Chichilnisky 1997a,b,c, 1998). However, a key issue is the distribution, use, and applicability of the property rights for licenses.

It is clear that a system of licenses on knowledge products (such as operating systems for software, biological information, and how-to-do-it systems) could preserve or even worsen today's uneven distribution of wealth in the economy, because the knowledge economy has a built-in incentive for the creation of monopolies. Indeed, any knowledge-based corporation is a "natural monopoly," that is, the cost of duplicating knowledge products (such as software) is very small, so the larger the firm, the lower its costs. That is an extreme case of "increasing returns to scale," wherein larger firms have an advantage over their smaller competitors and can deter entry by newer and smaller competitors. Such natural monopolies are characteristic of the knowledge society. How to avoid their effects in concentrating welfare in the hands of a very few?

The system of property rights proposed here takes into account those possibilities. It establishes how the distribution of licenses in competitive markets is crucial in achieving efficient solutions. It shows that markets in knowledge operate differently from the standard markets because knowledge is a privately produced public good. The solution proposed here is a distribution of property rights through licenses that is negatively correlated with the property rights of private goods.

How will such a system of property rights become accepted? There is a parallel with the introduction of laws to ensure fair trade, to which natural monopolies have offered much resistance, but which were eventually adopted by society as a whole (Shulman 1999). There are substantial economic incentives for corporations to accept fair trading and the system of property rights that I propose, although it is clear that more economic thinking and business education are needed before acceptance becomes widespread. Producers that benefit from increasing returns to scale could benefit from a system of licenses in which the lower-income segments of the population are given proportionately more rights to use knowledge than the rest. This would expand the market for their products and thus favor them. Consider as an example the case of subsidized worker-training schemes. Because knowledge is so important for the productivity of society as a whole and produces positive "externalities" on all producers, there is an incentive to develop a skilled pool of workers. Corporations know that skilled workers are essential to the success of knowledge industries.

To reach an efficient market solution, namely one that cannot be improved so as to make everyone better off, lower-income traders (individuals or nations) should be assigned a larger endowment of property rights in the use of knowledge (Chichilnisky 1997a,b,c, 1998). In practice, a larger amount of licenses to use knowledge are assigned to such lower-income countries or groups.

The regime that I propose is new but realistic. Similar systems are already in place in most industrial societies within educational systems. For example, school subsidies offer lower-income groups preferential prices in educational services. The US federal government auctions off the use of airwaves in such a way that members of minority groups and women are given substantial discounts (in some cases, of 40%) when they participate in those auctions. In the United States, Microsoft has introduced licensing regimens for some of its products that benefit disproportionately the lower-income groups. More examples of this nature can be found in Shulman (1999), who also advocates compulsory licenses without however offering an economic analysis of distributional issues or efficiency.

LICENSES: WE MAKE IT, WE TAKE IT BACK

The system of property rights proposed here, although unique in its economic formulation, is reminiscent of a development that is already taking place in the corporate world, a development that is also connected with environmental issues that have a public-good aspect: the disposal of materials involved in heavy industrial products, such as vehicles and electronic equipment. Leasing vehicles and electronic equipment, a thriving business, hardly existed 20 years ago. One of the largest packaging companies in the world, Sonoco Products Co., started taking its used products off customers' hands after CEO Charles Coker made a pledge in 1990: "We make it, we take it back." The policy has already been adopted by the car industry in Germany, where, because of environmental concerns, car manufacturers are responsible for disposing of vehicles that customers return at the end of their useful life. Another example is in the floor-covering industry: Ray Anderson, CEO of Atlanta-based Interface, the largest maker of commercial carpeting, has set up as a goal to create zero waste while making a healthy profit, and the company takes back its products when they have been used to recycle them. What all of these examples have in common is that they perceive the businesses' mission to be the sale of services, not products. For example, selling viewing services rather than television sets, selling transportation services rather than vehicles, and selling the comfort and visual services that carpets provide rather than the carpets themselves. Licensing gives the producers an incentive to minimize waste and environmental damage—for example, the waste produced by wrapping or by defunct car bodies—because they will be responsible for them. The businesspeople see licensing services as the way to the future, particularly when consumers must pay for the disposal of industrial waste.

Implicit in the new system of property rights is the idea of licensing the use of services rather than owning the products that deliver the services. The analogy with licensing is therefore clear.

Knowledge, as we saw above, has much in common with environmental assets: it is a privately produced public good. Knowledge products have been licensed for many years, although case by case and without securing the competitiveness of the market for licenses and the distribution of property rights that would ensure efficient outcomes. In this sense, the new developments in industry reported here move in the same direction as the system of property rights involving licenses. The new system of property rights that is proposed here can be thought of as an improvement in, an institutionalization of, and an economic formalization of licensing and leasing systems that have recently emerged in advanced industrial economies.

A PROPERTY-RIGHTS REGIME FOR BIODIVERSITY

The Convention on Biodiversity faces a controversial issue with respect to property rights to the knowledge contained in biodiversity samples obtained from developing nations. The pharmaceutical industry faces difficult ethical and business issues on how to involve and compensate developing countries and how to price newly discovered drugs on which much R&D money has been spent but that should be available as widely as possible (such as newly found AIDS medication). The regime suggested above can deal with those issues because it ensures the widest possible use of knowledge while providing compensation for the discoverer and developer. In essence, patents would be replaced by long-lived compulsory licenses on the use of the implicit knowledge that would be traded in competitive markets. This regime would expand maximally the use of the products without depriving the creator of due rewards. Initial fixed costs could be recovered from higher-income groups through the appropriate use of initial allocations that favor low-income groups.

HUMAN IMPACTS OF PROPERTY RIGHTS TO KNOWLEDGE

The rules that govern the use of knowledge in society are important because they can lead to threats to as well as opportunities for human development. These rules have an effect both directly and through changes in the patterns of consumption of goods and services. They can determine the impact of human societies on the environment and on inequalities across the world economy. The way we use and distribute knowledge casts a very long shadow on human societies.

A historical comparison helps to explain the process. In agricultural societies, the way humans organized the ownership of land, which was the most important input to production, led to such social systems as feudalism. Ownership of land had a major impact on human welfare and on economic progress. Similarly, in industrial societies, the way humans organize the use of capital, the most important input of production, led to different social systems, such as socialism and capitalism. Indeed, those two systems are defined by their rules on ownership of capital: In socialism, ownership is in the hands of the governments or other public institutions; and in capitalism, capital is in private hands. Property rights to capital have mattered a great deal and have even led to global strife in most of this century.

Because capital is the most important input of production in industrial society, it is clear that property rights to capital had an enormous impact on the organization of society, on economic progress, and on people's welfare. Similarly, in the knowledge society, the way humans organize the use of knowledge, its most important input to production, will determine human welfare and economic progress across the world. Human institutions that regulate the use of knowledge, such as through property rights and markets for knowledge, will become increasingly important. As we saw, knowledge is a different type of commodity from land or capital: it is a privately produced public good. Markets with public goods—and other economic institutions, such as property rights to public goods—are still open to definition and require much economic analysis. Markets themselves will operate differently in the knowledge economy because the nature of the goods traded will be different. There will be new challenges and new opportunities for economic thinking and organization.

THE ECONOMIC IMPACT OF KNOWLEDGE-INTENSIVE VS. RESOURCE-INTENSIVE GROWTH

To focus our thoughts, it is useful to distinguish between two patterns of economic growth, two extreme cases between which is a spectrum of possibilities: economic development that is knowledge-intensive and economic development that is resource-intensive. The former means achieving more human welfare with less material input; the latter means achieving more production through more material use. These two categories were introduced in Chichilnisky (1995a, 1994b).

There are excellent historical examples of the two patterns of development and of the differences they induce in economic growth. East Asian nations approximate the knowledge-intensive paradigm, whereas Latin American and African countries fit well the pattern of resource-intensive growth. On the whole, knowledge-intensive development strategies succeeded, and resource-intensive development patterns did not. I studied the historical patterns, focusing on East Asian nations that are now called the Asian Tigers (including Japan, Korea, and Taiwan) and later those called the Small Tigers (such as Singapore, Philippines, Hong Kong, and Malaysia) Chichilnisky (1997a). Those nations focused on exports of technology-intensive products, such as consumer electronics and technologically advanced vehicles, and overturned the traditional economic theory of "comparative advantages." In contrast, Latin America and Africa followed a traditional resource-intensive pattern of development and lost ground.

The most dynamic sectors in the world economy today are not resource-intensive; they are knowledge-intensive, such as software and hardware, biotechnology, communication, and financial markets (Chichilnisky 1994b, 1995a, 1997a,b,c, 1998). These sectors are relatively friendly to the environment. They use fewer resources and emit relatively little CO_2. Knowledge sectors are the high-growth sectors in most industrialized countries.

Some of the most dynamic developing countries are making a swift transition from traditional societies to knowledge-intensive societies. Mexico produces computer chips, India is rapidly becoming an important exporter of software, and Barbados has unveiled a plan to become an information society within a generation (Fidler 1995). Those policies are an extension of the strategies adopted earlier by Hong Kong, the Republic of Korea, Singapore, and Taiwan, which have achieved extraordinary success over the last 20 years by relying not on resource exports, but on knowledge-intensive products, such as consumer electronics.

One lesson of history is clear: not to rely on resource exports as the foundation of economic development. Africa and Latin America must update their economic focus. Indeed, the whole world must shift away from resource-intensive economic processes and products. If they do, smaller quantities of minerals and other environmental resources will be extracted, and their prices will rise. That is as it should be because today's low resource prices are a symptom of overproduction and inevitably lead to overconsumption.

Not surprisingly, from an environmental perspective one arrives at exactly the same answer: higher resource prices are needed to curtail consumption. Producers will sell less, but at higher prices. That is not to say that everyone will gain in the process. If the world's demand for petroleum drops, most petroleum producers will lose unless they have diversified into other products that involve less use of resources and higher value. Most international oil companies are investigating this strategy. Indeed, British Petroleum and Shell are already following such policies. Monsanto is doing the same within the chemical industry.

The main point is that nations do not develop on the basis of resource exports. At the end of the day, development can make all better off. The trend is inevitable, and the sooner one makes the transition to the knowledge revolution, the better. The data and a conceptual understanding of how markets operate lead to the same conclusion. Economic development cannot mean, as in the industrial society, doing more with more. It means achieving more progress with less use of resources.

PEOPLE-CENTERED DEVELOPMENT: OPPORTUNITIES AND THREATS

The knowledge revolution could develop in different ways, depending on how our institutions and policies unfold. As already explained, knowledge has the capacity to amplify current discrepancies in wealth because knowledge sectors can lead to natural monopolies such as those due to the adoption of operating systems (Microsoft's Windows is a case in point) or other standards. Knowledge sectors could amplify the differences in wealth between the North and the South. If that occurs, the low prices of resources from developing countries will persist, because they result in part from the necessity to export at low prices in a difficult international market climate. It has been shown that with current institutions of property rights, anything that leads to more poverty will lead to increased resource exports from developing countries (Chichilnisky 1994a).

However, knowledge sectors will flourish in nations that have skilled labor. Several developing nations are or soon could be in that position; examples are the Caribbean area and Southeast Asia and many areas in Latin America (Harris 1994).

The main issues here are

• abandonment of the resource-intensive development patterns that those nations have followed for the last 50 years, with the support and encouragement of the Bretton Woods institutions, such as the World Bank and the International Monetary Fund; and
• establishment of the institutions (property rights and financial markets) that could lead them to overcome the mirage of resources as a "comparative advantage," help avoid the heavy stages of industrialization, and move directly ("leapfrog") to the knowledge society.

Heavy accumulation of capital (financial or physical) is not needed for most knowledge sectors. Indeed, most new technologies were developed in small firms within the United States (the proverbial "garages" in Silicon Valley), and software production in developing nations is labor-intensive and requires relatively little capital. Bangalore, a typical example, became in 10 years one of the world's most active exporters of software; it now exports US$2 billion worth per year. What is needed is good managerial ability and highly skilled labor of the type that does not require expensive machinery or heavy capital investment in plants.

REFERENCES

Brundtland GH. 1987. The UN world commission on environment and development. Oxford UK: Oxford Univ Pr.

Chichilnisky G. 1993a. The abatement of CO_2 emission in industrial and developing countries. OECD/IEA conferences on the economics of climate change, published in OECD: The Economics of Climate Change (ed. Jones T), Paris France, June 1993, p 159-170.

Chichilnisky G. 1994a. North-South trade and the global environment. American Economic Review, Bol. 84, NO. 4, Sept 1994, p 427-434.

Chichilnisky G.1994b. Trade regimes and GATT: resource intensive vs. knowledge intensive growth. Economic Systems merged with Journal of International Comparative Economics, special issue on globalization of the world economy, CIDEI conference, Rome Italy, 1994, 20, 1996, p 147-181.

Chichilnisky G. 1995a. Strategies for trade liberalization in the Americas, in Trade Liberalization in the Americas. Interamerican Development Bank (IDB) and United Nations Commission for Latin America and the Caribbean (ECLAC), Washington DC.

Chichilnisky G. 1995b. Global environmental markets: the case of an international bank for environmental settlements. Proceedings of the Third Annual World Bank Conference of Effective Financing for Environmentally Sustainable Development. World Bank, Washington DC. Oct 6 1995.

Chichilnisky G. 1996. Environment and global finance: the case for an international bank for environmental settlements. UNESCO-UNDP paper No. 10, Office of Development Studies (ODS) UNDP, New York, NY, 10017, Sept 1996.

Chichilnisky G. 1995-1996. The economic value of the earth's resources. Invited perspective article, Trends in Ecology and Evolution (TREE), 1995-1996, p 135-140.

Chichilnisky G. 1996b. The greening of Bretton Woods. Financial Times, section on economics and the environment. 10 January 1996, p 8.

Chichilnisky G. 1997a. The knowledge revolution: its impact on consumption patterns and resource use. Human Development Report, United Nations Development Program (UNDP) New York, NY, November 1997.

Chichilnisky G. 1997b. Updating property rights for the knowledge revolution. John D and Catherine McArthur Lecture, Program on Multilateralism, Institute for International Studies, Univ of California, Berkeley. Nov 3, 1997.

Chichilnisky, G. 1997c. The knowledge revolution. New Economy. London: The Dreydon Pr. p 107–11.

Chichilnisky, G. 1998. The knowledge revolution. J Int Trade Eco Devel 7(1):39–54.

Fidler S. 1995. An information age society is booming. Financial Times, 26 April 1995.

Harris DJ. 1994. Determinants of aggregate export performance of Caribbean countries: a comparative analysis of Trinidad & Tobago. Department of Economics, Stanford Univ, Sept 1994.

Shulman S. 1999. We need new ways to own and share knowledge. The Chronicle of Higher Education, Feb 19, 1999, p A64.

World Development Report. 1992. Development and the environment. Oxford UK: Oxford Univ Pr.

WRI, UNEP, UNDP [World Resources Institute, United Nations Environment Program, United Nations Development Program]. 1995. A guide to the global environment. Oxford UK: Oxford Univ Pr.

P A R T

7

INFRASTRUCTURE FOR SUSTAINING BIODIVERSITY— SOCIETY

BIODIVERSITY:
A WORLD BANK PERSPECTIVE

ISMAIL SERAGELDIN

Special Programs, The World Bank,
1818 H Street NW, Washington, DC 20433

WE LIVE in a time of unprecedented assault on biodiversity and natural resources at global, national, and local levels. The battle for the environment is being fought between growing populations and the need to conserve natural systems in countless arenas. Solutions are attainable, but it will require our genius, commitment, and ability to cooperate if we are to secure a future that generations to come can celebrate, instead of looking back and condemning us for opportunities lost, challenges forgone.

From the World Bank's point of view, however, that does not translate only into protection of pristine environments and conservation of a rare plant or animal, important as these might be. Rather, it is about the maintenance of life-support systems and people. It is about recognizing the need to conserve resources and manage them sustainably so that people have access to clean air, clean water, and fertile soils both now and in the future. Today, such access is denied to much of mankind.

At the global level, we face the pervasive reach of poverty, uncertainty over food security and the resource base, and increasingly diminished if not lost natural habitats and ecosystems. Biodiversity is being eroded at an unprecedented rate, and we can only guess its ultimate impact. Of the estimated 10–100 million species on the planet, only 1.4 million have been named. Fungi are the least known (only 69,000 of the 1.6 million thought to exist have been described) and we can only imagine the complexity and wealth of the estimated 8 million arthropods. However, bacteria are the " black hole" of systematics, with only some 4,000 recognized. In a recent study in Norway, 4,000–5,000 species (virtually all new to science) were indicated among the 10 billion organisms to be found in each gram of forest soil.

Humanity's call on the food base is precarious. Although staple cereal and root crops will continue to feed humanity for some time to come, the jettisoning of many useful plants will bring unnecessary costs. The decrease in the number of species used in forestry and in animal husbandry has also narrowed the genetic base, greatly reducing the options for adapting to change.

We continue to struggle in assessing the economic values of environmental assets, especially biodiversity. Methods are being developed to introduce conservation practices in the marketplace and to reduce the subsidizing of the mining of natural systems—full-cost accounting, green taxes, economic incentives for conservation, and internalization of environmental externalities. New ways are being used to measure well-being by looking at the contribution of natural human and social capital, not just human-made capital, which is usually considered in financial and economic accounts. Recent findings reinforce the importance of the natural-resource base of all economies and the fundamental role of human resources in determining a nation's wealth and, in turn, the opportunities for welfare gains for a nation's population.

It is particularly sobering to contemplate the pervasive influence of humanity on the natural environment and the threats posed to ecosystems: marine fisheries are being harvested to extinction, land transformation and water use are pressuring every ecosystem, and modified rates of nitrogen fixation and CO_2 concentrations are altering global climate. These and other human effects pose substantial threats to both sustainable development and the very quality of life.

The major causes of biodiversity loss are the fragmentation, degradation, or loss of habitats (through conversion by agriculture, infrastructure, or urbanization), overexploitation of biological resources, the introduction of nonnative species, pollution, and climate change. It is estimated that extinction rates of plants and vertebrates are some 50–100 times higher than the expected natural rate and that future extinction rates will be substantially more than 1,000 times the natural rate (Reid and others 1992). For some groups of plants and vertebrates, 5–25% of identified species are already listed as threatened with extinction. The result might induce profound changes in many ecosystems and render them much less useful to people even if not less complex ecologically.

The deforestation of tropical rain forests, the greatest cause of species extinction, is expected to continue. Some 50% of the world's species (estimated at 10–100 million) are harbored by rain forests, and the current rate of loss might exceed 50,000/year, 137/day, or 6/hour. The loss of old-growth forest remains a major concern in many temperate countries.

Sound management of the earth's precious water resources constitutes the greatest challenge to sustainable development and the conservation of freshwater biodiversity. Freshwater fish are the vertebrate group that has suffered the highest extinction rates in both tropical and temperate regions. The productivity of freshwater ecosystems and their economic benefits are well known; if not properly managed, the competing demands of water, increasing pollution, the alteration of the hydrologic cycle, and the introduction of alien species will compromise the ability of freshwater ecosystems to sustain human livelihood.

Marine biodiversity is also experiencing overexploitation, habitat loss, and pollution; indeed, overfishing is the greatest threat to marine biodiversity and ecosystems. Protection of marine biodiversity is critical because the marine environment has greater diversity at higher taxonomic levels than land—coral reefs harbor over 1 million species of plants and animals and constitute the largest untapped source of bioproducts.

Change and disturbance are essential features of ecosystems, but ecologists view the survival of complex systems as depending on connectivity and interdependence among their parts and on feedback among related processes. This focus is helping lead to partnerships and to the understanding and building of motivational structures to achieve desired ends. Thus, biodiversity conservation and management are not just ecological concerns; for many countries, they are also intrinsic to socioeconomic development, particularly for the poor. Biological resources provide the most important contributions to livelihoods and welfare: food, medicines, health, income, employment, and cultural integrity. Over 80% of the world's population depends partly on traditional medicines and medicinal plants, and some 60% of plant species (35,000) have potential medicinal value. About 7,000 compounds have been extracted from plants, leading to products as varied as aspirin and birth-control pills; the search for more has never been greater.

Of the thousands of plant species deemed edible for humans, some 20 produce the vast majority of the world's food. Staple crops—such as wheat, maize, rice, and potatoes—are used to feed more people than the next 26 crops combined. Likewise, sheep, goats, cattle, and pigs supply nearly all land-based protein for human consumption.

The same process of specialization is evident for varieties within species—humans are increasingly reliant on a narrow range of species and then on specific varieties of these species. Consequently, biodiversity conservation is equally concerned with sustaining greater varieties of specialized and nonspecialized species. To meet that challenge, two approaches are being adopted: ensuring an adequate supply of genetic diversity for such industries as agriculture and medicine, and protecting unconverted habitats for the supply of genetic diversity.

Conserving biological diversity needs to address complex issues that call for a wide range of responses across many private and public sectors. All responses are necessary, with adjustments for local conditions: in situ conservation, ex situ conservation, intellectual-property rights, indigenous knowledge, human and institutional capacity, access to technology, equitable sharing of benefits, morals and ethics, and biosafety and risk. Information on those issues is becoming more readily available, and this will help to address such central problems as limits to the flow of germplasm (particularly of processed products), the debate over intellectual-property rights, and trade rules. Basic inventory and fundamental research work should be carried out simultaneously with field action, the two forms of activity reinforcing each other.

High-yielding crop varieties produced during the "Green Revolution" helped to avert a food crisis in the 1960s. It has continued to save land, and its influence is still spreading, but a huge agenda remains. More genetically diverse new crop varieties are needed, and we need to adopt integrated pest management to minimize the use of pesticides. Likewise, on-farm water and nutrient management combined with traditional wisdom will produce efficiencies for farmers and maintain the health and productivity of agricultural systems. And the promise of and obstacles to biotechnology continue a lively debate, but we can be confident that it too will play a seminal role in securing food on a more sustainable basis, recognizing the mutual interest of the material-rich states and the biodiversity-rich states in the development and conservation of the remaining biological diversity.

The World Bank is the largest financier of targeted environmental projects, with an active portfolio of more than 170 projects at a funding level of $15 billion. Lending in biodiversity conservation itself has grown to $956 million, involving 101 projects in 56 countries. Investment has leveraged an additional $536 million from borrowing governments and donors, bringing the total commitment since 1989 to $1.34 billion. In addition to projects and project components with specific biodiversity objectives (the biodiversity portfolio), the bank has supported environmental projects that can have a favorable, although indirect, effect on biodiversity. Of these "environmental" projects, the ones aimed at improving natural-resource management ("green" projects) and those designed to strengthen environmental institutions ("institutional" projects) can help to conserve biodiversity through improved natural-resource management and development of appropriate incentives and policies.

The emphasis on sustainable economic development, the better valuation of renewable natural resources, strengthening of national institutional capacity, and improvement in project preparation and implementation will all benefit the conservation and use of biodiversity. It is clear that biodiversity will not be conserved without consideration of the broader context, but improving the management of biological resources in general will not prove sufficient. Biodiversity can and should be addressed as a distinct problem although it is related to the degradation of biological resources.

Sustainable use and biodiversity conservation also require understanding of the social and economic contexts. In the case of the rural poor, biological resources are often the most important source of economic and social well-being in the form of food supplies, medicine, shelter, income, employment, and cultural integrity. Successful biodiversity conservation also depends on sound policies and effective institutional and social arrangements.

A wide range of national policies, laws, and regulations can create "perverse" incentives that discourage conservation even as other policies are intended to provide incentives to conserve. For example, the conversion of natural areas and loss of biodiversity have often been accelerated by economic policies that encourage production for export markets, promote population resettlement, or open remote areas to road construction and logging. Policies aimed at increasing agriculture, forestry, fisheries, and energy and industrial production can have similar effects. Appropriate policies provide the basis for national development and for meeting the economic needs of people, but inappropriate policies can result in unsustainable and inefficient natural-resource use and contribute unnecessarily to the loss of biologically important natural habitats and species. Policies related to land tenure, forestry, and agriculture are particularly critical in this respect. Diverse experience has shown that the role of institutions in conservation is complex and taxing. Top-down conservation has seldom been effective except when large budgets are available for enforcement and society is willing to accept a rather undemocratic conservation process. Giving responsibility to local government and nongovernment organizations appears to create both opportunities and potential problems. To take advantage of the former while avoiding the latter, it seems that a cluster of arrangements must be made as a whole if conservation is to work well in an institutionalized setting. These arrangements include provisions for local participation, capacity-building, and incentive structures.

Decentralization can increase local responsibility for biodiversity conservation, making it more relevant and useful to local people. Reforms that a country might make affecting self-regulation, tenure, and accountability will help to ensure that people who decide how to use biological resources are directly affected by the consequences of their decisions. By shortening the feedback loop between a decision and its effect, such reforms will reward cautious decision-making. In addition, changes that give authority specifically to people living in the managed environment encourage decisions that are responsive to local conditions. If other local stakeholders are encouraged and enabled to question the decisions, responsibility will be promoted and a strong force for good governance will have been created.

The tools that can be used to conserve biodiversity—the protection of critical ecosystems (in situ measures) and such entities as arboretums, aquariums, botanical gardens and zoos, and seed and gene banks (ex situ measures)—all provide enormous benefits to humankind. Each conservation tool has its place in a comprehensive strategy for conserving biodiversity, including meeting human needs and maintaining the greatest possible numbers of species and genes.

Most national governments have established legal means for protecting habitats that are critical for conserving biological resources; the responsibility is often shared by public and private institutions. Although accomplishments have been impressive, the amount of protected habitat and ecosystems needs to be increased substantially if these areas are to ensure the long-term conservation of the world's biodiversity. However, such protected areas will succeed only if they are effectively managed and if the management of the surrounding areas is compatible with the objectives of the protected areas. That will typically mean making protected areas parts of larger regional schemes to ensure biological and social sustainability and to deliver appropriate benefits to neighboring populations.

Ex situ conservation programs supplement in situ conservation by providing for long-term storage and analysis, testing, and propagation of threatened and rare species of plants and animals and their propagules. They are especially important for wild species whose populations are severely reduced, serving as a backup to in situ conservation, as a source of material for reintroductions, and as a major repository of genetic material for future programs of breeding of domestic species. Some ex situ facilities—notably zoos and botanical gardens—offer important opportunities for public education and contribute substantially to taxonomy and field research.

Many of the current responses to the world's biotic impoverishment have been supported by international conventions that have fostered cooperation and partnerships in conserving biodiversity. These conventions, especially the Convention on Biological Diversity (CBD), represent unprecedented opportunities for the development of institutions concerned with fostering environmentally sustainable development. Posing unique intellectual challenges as it does, the CBD provides perspectives on a number of disciplines—its biological foundation is partnered by economics, sociology, and other social sciences to bring innovation and integration and to facilitate consensus-building. It will also help to define a systematic approach to encouraging investment in biodiversity.

Current approaches to sustainable development are still rudimentary. A rough-and-ready set of initiatives is in place, the development cycle is undergoing change (from a project orientation to one of listening, piloting, assessing, and mainstreaming), new partnerships are emerging, and the increasing accessibility to information is challenging the ownership of decision-making. But promising though these developments are, we must be sure of their selective and rigorous application.

Progress has always been heralded by paradigm shifts that seemed somehow difficult and dangerous, but moved the world forward into new realms of freedom and prosperity. We need to promote a paradigm shift in how we think about development—we need to think holistically, and we need to consider what is best for the common good. We need to do that for the poor and the marginalized of the world. We need to do it for the women who are carrying the burden of continuing degradation and discrimination. We need to do it for the future generations for whom we are but passing stewards of this globe.

SELECT BIBLIOGRAPHY

Brown K, Pearce D, Perrings C, Swanson T. 1993. Economics and the conservation of global biological diversity. Washington DC: Global Environment Facility, World Bank.

Kottelat M, Whitten A. 1996. Freshwater biodiversity in Asia, with special reference to fish. World Bank Tech Pap 343. Washington DC: World Bank.

Lambert J, Srivastava J, Vietmeyer N. 1997. Medicinal plants: rescuing a global heritage. World Bank Tech Pap 355. Washington DC: World Bank.

McNeely JC, Miller KR, Reid WV, Mittermeier RA, Werner TB. 1990. Conserving the world's biological diversity. In cooperation with the International Union for Conservation of Nature and Natural Resources (The World Conservation Union/IUCN), World Resources Institute, Conservation International, World Wildlife Fund-US, and the World Bank. Gland, Switzerland and Washington DC: IUCN and the World Bank.

Pagiola S, Kellenberg J, Vidaeus L, Srivastava J. 1997. Mainstreaming biodiversity in agricultural development: toward good practice. World Bank Environ Pap 15. Washington DC: World Bank.

Reid W, Barber CV, Miller KR. 1992. Global biodiversity strategy: guidelines for action to save, study, and use earth's biotic wealth sustainably and equitably. World Resources Institute, The World Conservation Union (IUCN), and the United Nations Environment Programme in consultation with the Food and Agriculture Organization of the United Nations and the United Nations Educational Scientific and Cultural Organization. Washington DC: World Resources Inst.

Rodenburg E, Tunstall D, van Bolhuis F. 1995. Environmental indicators for global cooperation. GEF Work Pap 11. Washington DC: World Bank.

Serageldin I. 1996. Sustainability and the wealth of nations: first steps in an ongoing journey. Envir Sust Devel Stud Monogr Ser 5. Washington DC: World Bank.

Srivastava JP, Smith NJH, Forno DA. 1997. Biodiversity and agricultural intensification: partners for development and conservation. Envir Sust Devel Stud Monogr Ser 11. Washington DC: World Bank.

Thrupp LA. 1998. Cultivating diversity: agrobiodiversity and food. Washington DC: World Resources Inst.

World Bank. 1998. Biodiversity in World Bank projects: A portfolio review. Washington DC: World Bank.

CREATING CULTURAL DIVERSITY:
TROPICAL FORESTS TRANSFORMED

OLGA F. LINARES

Smithsonian Tropical Research Institute,
Apartado 2072, Balboa, Ancón, Rep. de Panamá

TODAY, there is "an increasing realization that cultural [diversity] and biological diversity are intimately and inextricably linked" (McNeely and others 1995:767). The enormous variety that underlies the structures, beliefs, knowledge, and cultural practices of peoples around the world is a unique and valuable reservoir of environmental knowledge and know-how. During millennia of careful observation and experimentation, human groups have developed different uses for the plants and animals that make up the diverse ecosystems of the world. Distinct cultural patterns have emerged, have become specialized, and ultimately have changed in response to coevolution, coexistence, and mutual transformation along a nature-culture continuum. These cultural lifeways are increasingly threatened, as are the biological systems that support them.

This essay explores the many ways in which indigenous peoples relate to each other and to components of the ecosystems in which they play an essential role.[1]

---

[1] Indigenous peoples are members of communities that to a large extent follow their own cultural rules and their own social and economic practices and often also elect their own local leaders. *Indigenous* has been applied mainly to small-scale societies and often to New World (Amerindian) groups. The term seems to me less apt when applied, for example, to such rural farmers of Africa and Malaysia as the Ibo shifting cultivators of Nigeria or the Batak agroforesters of north Sumatra. Both those peoples are numerous and form part of large, semiautonomous political entities (they form nations within modern states). Moreover, *indigenous* is inapplicable to temporary or permanent migrants. Because other alternatives, such as *native* and *tribal* are even more inappropriate, I will be using the term *indigenous* here to refer in general to relatively autonomous tropical peoples. When possible, either the name by which certain groups are known in the literature or, even better, the name that the people themselves use (their self-definitional label) should be used.

Small-scale communities differ within themselves and between each other along several dimensions—linguistic, ideological, social, and political—and in subsistence pursuits and modes of insertion into the modern global economy. I emphasize two interrelated aspects: human ecological and economic behavior, especially with respect to physical resources, and cultural constructs, which are the beliefs and attitudes that people have toward their natural surroundings. A truism worth repeating is that the mental and material systems humans have devised to survive and reproduce are, simultaneously, responses to the environment and ways of shaping its biological diversity for human use. Thus, scholars justifiably argue that nature and culture are indivisible and that the real subject matter of human ecology should be the analysis of socionatural systems (Bennett 1996). Doubtless, they are correct; all human existence presupposes a degree of ecological involvement. To facilitate making empirical generalizations and forging comparisons, I will focus this discussion on tropical areas of the New World, Africa, and Asia. These regions have particularly high rates of biological diversity and are inhabited by diverse rural peoples who have devised highly specialized and low-energy, as well as high-energy, adaptations to multiple resources.

THE TROPICS: BIOLOGICAL DIVERSITY

Tropical forests are among the most complex and diverse of terrestrial ecosystems, having the greatest number of dynamically interacting plant and animal species (Whitmore 1992). High temperatures, abundant rainfall, and fragile soils are their general characteristics, but tropical forests differ greatly from one another in terms of their composition, dynamics, and size. Commonly, a distinction is made between climax or mature rain forests that have ever-wet environments and monsoon or seasonal secondary forests that have a marked dry period, but this is an oversimplification. In reality, all tropical forests are dynamic, subject to constant processes of natural disturbance caused by a series of biotic factors (such as insects and vertebrates), abiotic factors (for example, tree falls, landslides, storms, and droughts), and anthropogenic disturbances (usually repeated and prolonged) (Denslow 1996).

About half the world's tropical areas are in the American Neotropics, including southern Mexico to Panama, the Amazon and Orinoco basins in northwestern South America, and central and coastal Brazil. Next in extent are the eastern tropics of the Indo-Malayan region, including Indonesia and continental Southeast Asia.[2] The smallest block of tropical rain forest is in western and central Africa, including the Congo Basin.

---

[2] The Indo-Malayan rain forest (also called the eastern rain forest) covers the western Ghats in India and the southwestern corner of Sri Lanka. It is centered on the Malay archipelago, in the phytogeographic region that botanists call Malesia (or Malaysia). The term includes peninsular Thailand, the Bismarck archipelago, and the northwestern corner of New Guinea (Irian Jaya). Furthermore, the Indo-Malayan rain forest extends beyond the Malay peninsula into Burma, Indochina, southern China, and Vietnam. Rain forest also covers Indonesia, most of New Guinea, and Borneo (Kalimantan). See Whitmore (1992 10, [figure 2.1 11, 213 [glossary], 223 [index]).

These large tropical regions are by no means homogeneous. For instance, they differ in species diversity, that is, in the number of plant species present. Of the 170,000 species of flowering plants in the tropics, for example, about half are in the Neotropics, 35,000 are in tropical Africa, and 40,000 are in tropical Asia; the rest are found in Madagascar and Malesia (Whitmore 1992:28). In all of Africa, there are only about 15 genera and 50 species of palms (Whitmore 1992:28), compared with 71 genera and about 800 species in the New World (Henderson 1990:2–3).[3] In the Indo-Malayan and Australasian regions, there are at least 93 genera of palms (Uhl and Dransfield 1987:550–3) and over 1,000 species. Although those three regions share plant families, they have few plant genera and even fewer species in common. The distribution of some kinds of vertebrates is similar. For example, 1,300 avian species exist in the Neotropics, 900 in the Asian tropics, and 400 in the African tropics (Myers 1992). With the exception of primates, fewer mammal species exist in Africa than in the other two regions.

Despite their diversity, some forests throughout the tropics are monospecific—that is, they are dominated by a single species of canopy trees—and can occur next to mixed-species, old-growth forests (Hart and others 1989; Hart 1990). Thus, it is possible for fruiting trees, including trees that might yield fruit edible for humans, to occur in large stands throughout particular forests. This phenomenon might have enabled forest peoples to survive in tropical regions during pre-agricultural times.

RAIN-FOREST ENVIRONMENTS AND "PURE" HUNTER-GATHERERS

During the last decade, scholars have vigorously debated the problem of whether foraging peoples (that is, hunter-gatherers) could have lived in mature forests without cultivating plants or domesticating animals or could have lived independently of their agricultural neighbors, with whom they exchanged forest products for food crops (Harlan 1995; Hladik and Dounias 1993; Piperno and Pearsall 1998:76–8). Those who argue against the possibility of foragers living independently portray the tropical rain forest as having limited food resources, especially wild starches and animals with adequate fat reserves (Hart and Hart 1986; Headland 1987; Bailey and others 1989). The Mbuti Pygmies of the Ituri Forest in Zaire (now the Democratic Republic of the Congo) are used as an example of how these constraints operate. It is said that in the Ituri Forest, important food-plant species, including yams, are rare, sparsely distributed, or seasonal—as are other products, such as honey, grubs, and caterpillars—and that the mammals hunted are lean most of the year. In fact, however, large areas of the Ituri Forest are monospecific, dominated by the edible species *Gilbertiodendron dewevrei*, which

---

[3] Genera and species numbers are constantly being revised and should be accepted with caution. For example, Corner (1966, 230, table 2) lists 255 genera and 2,009 species of New World palms; 24 years later, Henderson (1990:2–3) reduced these numbers drastically. Here, I have used Henderson's estimates.

produces sizable, starchy seeds. Although this tree yields for only 3 months of the year (October-December), other foods are abundant at other periods, including honey (May and June), termites (November), and fat animals (the dry season). Thus, seasonality might be important, but the seasons of abundant food tend to be staggered. Moreover, recent studies of the standing biomass of wild yams in the forests used by Pygmies have revealed year-round availability and considerable density (3–6 kg/hectare), certainly high enough for sparse populations to survive (Hladik and Dounias 1993). In fact, it has been estimated that even at a very low yam density of 2 kg/hectare, one Aka Pygmy camp of 26 persons foraging in a 2-km radius could feed itself on yams alone for 6 months; of course, other wild plant and animal foods are also available (Bahuchet and others 1991).

Other forest foragers, in Malaysia and the Philippines, also consume large amounts of wild yams. Some Pygmy groups, such as the Baka, encourage the regeneration of wild yams by carefully reburying the heads of the plants after harvesting them, a management technique that could have been used for other wild-forest resources as well (Dounias 1993). Other plant species in other continents also are conserved; for example, palms are protected and tended in Amazonia, and the sago palm is carefully pruned by some foragers in Southeast Asia.

Recent excavations of 10 archaeological sites in the Ituri Forest of the northeastern Congo Basin confirmed that hunter-gatherers who exploited wild vegetable-oil resources were living in the African tropical rain forest in the 11th millennium BC, during Pleistocene times (Mercader 1997).

Today, hunter-gatherers might not consume as much wild food, simply because it is less work to obtain cultivated foods through trade. Thus, Efe Pygmy hunter-gatherers and Lese farmers in the Ituri Forest have become economically interdependent. The Efe exchange wild meat, honey, medicines, and their labor for crops that the Lese raise—such as manioc, plantains, rice, and peanuts—and for such items as metal tools, cloth, and pots (Wilkie 1988). Despite this symbiosis, the Efe have maintained their separate ethnic identity and cultural ways, if not their language. They have had a limited effect on the forest, scarcely more permanent or disruptive of ecosystem functioning than are the natural processes of forest disturbance.

The same might not be true of all hunter-gatherers. The aborigines that once inhabited Kangaroo Island in Australia, for example, had a marked impact on the forest. Well before the Europeans arrived, they had transformed naturally occurring thickets on the mainland into open woodlands (Harlan 1995). Elsewhere in Australia, native peoples flooded forests and built ditches to increase the abundance of wild plants and fish. They also dug up yams of the genus *Dioscorea* so intensively that the churned-up fields "resembled plowed fields" (Harlan 1995:11). Other "advanced" hunter-gatherers planted seeds, fertilized with ashes, settled in villages, and reached high population densities without domesticating either plants or animals. Only when and if their manipulative practices involved the deliberate selection and enhancement of useful traits in plant populations—altering their genetic makeup—can we talk of plant domestication. The initial stages of this process are known as horticulture, and its later stages as fully developed agriculture.

ANCIENT AGRICULTURAL DEVELOPMENTS

Continental differences in the availability of plants and animals suitable for domestication presumably led to very different early patterns of food procurement and production in different regions of the world (Diamond 1997). Actually, a very small number of plant species provide the bulk of the food consumed by the world's population. Common opinion has it that none of the major crops originated in tropical forests—that most economically important plants came from areas with low species diversity, such as the Middle East (or southwestern Asia), Eurasia, Mesoamerica, the Andes, and North Africa. Grains belonging to the grass family (for example, wheat, maize, rice, millet, and sorghum) and legumes (for example, beans, peas, peanuts, or groundnuts) all were cultivated first in independent centers of domestication that had marked dry and wet periods. Moreover, the five major big herbivores that were domesticated in the Old World had the same original distributions as the staple plants: sheep and goats from western Asia, and cattle, pigs, and horses from Eurasia and North Africa.

Recent evidence suggests, however, that the tropics did not lag behind other centers of early plant domestication (Friedberg 1996). In the highlands of New Guinea, native species of taro, bananas, and yams might have been domesticated by 9,000 years ago (Bayliss-Smith 1966:507–8; Golson 1997). In the tropics of southwestern Ecuador, Colombia, and Panama, squash (*Cucurbita*), maize, bottle gourd, avocado, and lerén (Sp.), or *Calathea*, a minor root or tuber crop, and perhaps also *Maranta* (arrowroot) were being planted by horticulturists between 10,000 and 7,000 years ago (Piperno and Pearsall 1998:182–227). In the southern Guianas, northeastern Brazil, and the Orinoco Venezuelan region, indirect evidence suggests that manioc and sweet potatoes also might have been cultivated early (Piperno and Pearsall 1998:230–2 table 4.5). Thus, in those tropical areas and perhaps in other regions, such as tropical Africa, that are less well known archaeologically, food production could have been more precocious than was thought previously.

In any case, complex centralized states and stratified chiefdoms eventually also flourished in those tropical areas. Some examples are the ancient Maya civilization of Mesoamerica, the chiefdoms of Central America and northwestern South America, the Mon Khmer states of Cambodia, the African forest kingdoms of Ghana and Nigeria, and the Polynesian chiefdoms of Hawaii and Tahiti. When Europeans reached the Old World and New Old World tropics, they encountered an amazing array of diverse peoples whose cultural accomplishments rivaled those of nontropical indigenous groups. Many of those cultures did not survive the diseases and destruction wrought on them by the newcomers. Thus, present conditions reflect poorly the great diversity that once existed among tropical-forest societies.

PEOPLE OF THE TROPICAL FORESTS

As I already indicated, the biological diversity of tropical forests is not the same around the world. Has that affected the number of societies that various forest ecosystems could support? A crude estimate, based on the number of ethnographically described groups that inhabit or in the recent past inhabited the main tropical areas of the world, reveals that it has not. Tropical America and tropical Africa are home to 434 and 445 ethnic groups, respectively; Southeast Asia and the Pacific (excluding Australia) harbor an additional 539 groups (Price 1990). Although humans have developed broadly similar types of ecological and economic adaptations in all three regions, a great deal of cultural diversity occurs at the local level. For example, the members of 46 households in a single Amazonian village (Santa Rosa, on the Ucayali River of Peru) practice 12 distinct types of agriculture and employ 39 strategies of resource use that they constantly modify over the short term (Padoch and de Jong 1992). In all these cases, cultural diversity can occur at inter-ethnic as well as intra-ethnic levels.

Hunter-Gatherers

Let us return to the so-called Pygmies of central Africa. At least 10 ethnolinguistically distinct populations of these foragers are found unevenly distributed throughout the Congo Basin and adjacent areas. They differ markedly in subsistence and settlement patterns, as they do in other cultural aspects (Hewlett 1996). For example, the Efe hunt with bows, the Mbuti and Aka with nets, and the Baka with spears. Among the net hunters, female Mbuti participate in the hunt, but female Aka do not. Whereas the Efe and the Baka spend 4–5 months a year in the forest and camp close to villages, the Mbuti and Aka spend as long as 8 months in the forest, camp far away, and eat less food from the village. Although all these groups rely to some extent on cultivated foods today, they still make extensive use of diverse forest resources. Thus, the Mbuti of the Ituri Forest use more than 100 species of plants and over 200 species of animals for food, even though a much smaller number of species provide the bulk of their diet. In fact, the four distinct groups of Mbuti foragers have different cultural preferences for different foods (Ichikawa 1993). Research on these African foragers therefore suggests that as much cultural variability exists within the same ethnic group as between groups: the locus of diversity is not only cross-cultural but also intracultural. This diversity is only loosely related to the specific resources at hand. It also must be explained with reference to particular historical experiences that have shaped social processes, such as the systems of belief, the technologies used, and the division of labor by gender.

Turning now to Southeast Asia, a few distinct groups of hunter-gatherers remain in the tropical forests of Malaysia, Thailand, the Philippines, Sumatra, and Borneo. On the island of Borneo, for example, live the Penan, who hunt wild boar and other animals and collect a wide variety of plants for food, especially the starch of the sago palm, which they prune regularly; they also use plants for shelter and craft materials (Hutterer 1998). Like most Southeast Asian and African foragers, the Penan rely on exchanges with their agricultural neighbors, trading mats and baskets for rice. Like many other foragers, they are struggling to save their forests. Other Southeast Asian groups exchange wild meat, resins, beeswax, medicinal plants, and other forest products with agriculturalists, even though they have the resources and know-how to survive solely on wild food species. Hence, their adaptations to the forest are not only highly variable, but also different from those of their pre-agricultural ancestors thousands of years ago.

Amazonia is a third large tropical area where very diverse groups of humans still rely heavily on wild forest products that are hunted, gathered, and fished, in addition to the practice of slash-and-burn cultivation. Their knowledge of the forest is vast and accurate. It is believed that the Yanomami of Venezuela could have subsisted on wild products alone, provided that they remained numerically small, mobile, and able to exploit the diversity of microenvironments in their habitat (Good 1995). The Yuqui of lowland Bolivia might have remained true foragers until relatively recently, exploiting the patchwork dynamics of the forest through constant mobility, overlapping sexual roles, and active sharing of information (Stearman 1995). Their fine-tuned knowledge of the fruiting phenology of plants and the feeding behavior of animals sees them though periods of resource scarcity. Their Tupi-Guaraní relatives, the neighboring Siriono, were also primarily a trekking society before they became sedentary in the 1940s and 1950s. Although wild game is still very important in their diet, the Siriono make their camps on artificial mounds that were built up by previous horticulturalists in the midst of seasonally flooded lowlands (*bajuras* Sp.; Holmberg 1960). On these abandoned mounds, the Siriono gather the semidomesticated palms and fruit trees that had been planted by their predecessors (Balée 1995). Hence, like most foraging groups today, the Siriono no longer rely entirely on wild products from the forest.

Swidden or Slash-and-Burn Agriculture

Despite enormous variation, swidden cultivation (the temporary clearing of forested land to grow crops)—also known as shifting cultivation, long-term fallow cultivation, and so on—still prevails in parts of the tropics around the world where population densities are relatively low and land is available for rotational forms of agriculture. The system is essentially the same everywhere it is practiced. During the dry season, the forest (usually secondary) is cleared, and trees are felled. Then the vegetation is burned just before the rains begin, and various plants are planted on the ashes in a manner that generally imitates the wild vegetation they replace (Harris 1972). In fact, "by substituting a diverse assemblage of cultivated plants for the wild species of the forest this type of polycultural conuco stimulates much more closely than monocultural plots do the structure and dynamics of the natural forest ecosystem" (Harris 1971:481). The same parcel may be cultivated for 2–3 years and then lie fallow for 5–20 years to restore its fertility. Not only soil depletion, but also weed growth and insect pests can force farmers to clear new land. The need for fallow periods requires that large tracts of land be held in reserve. In terms of labor input per unit time, however, swidden cultivation is often more productive than more labor-intensive methods of permanent cropping.

Those general statements aside, it is important to emphasize that people use a great array of planting techniques, crop combinations, and rotational practices in their swidden systems, even within the same general area. In West Africa, for example, the Sakata of the Democratic Republic of the Congo clear and burn plots in the forest, sparing such economically useful trees as the kola and the oil palm (Grove and Klein1979). They then plant manioc or cassava, maize, plantains, and bananas on mounds and abandon each plot after 3 years of use. They also grow vegetables, sweet potatoes, and other crops in gardens near their dwellings. In contrast, the Zande, who also live in the Congo, plant groundnuts and maize, finger millet, sorghum, and other minor crops. Although they might plant manioc in the third year, they are more dependent on grain crops than the Sakata are. It is difficult to find swidden farmers anywhere in tropical Africa who do not grow commercial crops as well, on a more permanent basis, such as oil palms, cocoa, coffee, and tea in the highlands. Growing single crops (monocrops) for the export market can increase the number of diverse groups that can live in a given region, but it can also reduce considerably the diversity of crops grown for subsistence purposes.

Swidden cultivation is still practiced widely in some tropical areas of the Indo-Malayan region (Aubaile-Sallenave 1997; Spencer 1966). Well known among swidden agriculturists are the Hanunóo of Mindoro in the Philippines, who grow or grew at least 430 cultivars, 40 or more of which can be planted in the same swidden plot, in parcels that they cultivated for 3 years and allowed to lie fallow for 8 years (Conklin 1957). Other, less well-known groups, like the Gidra of Papua New Guinea, are also swidden cultivators. The Gidra who live in inland villages rely on starch from the sago palm and meat from wild animals, whereas those who live in riverine villages rely more on garden crops and fishing (Ohtsuka 1996). Until recently, the former adaptation was the more successful of the two, but the sale of garden crops and the adoption of modern fishing technologies has conferred advantages to the riverine adaptation; this is another example of how intraethnic diversity can be created by outside influences.

The Kuikuru of central Brazil not only can name 191 trees, but also display an intimate knowledge of the multiple uses of 138 of them—many of them palms—including their role in feeding the animals that they hunt (Carneiro 1988). The Kuikuru cultivate 11 varieties of bitter manioc plus maize and several other food crops in swidden plots that average 0.61 hectare (1.5 acres), which they carefully plant and weed for 3 years, then abandon for as long as 25 years (Carneiro 1961). Kuikuru gardens produce 4–5 tons of manioc tubers per acre per year. Enough forest is available for clearing within walking distance of any village for settlements to be permanent. When the 150 or so inhabitants of a village change location, it is not for ecological reasons but for internal social pressures, most often disputes. Thus, as long as the population remains relatively small and the forest large, the swidden systems of the Kuikuru and some other Amazonian groups do not necessarily destroy the natural vegetation, even though they inevitably alter the species composition of the forest.

That does not mean, however, that all Amerindian tropical groups were equally well adapted to their environment. The Trumai, who lived along the Upper Xingu River in Brazil, not far from the Kuikuru, were much less successful. When they were first contacted by Europeans in the late 1930s, the total population was only 43 (25 in 1955 and possibly none today), and their tiny manioc gardens, of less than 0.2 hectare (0.5 acre) each, were barely large enough to feed their households (Murphy and Quain 1955). Constantly under attack from their neighbors, and lacking leadership and strong kin ties, their community was breaking down despite their not having had face-to-face contact with Western society. Although introduced diseases had reduced their numbers, the recentness of their move into the region from the Southeast was the principal factor in their demise. Ecological conditions could not have differed greatly between the two regions, but the Trumai had not yet formed alliances through marriage and political ties with neighboring groups that would have permitted them to live peacefully in this ethnically diverse area. Hence, the particular social history of an individual group, including its relations with its neighbors, and not only environmental constraints or direct contact with nonindigenous peoples, can contribute substantially to the shape of its future.

INDIGENOUS FORMS OF AGRICULTURAL INTENSIFICATION

Agricultural production can be increased by applying ever larger amounts of labor to improving small parcels that are cropped permanently, rather than by enlarging the amount of land that is cultivated. Many agricultural peoples in the tropics, including swidden farmers, also practice some form of more intensive permanent cultivation. Frequently, they make small, permanent house gardens (also called home or dooryard gardens), in which they plant a diversified mixture of trees, vines, bushes, grains, root crops, medicinal plants, and spices, and fertilize them with kitchen debris and animal dung. House gardens play an important ecological and economic role. For example, among the Ibo of eastern Nigeria, who are short of land, a compound garden is a diverse plant community that can include 60 species, including tubers, vegetables, maize, small and large trees, and palms (Ruthenberg 1976). Although it occupies scarcely 2% of the land farmed by the Ibo household, the garden produces half the crops consumed. In Java, the house garden is also a complex and dynamic ecosystem made up of tuberous plants at ground level, bushes and small trees (such as papaya and banana) at the middle levels, and tall fruit trees at the upper level. A closed canopy helps to control weeds and lessens erosion, and the decaying vegetation produces fertilizer, imitating natural-forest dynamics and causing minimal environmental degradation. Here, anywhere from 15 to 75% of the land may be dedicated to gardens that provide more than 40% of the caloric requirements of the household and more than 20% of its monetary income (Stoler 1978). On the other extreme are the groups of rural Jola in southern Senegal who do not make house gardens at all or, if they live near towns, grow introduced, foreign vegetables, mostly for sale in the market rather than for household use.

Agroforestry is a variant of the house-garden option; by incorporating trees into the agricultural landscape, it also reproduces the structure and dynamics of the natural forest. In Indonesia, for example, fruit trees of local forest species often are cultivated, as are bamboo, useful fibers, and so forth, all of which are only slightly modified genetically (Michon and Bompard 1987). Indeed, many of the species in these special forests are protected and tended but not necessarily planted deliberately. These systems often surround the village, linking the agrarian landscape with the natural forest. In southern Sumatra, agroforestry accounts for more than half the territory and is based on multiple species of trees—in some areas more than 300. This culturally created forest might have a greater biomass than the "natural" forest has (400–700 trees/hectare, compared with 500 trees/hectare, respectively), and they are similar in density and structure. The wood, resins, fibers, and so forth from these trees bring monetary revenue, and the fruits are used as complementary food.

In central Sumatra, planting trees, or favoring their spontaneous regeneration, legitimizes a farmer's rights to productive land. Customary rules dictate that "the land near the lake belongs to those who make it fruitful" (Aumeeruddy 1994:23). If productive trees—including commercial timber species, coffee, and cinnamon—are not grown in these highly diverse agroforest gardens, the land will be taken back by local community authorities, namely the customary chiefs, and assigned to others. In pioneer fronts, however, collective control of scarce resources is weak or nonexistent. Wealthy farmers are cutting down the forest to plant profitable cash crops. Monocrops are now being grown on hillside areas where a wise and carefully managed complex agroforestry system would have prevented the rampant erosion that is now menacing the fragile soils. This is one more example of how profit can compromise future productivity.

Among the Chiripa, a Guaraní people living in Paraguay, agroforestry has taken a different course than in Indonesia. The Chiripa integrate subsistence gardening and hunting with commercial harvesting from the forest of *yerba mate* leaves, which are used to make a kind of tea that is drunk widely by Paraguayans. "Rather than simply harvesting foliage, however, *yerbateros* have developed techniques that protect the standing trees and promote the growth of new ones" (Reed 1995:27). Stages in the production cycle—first gardens, then fallows that are still managed for food, and finally trees—replicate the natural succession of tropical ecosystems. Unlike their nonindigenous, *mestizo* neighbors who work as hired labor in such enterprises as logging under a coercive patronage system, Chiripa *yerbateros* belong to independent communities that comprise nuclear families integrated through bilateral kin ties and affiliation to elderly religious leaders. These enduring social institutions are not simply defined by productive relations; they are the principal explanation of why and how the Chiripa have survived as an ethnic group. Thus, agroforestry can allow indigenous peoples to participate in the national economy without irreversible damage to the environment, provided that they have the right social institutions in place.

In some societies, the entire farming system can rely on intensive techniques. The Kofyar of the Jos Plateau in Nigeria enhanced the natural productivity of the soil by making permanent, terraced homestead fields where crops are intermixed and heavily fertilized with dung from corralled goats. With no more than simple tools, small, independent Kofyar households can grow most of the family food (Netting 1968). The Jola of Senegal and other peoples living on the swampy coastal lands of the Upper Guinea coast cultivate wet (irrigated) rice in permanent diked paddy fields that are annually transplanted on with a single crop (Linares 1981). An individual family owns parcels in all the important sections of the rice fields, improving soil fertility by careful tilling and controlling water quality by diking and draining. Through centuries of careful experimentation, the Jola have developed multiple varieties (landraces) of the West African rice species *Oryza glaberrima*, to which they have added introduced varieties of the Asian species *Oryza sativa*, thus staggering harvest time and other labor requirements and spreading the risk of failure if precipitation is insufficient. As among the Kofyar, land among the Jola is privately owned but inalienable, and production is organized on the basis of independent nuclear households that exchange labor rather than extended households or larger lineages (Linares 1992). These social organizational features can, in fact, be shared by other intensive farmers elsewhere.

The rice economies of East and Southeast Asia vary in the intensity with which they use land, labor, and capital, but most farmers grow at least two crops a year, using household or family labor, exchange organizations, and irrigation societies (Bray 1986). Rice cultivation is used as the basis of economic diversification into commercial cropping and manufacturing. The combination of rice, fish, and silk production creates an enriched, diversified ecosystem capable of sustaining very high population densities (as in Java with 600 persons/km^2).

CONCLUSIONS

The examples discussed above suggest that most tropical forests have been inhabited by humans for a long time; to one degree or another, these forests are anthropogenic, having been transformed through human agency. Everywhere, indigenous groups have developed diverse and ingenious techniques to incorporate the biological diversity inherent in tropical forests into cultural patterns of resource use. Regardless of whether they are hunter-gatherers or intensive agriculturalists, some of their practices have had little effect on the environment and others have greatly modified it. In all instances, the particular lifeways that have emerged are a product of historical processes of cumulative social change and continuing adaptation. Culture is not in any simple way determined by nature, but rural economies are doubtless forged in the mutual interaction of humans with the diverse ecosystems that they occupy. In the process of engaging nature, indigenous farmers have created hundreds of varieties of cultivated plants (landraces), thus increasing food security through plant genetic diversity that confers resistance to pests, pathogens, and adverse climatic conditions. That is only one of the many ways in which local peoples have actually increased diversity.

Clearly, then, indigenous peoples have the capacity to transform tropical rainforest environments without destroying their biodiversity. "Cultural knowledge leads to different land-management practices that increase biological diversity—protection of sacred forests, building and maintaining hedgerows, planting a diversity of crops and varieties, and protecting plants in the forest" (Brush 1996:2–3). Such practices are generally sustainable as long as population numbers are kept down and land continues to be plentiful or as long as access by densely settled peoples to scarce resources, such as fertile soils, is carefully managed for the common good. Even under ideal conditions, however, examples of tropical forest peoples who misuse resources can be cited: in the Amazon, they overexploit game and fish populations; in Northern Luzon, they deforest (Lawless 1978); in the

Upper Guinea coast, during years of drought, they cut down and burn palm groves to grow rice (Beye and Eychenne 1990). But those instances do not add up to a worldwide systematic and massive assault by native peoples on their resource base. Before the 1950s, deforestation rates in the world's low-latitude tropical forests were "of negligible proportions" (Jones and Hollier 1997:317). Since then, conditions have changed; the world's tropical forests and the people living within them are increasingly under threat from overpopulation; from land-hungry peasants, unskilled migrants, loggers, miners, and cattle ranchers; from government projects to build roads and dams; and from commercial plantations and crop monocultures. Cultural diversity is being reduced even faster than biological diversity. Within the next century, 90% of the world's languages might disappear (Krauss 1992; Maffi 1998). It is estimated that in Brazil alone there were 230 indigenous cultures in 1900 but only 87 in 1957 (Sponsel 1995). With the loss of lives goes the loss of cultural knowledge about the forest and the myriad beneficial uses to which people can put its plants and animals as food, medicines, dyes, fibers, industrial materials, and so forth. But the forest itself is also disappearing fast. Close to 3 million hectares of Amazon forest are being cut down every year. Between 1990 and 2020, tropical deforestation might wipe out 5–15% of the world's 10 million species of plants and animals, or a yearly loss of 15,000–50,000 species (Reid and Miller 1989:37). If forest alterations by logging and surface fires are taken into account, however, the present rate of annual deforestation in Brazil's Amazonian region may be underestimated by a factor of 35–50% (Nepstad and others 1999). Doubtless, we are facing cultural and biological extinction rates of unprecedented magnitude.

There are no easy, blanket solutions to halt this destruction, for it is rooted in intractable socioeconomic problems having to do with overpopulation, poverty, neglect, exploitation, and commercial greed. Added to these is the precarious and ambiguous juridical status in which indigenous groups in the African, Asian, and American tropics find themselves (Grenand 1993). What seems evident, however, is that whatever diversity is inherent in tropical forests can be protected only by using diverse means and methods, to be applied alone or in combination, often case by case, with the full participation and empowerment of the local populations affected. In most instances, farmers must be compensated for safeguarding, in situ, crop genetic diversity (Orlove and Brush 1996; Wilkes 1991). In other instances, new forms of gaining a livelihood must be found for people living in protected areas; and educational opportunities must be extended to them (Redford and Mansour 1996). In many cases, local populations must be granted secure land rights before they are willing to conserve their patrimony. Intellectual-property protection and prospecting contracts for indigenous communities might work in some cases (Brush and Stabinsky 1996; Greaves 1994). Nonetheless, caution should be exercised in this connection (Cleveland and Murray 1997). Governments in the developing world must enforce legislation that respects the rights of poor rural peoples and should offer them new incentives to develop productive and profitable farms. The increased use of such arrangements as debts for nature swaps and restrictive measures—such as prohibiting the exploitation of particular trees for timber, imposing selective tariffs, taxing extractivist activities, and outlawing illicit mining—should also help to ensure a wise use of the forest. And among the citizens of the industrial world, a less rapacious attitude toward the resources of their southern neighbors should be encouraged. In every instance, the general rule should be "to put more faith in the rural population, the people whose way of life depends on how well they manage their biological resources" (McNeely and Ness 1996, p 64). Everywhere, but crucially in the world's tropical forests, fulfilling cultural needs and conserving biodiversity must proceed hand in hand.

REFERENCES

Aubaile-Sallenave F. 1997. L'Asie du sud-est: introduction régionale. In: Joiris DV and de Laveyele D (eds). Les peuples des forêts tropicales: systèmes traditionnels et développment rural en Afrique équatoriale, grande Amazonie et Asie du sud-est. 70–175. Civilisations (special issue) XLIV(1–2).

Aumeeruddy Y. 1994. Local representations and management of agroforests on the periphery of Kerinci Seblat National Park, Sumatra, Indonesia. People and plants working paper No 3. Paris France: UNESCO.

Bahuchet S, McKey D, de Garine I. 1991. Wild yams revisited: is independence from agriculture possible for rain forest hunter-gatherers? Hum Ecol 19(2):213–43.

Bailey RC, Head G, Jenike M, Owen B, Rechtman R, Zechenter E. 1989. Hunting and gathering in tropical rain forest: is it possible? Amer Anthropol 91(1):59–82.

Balée W. 1995. Historical ecology of Amazonia. In: Sponsel LE (ed). Indigenous peoples and the future of Amazonia: an ecological anthropology of an endangered world. Tucson AZ: Univ of Arizona Pr. p 97–110.

Bayliss-Smith T. 1996. People-plant interactions in the New Guinea highlands: agricultural heartland or horticultural backwater? In: Harris DR (ed). The origins and spread of agriculture and pastoralism in Eurasia. Washington DC: Smithsonian Inst Pr. p 499–523.

Bennett JW. 1996. Human ecology as human behavior. London UK: Transaction Publ.

Beye M, Eychenne D. 1990. La palmeraie de Casamance; quel avenir?...les paysans parlent. Série études et recherches no 105. Dakar, Senegal: Enda.

Bray F. 1986. The rice economies: technology and development in Asian societies. Oxford UK: Basil Blackwell Ltd.

Brush SB. 1996. Whose knowledge, whose genes, whose rights? In: Brush SB, Stabinsky D (eds). Valuing local knowledge: indigenous people and intellectual property rights. Washington DC: Island Pr. p 1–21.

Brush SB, Stabinsky D (eds). 1996. Valuing local knowledge: indigenous people and intellectual property rights. Washington DC: Island Pr.

Carneiro RL. 1961. Slash-and-burn cultivation among the Kuikuru and its implications for cultural development in the Amazon basin. In: Wilbert J (ed). The evolution of horticultural systems in native South America: causes and consequences. Caracas Venezuela: Editorial Sucre p 47–67.

Carneiro RL. 1988. Indians of the Amazonian forest. In: Denslow JS and Padoch C. (eds). People of the tropical rain forest. Berkeley CA: Univ California Pr. p 73–86.

Cleveland DA, Murray SC. 1997. The world's crop genetic resources and the rights of indigenous farmers. Curr Anthropol 38(4):477–515.

Conklin HC. 1957. Hanunóo Agriculture. A report on an integral system of shifting cultivation in the Philippines. FAO forestry development paper No 12. Rome Italy: FAO.

Corner EJH. 1966. The natural history of palms. Berkeley CA: Univ California Pr., p 230, table 2.

Denslow JS. 1996. Functional group diversity and responses to disturbance. In: Orians GH, Dirzo R, Cushman JH. Biodiversity and ecosystem process in tropical forests. Berlin Germany: Springer-Verlag. p 127–51.

Diamond J. 1997. Guns, germs, and steel: the fates of human societies. New York NY: WW Norton.

Dounias E. 1993. Perception and use of wild yams by the Baka hunter-gatherers in south Cameroon. In: Hladik CM, Hladik A, Linares OF, Pagezy H, Semple A, Hadley M. Tropical forests, people and food: biocultural interactions and applications to development. Man in the Biosphere Series, Vol 13. Carnforth UK: Parthenon Publ Grp. p 621–32.

Friedberg C. 1996. Forêts tropicales et populations forestières: quelques repères. Nat-Sci-Soc (4)2:155–67.

Golson J. 1997. From horticulture to agriculture in the New Guinea highlands. A case study of people and their environments. In: Kirch PV and Hunt TL (eds). Historical ecology in the Pacific Islands: prehistoric environmental and landscape change. New Haven CT: Yale Univ Pr. p 39–50.

Good K. 1995. Yanomami of Venezuela: foragers or farmers—which came first? In: Sponsel LE (ed). Indigenous peoples and the future of Amazonia: an ecological anthropology of an endangered world. Tucson AZ: Univ Arizona Pr. p 113–20.

Greaves T (ed). 1994. Intellectual property rights for indigenous peoples, a sourcebook. Oklahoma City OK: Society for Applied Anthropology.

Grenand P. 1997. Situation des peuples indigènes des forêts tropicales: systèmes traditionnels et développement rural en Afrique équatoriale, grande Amazonie et Asie sud-est. Civilisations (special issue) XLIV(1–2) 32–5.

Grove AT, Klein FMG. 1979. Rural Africa. Cambridge UK: Cambridge Univ Pr.

Hames R. 1991. Wildlife conservation in tribal societies. In: Oldfield ML, Alcorn JB (eds). Biodiversity: culture, conservation, and ecodevelopment. Boulder CO: Westview Pr. p 172–99.

Harlan JR. 1995. The living fields: our agricultural heritage. Cambridge UK: Cambridge Univ Pr.

Harris DR. 1971. The ecology of swidden cultivation in the Upper Orinoco rain forest, Venezuela. The Geographical Review LXI(4):475–95.

Harris DR. 1972. Swidden systems and settlement. In: Ucko PJ, Tringham R and Dimbleby GW (eds). Man, settlement, and urbanism. London UK: Gerald Duckworth. p 245–62.

Hart TB, Hart JA, Murphy PG. 1989. Monodominant and species-rich forests of the humid tropics: causes for their co-occurrence. Amer Natural 133(5):613–33.

Hart TB. 1990. Monospecific dominance in tropical rain forests. Trends Ecol and Evol 5(1):6–11.

Hart TB, Hart JA. 1986. The ecological basis of hunter-gatherer subsistence in African rain forests: the Mbuti of eastern Zaire. Hum Ecol 14(1):29–55.

Headland TN. 1987. The wild yam question: how well could independent hunter-gatherers live in a tropical rain forest ecosystem? Hum Ecol 15(4):463–91.

Henderson A. 1990. Flora Neotropica: Arecaceae. Part 1: Introduction and the Irarteinae. New York Bot Gard Monogr 53: 2–3.

Hewlett B. 1996. Cultural diversity among African Pygmies. In: Kent S (ed). Cultural diversity among twentieth-century foragers: an African perspective. Cambridge UK: Cambridge Univ Pr. p 215–44.

Hladik A, Dounias E. 1993. Wild yams in the African forest as potential food resources. In: Hladik CM, Hladik A, Linares OF, Pagezy H, Semple A, Hadley M. Tropical forests, people and food: biocultural interactions and applications to development. Man in the Biosphere Series, Vol 13. Paris France: UNESCO; and Carnforth UK: Parthenon Publ. p 163–76.

Holmberg AR. 1960. Nomads of the long bow: the Siriono of eastern Bolivia. Chicago IL: Univ Chicago Pr. p 5–6.

Hutterer KL. 1988. The prehistory of the Asian rain forests. In: Denslow JS, Padoch C. People of the tropical rain forest. Berkeley CA: Univ California Pr. p 63–72.

Ichikawa M. 1993. Diversity and selectivity in the food of the Mbuti hunter-gatherers in Zaire. In: Hladik CM, Hladik A, Linares OF, Pagezy H, Semple A, Hadley M. Tropical forests, people and food: biocultural interactions and applications to development. Man in the Biosphere Series, Vol. 13. Paris France: UNESCO; and Carnforth UK: Parthenon Publ. p 487–96.

Jones G, Hollier G. 1997. Resources, society and environmental management. London UK: Paul Chapman. p 317.

Krauss M. 1992. The world's languages in crisis. Language 68:4–10.

Lawless R. 1978. Deforestation and indigenous attitudes in Northern Luzon. Anthropology (2)1:1–13.

Linares OF. 1981. From tidal swamp to inland valley: on the social organization of wet rice cultivation among the Diola of Senegal. Africa 51(2):557–95.

Linares OF. 1992. Power, prayer and production: the Jola of Casamance, Senegal. Cambridge UK: Cambridge Univ Pr.

Maffi L. 1998. Language: a resource for nature. Nature Res 34(4): Oct-Dec 1998. Maffi L, Anthropol News Feb 1997: p 11.

McNeely JA, Ness G. 1996. People, parks, and biodiversity: issues in population-environment dynamics. In: Dompka V (ed). Human population, biodiversity and protected areas: science and policy issues. Washington DC: AAAS. p 19–70.

McNeely JA, Gadgil M, Leveque C, Padoch C, Redford K. 1995. Human influences on biodiversity. In: Heywood VH (ed). Global biodiversity assessment. Cambridge UK: Cambridge Univ Pr. p 711–821.

Mercader J. Bajo el techo forestal: la evolución del poblamiento en el bosque equatorial del Ituri, Zaire. PhD dissertation. Madrid Spain: Universidad Complutense.

Michon G, Bompard JM. 1987. Agroforesteries indonésiennes: contributions paysannes à la conservation des forêts naturelles et de leurs ressources. Rev d'Ecol (la Terre et la Vie) 42:3–37.

Murphy RF, Quain B. 1955. The Trumaí indians of central Brazil. Seattle WA: Univ Washington Pr.

Myers N. 1992. The primary source: tropical forests and our future. New York NY: WW Norton. p 59.

Nepstad DC, Veríssimo A, Alencart A, Nobre C, Lima E, Lefebvre P, Schlesinger P, Potter C, Moutinho P, Mendoza E, Cochrane M, Brooks V. 1999. Large-scale impoverishment of Amazonian forests by logging and fire. Nature 398(6727):505–8.

Netting R.McC. 1968. Hill farmers of Nigeria: cultural ecology of the Kofyar of the Jos plateau. Seattle WA: Univ Washington Pr.

Ohtsuka R. 1996. Long-term adaptations of the Gidra-speaking population of Papua New Guinea. In: Ellen R and Fukui K (eds). Redefining nature: ecology, culture and domestication. Oxford UK: Berg. p 515–30.

Orlove BS, Brush SB. 1996. Anthropology and the conservation of biodiversity. Ann Rev Anthropol 25:329–52.

Padoch C, de Jong W. 1992. Diversity, variation, and change in ribereño agriculture. In: Redford KH, Padoch C (eds). Conservation of neotropical forests: working from traditional resource use. New York NY: Columbia Univ Pr. p 158–74.

Piperno DR, Pearsall DM. 1998. The origins of agriculture in the lowland Neotropics. San Diego CA: Acad Pr. p 76–8, 230–2, table 4.5 on p 204–5.

Price DH. 1990. Atlas of world cultures: a geographical guide to ethnographic literature. Newbury Park CA: Sage Publ. p 156.

Redford KH, Mansour JA (eds). 1996. Traditional peoples and biodiversity conservation in large tropical landcapes. Washington DC: The Nature Conservancy.

Reed RK. 1995. Prophets of agroforestry: Guaraní communities and commercial gathering. Austin TX: Univ Texas Pr. p 27.

Reid WV, Miller KR. 1989. Keeping options alive: the scientific basis for conserving biodiversity. Washington DC: World Resources Inst. p 37.

Ruthenberg H. 1976. Farming systems in the tropics. 2nd ed. Oxford UK: Clarendon Pr.

Spencer JE. 1966. Shifting cultivation in Southeastern Asia. Publications in Geography. Vol. 19. Berkeley CA: Univ California Pr.

Sponsel LE. 1995. Relationships among the world system, indigenous peoples, and ecological anthropology in the endangered Amazon. In: Sponsel LE (ed). Indigenous peoples and the future of Amazonia: an ecological anthropology of an endangered world. Tucson AZ: Univ Arizona Pr. p 263–93.

Stearman AM. 1995. Neotropical foraging adaptations and the effects of acculturation on sustainable resource use. In: Sponsel LE (ed). Indigenous peoples and the future of Amazonia: an ecological anthropology of an endangered world. Tucson AZ: Univ Arizona Pr. p 207–24.

Stoler A. 1978. Garden use and household economy in rural Java. Bull Indones Econ Stud 14(2):85–101.

Uhl NW, Dransfield J. 1987. Genera Palmarum: a classification of palms based on the work of Harold E. Moore, Jr. Lawrence KS: Allen Pr. p 550–3.

Whitmore TC. 1992. An introduction to tropical rain forests. Oxford UK: Clarendon Pr. 3rd ed p 10–11, 28, 213 (glossary), 223 (index).

Wilkes G. 1991. In situ conservation of agricultural systems. In: Oldfield ML, Alcorn JB (eds). Biodiversity: culture, conservation and ecodevelopment. Boulder CO: Westview Pr. p 86–101.

Wilkie D.S. 1988. Hunters and farmers of the African forest. In: Denslow JS, Padoch C (eds). People of the tropical rain forest. Berkeley CA: Univ California Pr. p 111–26.

ENDANGERED PLANTS, VANISHING CULTURES: ETHNOBOTANY AND CONSERVATION

PAUL ALAN COX
National Tropical Botanical Garden,
P.O. Box 340, Lawai, Hawaii 96765

ENDANGERED PLANTS

WE LIVE in a time of mass extinction of biological species. Although there is a public outcry over the demise of well-known species, such as whales or condors, the relentless extinction of species affects all taxonomic groups. May and others (1995) reported that 485 animal species and 585 plant species are known to have become extinct since 1600. Although this is an extremely high level of extinction—for plants, the average rate of extinction is 0.5% of all species per century—what is more alarming is the increase in these rates—half of all those extinctions occurred within the last century. As a result, the period from inception to demise of a bird or mammal species has been reduced from 5–10 million years to about 10,000 years (May and others 1995).

The creation and extinction of biological species is, of course, a natural process that occurs over evolutionary time. Those few living relics of earlier geological periods, be they coelacanths deep in the oceans or ancient conifers hidden in Australian valleys, are quite properly regarded as objects of curiosity. It is not the fact of extinction, but the acceleration of extinction that concerns conservation biologists. If present trends continue, future generations will inherit a planet of greatly reduced biological diversity. Although this is not the first mass extinction caused by people—one need think only of extinctions of birds in Hawaii or Pleistocene extinctions of mastodons in North America (Martin and Wright 1967; Pimm and others 1995)—ours is the first known human-induced mass extinction of plants. Botanists are particularly troubled about the extinction of plant species, because, with only a few exceptions, plants form the foundation of all known eco-systems. Plants also constitute the major sources of food, medicine, and building materials throughout the world and have strongly influenced the trajectory of human civilization (Balick and Cox 1997).

The International Union for the Conservation of Nature (IUCN) currently lists 12% of the world's plant species as threatened with extinction. This figure is al-most certainly an underestimate, in that the IUCN lists only species known to science. With the rapid destruction of tropical rain forests, ecosystems in which roughly one new plant species is discovered for every hundred collected, many unknown plant species are disappearing. If current rates of extinction continue, nearly half of all plant species worldwide will disappear in 3,000 years (May and others 1995).

What has driven this high rate of extinction? The oft-cited ultimate causes are deforestation, pollution, and growth of the human population, but little is known about the proximal processes that lead to extinctions of plants. Sometimes the fate of an entire species can hinge on small things. We are just beginning to understand how the loss of small insects, birds, flying mammals, and other polli-nators and seed dispersers can lead in turn to extinctions of plants (Bond 1995; Buchmann and Nabhan 1996). Temple (1977), for example, argued that the extinction of dodos (*Raphus* spp.) in the Mascarene Islands led to a lack of dis-persal and germination of seeds for *Sideroxylon* trees. At this point, no new seed-lings of *Sideroxylon* are being produced in nature.

In oceanic islands that have limited guilds of pollinators, loss of pollinators can affect plant assemblages dramatically, affecting major structural components of the rain forest and creating cascades of linked extinctions. In Samoa, more than half of all canopy-level trees depend on flying foxes of the genus *Pteropus* for pollina-tion (Banack 1998). When the flying foxes began to disappear because of com-mercial hunting and destruction of habitat, biologists became concerned that their loss could lead ultimately to loss of the Samoan rain forest (Cox and others 1991). An urgent appeal was made to the 108 signatory nations of the Convention on International Trade in Endangered Species, who responded by banning interna-tional traffic in Samoan flying foxes (*Pteropus samoensis*). The US Congress also used the finding about the importance of flying foxes as pollinators as a justifica-tion for granting national-park status to several areas in American Samoa, a US territory (Cox 1997a).

Many extinctions of pollinators occurred before protective legislation had been envisioned. In Hawaii, the native Hawaiian birds that had pollinated the flower-ing vine *Freycinetia arborea* became extinct in the late 19th century, but pollina-tion was continued by the Japanese white-eye, *Zosterops japonica*, an introduced species (Cox 1983).

Another cause of extinctions of plants, particularly in oceanic islands, is the in-troduction of exotic species. Introductions of beneficial plants to islands seem to be the exception rather than the rule, so it should be no surprise that half the plants (263 of 553) on the endangered species list in the United States are from Hawaii. Recent exotic plant introductions to Hawaii, such as *Clidemia*, pose grave threats to native plants, particularly in the aftermath of hurricanes or forest fires, a pattern that occurs throughout Oceania. In Samoa, introduced weeds, such as *Mikania micrantha*, have slowed dramatically the regeneration of native forests from hurricanes (Elmqvist and others 1994). This vulnerability of island flora to foreign weeds long has been known.

> From the extraordinary manner in which European productions have recently spread over New Zealand and have seized on places which must have been previously occupied, we may believe, if all the animals and plants of Great Britain were set free in New Zealand, that in the course of time a multitude of British forms would become thoroughly naturalized there, and would exterminate many of the natives. Yet the most skilful naturalist from an examination of the species of the two countries could not have foreseen this result [Darwin 1859].

Although a skillful naturalist might not have foreseen the extirpation of native New Zealand's species as a result of the introduction of exotic competitors, Darwin learned in New Zealand that native Maoris predicted not only biological extinctions but also cultural extirpations. "As the white man's rat has driven away the native rat, so the European fly drives away our own, and the clover kills our fern, so will the Maoris disappear before the white man himself" (Crosby 1986). Clearly, the Maoris foresaw early the link between biological extinction and cultural loss.

VANISHING CULTURES

A variety of animal species ranging from social insects to chimpanzees can be said to have societal structures complete with communication systems, but human cultures are distinguished by the complexity of the languages used. The ability to use language, symbolic systems of vocalization that have sophisticated grammar and syntax, is one of the characteristics of our species. As a species, we have used language for at least 40,000 years and perhaps far longer.

Just like biological species and populations, languages vary in range and size. Some, like Mandarin, are spoken by millions and even billions of people, while others, such as the Eyak language of Alaska, are limited to one or two living individuals (Krauss 1992). Just like species, languages originate in different ways but eventually become extinct or significantly altered. Within the last century, Bishlama, a pidgin language spoken in Vanuatu, has arisen, whereas Dalmatian, a Romance language, ended when the last native speaker died. The current lingua franca of international commerce and scholarship is English, like Latin, Arabic, and Greek before it; but if the historical pattern of change continues, this role in the future likely will be accorded to some other tongue, such as Mandarin, Hindi, or Japanese.

Many parallels exist between languages and species. Just as museums catalog Tasmanian tigers or passenger pigeons, so do linguists attempt to classify and display extinct languages. Hattic, Sumerian, and Etruscan once were spoken widely among flourishing populations in Anatolia, Mesopotamia, and Northwestern Italy, respectively, but these languages now survive only in clay tablets, stone inscriptions, and ancient scrolls. Like European bison, which are extinct in the wild but protected in zoos, Latin, Sanskrit, and other "dead" languages are cared for lovingly by scholars.

The analogy between endangered species and endangered languages is not perfect—unlike biological species, languages once extinct can still be revived—but the correspondence is closer than might be thought. As the number of species is an indicator of the planet's ecological health, so is the number of languages a manifestation of the world's cultural diversity. As the declining number of species alarms biologists, so does the vanishing number of languages dismay linguists. Krauss has estimated that of the 6,000 languages present at the beginning of our century, half have disappeared. By considering languages to be "endangered" if they are no longer being learned by small children, Krauss (1992) believes that the coming century "will see either the death or doom of 90% of mankind's languages."

As the rate of extinction of species is but a crude index to the actual loss of the world's genetic diversity, so is the rate of disappearance of languages only a rough estimate of the world's vanishing cultural diversity. If we accept that both the biological and cultural diversity of the world are imperiled, then it seems that ethnobiology, the study of the interaction between human culture and biodiversity, will assume increasing importance in the future.

ETHNOBOTANY

Given the reliance of human cultures on biodiversity, it should not be surprising that the individual who invented the binomial system of nomenclature that underlies modern assessments of biodiversity also invented the field of ethnobotany, which provides the intellectual foundation for assessing interactions between cultural and biological diversity. In 1732, a young botanist in Uppsala, Sweden, became consumed with wanderlust. Unlike other students of his time, he decided to travel not to the academic centers of Europe, but to learn directly from indigenous peoples. "I set out alone from the city of Uppsala on Friday, May 12, 1732, at eleven o'clock," Carl Linnaeus wrote in his journal with characteristic precision, "being at that time within half a day of twenty-five years of age." Equipped with only 400 copper dalers from the Swedish Royal Society, a plant press, a hand lens, a fowling piece, and a change of clothes, Linnaeus began a 5,000-kilometer, 5-month-long journey to the land of the midnight sun. As the Galapagos were well surveyed before Darwin's arrival, so had Lapland been well mapped and explored before Linnaeus commenced his journey. The scientific significance of the travels of both Darwin and Linnaeus stemmed not from the novelty of the itineraries, but from the originality of the questions that they pursued.

The complete record of Linnaeus's journey to Lapland can be found only in his foolscap diary, which is carefully protected in the vault of the Linnaean Society in London. Written in Swedish and Latin, filled with sketches and notes, the handwritten travel diary of Linnaeus was never intended for publication. An abridgment of his diary was not published until 33 years after his death, first in

English, then, 78 years later, in Swedish. The diary is of enormous historical importance. "I here made the following observations relative to the remedies used by the Laplanders," Linnaeus penned on July 4, 1732, at the beginning of the first recorded interview of an indigenous healer by a trained botanist.

Previous botanists, such as Rauwolf and Rumphius, had returned from distant lands with accounts of the use of plants by local peoples, but Linnaeus's journey to Lapland in 1732 was the first time that a trained botanist had traveled to another land with the express purpose of interviewing indigenous people about their use and perceptions of plants. Although the term ethnobotany was not coined until a century and a half later by Harshberger (Balick and Cox 1997), the ethnobotanical field techniques pioneered by Linnaeus continue to provide evidence of the biological sophistication of indigenous peoples. Although it waned in the 1970s, ethnobotany has become reinvigorated and popularized. Ethnobotanical research methods vary (Martin 1995), but on one point nearly all modern ethnobotanists agree: indigenous knowledge about plants and animals is vanishing throughout the world.

The historical importance of ethnobotany in drug-discovery programs, coupled with new screening techniques, has generated an explosion of interest not only in the ethnobotanical approach to drug discovery, but also in related issues of indigenous intellectual-property rights (Cox 1990, 1995, 1997b; Cox and Balick 1994; Greaves 1994; Reid and others 1993). The potential importance of using indigenous knowledge to unlock the benefits of biological diversity is what led the international community to include preservation of traditional knowledge in the Convention on Biological Diversity drafted in Rio de Janeiro.

CONSERVATION

Linnaeus not only invented modern botanical nomenclature and ethnobotany; he also served as a pioneer of conservation. "I do not know how the world could persist gracefully if but a single animal species were to vanish from it," Linnaeus wrote in his journal. Today many people around the world share Linnaeus's view of the importance of conservation. Perhaps one of the most important manifestations of that sentiment in recent years was the Convention on Biological Diversity, commonly known as the Rio Treaty, which now has been signed by 161 different nations.

Although this convention emphasizes international responsibilities to protect the environment, its article 8j discusses the need to conserve traditional knowledge. This article mandates that, subject to national legislation, each signatory nation will "respect, preserve, and maintain knowledge, innovations, and practices of indigenous and local communities embodying traditional lifestyles relevant for the conservation and sustainable use of biological diversity and promote their wider application with the approval and involvement of the holders of such knowledge, innovations, and practices and encourage the equitable sharing of the benefits arising from the utilization of such knowledge, innovations, and practices." The parties to the convention thus commit to three major obligations:

- to respect, preserve, and maintain traditional knowledge;
- to promote wide application of traditional knowledge, with the approval and the involvement of the holders of such knowledge; and
- to encourage equitable sharing of benefits from traditional knowledge.

Properly implemented, article 8j of the convention can be a powerful tool in protecting both biological and cultural diversity. Although as some have pointed out (Glowka and others 1994), a narrow reading of article 8j could suggest that these obligations can be obviated through national legislation, but originally this provision sought to avoid inadvertent conflict with national laws that were in place before the convention was signed, such as proscriptions against administration of traditional ordeal poisons.

Different nations can point to different ways that they are implementing article 8j. In Japan, skilled practitioners of traditional knowledge are considered "living treasures," a rather explicit demonstration of the respect for traditional knowledge required under article 8j. This concept could be expanded in other countries to include weavers, healers, shipwrights, or others who serve as custodians of traditional knowledge. Sweden has sought to preserve and maintain traditional knowledge by launching a national survey of folk knowledge about Swedish plants and animals. The resulting multivolume work will be published by the Swedish Biodiversity Centre in Uppsala and a consortium of Swedish museums and universities. Regional museums in particular have demonstrated an important ability to involve Swedish citizens beyond the confines of traditional academe. Thailand has sought to promote wide application of traditional knowledge by educating its citizens about traditional Thai medicine. Mahidol University has produced a series of informative books and a filmstrip designed to be shown in schools and to community groups. Belize seeks to encourage equitable sharing of benefits from traditional knowledge by granting oversight of a rain-forest preserve to an organization of traditional healers. Proceeds from a book on traditional medicine and from a line of "rain-forest remedies" are used to provide pensions to healers (Balick and Cox 1997).

Although the US Congress has yet to ratify the Convention on Biological Diversity, indications are clear that it supports the provisions of article 8j. Congress required that a new national park in American Samoa be managed with the input of an advisory council of village chiefs. Nominees to the council convened in 1998 at the National Tropical Botanical Garden in Hawaii to discuss rules for the park. In addition, the US House of Representatives recently passed the Tropical Forest Conservation Act of 1998 (HR 2870), which not only facilitates the exchange of international debts for conservation of rain forests, but also requires consultation with indigenous peoples in the use of funds for conservation.

THE ROLE OF BOTANICAL GARDENS IN ETHNOBOTANY

Clearly, all these efforts depend on the ability to document traditional and indigenous knowledge and to identify the custodians of such knowledge. This in turn highlights the need to provide far broader opportunities for ethnobotanical training in the future. Yet, with the exception of a few institutions, such as the Autonomous National University of Mexico, universities have been slow to fill the breach. Although interest in ethnobotany has expanded rapidly among students, universities have been slow to present ethnobotany among the traditional academic disciplines. Pharmacognosy courses, which can provide a stepping stone to ethnobotany, long since have disappeared from most pharmacy schools, and chairs in ethnobotany are rare in liberal-arts institutions. As a result, meetings of the Society for Economic Botany or the International Society for Ethnopharmacology are often attended by students of anthropology, botany, and chemistry who lack mentors in ethnobotany at their institutions. The minimal requirements for a program in ethnobotany include access to a well-curated herbarium and a good library, which is why attempts by the private sector (such as those by firms that produce herbal supplements or pharmaceutical firms) to launch research programs in ethnobotany so often fail.

Universities are not the only public institutions that have significant herbaria and libraries, however. Botanical gardens, particularly those with significant living collections, offer an untapped but potentially important resource for training ethnobotanists. In the summer of 1998, a pilot course in tropical botany and ethnobotany was cosponsored by the National Tropical Botanical Garden and the Swedish Biodiversity Centre. Students from Canada, Estonia, Ethiopia, Korea, Russia, Singapore, Sweden, and Tanzania were offered the opportunity to be trained by ethnobotanists, plant systematists, and resident Polynesian weavers and healers in a large species-diverse garden and its associated preserves. The course is expected to be offered annually to students from around the world, in addition to a specialist course in ethnobotany begun in 1999.

CONCLUSION

Both plant species and indigenous knowledge are disappearing at an alarming rate. The loss of plant species is particularly acute in oceanic islands, where as much as 50% of island flora is endangered. Folklore about plants might be disappearing even faster than the species themselves, as suggested by the loss of indigenous languages: half have disappeared in this century, and, of the remaining languages, 80% are endangered. Ethnobotany, which deals with the relationship between biological and cultural diversity, can play a crucial role in helping nations meet the obligations under article 8j of the Convention on Biodiversity to respect, maintain, and preserve traditional knowledge. Although universities have been slow to meet the demand for ethnobotanical training, botanical gardens offer a unique setting for students and custodians of traditional knowledge to meet and discuss strategies for protecting both species and cultural biodiversity. Most countries of the world have botanical gardens, and those gardens should work now to fill the breach in ethnobotanical training, emphasizing the relationship between endangered plants and vanishing cultures.

REFERENCES

Balick MJ, Cox PA. 1997. Plants, people, and culture. New York NY: WH Freeman.

Banack SA. 1998. Diet selection and resource use by flying foxes (Genus *Pteropus*). Ecology 79(6):1949–67

Bond WJ. 1995. Assessing the risk of plant extinction due to pollinator and disperser failure. In: Lawton JH, May RM (eds). Extinction rates. Oxford UK: Oxford Univ Pr. p 131–46.

Buchmann SL, Nabhan GP. 1996. The forgotten pollinators. Washington DC: Island Pr.

Cox PA. 1983. Extinction of the Hawaiian avifauna resulted in a change of pollinators for the ieie, *Freycinetia arborea*. Oikos 41:195–9.

Cox PA. 1990. Ethnopharmacology and the search for new drugs. In: Chadwick DJ, Marsh J (eds). Bioactive compounds from plants. New York NY: J Wiley. p 4–55.

Cox PA. 1995. Shaman as scientist: indigenous knowledge systems in pharmacological research and conservation biology. In: Hostettmann K, Marston K, Maillard M, Hamburger M (eds). Phytochemistry of plants used in traditional medicine. Oxford UK: Clarendon Pr. p 1–15.

Cox PA. 1997a. Nafanua: saving the Samoan rain forest. New York NY: WH Freeman.

Cox PA. 1997b. Indigenous peoples and conservation. In: Grifo F, Rosenthal J (eds). Biodiversity and human health. Washington DC: Island Pr. p 207–20.

Cox PA, Balick MJ. 1994. The ethnobotanical approach to drug discovery. Sci Amer 270(6):82–7.

Cox PA, Elmqvist T, Pierson ED, Rainey ED. 1991. Flying foxes as strong interactors in South Pacific island ecosystems: a conservation hypothesis. Cons Biol 5:448–54.

Crosby AW. 1986. Ecological imperialism and the biological expansion of Europe 900–1900. Cambridge UK: Cambridge Univ Pr.

Darwin C. 1859. On the origin of the species by means of natural selection; or, the preservation of favoured races in the struggle for life. London UK: J Murray.

Elmqvist T, Rainey WE, Pierson ED, Cox PA. 1994. Effects of tropical cyclones Ofa and Val on the structure of a Samoan lowland rain forest. Biotropica 26:384–91.

Glowka L, Burhenne-Guilmin F, Synge H, McNeely JA, Gündling L. 1994. A guide to the Convention on Biological Diversity. Glan Switzerland: The World Conservation Union.

Greaves T (ed). 1994. Intellectual property rights for indigenous peoples, a source book. Oklahoma City OK: Society for Applied Anthropology.

Krauss M. 1992. The world's languages in crisis. Language 68:1–10.

Martin G S 1995. Ethnobotany: a method manual. London UK: Chapman & Hall.

Martin PS, Wright HE Jr (eds). 1967. Pleistocene extinctions: the search for a cause. New Haven CT: Yale Univ Pr.

May RM, Lawton JH, Stork NE. 1995. Assessing extinction rates. In: Lawton JH, May RM (eds). Extinction rates. Oxford UK: Oxford Univ Pr. p 1–24.

Reid EA, Laird SA, Meyer GA, Gámez R, Sittenfeld A, Janzen DH, Gollin MA, Juma C. 1993. Biodiversity prospecting: using genetic resources for sustainable development. Washington DC: World Resources Inst.

Pimm SL, Moulton MP, Justice JL. 1995. Bird extinctions in the Central Pacific. In: Lawton JH, May RM (eds). Extinction rates. Oxford UK: Oxford Univ Pr. p 75–87.

Temple SA. 1977. Plant animal mutualism: coevolution with dodo leads to near extinction of plant. Science 197:885–6.

RELIGION AND SUSTAINABILITY

JAMES PARKS MORTON
Interfaith Center of New York,
38 East 30th Street, New York, NY 10016

Two of my environmental gurus, and also close friends, died this last year, just before Christmas. Carl Sagan and Laurens van der Post could not be more different, but they were equally passionate stargazers. Science and religion agree that we humans and the planet itself are made of stardust, so I offer these words on religion and sustainability in grateful memory of those remarkable latter-day astrologers.

Let me begin with a story about Sir Laurens that Larry Hughes told at van der Post's funeral in London on December 20, 1996. Larry was his American publisher of 37 years and 25 books, and the story took place on Laurens's second trip to New York, in 1961.

It was a summer evening and I was accompanying Laurens to a supper party. When we entered the large lobby of the building in which our host had an apartment, we saw three or four people excitedly running around trying to catch a pigeon which had flown in through the front door, but couldn't find its way out. Immediately, but in a very quiet way, Laurens took charge. He directed that someone open a back door which led out to a garden and that all of us stand absolutely still. Within minutes the pigeon flew down to the lobby floor and from about 10 feet away stood inspecting Colonel van der Post with that side-to-side movement that pigeons employ when sizing up a situation. Then the pigeon took one last look at this smartly dressed stranger, turned, and flew out the open back door into the garden. Am I imagining that Laurens spoke to this bird? I believe he did because I remember thinking 'here we have Montgomery of Alamein and Francis of Assisi rolled into one' (Hughes 1996).

Later, as I came to read and learn more about Laurens van der Post, I realized that communication with animals was a very natural part of his life. Can we forget his persuasive chat with that stubborn camel or his meeting with that tiger on a jungle trail, or his marvelous portrayal of Blady the horse, or of Mantis or Hintze, the African ridge-back? As the hunter in *The Hunter and the Whale* explained: "One of the reasons why nature, and animals in particular, were so important to us today was because they are a reminder that we could live life not according to our own will, but to God's' (van der Post 1987).

Such is our context this afternoon: pigeons, humans, aquifers, apartment buildings and rain forests, stubborn camels, and stars—in particular, the structural interdependence and interconnectedness of all creatures. "Inter" is today's buzzword for us because it is the necessary qualifier for everything that touches both sustainability and religion: interrelatedness, interdisciplinary, intercontinental, intergenerational, interracial, intercultural, interspecies, interfaith—all interdependent, all interconnected. No man—or woman—is an island, especially in the age of the Internet.

I want to give you my particular twist in defining religion and sustainability—that specific environmental subset of ecology. And then, like every preacher, I will briefly outline what I will say, then say it, and then tell you what I have said.

DEFINING RELIGION AND SUSTAINABILITY

The etymology of the word *religion* is so utterly fundamental and simple that it surprises many people. It comes from the Latin verb *religare,* which means to connect, to join together, to assemble, to create connectedness, to create community. By definition, then, being religious means being inclusive, perhaps even being compulsive about the idea that no one, indeed nothing in Heaven or on Earth, is left out. This primary "action" definition of religion as "connecting" is my particular slant in contrast with defining religion as believing such-and-such. "Believing" comes much further down the line. Mine is the more primordial meaning of religion as community-building and maintenance—getting all the people together, keeping them together, and celebrating cosmic togetherness. I will speak of several generic and universal rituals of connecting that people all over the world have practiced since time immemorial and also of their manifold variations and often striking differences. But my point is to celebrate diversity in religious practices in the same way we celebrate the unique gifts that different trees bring to the forest and thereby save us from missing the forest for the trees or vice versa. Religion and ecology both deal with individual persons and individual trees but always in the context of connecting the whole creation, the whole forest.

In fact, this fundamental etymological closeness of religion and ecology has led several of my environmental pals to claim with a grin that ecology is the flip side of religion, maybe even *the* religion of the new millennium. We could do worse.

But if so-called secular ecology defines itself as the study of connections—of how biological systems connect with each other and with their larger environment—when we come to the definition of sustainability, we plunge once again into the incense-laden atmosphere of religion. Sustainability is the whistle-blower of ecology, the hard-nosed safeguard and guarantor of harmony and balance within a given "connectedness," lest it become overwhelmed by one or a combination of factors. The precise mission of environmental sustainability is to monitor and turn around threats to harmony and balance due to overconsumption, overproduction, and overreproduction and, equally important, due to disrespect for human rights and disregard for Earth's regenerative capacities. But whistle-blowing is also a cry for justice and compassion that overlaps with another function of religion, stretching from the Hebrew prophets to the two Martin Luthers on to Sadat and Rabin and Mandela and Rachel Carson and Rosa Parks and Mother Teresa.

In short, with the ancient rituals and connecting rhythms of the world's religions basically in synch with the music of modern-day ecology and sustainability, have we not just about got it made? Are not religion and ecology almost two sides of the same coin?

WHAT WENT WRONG?

Often a situation comes into focus instantly if we reverse gear or look at a negative photo print. For a few minutes, let us define "bad" religion and "non"-sustainability. Immediately we see the drive to all-inclusiveness and interconnectedness turn into its opposite—to exclusiveness, to one community splitting into factions, to communion turning into excommunication, to Us versus Them, to words and credos speaking louder than actions and behavior. And on the environmental side, is not nonsustainability inevitable if life is driven by economic determinism and its three bedfellows of maximal production with cheapest resources, psychological advertising, and unlimited consumption? The Me Generation becomes the me cosmos, and mono crop is king. Exit biodiversity.

Time out for the depressing light of reality: If we ask Mr. and Mrs. America and the families of Japan and Europe—we'll hold off asking Africa, the Far East, Latin America, and China for the moment—if we ask these "developed" folks where in their lives they honestly rate the importance of religion and the crisis of the environment, I think the answers to both will be lukewarm. Of course, positive polls can be cited about "belief in God," increased church attendance here and there, and even the recent inclusion of the word *environment* in national surveys of important issues facing humanity. But what seems lacking is any widespread dimension of urgency or immediacy. Instead, what appear to be dominant characteristics of modern religions in developed countries are their privatized and sectarian nature and their being optional, as in sports or collecting. In no sense today is religion recognized as a generic and given part of basic human reality, like breathing, eating, sex, communication, tools, and art. To put it simply, religion in recent years has been severely trivialized.

Similarly, most people's concern about environmental sustainability is on the back burner, but for different historical reasons. Since the 18th century, what we in the West have thought of as the environment has been largely subsumed under the abstract category of Nature with a capital N and therefore eternal, invisible (apart from 18th-century landscape painting and new science), and just there as a stage and resource for the human project. Sustainability, let us remember, was virtually unheard of until 1992 and Rio de Janiero; and for most people it is still an unknown word.

The history of abstract Nature, structurally divorced as it was from concrete humanity, is similar to the history of light. That is why Marshall McLuhan put a photograph of a very concrete light bulb on the cover of his 1966 book *The Medium Is the Message* to awaken people to the all-pervasive, and therefore invisible, medium of light.

But it took an additional 22 years for the abstract environment to become visible and concrete for most people. Indeed, it took a series of four blockbuster events beginning in the sweltering hot summer of 1988 to accomplish this visibility. First, that lonely garbage barge as it wandered the oceans seeking a place to dump its cargo; second at the peak of the summer's heat wave, the closing of New York's public beaches because of the piles of smelly washed-up trash, hospital wastes, and orange peels; third, as shown in *Newsweek's* August 1988 cover drawing of the modern nuclear family (mom, pop, junior, and sis) sweating like pigs under a bell jar, "the greenhouse effect"; and fourth, as shown in *Time's* man-of-the-year December 31 issue with its lurid cover titled "Planet of the Year," Christo's wrapped, deflated beach ball sporting the moonshot image of Earth, washed up and sagging on the shore.

GENESIS OF A NEW GLOBAL ENVIRONMENTAL FORUM

Oxford 1988

Now, back to a sense of urgency and immediacy. Even with the creation of Earth Day in 1970 and the valiant slugging efforts throughout the 1970s and 1980s of the environmental groups and of those relatively few scientists and activists and statesmen whom we all know, environmental concerns didn't begin to become priorities at the American breakfast table until 1988, for a variety of unplanned, serendipitous convergent reasons. In addition to the four media events of that hot 1988 summer just mentioned, one of the surprising alliances of 1988 was the remarriage of science and religion. Of course, as in most marriages today, there had already been a number of one-night stands and some closeted attempts at cohabitation. For example, Gregory Bateson, Rene Dubos, Margaret Mead, Carl Sagan, James Lovelock, and 2 dozen other environmental scientists and scholars had regularly preached at New York's Cathedral of St. John the Divine beginning in the 1970s. But I was considered an oddball in the tradition of my predecessor Dean James Pike and the famous Red Dean of Canterbury. According to hallowed English tradition, Anglican Cathedrals are often known as "Royal Peculiars."

At any rate, because of this decade-long practice of using *the pulpit* of St. John the Divine as an open forum for environmental issues as religious issues, I had been approached in 1985 by a delegation of three wise men, headed by Ambassador Angier Biddle Duke, and including Claus Nobel and Akio Matsumura. All three had worked for several years with global population issues involving parliamentarians and UN agencies. Now they had a new agenda and wanted my help to expand their population concern to include the full spectrum of environmental issues and to bring religious leaders worldwide into association with their band of global parliamentarians—in short, to create a new global forum *combining* church and state, an obvious no-no both to the UN and to all so-called modern states.

I replied that I thought this was just what the world needed, especially if we invited major scientists to join the plot from the beginning. They agreed, and I immediately called my friend Carl Sagan for help.

We organized ourselves that year as a nonprofit called the Global Forum for Spiritual and Parliamentary Leaders and set about planning our first meeting for spring 1988 in Christ Church College, Oxford. We all had agreed that it should take place in the most kosher setting possible. It was to be under the formal patronage of the archbishop of Canterbury on the religious side, joined by the Dalai Lama, Cardinal Koenig of Vienna, Mother Teresa, and the high priest of the African rain forest and on the political side by two senior senators, Sat Pal Mittal of India and Manuel Ulloa of Peru. Most astonishing was the arrival of a delegation from the Soviet Union that included archbishops, rabbis, imams, cosmonauts, the president of the Supreme Soviet, and Gorbachev's chief nuclear adviser, Evgeny Velikov, who headed up the Soviet Academy of Sciences. In all, 300 persons worked together under the huge banner of the planet seen from the moon: 100 sitting parliamentarians, 100 spiritual leaders, and 100 scientists, artists, and journalists. We were together for 4 full days; major addresses were by given by Sagan, Velikov, James Lovelock, Father Tom Berry, Mother Teresa, and the Dalai Lama; and we closed with a banquet at Blenheim Palace, where Dr. King's right-hand organizer, the Rev. C. T. Vivian, offered the blessing, and Carl and I chatted over port about atheism—our unending conversation.

I have treated the 1988 Oxford meeting at some length for two reasons. First, it really was the first major public viewing of environmental science, religion, and politics as partners in the common enterprise of living sustainably on Earth—a solemn return to the way all the world had operated until the modern era. But second, and perhaps more important, the Oxford meeting had a substantial impact in the United States on the forthcoming remarriage of environment and religion. The two marriage brokers were Carl Sagan and his Russian colleague, Evgeny Velikov, who reported to Gorbachev that a meeting just like Oxford, only much bigger, should take place in Moscow as soon as possible. They told Gorbachev that it would re-educate Russian scientists, politicians, and religious leaders about how to work together to rehabilitate the dangerously degraded "Chernobylized" Russian environment. Gorbachev gave an immediate go-ahead (he was just approaching the zenith of his power in 1988) and scheduled a 5-day meeting for the second week in January 1990, when all his new election and restructuring procedures would be in place.

But while the Moscow preparations were zooming ahead, things were moving at a snail's pace in the American religious community with respect to any recognition of crisis, let alone action, on issues of sustainability, even after the hot summer of 1988 with its garbage-strewn beaches and *Newsweek* and *Time* covers. I thought that the obvious course to follow was to go to the very top religious leaders with a blue-ribbon delegation of respected environmental true believers. Several polite meetings with the highest muckety-mucks actually took place, but eyes glazed amid confusion about the meaning of environmental stewardship. Were we talking about fund-raising stewardship as in churches and synagogues?

I was exasperated, and so was Paul Gorman, my sidekick and environmental partner at the cathedral. What next? We called Sagan and out of our shared frustration, a brilliant, totally different strategy was forged: To convert religious leaders, we decided to "lead with the enemy" and have Sagan recruit a small army of the world's most distinguished scientists, who would implore the religious leaders to join them in their urgent appeal to save the world. A one-paragraph cover letter signed by four impeccably placed national religious leaders paved the way and was followed by an impassioned appeal signed by 34 impeccable scientists with Sagan's name at the top of the list (Sagan and others 1988). And the miracle happened. All the top American religious leaders, deeply flattered, immediately signed up, and we had an instant Joint Appeal of Science and Religion sent to every minister, rabbi, and priest in the country (Hurst and others 1988).

As it turned out, our Machiavellian plans were providentially timed—the appeal letter had been mailed in November 1989, and replies arrived like Christmas cards just in time for presenting to the January 1990 meeting in Moscow.

Moscow Global Forum

Indeed, the Moscow Global Forum of Spiritual and Parliamentary Leaders was the full flowering of the seed planted at Oxford in 1988. Some 1,200 people sat for 5 full working days under the banner of planet Earth seen from the moon; the meeting was opened by UN Secretary General Perez de Cuellar, followed by keynotes from Elie Wiesel and Jim Grant of UNICEF; Senators Al Gore and Clayborne Pell and Undersecretary of State for Global Affairs Tim Wirth all pushed the emergency button; 15 American 8th-graders attended with their 15 Moscow 8th-grader hosts; and a delegation of 55 indigenous religious leaders came from five continents. The expected tradeoff also occurred: nine Chinese delegates came on condition that the Dalai Lama not be invited. But on the last day, Carl Sagan and Evgeny Velikov asked all the assembled world religious leaders to join their American confreres and add their names to the "Joint Appeal of Religion and Science," bringing the total to 300. And every session was opened by either a prayer or a chant or a moment of silent meditation from the religious traditions of the planet.

The meeting was extraordinary in its urgency and spirit. Artists made remarkable contributions, and vast amounts of vodka were joyously consumed in the midst of bureaucratic Moscow's predictable nonfunctionality: the telephone system, faxes, office supplies, and literally tons of fresh food all had to be imported from Finland and Frankfurt. But what a pivotal moment! Of the 1,200 attendees, 800 were from Russia; and for the first time since the revolution, religious leaders, elected public officials, scientists, news reporters, and artists mingled freely and in small buzz groups opened their hearts about a world after Chernobyl with its contaminated breast milk, poisoned rivers and aquifers, and deformed animals and insects.

The final session, on Friday afternoon, was held in the Kremlin in the ceremonial Hall of the Soviets, a climax past imagining. On the wall behind the podium stood a colossal 12-ft statue of Lenin, the only decoration in this exceedingly austere chamber, and on the rostrum were eight chairs: Gorbachev in the center flanked by Archbishop Pitorim, Angie Biddle Duke, Senator Manuel Ulloa, Akio Matsumura, Soviet Foreign Minister Eduard Shevardnadze, Carl Sagan, and me. To the astonishment of all 1,200, the session began when a skinny saffron-robed swami from India, who happened also to be a well-known microbiologist, mounted the podium, rapped the floor with his walking stick, and slowly began to chant "om." The entire room joined in the om-ing, and I wondered what Lenin thought from his lofty perch above us.

Gorbachev's speech was like Martin Luther King's 1963 "I Have a Dream" call to arms. He acknowledged the environment as the major global crisis before us, now that nuclear arms were coming under control, and then apologized very candidly for Russia's major role in creating a polluted world and pledged his personal leadership to meet the challenge. His final words were a very practical proposal. The great humanitarian achievement of the 19th century, he reminded us, was the creation of the Red Cross to alleviate human suffering from natural and human-made disasters. What the world needs today, he concluded, is the creation of an international Green Cross to heal the wounds of the degraded and ravished natural environment and to restore harmony to the total created order.

A final comment about the extraordinary timing of the Moscow meeting in January 1990: In the very same week that Gorbachev made his speech on Friday afternoon, the citizens of the Baltic states threatened to leave the Soviet Union if they were not granted their political independence. I remember Gorbachev's impassioned face on Russian television on Tuesday and Wednesday nights, imploring the crowds in the streets of Vilna; and on Thursday, we were told that the president very much wanted to keep his appointment with us, but that we must recognize the unpredictability of his schedule. After Friday afternoon *actually happened*, with the Green Cross buzzing in our brains, along with the prospect of a vodka and caviar reception with Gorbachev, the two dozen or so Jewish members of our ranks, Sagan included, gathered in a basement room of the Kremlin Hall of the Soviets and said prayers for the beginning of the Sabbath—certainly the first time *that* ceremony had ever taken place in *that* building. More miracles of timing.

Perhaps no one will ever know how important the Moscow meeting was, but several specific results are important for our brief consideration of religion and sustainability. First, the Global Forum of Spiritual and Parliamentary Leaders barged ahead and asked Gorbachev whether he would agree to be president if we did all the leg work of organizing his International Green Cross. Those negotiations took the better part of the next 2 years but provided the impetus for our planning a third global forum, to occur in Rio de Janeiro as part of the Environmental Summit in June 1992. That meeting in 1992 took place in Rio de Janeiro's original city hall, and all our by now faithful regulars chimed in—Al Gore, Tim

Wirth, Carl Sagan, Perez de Cuellar, Maurice Strong, the Grand Mufti of Damascus, and even the Dalai Lama, who just happened to be in Rio de Janeiro. The high point came when Senator Manuel Ulloa announced that Gorbachev had indeed agreed to be president of the newly forming International Green Cross, and we were to go ahead full speed in getting its organizational shape together in preparation for an inaugural meeting to take place a year hence, in 1993 in Japan. Keep in mind that at this point Gorbachev's domestic problems were coming to a boil, and his own political future was very unclear.

Moscow's second direct result for religion and sustainability was the tremendous affirmation it gave to the newly forming "American Joint Appeal of Religion and Science for the Environment." Paul Gorman at the cathedral made his priority the organizational task of transforming enthusiastic responses to an impassioned letter into a functioning program. Once again, major input came from Carl Sagan, Al Gore, and Tim Wirth; and by April 1990, the joint appeal came into existence as an organization, with Paul Gorman as its director, I as chairman, and offices based at the cathedral—just 3 months after Moscow.

THE AMERICAN RELIGIOUS COMMUNITIES' COMMITMENT TO GLOBAL CONCERNS

In the following June (1991), the new joint appeal held its first conference, beginning at New York's Museum of Natural Science and continuing at the cathedral, where 24 top religious leaders met—Jewish, Roman Catholic, Evangelical, mainline Protestant, and Eastern Orthodox leaders, and executives of the historically black churches—to be briefed by no less than Peter Raven, E.O. Wilson, James Hansen, Sherwood Rowland, Henry Kendall, Anne Whyte, Beverly Davison, Steven Jay Gould, Ann Druyan, and, of course, Sagan, Gore, and Wirth. At the end of the second day, the 24 religious leaders issued a powerful public statement committing the American religious community to solid environmental concerns.

Urgency at last was truly the name of the game. In March 1992, the Joint Appeal held a consultation with top leaders of the Jewish community (with Sagan, Gore, and Wirth again as principal teachers); and in May 1992, the Joint Appeal made its maiden trip to Capitol Hill with its "Mission to Washington." First, 50 heads of religious denominations were lectured by 50 scientists (the same faithful soldiers), and then in pairs they took on representatives, senators, and finally a joint congressional committee.

That night, after it was all over and the exhausted triumphant faithful sat down to strong drinks and dinner at a favorite Italian hideaway of Gore's, the second baby from the remarriage of religion and the environment was conceived. The question was, What next? And by the time the party had broken up, it had been decided to go for broke: seriously to take on the American religious establishment—obviously as partners, not adversaries—with a powerful up-front goal— guaranteeing that for local churches and synagogues the environmental crisis would have a clear priority for prayer and meditation, for study and proclamation and public action. In short, a central *religious* issue in the same sense that justice, peace, and poverty had become intrinsic religious concerns, not just "secular" issues. At long last, it looked as though the old sacred-secular standoff could be put to rest and that *all of creation* could be seen as holy and the very stuff of religion. Could it be that our starting point of religion and ecology as flip sides of each other, as the twin practitioners of connecting everything, just might be gaining ground?

National Religious Partnership for the Environment

That was all in May 1992; the next month, off we went to Rio de Janeiro and the environmental summit with its new word, sustainability. By fall of 1992, we were laying the groundwork for the new child of the Joint Appeal, a brand new baby to be called the National Religious Partnership for the Environment and to be composed of four religious partners: the Roman Catholic Church, the three denominations of Judaism, the National Council of Churches (including all the mainline Protestant, Orthodox and black churches), and the Evangelical Christians (including the Southern Baptists and all the Pentecostal churches). That is a huge mouthful, but the point is that its structure included virtually all the Christians and Jews who together make up the majority of Americans.

What made it politically important was its organizational structure with a small governing board composed of the top brass of the four religious partners—the folks who control the denominational budgets and make the policy decisions. Carl Sagan and Henry Kendall from MIT were also on the governing board maintaining the strong link with the scientific community, I continued to serve as chair, and Paul Gorman was president of the new organization, still with its offices at the cathedral. This trim structure proved excellent for fund-raising, the fact that Gore was in the White House did not hurt, and in 1 year we raised enough millions from major foundations to assure *each* of the four partners an annual grant of $250,000 for *each* of 3 years to be used by their *own staff* in their *own style* to make environmental sustainability come alive for their own religious tradition. The bottom line is that today we are involved directly with 50,000 local parish churches and synagogues. It is a major foot in the door. Again and again, we receive deeply moving testimonies of something cooking with kids in Sunday school, of how a certain team of interfaith activists turned around a certain city's incinerator policy, of extraordinary sermons and study groups and retreats and liturgies that have literally changed people's lives. It has just begun, and the consortium of foundations has already renewed grants for another 3 years.

But there is a negative side. Inertia remains all too real. The reality of unsustainable lifestyles is still our daily bread—James Lovelock's unholy trinity of cows, cheeseburgers, and chain saws. Sagan is dead, and the national environmental agenda has taken a tough political beating in spite of Gore and Wirth. Gorbachev's Green Cross has gone nowhere, although his very expensive State of the World Forum has kept the environmental flicker somewhat alive. In late June 1997, the UN convened its special summit session to review progress on the environmental commitment that the nations had made at Rio de Janeiro five years before in 1992. *Rio plus 5* rather uneventfully came and went, although Steven Rockefeller's heroic work on creating a charter, with major input from the religious traditions of the world, is a hopeful sign that had its preliminary airing in draft form at the June UN meeting. We hope that it will be ready for adoption by the total General Assembly for the millennium. So fasten your seatbelts! My only dour reminder is the final statement in Elizabeth Dowdeswell's preface to the excellent 1997 UN Environment Program paperback *Global Environmental Outlook* (GEO-1), prepared for *Rio plus 5*: "We know that the knowledge and technological base to solve the most pressing environmental issues are available. However, the sense of urgency of the early 1990s is lacking. Progress towards a sustainable future has simply been too slow."

How troubling it is once again to return to a lack of urgency! We already know that both ecology and religion are about the connections that include every iota of creation. And we know that both sustainability and religion are whistle-blowers, prophets, and trouble-makers when situations become nonsustainable either for human justice or for Earth's capacity to regenerate itself and remain livable. But it is urgency that alone seems to be the lifeblood, the spiritual linchpin that can make ecology truly sustainable and make religion truly religious. Indeed, if history can be our teacher, it seems that only a giant environmental disaster equivalent to Hiroshima or a global stock-market crash could produce the urgency necessary to make it crystal clear that the present situation is genuinely life-threatening. There is historically only one alternative that is *not* catastrophic and that is capable of generating the urgency to turn people around. And that is a profound spiritual awakening.

I do not mean the urgency of shouting preachers or millennarian threats on television. A spiritual awakening can stem from any source or any combination of sources—political, scientific, artistic, religious, media, cyberspace. Because this awakening must be spiritual, I cannot exclude any possibility. But whatever the sources, it must produce a profound urgency that turns our whole lifestyle and life orientation upside down for the long haul—not for 2 years but for 2 millennia. It must be an urgency that converts us and makes us gladly adopt a positive asceticism that can literally preserve Earth and all life on it. We can all learn these ascetic disciplines from folks who have been practicing them for years, for centuries, and who are willing to teach us how—from Buddhists and Benedictines, from Quakers and Jains and Jews and Hindus and Sufis and medicine men and women. Spiritual poverty was not grim or sentimental for St. Francis, but vital and urgent, full of guts and joy. Urgency gives us an earthy sense of humor with human tears and with the sincere expectancy of surprises from many quarters. Spiritual urgency makes us capable of being at home in any situation.

The disciplines of positive asceticism that I am talking about can awaken us to the basic spiritual orientations of urgency and immediacy and vitality. They can open our eyes to the fact that most in life is not either-or, but a necessary combination of opposites like body and spirit, and that so-called spiritual disciplines in fact deal with physical breathing and diet and mantras and prostrating and sex. Positive asceticism can make us see that it is spiritually urgent for everyone—literally everyone—to experience both city living and real wilderness; that it is likewise urgent for everyone to understand intellectually the structures of stasis and kinesis, of crystals and gases, of classicism and romanticism; and that everyone's life is made up of the opposites of enthusiasm (literally *en-theos* or god inside us) and abstention, of Easter and Ramadan, of Yom Kippur and Divali.

Spiritual asceticism teaches us that nothing is more vital than the urgency of rhythm. Rhythm itself is the basis of all religious practice, coming from our private internal and inescapable rhythm of the heartbeat—hence, the primordial urgency of the shaman and his drum, of music as the necessary handmaid of religion with tambourines and thundering organs and deeply monotonous unvarying chants. These rhythms sustain us and get us through life. Weekly rhythms, feasts and fasts, the rhythm of the seasons with solstices and equinoxes, spring and fall, the summer powwow and the piercing sundance—they make our blood circulate.

As a New Yorker, I have come to be sustained by our cathedral's annual environmental extravaganza for the Feast of St. Francis. Every year on that first October Sunday, everyone brings an animal to church. Paul Winter performs his Missa Gaia with full band, with African dancers and drummers, with a choir of 600 and the recorded voices of timber wolf and whale, and with legendary sermons by Carl Sagan and Al Gore. The climax comes with a procession down the aisle led by an elephant, followed by llama, horse, cow, sheep, owls and macaws, hamsters and tortoises, New York rats and cockroaches, a tree, a treasured moon rock and meteorite, and finally a glass vial containing 500 trillion blue-green algae. Last year, 6,000 two-leggeds were inside, and 1,000 had to listen from loudspeakers in the garden. It is urgent and immediate, vitality itself, deeply sustaining, deeply environmental, deeply religious.

Religion in this most basic primordial form can raise us to ecstasy, where we can experience our connectedness to everything that is and where we can see the true meaning of urgency and the necessary cost of a universal spiritual asceticism for the long haul that will make catastrophe unnecessary and instead make a sustainable future for the planet possible.

My penultimate comment is personal. Since 1972, 1 have worked to reveal the primordial rhythms and vital connectedness of creation itself as the essence both of religion and of environmental sustainability. I have worked from a base that is necessarily limited—American, Christian, Anglican, Episcopalian. We have done some good things, and I think the cathedral is still a useful model 25 years later.

The Interfaith Center

But starting in January 1997, a small group of us have been approaching the same reality of sustainability and religion from the other end of the stick. We started a new project in New York called the Interfaith Center, where our base is not one religion but many of the world's major religions working together—Hindus and Moslems, Sikhs and Buddhists, Jews and Christians, Taoists and Shintos, Jains and indigenous traditions, and so on—in short, the de facto religious picture of New York City, which is only a mild exaggeration of the religious mix in Toledo or Miami or Los Angeles. Today, there are more Moslems than Presbyterians in Houston, Texas, and more mosques than Anglican churches in Birmingham, England. All life today is urbanized and every city worldwide is increasingly an implosion of the planet's religious, racial, and cultural diversity into new demographic containers of connectedness. To help make this given unavoidable physical connectedness also spiritual, sustainable, vital, and urgent is the task that our new Interfaith Center has set for itself. We all know that this kind of compression of different traditions is causing increasing conflict everywhere. But we believe that conscious cultivation of interfaith respect based on interfaith exposure and cooperation is the way to go.

So we are about developing curricula for 3rd- and 4th-grade public-school kids to learn the stories and songs of the world's religions just as they are learning about the different continents and their peoples. Just think of the pictures the kids can color! We will work heavily with the whole gamut of religious art and music and poetry and dance and drama. It will be deeply cultural. And we will train rabbis, imams, ministers, and priests to become skilled in conflict resolution. We will work closely with the UN and its agencies. We will have a Web site and also an interfaith gift shop and bookstore. We really want to do everything to make our given connectedness visible and understandable in all its beautiful diversity so that everyone can learn to appreciate both the forest and the trees, both the pigeons and the stars.

Now let me end where I began with our two stargazers, Laurens van der Post and Carl Sagan. On the fourth advent Sunday in December for the last 5 years, Laurens has preached at the cathedral and included in his sermon the story about the Kalahari bushman from his book *A Mantis Carol*.

Van der Post answers a woman's question about the reasons that a bushman dances. There are two different dances, he tells her: the Dance of the Little Hunger and the Dance of the Great Hunger.

> The first one is of the physical hunger the child experiences the moment he is born and satisfies first at his mother's breast, and which from then on stays with him for the rest of his life on earth. But the second dance is the dance of a hunger that neither the food of the earth nor the way of life possible upon it can satisfy. It is the dance of the Bushman's instinctive intimation that man cannot live by bread alone, although without it he cannot live at all; hence the two.

> Whenever I asked them about this great hunger, he writes, "they would only say not only we dancing, feeling ourselves to be raising the dust which will one day come blown by the wind to erase our last spoor from the sand when we die, lest others coming and seeing our footsteps there might still think us alive, not only we feel this hunger, but the stars too, sitting up there with their hearts of plenty, they too feel it and feeling it tremble as if afraid they would wane and their light die, on account of so great a hunger.

When we know that the stars too share our hunger, then life on Earth can really become sustainable. Because we will know its deep urgency in our bones and in our blood.

REFERENCES

Hughes L. 1996. Funeral address for Laurens van der Post. London UK: 20 December 1996.
Van der Post L. 1987. The hunter and the whale: a story. London UK: Chatto and Windus.
Van der Post L. 1994. A mantis carol. Washington DC: Island Pr.

REACHING THE PUBLIC: THE CHALLENGE OF COMMUNICATING BIODIVERSITY

JANE ELDER

The Biodiversity Project,
214 N. Henry Street, Suite 203, Madison, WI 53703

JOHN RUSSONELLO

Belden & Russonello,
1250 I Street, NW, Suite 460, Washington, DC 20005

SCIENTISTS TELL US that we stand on the brink of a great wave of extinction, unparalleled since the demise of dinosaurs. This time our own species is the driving cause. The planet stands to lose an untold storehouse of genetic information. Unique and miraculous expressions of Creation will be erased forever. Lost, too, will be precious threads in Earth's complex tapestry of life called biological diversity. We do not know which are the critical threads that hold together the magical system of oxygen, water, nutrients, food webs, and climate that sustain life on Earth, but when each loss is permanent, there is no turning back.

Scientists are worried about the future survival and well-being of humans on a planet whose life-support systems are being eroded and changed so rapidly. But where is the public outcry, the mandate for action to stem the loss of biodiversity? We have learned from the debate on global climate change that even when there is widespread and convincing scientific evidence of impending environmental danger, the public does not rise automatically to demand a political response.

Americans have lived with messages about environmental Armageddon since the first Earth Day, and they continue to be bombarded with fearful messages ranging from water pollution to destruction of the rain forests. How do we reconnect the American public with the natural world and engage its involvement in stemming the biodiversity crisis? In this paper, we discuss the context of biodiversity as a public issue and how Americans perceive it, and we recommend approaches for increasing public awareness and action.

BIODIVERSITY: CONCEPT AND CONTEXT

Scientists coined the term *biodiversity* in 1986 to describe the diversity of life and life systems on Earth and as a way to focus growing concern and expertise within the scientific community on the high rate of extinctions throughout the world. After more than a decade of use within scientific and environmental circles, *biodiversity* is commonly defined as the diversity of genes, individuals, species, and habitats on Earth. This definition, however, fails to convey the concepts of interconnectedness and interdependence and ecological processes, which most conservation biologists also associate with the term. In the environmental community, the term *biodiversity* has been adopted as a shorthand description for the variety of species in an ecosystem, a definition that frustrates those who seek to convey a richer and more complex meaning with the word.

In retrospect, we can only wish that some linguists had been among the scientists involved in planning the first National Forum on Biodiversity in 1986. The word describes a scientific construct; it was not part of common English then, nor is it now. Biodiversity is both a challenging concept and a difficult word around which to design a public-education strategy. It requires explanation. It is simultaneously a cause and a scientific term (Takacs 1996), and it suffers from carrying both meanings. Terms like *clean water* and *safe sex* are elegantly simple and easy for the public to grasp, having common meanings and adjective-noun structure. *Biodiversity*, unfortunately, is not a user-friendly word. In an age of sound bites and slogans, the word provides us with a challenging starting point for public education.

Focus-group research commissioned by the Communications Consortium Media Center (CCMC) in 1995 (Belden and Russonello 1995) revealed that the word *biodiversity* communicates different types of life, but it does not imply other key concepts surrounding biological diversity, such as interconnectedness and ecological relationships. A more familiar term, *ecosystem*, was used by focus-group participants, who understood "eco" to refer to the environment and "system" to the interconnected parts.

Champions of the word *biodiversity* need to link the term to the other concepts that help define its implications. The 1996 CCMC Biodiversity Poll (Belden and Russonello 1996) showed that only one in five Americans has an awareness of the term *biological diversity*, but once it is explained to them, Americans overwhelmingly express support for the concept of protecting habitats and species, at least superficially. In addition, many Americans easily grasp the concepts of interconnectedness and interdependence of life. This alone is an encouraging foundation for public education.

A NEXUS FOR ACTION, OR A COMMON COMPONENT OF MULTIPLE ACTION AGENDAS?

Almost every environmental issue embraces some aspect of biodiversity, but most environmental-policy and activist groups do not focus their work through the lens of biodiversity. Biodiversity provides valuable scientific justification for protecting wilderness, large landscapes, endangered species, and many other long-standing objectives in the environment movement. As a result, biodiversity has been embraced widely in the environment community and invoked in countless debates as a new reason for protecting natural habitats and species. In spite of this enthusiasm, for a large portion of the environment community, protecting biodiversity is simply one more reason to achieve more traditional end points: saving places, stopping pollution, or moving a particular policy through the system. As one forest activist exclaimed at a Biodiversity Project working-group discussion on forests, "If biodiversity will help me save my forest, I'll talk about biodiversity. If it won't, forget it!"

Unlike "wilderness," for example, biological diversity lacks a driven, grassroots, quasi-religious base of activism with a clear policy agenda. Conservation of biodiversity, whether labeled so or not, is part of the overarching agenda of such organizations as the World Wildlife Fund, Defenders of Wildlife, The Nature Conservancy, and the National Audubon Society. However, no national "biodiversity coalition" or similar coalescing point exists. In the overall tapestry of the environmental agenda, biodiversity is the warp: It holds the fabric together but is not seen on the surface.

This hard-working common thread invites us to frame the educational message and strategies for conserving biodiversity in multiple ways across issues and agendas. For example, clear-cutting, destruction of wetlands, and toxic pollution of the food web are all biodiversity issues. In some instances, biodiversity itself is the issue, such as the UN Convention on Global Biodiversity. Arguably, the Endangered Species Act is almost exclusively a biodiversity issue. However, each of these large-scale policy debates embraces only a portion of the broader public debate that will determine the fate of biodiversity in the long run.

PUBLIC PERCEPTIONS AND ATTITUDES

In February 1996, the public-opinion firms of Belden & Russonello and R/S/M, under the auspices of the CCMC, conducted a national public-opinion poll on biodiversity and the environment to define more sharply the findings of the focus group conducted in 1995. The telephone survey was administered to 2,000 adults in late February and early March 1996.

In the last 10 years, the public has consistently supported government action to protect the environment, and the 1996 poll indicated that large majorities are in favor of maintaining strong clean water (85%) and endangered species (76%) acts. There is, however, a limit to the public's approval of government action. Support for the environmental position drops off on issues that juxtapose competing values, such as an individual's private-property rights versus protection of public resources like wetlands or endangered habitats.

The 1996 survey showed that majorities of Americans are aware that species are being lost (69%) and that humans are the cause (59%), but appreciation for biological diversity proves to be superficial when such countervailing pressures as jobs, property, or human convenience are introduced. In the poll, 87% of Americans expressed support for maintaining biological diversity, that is, preventing the extinction of plants and animals. However, this broad support can be eroded quickly; 48% of the public said that protecting jobs is more important than saving habitat and 49% that it is acceptable to eliminate some species of plants and animals. This decline in support for maintaining biodiversity when other priorities come into play tells us that the major task for environmentalists is not simply to reinforce the facts about loss, but to demonstrate clearly why the loss is so important to our lives and our world. The poll offered some insights for communication about the effect of the loss of biological diversity. The educational messages that register the most concern are those related directly to dangers to human health and threats to habitats and ecosystems that clean our air and water. Beyond these human-centered reasons, educational messages that appeal to the appreciation and enjoyment of places in nature are also of broad concern to the public: loss of ancient forests that cannot be replaced, places of natural beauty, and recreational areas. Other areas of high public concern are the elimination of possible new medicines to cure diseases and the loss of jobs in fishing and tourism owing to a loss of biodiversity.

Making sense of the clash of public concerns over the environment requires an understanding of the values that underlie attitudes. The poll identified responsibilities to family and saving Earth for future generations as the most widely held values that form attitudes toward the environment. Other values—such as respecting God's Creation, aesthetics, personal use and enjoyment, patriotism, and a belief in nature's rights—were fundamental to segments of the public but not as broadly held as responsibility to family and future generations.

Thus, the 1996 survey revealed that Americans will be most responsive to messages about biodiversity that address the values of family, responsibility to future generations, and, for some audiences, respect for God's Creation. Education about the meaning of biodiversity for humans and the value of habitats will be a key to building greater public commitment to maintaining biodiversity in the future.

WHAT SHOULD AN "AWARE" PUBLIC KNOW?

Early in its development, the Biodiversity Project identified three fundamental educational goals for building broad public support for policies, practices, and personal behaviors that will maintain biological diversity. These are to increase comprehension of biodiversity issues, to heighten recognition of the threats to biodiversity, and to generate public support for policies and actions to reduce those threats.

Although the public has a broad appreciation for protecting the web of life, public-opinion research shows that this appreciation is shallow. If Americans do not understand the basic components of the living environment and the policies that influence that environment, they will have difficulty recognizing or truly caring about what diminishes biological diversity. Moreover, if individuals do not comprehend their particular connection to living systems and species, they are less likely to be motivated to care or act.

Our perceptions are shaped by what we experience in life, and to many Americans biodiversity (or the balance of nature) is only a concept, not something that can be seen or experienced. Yet the workings of biodiversity are all around and within us. Educators and communicators need to be able to illustrate biodiversity so that it can be seen and understood. For example, if Americans cannot distinguish between a pine plantation and a healthy natural forest, they will have difficulty grasping the value of biodiversity.

We also face the challenge of reconnecting the American public to the natural systems and species in their home ecosystems. If biodiversity is only something that happens "out in nature," we will lose the ability to motivate many Americans to change public policies and behavior as consumers. By linking biodiversity and habitat to our air, water, food, and so forth, we can make essential connections to the local landscape and living systems that are in our daily experience as well as provide a basis for making global connections: We can make biodiversity tangible.

A general cognizance that humans are the primary cause of extinctions and loss of habitat is insufficient to help the public recognize and respond to threats to species and critical habitats. The public must become aware of and knowledgeable about specific causes of loss of biodiversity and the actions that individuals and society can take to address these problems.

Environmentally sensitive policies and community practices will come about only if the public can be engaged to support positive changes. To translate concern and awareness into action, the public needs to understand what it can do and then be inspired and empowered to take action. Americans have become more mistrustful of government institutions, and they are searching for solutions and actions that individuals and communities can take themselves. These individual actions need to have some direct (or easily understood indirect) effect on conservation and need to expand beyond practices like recycling, which are already widespread.

At the same time, the public needs to participate more in major policy decisions to offset the pressures that are driving environmentally damaging policies. Activists need to address policy with attention to values and the public's primary concerns about the environment. Jargon or technical language and government processes themselves present barriers to communication with the public. In summary, we propose the following educational objectives as a starting point for framing a strategy to increase public awareness and involvement:

- Help Americans recognize biodiversity in their everyday experiences.
- Help the public understand its dependence on nature.
- Raise fundamental ecological literacy.
- Help the public understand the specific effects of humans on biodiversity.
- Help the public understand its capability to act to conserve biodiversity.
- Motivate the public to act to conserve biodiversity.

ENGAGING THE PUBLIC

Public-education strategies for biodiversity need to be designed on the basis of widely accepted, solid scientific grounds to sustain their credibility and on the basis of widely held values and concerns that will engage the public. We recommend taking a broad approach to education about biodiversity, which uses the tremendous energy and activity in the broad range of issues in which biodiversity is a key element, rather than seeking to raise the profile of biodiversity as a stand-alone concept. Many issues can raise the profile of biodiversity when it is an obvious element, such as protection of forests, wetlands, and marine fisheries. For other issues, such as suburban sprawl and climatic change, in which it is not so obvious, we will need to direct attention to the idea of biodiversity.

Regardless of the issue, the dialogue for education should begin with easily grasped concepts rather than with scientific statistics or pronouncements of impending doom. Aldo Leopold cautioned that it is important to keep all the parts (Leopold 1978), and this is perhaps the fundamental principle from which strategies for public education about biodiversity can begin. This and similar principles, such as the value of keeping all the "connections," can provide a framework for public awareness, through which the public can evaluate and respond to rapidly changing information and debates about policy. The concepts are far more important than the word "biodiversity." We should use familiar terms like "nature," "web of life," and "ecosystem" to introduce biodiversity to the public.

Ultimately, conserving the diversity of life on Earth will require action on global and individual levels and on many levels in between. We need to reach Americans as parents, as consumers, as participants in their communities, and as citizens of Earth. Education about biodiversity needs to be well grounded in science, but scientific information must be translated in a manner that can resonate with the public. Moreover, it is not enough for the public to be aware of the loss of biodiversity and threats; the public must be given the means and the motivation to participate in the decisions that will form the basis of conservation of biodiversity. Providing the motivation for citizens to participate as players in a democratic society is perhaps our greatest challenge.

A clear and widely embraced domestic policy and action agenda for biodiversity, with tangible goals and objectives, is one step that would help engage the public; it provides a starting point for solutions, and it provides benchmarks for charting progress. Organizing an agenda-setting dialogue among leading scientists and nongovernment organizations could provide a forum for exchanging ideas and developing such an agenda and could serve as a test of using biodiversity as a nexus for advancing policy on many issues.

WHO CAN DO THE JOB?

To raise public awareness to a level at which conservation of biodiversity is integrated into public policy, consumer behavior, and corporate accountability, we must move beyond traditional public-education efforts that are linked to a specific policy for a short term. Legislative and regulatory mechanisms alone will not save sufficient habitat and species to sustain biodiversity, and, even if they could, we currently lack a popular mandate to enact the policies that would be effective. Thus, we must embrace new strategies that reach citizens at the following levels:

• values and ethics through religious, cultural, and community institutions;
• fundamental literacy and critical thinking, through formal and informal educational venues; and
• awareness and participation in social and political issues through news and popular-cultural media and through consumer and health education.

The responsibility to carry this agenda forward falls on a broad array of institutional communities within the biodiversity-conservation family. It includes the policy and advocacy groups, the land trusts and conservancies, recreational-user groups, environmental educators in many sectors, scientific and academic leaders, the grant-making community, organized religion, health and medicine, Hollywood, Madison Avenue, the news media, and even corporations. Forging a partnership at this scale is untested and unprecedented in circles of environmental education and policy, but the scope of the task begs for a concerted effort.

The fraying tapestry of life on Earth respects neither human institutions nor the challenges of working across traditional boundaries of specialty, discipline, and expertise. At the Biodiversity Project, we are persuaded by the daily, if not hourly, disappearance of species and by the rapid destruction of habitat throughout the globe that the urgency of this issue demands creative new responses. Future generations will not forgive our hand-wringing at how large the task is; they will thank us only if we rise to this challenge of survival and embrace our partnership with the diversity of life on Earth.

REFERENCES

Belden N, Russonello J. 1995. Communicating biodiversity: summary of focus group research findings conducted for the Consultative Group on Biological Diversity. Available from the authors.

Belden N, Russonello J, Breglio V. 1995. Human values and nature's future: Americans' attitudes on biological diversity. A public opinion survey analysis conducted for the Communications Consortium Media Center. Available from the authors.

Elder J, Farrior M. 1997. Report of the Biodiversity Project Message Development Working Group on Forest Ecosystems. Workshop held 15 April 1997. Available from the authors.

Leopold A. 1978. Sand County almanac. New York NY: Ballantine Books. p 190.

MacWilliams, Cosgrove, Snider, Smith, Robinson. 1996. Final presentation, Mississippi River Project, a private study prepared for the McKnight Foundation, February 1996.

Takacs D. 1996. The idea of biodiversity. Baltimore MD: Johns Hopkins Univ Pr.

CENTER FOR ENVIRONMENTAL RESEARCH AND CONSERVATION (CERC): A NEW MULTI-INSTITUTIONAL PARTNERSHIP TO PREPARE THE NEXT GENERATION OF ENVIRONMENTAL LEADERS

DON J. MELNICK*
MARY C. PEARL†
Center for Environmental Research and Conservation,
Columbia University, New York, NY 10027
*Departments of Anthropology and Biological Sciences,
Columbia University, New York, NY 10027
†Wildlife Preservation Trust International,
1520 Locust Street, Suite 704, Philadelphia, PA 19102

SOLUTIONS TO THE crisis of biodiversity loss will be as complex as the forces that led to it. Therefore, the serious task of educating and training current and future leaders of the public and private sector must include all areas of environmental management and conservation that are critical to the health of the ecological-support systems on which we rely. The fields of environmental management and conservation not only are complex and multidisciplinary, but also extend beyond the realm of the natural sciences.

We describe here the outcome of the first 5 years of a long-term effort to build an innovative and productive training and research consortium, the Center for Environmental Research and Conservation (CERC), through a new multi-institutional partnership of existing organizations. In New York, we have a unique opportunity because of the presence of several biodiversity-research institutions of international caliber complemented by an internationally renowned research university. Hence, we chose a strategy to bring them together: Columbia University, the American Museum of Natural History, the New York Botanical Garden, the Wildlife Conservation Society, and Wildlife Preservation Trust International.

SCOPE

Much has been written about how education and training in environmental conservation should be delivered (Jacobson 1995). Universities in the United States have been criticized because their professors and students alike are so narrow in their perspective that the universities fail to turn out graduates who can find employment in government agencies and nongovernment organizations (NGOs), much less make a substantial contribution to environmental conservation if they could find a job (Noss 1997). In fact, the need to broaden the definition of who should teach in this field, as well as who should be taught, was one motivator for the development of CERC.

It is clear, from the type of institutions and researchers that have been brought together to form the consortium, that we are not using a narrow definition of a research-university professor as the conveyor of knowledge. Our multi-institutional staff has a range of experience in academic institutions, research institutions, NGOs, and government agencies and as field-based practitioners. Equally diverse are our students, who are upper-level high-school students, undergraduates, graduate students, midcareer professionals, and those who already occupy positions of leadership in government and nongovernment institutions. Perhaps most important, we have sought the participation of both teachers and students who have diverse personal backgrounds; for example, those who come from nations of high or unique biodiversity and those from groups that are underrepresented in the academic and conservation communities in the United States.

By providing education and training programs that extend from high-school students to high-level environmental managers, CERC hopes first to generate interest among and identify those high-school and college students who have the greatest aptitude for environmental conservation. Second, we hope to provide unique opportunities to build the capacity of future environmental leaders at the level of graduate students and midcareer professionals. Finally, we hope to enhance the background knowledge of current environmental leaders and their staffs. We believe that such a broad-based approach is needed if we are to find solutions to the complex environmental problems we face now and those we will face in the future. If our five institutions can implement the consortium's goals successfully, then, in the process, we will have created a new model for conservation education and research in which the expertise of each institution is brought to bear on significant issues beyond the scope of any one institution.

STRUCTURE

The central administrative and education facilities of CERC are on the campus of Columbia University. Facilities include offices for administrative staff and faculty, a computer and student center, seminar and lecture rooms, a teaching and research greenhouse, and a planned, integrated set of teaching and research laboratories. In addition, students, visiting scientists, and teaching and research staff have a variety of laboratory, library, and computer facilities available on the same campus. It is at these facilities that many of the programs we will describe are based, but the true strength of the CERC consortium is realized in the activities, facilities, and expertise of the staff drawn from the five consortium-member institutions.

The American Museum of Natural History has one of the world's most extensive collections of animal species, and its staff have world-renowned expertise in identifying, cataloging, and systematizing animal diversity. Through the use of collections, individual research, and participation in field projects organized by individual curators and the Museum's new Center for Biodiversity, the museum staff offer an array of opportunities for animal-diversity research.

The New York Botanical Garden (NYBG), which has the largest herbarium in the Western Hemisphere plus a botanical library of more than a million books, journals, and other items, is the most comprehensive botanical-research center on a single site in North America. Its participation in the consortium is through its two key research divisions, the Institute of Economic Botany and the Institute of Systematic Botany.

The Wildlife Conservation Society, which was founded as the New York Zoological Society, has the largest field staff of any international conservation organization based in the United States. It conducts more than 250 field projects in more than 50 countries throughout Latin America, Africa, and Asia. The Bronx Zoo in New York—which has experts in captive breeding, veterinary medicine, school-curriculum development, and data analysis—complements these field programs.

Wildlife Preservation Trust International (WPTI), working through local conservation scientists and educators, conducts interdisciplinary, small-scale projects at the grassroots level in Latin America, the Caribbean, Asia, and Africa. WPTI works to protect threatened species and their habitats in areas where human pressures and human-wildlife conflicts exist in highly diverse or unique ecosystems. To train local conservation professionals, WPTI provides on-site backup and formal courses.

Columbia University in New York City is one of America's top research universities and the most internationally oriented. CERC is integral to the university's Columbia Earth Institute (CEI), which brings together scientists from a broad range of natural- and social-science disciplines to understand better how the earth works and to mitigate the negative effects of human activities on natural systems. In addition to CERC, eight divisions of the CEI or the university itself have direct relevance to CERC's mission.

• The School of International and Public Affairs offers a Master's of International Affairs concentration in environmental policy and collaborates in the environmental-policy certificate offered by CERC to its PhD students.

• The Lamont Doherty Earth Observatory supports graduate education in the earth sciences and is the home of the new International Research Institute for Climate Prediction.

• The Goddard Institute for Space Studies was the first to identify global warming.

• The Biosphere 2 Research Center (in Arizona) provides opportunities for conducting closed-system ecological research and relating it to other atmospheric research.

• The Black Rock Forest, which is run by a consortium that includes Columbia University and the American Museum of Natural History, is a temperate research forest 50 miles from New York City that has an array of habitats, including a patch of rare primary, climax vegetation.

- The Rosenthal Center for Alternative/Complementary Medicine is an institute that investigates the relationships between traditional health practices, ethnobotany, and related areas of environment, biodiversity, and economic development.

- In the School of Public Health, the Division of Environmental Health Sciences is involved in a number of international projects that focus on the relationships between environmental change, climate variations, and emerging diseases.

- Finally, the Program for Information and Resources is composed of a group of applied mathematicians and economists who study the relationship between environmental changes and economic trends.

All these schools, centers, and institutes offer a broad array of expertise to our students and visiting scientists, and they provide a variety of powerful resources for all the member institutions of CERC.

GOVERNANCE

To ensure the participation of all member institutions of CERC, we have created joint committees to assist in the development of our education, training, and outreach programs, to evaluate small-grants proposals, to identify potential affiliate centers throughout the world, and to handle general issues of interinstitutional integration and consortium policy. In addition, undergraduate and graduate students are mentored by both Columbia faculty, some hired specifically for CERC's educational programs, and adjunct faculty, consisting of selected staff members from all five CERC institutions.

PROGRAMS

The programs of CERC are in three areas: degree-granting education, professional training and public outreach, and interdisciplinary research. In all areas, the paramount goal is to provide opportunities for individuals to improve their ability to assess the effect of human activities on natural environments and the services they supply and to help develop the scientific, economic, and political means to mediate effects that adversely affect important ecosystems and the species they contain.

Degree-granting Education

Undergraduate Program. CERC staff have designed a new interdepartmental undergraduate major, Environmental Biology. Graduates of this major have a strong foundation in the life sciences and an exposure to relevant fields in the social sciences, such as economics and anthropology. This major also provides the necessary training for students who wish to pursue graduate studies, such as the new biodiversity-conservation-based Ecology and Evolutionary Biology PhD program offered by CERC, the Conservation Biology MA program offered by

CERC, and the social-science-based Environmental Policy master's programs offered by Columbia's School of International and Public Affairs (SIPA).

The undergraduate course of study includes a two-semester sequence of earth and environmental sciences, a two-semester sequence of molecular and organismal biology, and several other introductory courses in physics, chemistry, and quantitative methods. These introductory studies are followed by elective courses in such areas as environmental policy and economics, conservation and population biology, ecology and behavior, and evolution and genetics. The courses are taught by Columbia faculty and by adjunct faculty from CERC. These collaborations have borne productive fruit, including cross-cutting courses in ethnobotany and human ecology and in biodiversity loss and human disease.

Besides the novel policy—social-science component as part of a natural-science major—we require and facilitate a summer internship for each student major between the junior and senior years at CERC. In many cases, these internships are overseas and place students with researchers from one of the five CERC institutions. They also may be in New York, where they may involve collections or policy-related research. The purpose of these internships is to expose students to practical biodiversity-conservation research and to help them focus their interests and goals for their future careers.

Master of Arts Program in Conservation Biology. The 2-year, stand-alone MA program emphasizes the biological sciences but includes a basic foundation in environmental policy. After taking specially developed MA core courses in the natural science of conservation biology and the social science of environmental policy, students have the option of tailoring their remaining coursework to follow either an academic or a professional track. The academic track is designed for students who wish to continue on to a PhD program, and the professional track is for students pursuing positions with nongovernmental organizations, government agencies, or consulting firms concerned with the conservation of biodiversity or environmental protection.

In addition to the required coursework, all students must complete an internship and a thesis. This experience provides the practical experience necessary to pursue a career in a field related to natural-resource conservation. For the academic track, the internship, done during the summer after the first year of the program, is usually conducted in the field or in a laboratory at one of the CERC institutions; a thesis results from this research. Students in the professional track can do original research or conduct their internship with a conservation organization or government agency. In the latter case, a report or policy formulation is the expected focus of the thesis.

PhD Program in Ecology and Evolutionary Biology. The Ecology and Evolutionary Biology (EEB) program is designed to provide the broad scope of education needed to describe, understand, and conserve the diversity of life on Earth. This program offers specializations that are strictly biological (ecology, evolution, systematics, and population biology) or are at the interface of biology and human activities (ethnobiology). The aim is to prepare students to conduct ecological, behavioral, systematic, molecular, genetic, and other evolutionary-biological research and to formulate or implement biodiversity-related environmental policy. Graduates of the program likely will pursue academic careers as researchers and teachers or take professional positions in national- and international-conservation, environmental, and multilateral-aid organizations. It is also our hope that some of our graduates will fill public-sector positions in environmental ministries, national park systems, and other agencies that deal with environmental conservation and sustainable-development planning.

Students in this program take two core courses at the outset that include the basics of ecology, evolution, systematics, population biology, and genetics. This initial semester is followed by several semesters of electives in these areas and at least three laboratory or field-based research internships. After passing the qualifying examinations and an oral defense of their research proposal, students engage in original research. The CERC consortium is so rich with conservation-oriented research projects being conducted around the world that the opportunities for internships and research projects are enormous in scope and number.

Perhaps the truly unique aspect of this PhD program is that all the students must complete a certificate in environmental policy. This certificate program is designed to give candidates in the biological sciences a better understanding of the workings of the markets, policy, and law that affect the efforts to preserve biodiversity. The certificate program, co-organized with SIPA, includes the completion of six courses and one internship, participation in a problem-solving workshop, and preparation of one interdisciplinary research paper.

Outside the CERC program, we also offer a parallel certificate in conservation biology for social-science PhD students, to give candidates in the social sciences a strong foundation in areas of biology that will enable them to contribute as much as possible to the formulation of sound environmental policy.

Professional Training and Public Outreach

Training and outreach to nontraditional, non-degree-oriented students are often conducted by nonacademic institutions rather than universities, although universities harbor the research infrastructure and pedagogical resources to have a major effect on this audience. Recognizing this, CERC has established the Morningside Institute (MI) as a way of directing energy and resources to these nontraditional audiences, giving them the opportunity to fill in gaps in their academic background, gain new skills, and profit intellectually from the interaction and exchange of ideas with others in the CERC community.

The MI programs include career days, 1-day workshops designed to bring the staff of the five CERC institutions up to date on the latest areas of inquiry and the latest technology used in biodiversity conservation and environmental management. Formal training in these areas is relatively recent, and many individuals have learned their trade on the job. To enhance the ability of every member institution in CERC to work more effectively, we have offered workshops in such areas as environmental economics, conservation genetics, geospatial positioning systems and mapping, the use of computers and the Internet in environmental management, environmental ethics, environmental education, and ecotourism.

These workshops have been a great success in providing exposure to new fields and have provided an opportunity for interaction among people who have similar interests, which can lead to new research collaboration.

Another MI program is the annual Environmental Leaders' Forum (ELF), which was created to help high-level conservation managers from countries of high or unique biodiversity in Asia, Africa, Latin America, the Caribbean, and eastern Europe to develop strategies to carry out their individual mandates for conserving biodiversity. The curriculum includes sessions on strategic planning, on emerging techniques such as conservation genetics, in systematics research, in population biology, and in resource economics. The member institutions of CERC participate by presenting to the environmental leaders the types of research and training opportunities they offer. Each ELF culminates with a group statement on the status of biodiversity conservation and the critical needs of the developing world in this area.

Each participant becomes part of a growing communication network of environmental leaders around the world. An electronic newsletter has been developed to invite communications and observations from all former participants and to provide an opportunity for information exchange among high-level managers around the world. So far, 66 leaders from 29 countries have participated in the ELF program. We believe that ELF is an important instrument of training and communication for conservation managers. For example, a program of community use that was established by the head of Chitwan National Park in Nepal was presented at the ELF and likely will be considered by the environmental ministry in Cameroon, whose head officer also attended that forum.

The High School Summer Program is a 1-month intensive course for high-school juniors and seniors that includes lectures and practical field projects. During this course, teachers from the CERC faculty and staff engage students in grappling with critical biological and policy issues at a level that they can understand. They visit nearby protected areas and meet staff from the member institutions of CERC as part of the course. In this way, the information they get is given life, and they can begin to identify potential role models and career paths. We believe this program will become a major vehicle for generating interest in environmental conservation among high-school students in our region.

The Visiting Scholars Program is designed to allow selected scholars and environmental-resource managers to spend 1–6 months at CERC. It is intended to provide the intellectual environment necessary for an international group of practitioners to read widely, interact with other environmental scholars, and write on their concerns. The resources we have at the CERC institutions, particularly our libraries and computer systems, are resources we want to share for the completion of critical research and of the formulation of policy.

The Mid-Career Certificate in Conservation Biology is a two-semester sequence in the science, techniques, and policy of conservation biology that is designed for professionals who want to enhance their knowledge to perform their jobs better or to reorient their careers. Courses are held in the evenings and are taught by the faculty and research staff of CERC. Classes are designed to give participants a broad overview of the basic science of conservation, the techniques used to conduct research, and the social-science issues that are related to policy development. The program consists of four intensive courses that cover the theory and practice of conservation biology and resource use by humans. The brevity of the training program, compared with degree programs, still provides substantial coverage of key topics in conservation science, but it is extremely appealing to individuals from fields as diverse as law, business, finance, public service, and teaching. More recently, we have designed specific tracks in this program for high-school science teachers and international trainees.

INTERDISCIPLINARY RESEARCH

Small-Grants Program

To stimulate new cross-disciplinary, cross-institutional research in biological conservation, we have established a small-grants program to assist in the development of research teams and the collection of preliminary data that would stimulate research on a much larger scale. Projects this program has supported so far and the disciplines they have involved include developing models for trading in carbon-sequestration credits (agronomy, economics, and human ecology), preliminary research on the relationship between climatic variability and vector-borne diseases (climatology, ecology, and epidemiology), and a multifaceted approach to conservation of manatees, including basic biological monitoring and community-based ecotourism (marine biology, community development, and wildlife management).

Affiliate-Centers Program

We have set up affiliate centers in Indonesia and Brazil and are engaged in discussions with colleagues in Madagascar, Vietnam, and Belize. We hope to have 10 such affiliates around the world that can act as regional training centers and provide a means of identifying individuals who would benefit from one of the many New York-based programs of CERC. In its ideal form, the affiliate-center model involves a university, an NGO, and a government division. In the case of Indonesia, we have forged an agreement with the Center for Biodiversity and Conservation Studies and with the University of Indonesia. In Brazil, we have concluded an agreement with the Instituto de Pesquisas Ecologicas and with the government of São Paulo. In Madagascar, the agreement will be with the major national university and the government. In each case, a small annual budget is provided to the affiliate center to improve facilities, initiate innovative research, and assist students in the pursuit of their training.

CONCLUSION

The Center for Environmental Research and Conservation is an ambitious experiment to meld the many strengths in science and policy of its member institutions into a cohesive effort that can make a difference globally in the education and training of both current and future environmental leaders. It is a model worth trying in other regional centers of biodiversity research in the United States and abroad. Indeed, given the rapidity with which biodiversity is being lost, we, as a community of researchers and practitioners, need to move as quickly as possible to make the most of our collective resources. To do less would be difficult to comprehend and even more difficult to defend.

REFERENCES

Jacobson SK (ed). 1995. Conserving wildlife: international education and communication approaches. New York NY: Columbia Univ Pr. 302 p.

Noss RF. 1997. The failure of universities to produce conservation biologists. Cons Biol 11(6):1267–9.

NATURAL CAPITALISM

PAUL G. HAWKEN
Gate Five Road #20 South Forty,
Waldo Point Harbor, Sausalito, CA 94965

THE WORLD is entering a period of historical and economic discontinuity that will change our lives in radical ways. The discontinuity is brought about by a fundamental shift in the relationship between industrialism and living systems. Industrial systems have reached pinnacles of success and are able to muster and accumulate human-made capital on vast levels, but living systems, which are the sources of our natural capital, and on which we depend to create our industrial capacity, are all declining.

Humankind has a long history of destroying its natural capital, especially soil and forest cover. The entire Mediterranean region shows the effects of siltation, overgrazing, deforestation, and erosion or salinization caused by irrigation (Hillel 1991). In Roman times, one could walk North Africa's coast from end to end without leaving the shade of trees; now it is a blazing desert. Today, human activities are causing global decline in all living systems. The loss of 750 metric tons of topsoil per second worldwide and 5,000 acres of forest cover per hour becomes critical. Turning 40,000 acres a day into barren land—the present rate of desertification—is not sustainable, either (UNEP 1996). In 1997, more than 5 million acres of forest were destroyed by "slash-and-burn" industrialists in the Indonesian archipelago. The Amazon basin, which contains 20% of the world's freshwater and the greatest number of plant and animal species of any region on Earth, saw 19,115 fires in a 6-week period in 1998, five times as many as in 1995. In the oceans, the losses are similar. Our ability to overfish oceans with 30-mile-long lines results in 20 million tons of annual bycatch—dead or entangled swordfish, turtles, dolphins, marlin, and other fish that are discarded, pushed overboard, tossed back, or, in the case of sharks, definned for soup. The bycatch that is thrown overboard is the equivalent of 10 lbs. of fish for everyone on Earth (San Francisco Chronicle 1998). By now, almost all the world's fisheries are being exploited at or beyond their capacity, and one-third of all fish species (compared with one-fourth of all mammal species) are threatened with extinction. A 7,000-square-mile "dead zone"—the size of New Jersey—is growing off the coast of Louisiana. No marine life can live there, because nitrate runoff in the form of agricultural fertilizers borne by the Mississippi River has depleted supplies of oxygen. The growing marine desert threatens a $26-billion-a-year fishing industry (Yoon 1998). Each fire, each degraded hectare of crop and rangeland, and each sullied river or fishery reduces the productivity and integrity of our living planet. Each of them diminishes the capacity of natural capital systems to process waste, purify air and water, and produce new materials (Hawken and others 1999).

It is often assumed that environmental improvements are expensive—clean water, elimination of dangerous chemicals, efficient nonpolluting transportation, a pesticide-free food supply, preserving our ancient forests, providing for the health and safety of people in nonindustrialized nations. In fact, these and most other environmental improvements can be brought about at a profit, not a cost. To put it differently, the massive inefficiencies that are causing environmental degradation cost far more than the measures that would reverse them. In energy, transportation, forestry, building, and other sectors, mounting empirical evidence suggests that large savings can be achieved by radical, even paradigmatic, improvements in efficiency—not the constant marginal improvements that industry continuously seeks, but leap-frog changes in design and technology that presage a different economic system.

Industrialism was a system of organized mechanistic production that increased the productivity of human beings. It did not replace the system before it, but subsumed an agrarian society within a new framework of production and understanding. In the next century, as human population doubles and the resources available per person drop by one-half to three-fourths, a remarkable transformation of industry and commerce can occur. Through this transformation, society will be able to create a vital economy that uses radically less material and energy. This economy can diminish our use of resources and begin to restore the damaged environment of the Earth. These necessary changes can take place because they will promote economic efficiency, ecological conservation, and social equity. The change in business economics can be called natural capitalism. Natural capitalism recognizes the critical interdependence of the production and use of human-made capital and the maintenance and supply of natural capital.

Natural capitalism includes four distinct yet intertwined patterns of change. The first is a shift from an economy based on incremental improvements in human productivity to one emphasizing dramatic and in some cases radical gains in resource productivity—increases of a factor of 4–10, which means getting 4–10 times as much wealth from the same resources. That is a critical message because much of this productivity revolution is available at "negative cost", that is, profitably. Countries moving toward resource productivity will become stronger, not weaker, in their international competitiveness. The second is the use of biomimicry as the means and basis of redesign of industrial systems. Reducing the wasteful through-put of materials—indeed, eliminating the very idea of waste—can be accomplished by reimagining industrial systems on biological lines, changing the nature of in-dustrial processes and materials, enabling the constant reuse of materials in con-tinuous closed cycles and often the elimination of toxicity. The third is a funda-mental change in the relationship between producer and consumer—a shift from an economy of matter and things to one of *service and flow*. This describes a new perception of value, a shift from the episodic acquisition of goods as a measure of affluence to the continuous receipt of quality, utility, and performance. A fourth stage is a centuries-long reversal in ecosystem and habitat destruction wherein profitable investments will begin to maintain and *increase our pool of natural capi-tal*. All four are interrelated and interdependent, and all four generate numerous other effects in the environment, finance, resources, and society.

RADICAL RESOURCE PRODUCTIVITY

Radical resource productivity means getting the same amount of work or ser-vice from a product or process while using 75–90% less resources. That increases the value we can obtain from each unit of resource and will create vast new op-portunities for business and society. As a society, we have become extremely pro-ductive with respect to labor and capital. Companies and designers will be mak-ing natural resources—energy, metals, cars, water, forests, and oil—work 5, 10, even 100 times harder than before. Radical improvements in resource productiv-ity offer a new terrain for business invention, growth, and development. They are critical because resource productivity will eventually determine which countries and corporations succeed. It is a hopeful concept because it means we can in-crease worldwide standards of living while reducing the energy and materials we use and the impact of their use on the environment. This concept can help to dispel the misunderstanding that core business values and environmental wisdom are incompatible or at odds.

For the last two decades, there has been a quiet design revolution in products, materials use, and energy. There are cars on drawing boards that can cross the country on the equivalent of a tank of gas, buildings that can create more energy than they consume, plastics that can be reused for centuries. The list is long and somewhat technical. Reading about an air conditioner that uses 90% less energy might not fascinate the average citizen, but the fact that it is utterly quiet while dramatically reducing energy costs will be compelling. As you move through life, listen to the din of daily life, the city and freeway traffic, the airplanes, the gar-bage trucks outside your windows and remember this: Most noises are the signs of inefficiency and will disappear as surely as did manure from the streets of 19th century London. If not in a city, then one need only look from the window of a low flying plane to see the enormous devastation and waste of living systems throughout America and other lands. Either way, the signs are everywhere. For reasons that are essentially inevitable, industry will need to redesign everything it makes and does to meet this coming efficiency revolution and in the process greatly reduce its impact on living systems.

BIOMIMICRY

The present industrial system is like a person with a metabolic disorder. It eats too much and gets too little exercise. Our overmature industrial system runs on machines that require enormous heat and pressure, is petrochemically dependent and material-intensive, and requires large flows of toxic and hazardous chemicals that degrade the environment in unforeseen ways. Those industrial "empty calories" end up as pollution, acid rain, and greenhouse gases. The result is bloated amounts of waste that harm environmental, social, and financial systems. Despite the reengineering and downsizing trends that were supposed to sweep away corporate inefficiency, the overall industrial system is only about 1–2% efficient, probably less. (When economists refer to efficiency, they are usually measuring a process or outcome in terms of money—how much labor or other input costs compared with what was produced. Here, efficiency refers to resource efficiency, both material and energy. In the case of energy, it means how much work is accomplished by an input of energy. In the case of materials, it means the total material flow that is required to create a given product or service. Living systems are not affected by monetary calculations. What matters is how effectively we use the flow of energy and material resources to meet human needs. That is the only measure of efficiency that matters over the long term.)

Chemists, engineers, and designers are turning away from mechanistic systems requiring heavy metals, combustion, and petroleum and toward something closer to biological systems that require smaller inputs, low temperatures, and enzymatic reactions. They are moving from linear take-make-waste systems to closed industrial loops where technical nutrients, synthetic materials used in a prior product, become the raw material for successive production. In energy, this means the end of high-temperature, centralized power plants and the growth of small distributive sources feeding a grid. In transportation, it means hybrid-electric vehicles. In fuels, it means a continuing decarbonization of energy sources. In food, it will mean dramatic reductions in input of fuels and chemicals with increasing yields.

To create breakthroughs in radical resource productivity, chemists, materials scientists, process engineers, biologists, and industrial designers are reexamining the energy, materials, and manufacturing systems required to provide the specific qualities—strength, warmth, structure, protection, function, speed, tension, motion—required by products and end users. Business is rapidly switching to biomimicry and ecomimesis (imitating biological and ecosystem processes, respectively): replicating natural methods of production and engineering to produce chemicals, materials, and compounds and soon maybe even microprocessors. Some of the most exciting developments come from emulating nature's low-temperature, low-pressure, solar-powered assembly techniques, whose products rival anything made by humans. Janine Benyus's book *Biomimicry* points out that spiders make silk, strong as Kevlar but much tougher, from digested crickets and flies, without needing boiling sulfuric acid and high-temperature extruders. The abalone makes an inner shell twice as tough as our best ceramics. Trees turn sunlight, water, and air into cellulose, a sugar stiffer and stronger than nylon, and bind it into wood, a natural composite with a higher bending strength and stiffness than concrete or steel. We might never get as skillful as spiders, abalone, or trees, but smart designers are apprenticing themselves to learn the benign chemistry that natural processes have mastered (Hawken and others 1999).

Pharmaceutical companies are becoming microbial ranchers, managing feedlots. of enzymes; chemical companies are rearranging the genes in corn stalks to produce polymers as strong as nylon; biological farming, the precursor of tomorrow's industrial thinking, manages soil ecosystems to increase the amount of biota and life per acre by keen knowledge of food chains, species interactions, and nutrient flows, minimizing crop losses and maximizing yields; meta-industrial engineers are creating "zero-emission" industrial parks and their constituent tenants as an industrial ecosystem in which they feed on each other's nontoxic and useful wastes, just as farmers would intercrop, optimize yields, and nourish predators; and architects and builders are creating structures that process their own wastewater, capture light, create energy, and provide habitat for wildlife, all the while improving worker productivity, morale, and health. This revolution in thinking will cause high-temperature, centralized power plants to be replaced by smaller-scale, renewable power generation. In chemistry, it means an end to the witches' brew of compounds and nasty surprises invented in this century: DDT, PCB, CFCs, thalidomide, Dieldrin, xeno-estrogens, and so on. The 70,000 chemicals manufactured every year have ended up everywhere, as biophysicist Dana Meadows puts it, from our "stratosphere to our sperm", to accomplish functions that can be far more efficient with biodegradable compounds and naturally occurring toxins that imitate nature's assembly techniques. In transportation, ultralight hybrid-electric vehicles will replace carbon dioxide-spewing gas-guzzlers. There will be hydrogen fuel cells to power our cars (theoretically, 5,000 miles between fillups), with onboard 20-kw generating capacity as the utility of the future. There will be printable and reprintable paper that reduces printing-fiber use and forest impact by 90%. In materials, high-strength synthetics made of biodegradable or reusable engineered compounds will become common. Weeds will be grown to make pharmaceuticals and corn stalks to make biopolymeric plastics that are both reusable and compostable; bioremediation will be intensively used for cleanup; luxurious carpets will be made from landfill scrap. Not all those technologies will succeed, and some might have side effects that are unwanted and unexpected. Nevertheless, they and thousands more are lining up like salmon to swim upstream toward a world of radical resource productivity.

SERVICE AND FLOW

Beginning in the middle 1980s, Swiss economist Walter Stahel and German chemist Michael Braungart began to imagine a new industrial model that is now slowly taking shape. Rather than an industrial model wherein goods are sold, they imagined a *service economy*. This was not the often-discussed and conventional definition wherein service workers outnumber manufacturing workers. Their idea of a service economy is based on ecological models. In it, the concept of value undergoes a radical shift. In an industrial society, value is the selling price of a given product. In a service economy, value is measured by the flow of services received by the end user over some period. The industrial model is static and transactional. The service model is dynamic and relational.

Stahel's work focused on product life and durability. As a strategy to reduce the demand for resources and energy dramatically, Stahel proposed that manufacturers think of themselves not as sellers of products, but as providers of long-lasting, upgradable durables that provide customers with services. The product would remain the property of the manufacturer primarily because the focus would shift to the service needed by the user. In practical terms, instead of purchasing a washing machine, you buy the service of clean clothes. Just as in the use of a copying machine wherein you are charged for the number of copies rather than the machine, in the service economy products are valued by the quality and extent of the services they provide. The washing machine remains the property of the manufacturer. This would apply to computers, cars, and hundreds of other durable products that we now buy, use up, and ultimately throw away. The Carrier Corporation, a division of United Technologies, is now selling warmth and "coolth" to companies while retaining ownership of the equipment. The Interface Corporation is leasing carpets. Agfa Gaevert pioneered the leasing of copiers. Stahel's focus was on selling results rather than equipment, performance and satisfaction rather than motors, fans, plastics, or condensers.

In a service economy, the products are returned to the manufacturer, broken down, and then used to make new products. This concept of "cradle-to-cradle" was invented and first articulated by Stahel, who also named it "extended product responsibility" (EPR). EPR is now becoming a mandated or voluntary standard in European industry. The concept of an economy consisting of a flow of services rather than an amount of material products meshes extraordinarily well with biological concepts of ecosystem flows on which industry depends.

Braungart's model of a service economy focused not on product life, but on material cycles. Even if a product lasts longer, but the materials used cannot be reincorporated into new manufacturing or biological cycles, then society is still creating cumulative waste with its attendant problems of toxicity, worker ill health, and environmental damage. Braungart, working with architect William McDonough, proposed the *intelligent product system* wherein products that were not compostable would be designed so that they could be completely reincorporated into *technical nutrient* cycles of industry. In other words, all products would become the raw material of future products. Another way to look at Braungart and McDonough's concept is to imagine an industrial system with no landfills. If you knew that nothing that came into your factory could be thrown away and that everything you made would come back, how would you design the materials and products? That is precisely how Earth works. Braungart and McDonough's system is essentially an industrial system that mimics the nutrient cycles that maintain life on Earth.

INVESTING IN NATURAL CAPITAL

Businesspeople are familiar with the traditional definition of capital as accumulated wealth in the form of investments, factories, and equipment. But natural capital consists of the resources we use, both nonrenewable (such as oil, coal, and metal ore) and renewable (such as forests, fisheries, and grasslands). Although we usually think of renewable resources in terms of desired materials, such as wood or fish, their most important value is the *services* that they provide. Living systems feed us, protect us, heal us, clean the nest, and let us breathe. These services are related to but distinct from resources. They are not pulpwood, but forest cover; not food, but topsoil. They are the "income" derived from a healthy environment: clean air and water, climate stabilization, rainfall, ocean productivity, fertile soil, watersheds, and the less-appreciated functions of the environment, such as processing of waste, both natural and industrial.

A capitalistic system needs all three types of capital: financial capital in the form of money, investments, and monetary instruments; manufactured capital in the form of infrastructure, machines, tools, and factories; and natural capital in the form of resources, living systems, and ecosystem services. The industrial system is a transformation of natural capital in the form of energy, metals, trees, soil, water, and so on, into human-made capital: goods, highways, cities, transport systems, houses, food, and services, such as health and education. It was an ingenious system and continues to be especially now as computer and telecommunication technologies revolutionize our lives. A system based on natural capital recognizes the critical dependence between the production and use of human-made capital and the maintenance and supply of natural capital. Costanza and others, writing in *Nature* (15 May 1997), estimated that the flow of ecosystem services flowing directly into society from our stock of natural capital is worth $17–54 trillion a year. World GDP in 1998 is about $39 trillion. The approximate valuation provides some measure of the value of natural capital to the economy.

Former World Bank economist Herman Daly believes that humankind is facing a historic juncture in which, for the first time, the limit to increased prosperity is not human-made capital, but natural capital. For example, the limits to increased harvests of fish are not boats, but productive fisheries; the limits to irrigation are not pumps or electricity, but viable aquifers; and the limits to pulp and lumber production in many areas are not sawmills, but forests.

Historically, economic development has faced a number of limiting factors, including labor, energy resources, and financial capital. A limiting factor is one whose lack prevents a system from surviving or growing. If marooned in a snowstorm, you need water, food, and warmth to survive. The scarcest one is the limiting factor. Having more of one factor cannot compensate for the lack of another. Drinking more water will not make up for lack of clothing if you are freezing, and having more clothing will not feed you. Limiting factors cannot be substituted for one another. They are complements; as with the mountaineer marooned in a snowstorm, the scarcest complement is what must be increased if the enterprise is to continue.

The economy has faced limiting factors to economic development in the past—labor, energy resources, and financial capital. Industrial countries were able to continue to develop economically by increasing the limiting factor. It wasn't always pretty. Labor shortages were "satisfied" shamefully by slavery, as well as by immigration and high birth rates. Energy came from the discovery and extraction of coal, oil, and gas. Labor-saving machinery was supplied by the industrial revolution. Tinkerers and inventors created steam engines, spinning jennies, cotton gins, and telegraphy. Financial capital became universally accessible through central banks, credit, stock exchanges, and currency-exchange mechanisms. When new limiting factors intervene, everything changes, nothing works as before, and a restructuring of the economy occurs.

Daly (1994) believes that the current relationship between natural and human-made capital gives rise to the following propositions or principles:

1. If factors are complements, then the scarcest one will be the limiting factor. The question is, Which type of capital is scarcest, human-made or natural?

   Are cars or television sets scarcest? Or potable water, salmon runs, and old-growth forests? Business is already seeking to substitute human-made capital or services for natural capital or ecosystem services. Pure bottled water is the one of the best-selling beverages in the United States (2.95 billion gallons a year) (Hays 1998). There are even "oyu" (water) bars in Tokyo. But bottled water is not a substitute for freshwater flows. The act of manufacturing, storing, shipping, and selling bottled water uses natural capital rather than replacing it, as gasoline, trucks, steel, plastics, highways, ships, stores, lights, paper, and boxes are used to deliver what was once a free good. The more "pure water" is produced, the greater the loss of natural capital. Conversely, the more polluted water becomes, the greater demand for bottled water—a positive-feedback loop.

2. This proposition, according to Daly, gives rise to the thesis that the world is moving from an era in which human-made capital is the limiting factor into an era in which remaining natural capital is the limiting factor.

   There is no threshold point to verify the thesis. Although the complexity of living systems defies simplistic quantification, the *Nature* paper totaling the value of ecosystem services provides a perspective from which to understand the dynamics better. Knowing that freshwater tables are falling in China, Africa, India, and North America, that forest cover continues to shrink by about 17 million hectares per year, that topsoil losses are about 26 billion tons a year, and that thousands of lakes worldwide are biologically dead can become numbing. Seeing the problem in the context of the whole system makes clear the need to move toward upstream solutions—resource productivity, biomimicry, service-and-flow, and restoring natural capital.

   As natural capital becomes a limiting factor, we need to remind ourselves what income is. In 1946, J.R. Hicks defined income as the greatest amount of goods that a community can consume at the beginning of an extended period and still be able to produce the same or greater amount at the end of the period. That requires that the capital stock used to produce income—whether a soybean farm, semiconductor factory, or truck fleet—remain in place and complete. In the past, this definition of income was applied only to human-made capital because natural capital was so abundant. Obviously, it should also apply to natural capital. That means that to retain, let alone increase, income, we have to maintain stocks of both human-made and natural capital.

3. Economic logic requires that we maximize the productivity of the limiting factor in the short run and invest in increasing its supply in the long run.

This is common sense. If you have a distribution system and the roads are falling apart but you have abundant supplies of gasoline and trucks, you fix the roads. The only way to maximize natural-capital productivity is to change consumption and production patterns. Inasmuch as 80% of the world receives only 20% of the resource flow, it is likely that the majority will require more consumption, not less. The industrialized world will need to radically improve resource productivity, both at home and abroad, so that there does not have to be a reduction in quality of life.

4. When the limiting factor changes, behavior that used to be economic becomes uneconomic. Economic logic remains the same, but the pattern of scarcity in the world changes; the result is that behavior must change if it is to remain economic.

That last proposition does more than any other to explain the despair and excitement on both sides of the issue. On the environmental side, scientists are frustrated that business does not understand the basic dynamic involved in the degradation of biological systems. For business, it seems unthinkable, if not ludicrous, that you cannot extrapolate the future from the past and continue with present methods. In this intensely uncomfortable phase, people recognize, one by one, that economic activities that were once successful can no longer lead to a prosperous future. In itself, that recognition has caused polarization, frustration, anger, and name-calling. At the same time, it is already fueling the next industrial revolution.

The patterns of change that underlie natural capitalism appear to be the only known way to improve ecological health, create net economic growth, and provide meaningful employment in a world where one-third of the workforce—1 billion people and increasing—is marginalized, with no decent work or no work at all. It has been said that people are the only species without full employment. And we are also striving earnestly to make this ever more so, jettisoning people to create one more wave of short-term profits. The zeal to eliminate people is rooted in an obsolete industrialism designed for the bygone world of scarce people, general poverty, sparse technology, and abundant nature. The success of industrialism and capitalism has largely reversed those conditions. Today, continuing to deplete natural capital to make fewer people more productive and more people unemployed exhausts both the environment and society. Its logic is backward— using more of what we have less of (natural capital) to use less of what we have more of (people). The result is massive waste on three fronts: overstressed resources and hence deteriorating living systems, underworked or overworked (either way, harried and disrespected) people, and the expenditure of vast sums expended to try to cope with the costs of both.

Civilization in the 21st century is imperiled by three main problems: civil societies' dissolution into lawlessness and despair, the deteriorating capacity of the natural environment to support life, and the dwindling of the public purse needed to address these problems and reduce human suffering. All three megaproblems share a cause: waste. Its systematic correction is a common solution, equally unacknowledged yet increasingly obvious.

Natural capitalism is the key that unlocks the reversal of that waste. A manifold reduction in resource use can increase the overall level and quality of employment while dramatically reducing harm to the environment. The economy can grow, use less material, free resources for those who need them, and start to restore living systems. We should be laying off not productive people, but rather the wasted barrels of oil, gallons of water, pounds of metals, and acres of forest, thus regenerating natural capital, hiring more people to do so, and cutting total cost. Gradually and fairly rebalancing factor inputs to substitute increasingly abundant labor for increasingly scarce nature will help to heal society *and* Earth.

REFERENCES

Costanza R, Folke C. 1997. Valuing ecosystem services with efficiency, fairness, and sustainability as goals. In: Daily GC (ed). Nature's services: societal dependence on natural ecosystems. Washington DC: Island Pr.

Daly HE. 1994 Operationalizing sustainable development by investing in natural capital. In: Jansson A and others (eds). Washington DC: Island Pr.

Hawkens P, Lovins A, Lovins H. 1999. Natural capitalism: creating the next industrial revolution. New York: Little Brown.

Hays CL. 1998. Now, liquid gold comes in bottles. New York Times: Jan 20.

Hillel D. 1991. Out of the earth, civilization and the life of the soil. New York: The Free Pr.

San Francisco Chronicle. 1998. Accidental fishing called huge threat. May 21.

UNEP [United Nations Environment Programme]. 1996. Poverty and the environment: reconciling short-term needs and long-term sustainable goals. Nairobi: Mar 1 press release.

Yoon. 1998. A "dead zone" grows in the Gulf of Mexico. New York Times: Jan 20, p F1.

INFRASTRUCTURE
FOR SUSTAINING
BIODIVERSITY—
POLICY

LINKING SCIENCE AND POLICY:
A RESEARCH AGENDA FOR
COLOMBIAN BIODIVERSITY

CRISTIÁN SAMPER
Instituto Alexander von Humboldt,
Calle 37 #8-40 Mezanine, Santafe de Bogota, Colombia

THE CLOSE interaction between nature and human society has been the basis of life for cultures worldwide over many generations. Indigenous tribes, such as the Yukuna living along the Mirití River in the Colombian Amazonia, view their world as the conjunction of all biophysical, biological, and cultural elements. They have a "humanized" view of the forest, in which all the elements are closely connected, and they see themselves as the guardians of the spirits contained in plants, animals, and minerals (van der Hammen 1992).

In recent years, more and more people around the globe have been facing environmental problems as part of everyday life, and many of us have seen changes within our lifetimes. Access to clean water is increasingly difficult, the air in our cities is increasingly polluted, forests are being cut down, and some species are becoming increasingly rare or extinct (WRI 1996). As pressures on natural resources have increased and environmental degradation has become evident, public awareness has increased to an all-time high, and the interdependence of human society and our natural environment is widely accepted.

Environmental issues have become important in local, national, and international agendas, and decision-makers are facing the challenge of designing and implementing policies that achieve an adequate balance between environmental, economic, and social goals. Although much progress has been made in agriculture, transportation, and energy (Dower and others 1997), we are still seeing a steady decline in biological diversity worldwide.

One important reason for the decline is the gap that still exists between scientists and decision-makers. On the one hand, scientists are not providing the information that is required for the decision-making process at the right time or in the right language to be useful. On the other, decision-makers at all levels are not necessarily framing questions to scientists or providing the support that is needed to carry out research. In this paper, I describe the attempts made by scientists and decision-makers in Colombia to overcome this problem, and I present a research agenda for the conservation and sustainable use of bio-diversity.

THE EARTH SUMMIT AND THE CONVENTION ON BIOLOGICAL DIVERSITY

In June 1992, leaders of over 100 countries gathered in Rio de Janeiro as part of the UN Conference on Environment and Development (UNCED), also known as the Earth Summit. It was by far the largest gathering of decision-makers from around the world to discuss environmental issues—a clear recognition that these themes do not recognize political boundaries but require international coopera-tion. The results of the conference include Agenda 21, a global plan to halt and reverse environmental damage to our planet and to promote environmen-tally sound and sustainable development in all countries (Sitarz 1994). In addi-tion, three legally binding conventions were signed—on biodiversity, climate change, and desertification.

The Convention on Biological Diversity has been ratified by 173 parties and has become a global framework for decision-makers (see Juma, this volume). The convention defines biological diversity as "the variability among living organisms from all sources including, inter alia, terrestrial, marine and other aquatic eco-systems and the ecological complexes of which they are part, this includes di-versity within species, between species, and of ecosystems" (UNEP 1994). The convention has three main objectives: the conservation of biological diversity, the sustainable use of its components, and the fair and equitable distribution of benefits derived from its use. The last objective is far-reaching, ambitious, and difficult to achieve, but it is essential for future sustainable development.

The organization of the convention includes the Conference of the Parties (the highest ranking body), in charge of decisions that are legally binding on all par-ties. It also has a Subsidiary Body for Scientific, Technical, and Technological Advice (SBSTTA), in charge of analyzing relevant information on issues defined by the Conference of the Parties and making recommendations that are then offered for adoption by decision-makers. This scheme is intended to bridge the gap between science and policy, and it has allowed progress to be made on such issues as coastal and marine biodiversity, agricultural biodiversity, and capacity-building for taxonomy.

Many parties to the convention have adopted measures for its implementa-tion on a national level. Colombia has taken steps to implement the conven-tion, and I will examine the measures taken to strengthen scientific research on biodiversity to provide a stronger basis for designing policy and monitoring its effects.

THE BIODIVERSITY OF COLOMBIA

Colombia is among the countries with the richest biodiversity. With a land area of 1,140,000 km² (about 0.7% of the continental surface of the globe), it is home to over 40,000 plant species, over 1,815 bird species, over 604 amphibian species—more than 10% of the species of any of these groups.

Colombia's enormous richness can be attributed to its geological history and location. Its location near the equator, as a land bridge between North America and South America, allowed the migration of species between the continents. Many species, such as the oaks (genus Quercus), are widespread in North America, are found in the higher-elevation forests in Central America, and are in forests in the Andes of Colombia as far south as the border with Ecuador.

The geological history of Colombia has also played an important role in speciation and diversification. The oldest rock formations in Colombia are parts of the Guyana shield and are found as giants standing over the plains of the Orinoco and parts of the Amazonian region of Colombia. The Andes are more recent and split into three distinct ranges, with the eastern range stretching as far north as Venezuela. The Pacific coast of Colombia, known as the Chocó, has one of the largest rainfalls—some locations get more than 12,000 mm of rain annually—and is separated from other lowland forests by the Andes. This complex geography gives rise to over 140 biogeographic zones (Jorge Hernandez Camacho, unpublished).

THE INSTITUTIONAL STRUCTURE IN CHARGE OF COLOMBIA'S BIODIVERSITY

The environmental sector in Colombia was restructured as a response to the commitments of the Convention on Biological Diversity, ratified by Colombia in 1994. The result is a series of institutions and organizations that are collectively known as the National Environmental System. The highest-ranking body is the National Environmental Council, which is made up of representatives of the different ministries and government agencies and of the private sector, universities, and the civil society. This body is in charge of establishing general policy guidelines and facilitates cross-sectoral coordination.

The restructuring also led to the creation of the Ministry of the Environment, as a small entity in charge of supervising environmental policy and representing Colombian positions in international conventions and treaties related to the environment. Environmental control and management are decentralized in the new system and are in charge of regional autonomous corporations for sustainable development.

Most important for the purpose of this paper are the research institutes that are in charge of providing the scientific and technical support to the environmental system. The institute in charge of biodiversity research, named after Alexander von Humboldt, was established in 1995 as a joint venture of 24 partners, including the Colombian Ministry of the Environment, the Colombian Science Foundation, universities, and nongovernment organizations. This innovative institutional approach was designed to bring together the skills and experience of the public and private sectors and to bridge the gap between science and policy. The institute's mission is to promote, coordinate, and carry out research that contributes to the conservation and sustainable use of biological diversity in Colombia.

A CONCEPTUAL FRAMEWORK FOR BIODIVERSITY RESEARCH

The development of a biodiversity research strategy for Colombia requires a conceptual framework. The Convention on Biological Diversity itself has recognized several levels of organization, including genetic diversity, species diversity, and ecosystem diversity. Noss (1990) developed a useful framework to study biodiversity that recognizes those three levels of organization and three attributes that can be surveyed (composition, structure, and function). The result is a two-dimensional matrix that allows any combination of attributes at any level of organization.

In the framework presented by Noss (1990), *composition* refers to the identification of the components of biological diversity, such as species lists. *Structure* refers to the characterization of these components, including their relative abundance, for example, the types of ecosystems in a given area. By *function*, we mean the study of the dynamic nature of biodiversity in space and time, for example, monitoring allele frequency in a population over time or the effects of management practices on demography. It is not surprising that an analysis of biodiversity research over the last few decades shows that most work has been done on composition at the species level and very little on function at the genetic and ecosystem levels.

A helpful addition might be to include the human dimension and to evaluate the use of biodiversity at any level along a gradient of human intervention, from "pristine" habitats, through extractive systems, to highly transformed or even degraded areas. That would enable us to address such matters as the impact of logging on genetic diversity of nontimber forest products or the effects of wetland restoration on ecosystem services.

A STRATEGIC AGENDA FOR BIODIVERSITY RESEARCH IN COLOMBIA

The strategic plan for research in biodiversity in Colombia is designed to address the conceptual framework as a whole, identify gaps and weaknesses, and design actions to overcome them. The plan, developed in collaboration with 100 institutions and scientists nationwide, has six main objectives:

- to continue the inventory of biological diversity;
- to provide the scientific basis for the conservation of biodiversity;
- to develop new ways to use and value biodiversity;
- to study the effects of cross-sectoral policies and legislation on the conservation and sustainable use of biodiversity;

• to strengthen the national capacity to carry out scientific research and promote international cooperation; and

• to design ways to disseminate the results of research, especially to decision-makers.

Biodiversity Inventories

Although biological inventories have been carried out for the last 2 centuries, we still have little information on what biodiversity we have and where it is. Most of the biological collecting done since the journeys of Alexander von Humboldt and the botanical expedition led by Jose Celestino Mutis in the early 19th century has focused on vascular plants and vertebrates, especially birds and mammals. Invertebrates, fungi, and bacteria have received little attention, and overall we estimate that we probably know less than 10% of the species found in Colombia (figure 1). Research related to characterization at the genetic level is scarce, except for some species of importance for agriculture and health, although the cost and speed of molecular techniques are making these increasingly available to researchers worldwide.

The Alexander von Humboldt Institute has completed an exercise to determine the high-priority geographic areas for biodiversity inventories through a series of workshops involving leading scientists. The criteria to evaluate geographic priorities include species richness, endemism, current state of knowledge, and degree of threat, including such variables as extent of original habitat left, degree of fragmentation, rate of change, and existence of protected areas. Use of those criteria has led to the identification of areas that have top priority, primarily those with a

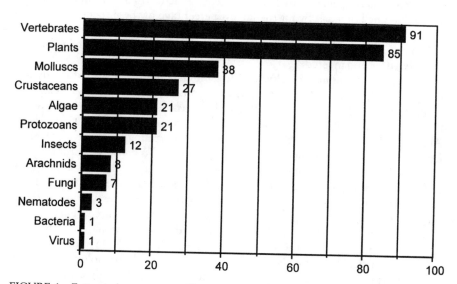

FIGURE 1   Estimated percentage of known species in taxonomic groups in Colombia.

combination of high diversity, high endemism, poor knowledge, and high degree of threat. The resulting maps are used to establish a set of geographic priorities that are used by institutions nationwide for inventories (Samper 1997).

In the research plan, the strengthening of biological collections nationwide and the repatriation of information to Colombia are very important. The 29 biological collections in the country house an estimated 1.7 million specimens. However, the collections are not always adequately curated, taxonomic identification is not always reliable, and the information is not readily available for studies in biogeography. Therefore, an important step is to support the exchange of material with national and international specialists and institutions, and a major effort is under way to computerize all collections in Colombia by the year 2000. An additional step is to establish agreements for the repatriation of information housed in museums and other biological collections abroad.

Conservation Biology

A second major line of work is related to research that directly contributes to the conservation of biological diversity at all levels. Research should address the direct causes of extinction, namely, habitat transformation, overexploitation, competition with alien species, and pollution and climate change (Heywood1995).

Research related to conservation should focus on a better understanding of the current status, monitoring, and trends of biological diversity, with emphasis on endangered or threatened taxa or habitats. Preliminary results of this work have resulted in a complete list of threatened plants of Colombia, including 620 species so far, according to the criteria used by the International Union for the Conservation of Nature (UICN 1994). We find that a major group of threatened plants consists of species with restricted geographic distributions and those commonly used by humans. By far the largest percentage of these species are orchids (29%) because of overexploitation for ornamental purposes and transformation of habitats (Calderón 1997). Some plant families that are used for timber are also threatened or endangered.

A recent survey of major ecosystems in Colombia has revealed that nearly one-third of the habitats have been altered or transformed as part of development (Ministerio del Medio Ambiente 1997). The most degraded ecosystems are, not surprisingly, those with the highest population pressures (table 1), such as the Andean cloud forests (26.5% of original cover remaining) and the tropical dry forests of the Caribbean lowlands (1.5% remaining). To conserve natural ecosystems and diversity, Colombia has set aside more than 9 million hectares in 45 protected areas, roughly 8% of the country. Although some ecosystems, such as the Andean and Amazonian forests, are well represented in the national park system, others, such as the Orinoco grasslands, are underrepresented. Furthermore, many of the areas lack the size or latitudinal gradients that would make them viable in the long term. In this context, the Alexander von Humboldt Institute is identifying critical areas for the establishment of new parks or biological corridors and is making recommendations on investment of limited resources to maximize the diversity preserved under in situ conditions.

TABLE 1   Current Status of Major Natural Ecosystem Types in Colombia

Ecosystem Type	Original Area, km²	Area Remaining, km²	Fraction Remaining, %
Tropical lowland humid forests	550,000	378,000	68.7
Tropical dry forests	80,000	1,200	1.5
Deserts and xerophytic vegetation	11,000	9,500	86.4
Andean cloud forests	170,000	45,000	26.5
Andean Páramo	18,000	18,000	100.0
Flooding forests (Amazonia)	36,000	36,000	100.0
Orinoco grasslands	113,000	105,000	92.9
Amazonian grasslands	14,000	14,000	100.0
Caribbean grasslands	3,500	1,000	28.6
Amazonian shrublands	7,500	7,500	100.0
Gallery forests	118,000	95,000	80.5
Wetlands	13,000	6,500	50.0
Mangrove forests	6,000	3,500	58.3

Source: Ministerio del Medio Ambiente 1997, based on Etter.

An additional strategy is to conserve components of biodiversity under ex situ conditions, such as germplasm banks and zoological and botanical gardens. The most important ex situ collections held in Colombia are related to genetic diversity of agricultural crops and livestock. The country has 16 registered botanical gardens, but they contain fewer than 5,000 plant species and no more than 5% of the threatened plants of Colombia. A major effort is under way to strengthen the role of botanical gardens in conservation of and research on endangered flora. However, in situ conservation is generally favored in the absence of a complete understanding of diversity and interactions.

One aspect that has received little attention in tropical ecosystems is the effect of alien species and living modified organisms on biodiversity. Research in other countries has shown that introduced species can make up an important fraction of local biodiversity, and in extreme cases, such as the islands of Hawaii, the total number of plants has doubled over the last 2 centuries. Some introduced species can be aggressive and more tolerant to environmental change and can therefore outcompete native species. The effect is especially severe in island and freshwater ecosystems. Over 140 species of freshwater fishes and crustaceans have been introduced into Colombian rivers and wetlands since the turn of the century and might have led to the extinction of several endemic freshwater fish species (Hernando Alvarado, unpublished data).

Use and Valuation of Biodiversity

Biodiversity has played a major role in the structuring of human populations. That can be clearly seen in the effects of crop and livestock exchange between continents in recent history and their effect on modern cultures (Hobhouse 1985;

Viola and Margolis 1991). Our livelihood ultimately depends on the direct benefits that we derive from biological diversity (for example, food) and ecosystem services (such as watershed regulation and air control).

Colombia is home to 81 ethnic groups that have interacted closely with their environment and in some ways shaped it over the centuries. The traditional knowledge of components of biodiversity, their ecology, and their natural history and of ways to manage resources is critical to our understanding of biodiversity. This knowledge is being lost at alarming rates, primarily as a result of the changes in cultures as they incorporate elements of western society. Some of the indigenous groups, such as the U'wa in the foothills of the Sierra Nevada del Cocuy and the Arhuacos in the Sierra Nevada de Santa Marta, have developed complex production systems that take into account seasonal variations and migrations along an altitude gradient that stretches from sea level to the timberline at 3,000 m (Franco 1997). Documenting these management practices and promoting the training of younger generations to preserve the knowledge have high priority.

The Convention on Biological Diversity is to some extent addressing a great paradox: the countries with the highest diversity are the ones with the least economic development. Those countries have legitimate interests in using biological diversity for their development in the 21st century, although the short-term economic benefits are often overestimated. It is important to provide a research basis that recognizes the roles of traditional and scientific knowledge. Preliminary results of our work indicate that the total value of goods and services derived from biodiversity in Colombia can be around $300 billion per year, 5 times the GNP (Mansilla and others, in press). Further research is required to determine the value of goods and services from biodiversity and to examine new uses of and markets for products.

Policy and Legislation

Research on biodiversity is too often left to biology and related disciplines, and little room is left for other fields of research. Therefore, a high priority in the research agenda is to strengthen policy research to evaluate the effects of cross-sectoral policies on the conservation and sustainable use of biodiversity. One clear example is the agrarian reform policy that was promoted during the 1960s and 1970s in Colombia, where "unproductive" land, defined as land that was not used for agricultural and livestock production, was redistributed to small farmers. The policy served as a disincentive for conservation, and the result was that many areas that had remnants of natural forest ecosystems were cleared to give way to pastures and crops. Not only has the policy been changed to be compatible with conservation of natural ecosystems, but also economic incentives for conservation of forest remnants have been established in recent years.

Another critical component is research on legislation at the international, national, and local levels and its effects on biodiversity goals. International conventions, such as the Convention on Biological Diversity, are increasingly important as we move toward a global economy. It is important to examine the relationship of legislative developments in related conventions, such as the negotiations under the Convention on Climate Change, the Food and Agricultural Organization, and the World Trade Organization. On the national level, the 1991 revision of the constitution of Colombia allowed for many environmental issues to be included. Additional developments have been made at the regional level, such as the agreement among the countries of the Andean Community (Venezuela, Colombia, Ecuador, Peru, and Bolivia) for a common regime for access to genetic resources, known as Decision 391 de la Junta del Acuerdo de Cartagena.

Training

It is no secret that the distribution of research capacities is severely unbalanced geographically and that many developing countries need to train scientists in many of the topics and areas described above. That is done in close collaboration with national and international universities, and our goal is to double the number of researchers in biodiversity in Colombia over the next 25 years. Specialized courses, scholarships, and internships will also play a major role in strengthening national capacity.

Communication and Information

One element that is often not considered in designing research programs is related to information management and delivery of the results in a manner that is useful for different audiences. Potential users include decision-makers, other scientists, the communication media, and the general public. Each audience has its own interests, background, and ways to receive information. A helpful exercise is to identify user groups, needs, and means.

The basis of all communication strategies is information, and such issues as database management are critical for research and decisions. Technological advances in hardware, software, and telecommunication are improving the exchange of information in developing countries. A number of initiatives, such as the clearinghouse mechanism of the Convention on Biological Diversity and the Inter-American Biodiversity Information Network, will strengthen database management and facilitate information exchange.

The results of scientific research on biodiversity are traditionally published by scientists in academic journals, and little effort has been made to deliver these results in other ways that make them readily accessible to decision-makers and the general public. Research on the natural history of plants and animals has served as the basis of an increasing number of documentaries that are featured on television networks around the globe. Strengthening the technical capacity for production and worldwide distribution of documentaries on Colombian biodiversity has high priority.

CONCLUSION

The actions described in this paper should strengthen capacity to carry out research that is strategically important for the conservation and sustainable use of biodiversity. The institutional developments undertaken in Colombia in response to the Convention on Biological Diversity are aimed at bringing together limited resources to address a common agenda and to help bridge the gap between science and policy.

REFERENCES

Calderón E. 1997. Plantas amenazadas. Boletín Bio No 1. Villa de Leyva Colombia: Instituto Alexander von Humboldt.

Dower R, Ditz D, Faeth P, Johnson N, Kolzoff K, MacKenzie J. 1997. Frontiers of sustainability: environmentally sound agriculture, forestry, transportation, and power production. Covelo CA: World Resources Inst; and Washington DC: Island Pr.

Franco R. 1997. Biodiversidad y sistemas tradicionales de producción en Colombia. Boletín Bio. Villa de Leyva Colombia: Instituto Alexander von Humboldt.

Heywood VH. 1995. Global biodiversity asessment. Cambridge UK: Cambridge Univ Pr.

Hobbhouse H. 1985. Seeds of change: five plants that transformed mankind. London UK: Papermac, MacMillan.

Mansilla H, Baptiste LG, Hernandez S, Cárdenas JC, Willis C. In press. La valoración económica de los servicios ambientales de la biodiversidad en Colombia. Villa de Leyva Colombia: Instituto Alexander von Humboldt.

Ministerio del Medio Ambiente. 1997. National biodiversity: policy for Colombia. Villa de Leyva, Colombia: Instituto Alexander von Humboldt.

Noss R. 1990. Indicators for monitoring biodiversity: a hierarchical approach. Cons Biol 4:355–64.

Samper C. 1997. Que tanto conocemos a Colombia? Boletín Bio No 2. Villa de Leyva Colombia: Instituto Alexander von Humboldt.

Sitarz D. 1994. Agenda 21: the Earth summit strategy to save our planet. Boulder CO: EarthPress.

UICN [Unión Internacional para la Conservcación de la Naturaleza]. 1994. Categorías de las listas rojas de la UICN. Gland, Suiza.

UNEP [United Nations Environment Programme] 1994: Convention on biological diversity. Geneva Switzerland: UNEP.

Van der Hammen MC. 1992. Managing the world: nature and society by the Yukuna of the Colombian Amazonia. Studies of the Colombian Amazonia IV. Bogotá Colombia: Tropenbos Foundation.

Viola HJ, Margolis C. 1991. Seeds of change. Washington DC: Smithsonian Inst Pr.

WRI [World Resources Institute]. 1996. World resources 1996–97: a guide to the global development. Washington DC: WRI, UNEP, UNDP . . . and the World Bank.

SUSTAINABILITY AND THE LAW:
AN ASSESSMENT OF THE
ENDANGERED SPECIES ACT

MICHAEL J. BEAN
Environmental Defense Fund,
1875 Connecticut Avenue, NW, Washington, DC 20009

IN THE UNITED STATES, the Endangered Species Act (ESA) has served for the last quarter-century as the final safety net against the loss of biological diversity. During that time, the list of legally protected species, subspecies, and populations has grown steadily, and it now numbers more than 1,200 in the United States. The goals of the ESA are to prevent the extinction of these and to recover them to the point where they are no longer in peril. The tools the act provides to achieve these goals are few: a duty of federal agencies to further the conservation of imperiled species and to avoid actions that jeopardize their continued existence, a prohibition against most commercial activities involving imperiled species, and a prohibition against collecting, killing, or otherwise "taking" them. Thus far, these tools have proved sufficient to arrest the decline of only a minority of imperiled species and to improve the status of an even smaller fraction of those. New and more diverse tools to address more effectively the threats to survival of species clearly are needed. These include positive incentives for private landowners and others to restore, enhance, and responsibly manage habitats for imperiled species; mechanisms to initiate conservation efforts toward species before they reach a point of crisis; and tools to broaden the focus of conservation efforts from individual species to assemblages of species in particular natural communities, habitats, or ecosystems. Recent efforts to fashion such tools administratively have offered promising results.

About 25 years ago, the modern ESA did not exist. Indeed, it had not even been conceived. Representative John Dingell was still 2 months away from introducing the bill that, with important changes, eventually would become the

ESA. The House passed Dingell's bill in only 8 months, by a margin of 390-12. The Senate's action was even speedier. New Jersey's Senator Harrison Williams introduced his bill on 12 June 1973, and the Senate approved it only 6 weeks later, by a vote of 92-0.

In only 8 months, bills were introduced in both houses, hearings were held, committee reports were written, debate was held, near-unanimous votes occurred, and a presidential signature was obtained on a bill that many regard as one of the most far-reaching and important environmental laws ever passed by any legislature in any nation.

How times have changed: The contrast with the situation today could not be sharper. Gridlock over the future of the ESA has nearly paralyzed Congress for several years. Congress last reauthorized the ESA in 1988 and was supposed to have done so again in 1992. It didn't. Not in 1992, 1993, 1994, 1995, 1996, or 1997, either. Only two of the last four Congresses even managed to report a reauthorization bill out of committee. Neither house has debated a reauthorization bill, although they have tried—with some success—to hamstring the endangered-species program by including budget cuts, narrowly targeted overrides, and moratoriums on listing in unrelated legislation.

The near unanimity of congressional opinion that prevailed in 1973 also has vanished. Congress is divided—deeply—over the future of the ESA. When the ESA has been brought up recently in congressional debate—as in recent debates over creation of a National Biological Survey, imposing a moratorium on adding further species to the endangered list, and exempting certain flood-related activities from the ESA's requirements—the debate has been rancorous, bitter, vitriolic, acerbic, and sometimes downright nasty. These divisions in Congress reflect similar divisions in society at large.

When will Congress get on with the business of reauthorizing the ESA? When it does, what will it do? Unfortunately, no one knows the answers to these questions.

Two things can be said with confidence, however. One is that the ESA has been a huge success. The other is that the ESA has been a huge disappointment.

Let us look at the successes first. When the ESA was passed in 1973, fewer than 50 whooping cranes survived in the wild; today, there are four times as many. The American alligator has recovered fully throughout the Southeast. In less than two decades, the bald eagle has increased its nesting population severalfold and its classification has been changed throughout the nation from endangered to threatened. Brown pelicans and peregrine falcons have increased their numbers and expanded their ranges. Numbers of Kemp's ridley sea turtles are increasing on their nesting beaches in Mexico and appear to be increasing in US coastal waters; in 1997, at least nine Kemp's ridley nests were found on the coast of Texas—apparent evidence that the "head-starting" effort begun here two decades ago may succeed yet. The northern Aplomado falcon once again occurs as a breeding species in the United States after an absence of nearly a half-century.

The list of similarly impressive results continues. Gray wolves have been reintroduced successfully into the northern Rockies, and red wolves into North Carolina. Soon, Mexican wolves are expected to be reintroduced into Arizona and possibly New Mexico. And in the Grand Canyon, California condors soar overhead, a sight that has not been seen there in a century or more, and one that few thought would be possible when the last handful of wild condors were taken into captivity a decade or so ago.

As encouraging and reassuring as these successes are, they are counterbalanced by a frustrating lack of success and continued decline of many other species. The ESA was not enough to save two species of fish in Texas, the Amistad and San Marcos gambusias; a bird in Florida, the dusky seaside sparrow; or a fish in Maryland, the Maryland darter. All are now apparently extinct. The Attwater's prairie chicken, although included on the first federal list of endangered species in 1967, has suffered a catastrophic decline, despite three decades of nominal protection, from more than 2,400 wild birds to only 42 in 1997. For the nation as a whole, 33% of the species that the ESA protects are declining.

How does one explain these disappointing results? Many observers offer one or more of the following explanations: the Fish and Wildlife Service does not have enough money, does not have enough backbone, and suffers from too much political meddling. If these are the sources of the problem, then the solutions are easy: give the service more money, stiffen its backbone, and halt the political meddling. It is wise to keep in mind, however, what H.L. Mencken (as quoted in Raspberry 1997) said of easy solutions: "There is always an easy solution to every human problem—neat, plausible, and wrong." In the case of the ESA, the solutions just enumerated are not so much wrong as they are incomplete. Yes, the service is woefully short of the financial and human resources it needs to do the job that Congress has assigned it. Yes, the Service often shows a remarkable propensity to cave in to pressure. And yes, that propensity to cave in has invited far too much political meddling in congressional administration of the ESA. Yet, although each of these assertions is undoubtedly correct, they do not provide a satisfactory explanation of the disappointments of the ESA.

What is missing from these explanations is the fact that the ESA does not give the Fish and Wildlife Service all the tools that it needs to conserve endangered species, particularly in states like Texas, Florida, and Hawaii, where a great many endangered species occur and where most of the land and most of the habitat that supports those species are privately owned. When dealing with private lands, the ESA gives the service only two tools. One is the authority to tap the Land and Water Conservation Fund to acquire land, and the other is the prohibition in section 9 against the "taking" of endangered species. The service long has interpreted this prohibition to extend to modification of habitat. When the Supreme Court in 1995 upheld that interpretation in the *Sweet Home* case, most conservationists breathed a sigh of relief.

The "taking" prohibition of section 9 can serve as a powerful hammer. The problem, however, is that many of the causes of decline of endangered species are not nails. Against them, a hammer is ineffectual. Contributing to the decline of most species on the threatened and endangered lists are the absence of natural disturbances like fire, the presence of introduced species, or both. Against these pervasive threats, the prohibition against taking endangered species or their habitats is largely ineffectual. Moreover, the prohibition is ill suited to restoring vanished habitats, reconnecting the pieces on a highly fragmented landscape, or countering the random events that drive isolated small populations into extinction. For species too, the world often ends not with a bang, but with a whimper.

After more than two decades, the weaknesses of the ESA are as evident as its strengths. One of its weaknesses is that it has led to a variety of unfortunate and unintended consequences. For example, as long as landowners believe that they will incur added regulatory restrictions on the use of their land if they do things that either attract endangered species or increase the number of such species on it, they are unlikely to do those things. Indeed, to avoid the possibility of such restrictions, landowners sometimes manage their land in ways that render its use by endangered species highly unlikely. That was the clear between-the-lines message of the National Association of Home Builders's recently published *Developer's Guide to Endangered Species Regulation* (Sauls 1996). In a chapter called "Practical Tips for Developers," the following frank advice appears:

> The highest level of assurance that a property owner will not face an ESA issue is to maintain the property in a condition such that protected species cannot occupy the property. Agricultural farming, denuding of property, and managing vegetation in ways that prevent the presence of such species are often employed in areas where ESA conflicts are known to occur. This is referred to as the "scorched earth" technique. The scorched earth management practice is highly controversial, and its legality may vary depending upon the state or local governing laws. But developers should be aware of it as a means employed in several areas of the country to avoid ESA conflicts.

Such practices are by no means confined to developers. In the Southeast, forestry consultants reportedly often advise owners of pine woodlands to cut their trees before they are old enough to serve as foraging habitat for red-cockaded woodpeckers. In the Northwest, commercial timber companies shorten harvest rotations beyond what otherwise would be economically rational to avoid having northern spotted owls take up residence on their property. In California's central valley, farmers plow fallow fields to prevent native vegetation and endangered species from reoccupying the fields. Sometimes, even the Fish and Wildlife Service finds itself advising landowners to avoid creating habitat for endangered species. Consider the following message I received from an environmentalist in California who had been promoting a habitat-enhancement effort on private land:

> Picture this: We are standing in a field, looking at a sediment basin that tends to stay wet part of the year and thus vegetate. The [Fish and Wildlife Service] biologist suggests that we remove vegetation from the basins to avoid creating habitat for and attracting red-legged frogs. Why? you ask. Why indeed; after all, the [service] is supposed to want habitat. The answer to this million-dollar question is that if we attract [threatened and endangered] species, then the [service] would slap some additional constraints on all the projects. So, in an effort to be helpful, [the service encouraged us] to avoid creating habitat. [Sigh]

These examples only demonstrate that the ESA is no exception to the common phenomenon of regulatory programs' spawning ingenious strategies on the part of the regulated parties to frustrate the regulatory purpose without technically violating the law. None of these actions violates the ESA, yet all of them virtually ensure that endangered species will not benefit.

This problem was made abundantly clear to me and my colleagues at the Environmental Defense Fund (EDF) while we were working to conserve the red-cockaded woodpecker in an area of North Carolina called the Sandhills. The Sandhills supports the second-largest remaining population of red-cockaded woodpeckers in the country; most of them are on Fort Bragg army base, but many are on nearby private lands. The numbers of these woodpeckers on private lands have declined steadily over the last two decades. Much of that decline has resulted from the lack of control of the hardwood understory, a task that Nature formerly performed with regular fires. An extensive system of roads now means that fires caused by lightning strikes will burn only as far as the next road and not cover the thousands of acres that formerly would have burned.

If landowners in the Sandhills would control the hardwood understory aggressively, rehabilitate some of the abandoned nesting cavities where woodpeckers once persisted for decades, install artificial cavities in suitably sized pines, let stands of pines that are nearly old enough to provide foraging habitat remain uncut for a few more decades, and protect those remaining very old trees that could serve as cavity trees, then the bird would be much better off there. The problem was that few landowners in the area could be persuaded to do these things, precisely because they feared the regulatory restrictions that would accompany the woodpeckers that would benefit from these practices. The result was a continuing, steady decline in the local population of woodpeckers, despite their nominal protection as an endangered species.

EDF decided to try something completely different. With generous support from the National Fish and Wildlife Foundation, we devised a new form of habitat-conservation plan. Unlike habitat-conservation plans used elsewhere, this one was intended to create, restore, and enhance habitat for endangered species. No immediate development activity or timber harvest that would result in incidental taking of any endangered species was anticipated. Incidental taking was authorized under this plan, but only in the future, only in those habitats that had been created, restored, or enhanced pursuant to the plan, and only if the participating landowner acknowledged and agreed not to diminish the baseline conditions that existed when he or she enrolled in the plan.

We called this idea "safe harbor". In return for a definite commitment to carry out specific management actions that were expected to benefit the woodpecker, the landowner was given protection—a safe harbor—from added regulatory restrictions beyond those which already applied to the land on the day he or she entered into the agreement. The aim of the Sandhills safe-harbor program was to accomplish something that no other strategy of the ESA had accomplished there: to halt and reverse the decline of red-cockaded woodpeckers and their habitat on privately owned land. We bet that enough landowners would be willing to beneficially manage enough habitat area for a long-enough period that we could accomplish a substantial improvement in the situation that would exist otherwise.

To date, the safe-harbor program in the North Carolina Sandhills has been received very well by landowners. Two dozen of them now are actively managing 25,000 acres of forest land for the benefit of an endangered species. More important, the benefits of this approach have been recognized by others, who are adapting it to their own circumstances. For example, the Peregrine Fund seized on the safe-harbor idea to expand dramatically a reintroduction effort for the northern Aplomado falcon. Along the Texas coast, coastal prairie habitat is being restored by ranchers and other landowners in a safe-harbor program for the Attwater's prairie chicken. In South Carolina, a statewide safe-harbor program for the red-cockaded woodpecker was launched in 1998. Even before it officially began, landowners signaled their intention to enroll some 80,000 acres of forest land in the program.

Safe-harbor agreements are not a panacea. For some species, they can contribute substantial benefits, both to species that have not benefited much from other strategies and to the attitudes of landowners toward the conservation of endangered species. Ultimately, however, what will be needed is a set of meaningful economic incentives to encourage landowners to carry out more broadly the active measures of management that are essential if the goal of recovering endangered species is to be accomplished (Eisner 1995). Cost-sharing programs, tax incentives, and creative contractual programs like those now being tried in Texas can add a new set of tools to the endangered-species toolbox. Without those new tools, the opportunity to reverse the slide toward extinction of many of our plant and animal species will be lost.

It is also necessary to begin directing attention and resources to declining species much earlier. By the time many species receive the nominal protection of the ESA, their numbers are already so reduced that their eventual recovery will be costly, protracted, and difficult, if it can be accomplished at all (Wilcove 1993). Despite the clear need for earlier action, the ESA has discouraged it in some respects. For much the same reason that landowners sometimes seek to manage their land so as not to attract endangered species to it, some landowners also seek to eliminate from their land species that have been identified as likely candidates for future addition to the endangered list. If they act quickly enough, before the government can accomplish the listing, they can avoid any restriction on the use of their land. Ironically, the identification of a species as a possible candidate for future listing can therefore accelerate the very factors that threaten it. To change this unfortunate dynamic, landowners need to be given clear incentives to help prevent species from being listed in the first place. One recently initiated approach is to authorize agreements between the government and landowners under which the landowner commits to do something that reduces threats to the species, in return for which the landowner receives an assurance that his or her obligations toward the species will be fixed for the duration of the agreement by the terms of the agreement. These "candidate-conservation agreements" are only now beginning to be used and offer the potential for a much more salubrious outcome than often has been achieved for declining species.

Finally, as the list of endangered species continues to grow, it has become increasingly clear that we need strategies for conservation that can address the needs of multiple species simultaneously. Such strategies can serve a wide array of interests. For the government, such strategies offer a means of more effectively stretching scarce funds for conservation, particularly because of the tendency of endangered species to be concentrated in a relatively few areas (Dobson 1997). For landowners, instead of having to deal seriatim with a steady parade of newly listed species, such strategies can increase certainty and reduce the cost of compliance. For the species themselves, strategies to maintain or restore the habitats and ecological functions necessary for the survival of natural communities likely offer a better hope of lasting success than do strategies that rely on artificial manipulation to sustain a species that is no longer capable of survival on its own. The continuing experience with "natural-community-conservation planning" under the ESA illustrates both the potential for good and the practical difficulty of taking a broader approach. The lessons now being learned from that experience are likely to guide the effort to balance goals of conservation with other societal goals in the coming decades.

REFERENCES

Dobson AP, Rodriguez JP, Roberts WM, Wilcove DS. 1997. Geographic distribution of endangered species in the United States. Science 275:550–3.

Eisner T, Lubchenco J, Wilson EO, Wilcove DS, Bean MJ. 1995. Building a scientifically sound policy for protecting endangered species. Science 268:1231–2.

Raspberry W. 1997. Neat, plausible—and right on. Quote from HL Menken. The Washington Post. 1Sep97:A21.

Sauls EG. 1996. Practical tips for builders. In: National Association of Home Builders developers guide to endangered species regulation. Washington DC: Home Builder Press. 107–14.

Wilcove DS, McMillan M, Winston, KC. 1993. What exactly is an endangered species? an analysis of the US endangered species list: 1985–1991. Conserv Biol 7:87–93.

GOVERNMENT POLICY AND SUSTAINABILITY OF BIODIVERSITY IN COSTA RICA

RENÉ CASTRO

Ministerio del Ambiente y Energía,
Piso 10, Calle 25/Av. 8–10, San José, Costa Rica

THE YEAR 1948 was a milestone in the consolidation of modern democracy in Costa Rica. A series of historical events took place that year: the breakdown of authoritarianism, the abolition of the army, the creation of mandatory and free education, and the acknowledgment of full citizenship of women and minority-group members and the restoration of all their rights, including the right to vote. Such outstanding events set the stage for the legitimization of equal opportunity as a permanent commitment of the state, and social investment continues to be the essential condition for development. Almost 50 years later and near the end of the millennium, Costa Rica shows significant success.

• In social terms, the average life expectancy is 76 years (3 years more than in the United States), and infant mortality is less than 12 per thousand; the literacy rate is 95%.
• In environmental terms, we have set aside more than 25% of our territory as national and other parks.
• In economic terms, we have achieved such indicators despite a per capita annual income of only $2,700 (although this is low, it is more than two times the regional average). Costa Rica ranks 33rd among nearly 150 nations, according to the human-development index of the United Nations, and ranks 52nd in terms of per capita income.

These outcomes indicate the high level of efficiency of social investment.

WHY THEN DOES A SOCIAL-DEMOCRATIC PRESIDENT REQUEST CHANGE?

The inauguration ceremony of the Figueres administration, a forum titled *From Forest to Society*, gathered together the government ministers and key persons from both Costa Rican society and friendly nations. The president's speech declared, "We have a relatively nice story in development at the end of a decade called the 'lost decade'. But then the world changed, [and it was] decided to end the Cold War and begin the construction of a global economy, without the permission of small Costa Rica." The new president continued, "Nowadays, as my country faces the challenge of entering and succeeding in the era of global economy, even with all the progress attained, Costa Rica stands no chance unless we shift our development paradigm to that of sustainable human development. Moving in that direction is not a high-tech decision; it is a policy decision."

Costa Rica's approach to sustainable human development places equal importance on the simultaneous achievement of the following three objectives:

• To consolidate macroeconomic balances to allow an increase in internal savings and to attract investments.

• To increase possibilities for strategic social investment, which means building the capacity and empowering the people to understand and take advantage of global economic opportunities through good health and adequate education. The Ministry of Public Education has formally included issues of environment and sustainable human development in the school curriculum and has implemented a bioliteracy project: 100% of public high schools will have a computer laboratory facility, and in 1998, 50% of the citizens of the 21st century will be obtaining a bilingual education.

• To construct an alliance with nature: reevaluating existing natural resources; searching for innovative, nondestructive uses; and creating the setting for groundbreaking business opportunities that cause significantly fewer environmental disturbances, all of which can affect the access of future generations to natural resources.

Only at the correct and responsible intersection of those three objectives does development truly become sustainable, and allow us to generate a virtuous circle of increased well-being for our people instead of a vicious circle of endemic underdevelopment.

This paper focuses on the use of our biodiversity, as well as its economic aspects, because it is a reality that an economically poor country cannot give itself the privilege of preserving 25% of its territory without obtaining an economic contribution of at least roughly the average benefits provided by the other land that is used for traditional patterns of production.

In the case of a tropical country like Costa Rica, having set aside a significant portion of the national territory to conserve wildlands has ensured the representation of a high percentage of its biological diversity in the remaining forest. Because of the state's inability to acquire new lands, a new category of private wild reserves was created, thus involving the civil society in the management of natural resources. That action gave way to the development of a network of private reserves, which comprises more than 120 reserves and more than 100,000 hectares; the percentage of protected wildlands has increased to nearly 30% of the total area of Costa Rica.

Does a protectionist approach to conservation represent an opportunity cost for the rural communities and the country's development? Discussion about this question helped us see the urgent need to change the financial framework of our conservation areas, and we now intend to implement a process to develop their self-sufficiency. From an economic perspective, protected wildlands should be considered producers of both direct and indirect environmental benefits and services, and these should be appreciated and valued adequately. From this standpoint, we undertook a careful revision of visiting and admission fees.

In the 1980s, the World Wildlife Fund predicted that if the high rates of deforestation in Costa Rica during recent decades continued, its forests would disappear in less than 40 years. Fortunately, the trend that followed has been quite different.

What prevented the prediction from being realized? One reason is that we increased the value of timber and have encouraged cultivation of trees. We also increased the efficiency of deforestation controls and restricted changes in land use forcing landowners to preserve some areas with forest cover. The World Wildlife Fund's prediction provoked a national reaction that led to the development of a plan for reforestation in 1986, an increase in the professional forestry capacity, and the offering of tax incentives and financial support to those who were willing to plant trees. The original instruments of the policy to reduce deforestation were refocused to correct its near-sightedness by eliminating distortions in other sectors' policies that previously had encouraged inadequate uses of land. Those actions implemented the national decision to stop the irreversible damage to biodiversity. Larger and larger sectors of the population organized themselves so that, in the following years, the national reaction was translated into a substantial decrease in the rate of deforestation and an increase in reforested areas, forest regeneration, and secondary growth (500,000 hectares).

During the last decade, we have recognized the economic importance of natural forest environmental services and their contribution to local and global societies. That has changed dramatically the framework of an effective fight against deforestation. One particularly important aspect of the national strategy of sustainable development is the adequate recognition of the benefits and services provided by the forest. Only a few of them are acknowledged in the marketplace; others, equally important, undergo the "tragedy of the commons." Table 1 summarizes the services provided by Costa Rican forests and identifies the level of beneficiaries.

To preserve the forests and their environmental services and to generate economic benefits to private owners and the whole country, we must make a practical and innovative effort to internalize the costs and benefits, which are recognized in theory but are ignored in practice by the marketplace. Costa Rica has decided to act in a pioneering and creative way, identifying products that can be

TABLE 1   Services Provided and Beneficiaries of Costa Rican Forests

Environmental Service	Beneficiaries		
	Local	Country	Global
Sustainable wood production	X		
Watershed protection (water for human consumption, irrigation, and hydroelectricity)		X	
Scenic value and ecotourism		X	
Biodiversity resources			X
Carbon sequestration			X

generated and developing markets that can help to reduce the external effects of deforestation and loss of habitats while compensating—directly and tangibly—those who, through adequate land use, maintain the vital life-support functions of the ecosystem and generate benefits to both our society and the world.

INBIO-MERCK CONTRACT

In 1992, the first INBio-Merck contract was approved and carried out successfully. In 1997, a group of major European pharmaceutical companies approached us with interest in creating similar agreements. They expressed surprise that the INBio-Merck contract already had undergone three additional rounds and that the original terms were no longer available. Biodiversity-prospecting agreements have since extended to other applied industries, such as perfume and natural pesticides. Our learning process over the last 5 years has shown us the usefulness of internalizing the benefits derived from our biodiversity resources to our society, which generates additional value to them, each time under improving conditions and benefits of the negotiations.

WATERSHED PROTECTION AND ELECTRIC POWER

Recently, a hydroelectric-power company based in Costa Rica and a group of rural landowners, supported by local nongovernmental organizations, entered a voluntary private agreement. It guarantees that the company will pay the landowners US$10 per hectare per year for providing protection to the watershed that the hydroelectric company depends on. Similar agreements are under way for other projects because the results have been a substantially improved water level and significantly reduced sedimentation and siltification in the dams. The benefits generated have resulted in minimal effect on consumers and constitute a step toward a win-win situation.

CARBON-FIXATION SERVICES

In 1995, we conducted a national inventory of the emission of greenhouse gases to comply with the mandates of the Climatic Change Convention. Article 30 of that convention anticipated the possibility of conducting joint activities between countries that are forced to meet reduced levels and countries that are not. We discovered in that article a way to internalize and create markets that repay those forest environmental services that are of global benefit.

A combination of energy-conserving activities (such as the increased use of renewable energy resources, the increased use of public transportation instead of private vehicles, the use of more efficient vehicles and other fuel options, and the planting of trees and protection of natural forests) can contribute to reduce pollution, particularly by carbon dioxide, and reduce the effects of climatic change. The Climatic Change Convention recognizes them as such, and Costa Rica has developed numerous pilot projects in each of those activities.

At the domestic level, the introduction of taxes on fuels and pollution generates more than $20 million a year, which is paid to owners of natural forests for the environmental services they provide to the country, including carbon fixation and watershed protection.

Domestic efforts should be followed by global efforts. In this sense, Costa Rica is taking the first steps in developing the global market. We have implemented activities jointly with several countries and businesses, and eight of 39 projects submitted to the convention's office take place in Costa Rica.

During the implementation stage of the eight projects, we have realized the advantage of having a portfolio of projects to reduce potential risks; lower the costs of designing, monitoring, and certification; and develop a unit capable of combining all these possible projects in the fields of energy and forestry. Thus, in 1996, the first "certified tradable offsets" (CTOs) and the first 200,000 tons of carbon dioxide were sold to the Norwegian government and private sector for $10 per ton. The resulting funds went to 238 landowners to finance reforestation projects.

One alternative use of land is raising cattle, and it is from this alternative use that the cost of the first CTOs was calculated. We needed to provide at least the same $50 per hectare per year that the landowner would have obtained from raising cattle. If the same activity were to take place in the United States (or if it were transformed into energy), the cost per ton would be 5–10 times higher than ours.

The estimated cost of fulfilling the Climatic Change Convention is $400 billion in 20 years to reduce emissions by 20%. The cost for the same effort could decrease to $150 billion if countries that have comparable advantages, such as fast-growth tropical forests, develop joint implementation activities. We suggest that the effort be global and that the savings of more than $250 billion be distributed fairly among all participating countries.

Taking into consideration the different domestic and global efforts, the United Nations State of the Nation Report shows that since 1996 the regeneration of forests, in addition to forest plantations and the increase in private natural forests, helped to meet the national demand for timber. That is, the net level of deforestation is zero. Furthermore, a net increase in rapid-growth forest in urban areas, such as Costa Rica's central valley, means a continued net increase in forest in the last decade from 178,000 to 206,000 hectares, verified through systematic field trips and geographic information systems.

FUTURE CHALLENGES

Is the recovery of our forests sustainable? Is it replicable in other countries? We believe so, but the effort will require both domestic and international actions. For example, domestically, we are developing water tariffs to include the economic and ecological costs associated with the future availability of water and with watershed protection. That will increase prices by 25–40%. We intend to allocate the difference as payments to owners of forest for the costs of environmental services that they incur in watershed-protection activities. In developing a global system of trade for carbon offsets, Costa Rica, in association with the Earth Council, has placed the first 4 million CTOs in the Chicago Stock Exchange, and we are ready to develop the global market.

By adopting similar means, the United States and other developed countries could reduce the costs of complying with the Climatic Change Convention, encourage environmentally sustainable world trade, support rural populations in tropical nations, and save millions of hectares of forest from becoming pasture or other agricultural lands, thus protecting biological diversity and the life-support functions that the forest provides.

Finally, the Climate Change Convention is undergoing negotiations for change. It is time for domestic and global action to save the rain forests. Costa Rica has implemented such projects and policies, and we have proof that success is possible; globally, however, we need to walk the walk as well as talk the talk.

NATIONAL SECURITY, NATIONAL INTEREST, AND SUSTAINABILITY

THOMAS E. LOVEJOY

Environmentally and Socially Sustainable Development, Latin America & Caribbean Region, The World Bank, 1818 H Street, NW, Washington, DC 20433 and Room SI 463, Smithsonian Institution, 1000 Jefferson Drive, SW, Washington, DC 20560

GLOBAL CHANGE is usually thought of—incorrectly—only in terms of physical change and primarily on a planetary scale (for example, climate change and ozone depletion). Often forgotten are local changes that lead to regional and global biodiversity loss—both direct changes (the "green" or conservation issues) and indirect changes (the "brown" or pollution issues) (Patrick 1961, 1962, 1964).

Biodiversity is thus affected by the aggregate of all environmental problems (brown and green) and, as a consequence, represents the global "bottom line". The loss of biodiversity proceeds in increments that often seem inconsequential, but there is virtual unanimity among scientists that given present trends, the planet is likely to be ravaged biologically with the loss of one-fourth to one-half of all species within a century (Heywood 1995). Most recently, in August 1999, botanists who were assembled in St. Louis, Missouri, for the International Botanical Congress issued a projection of a similar scale.

Given the scale of the problem and the fact that biodiversity is affected by so many other economic and political decisions, to be effective the conservation and scientific communities need to engage the foreign-policy community. If that is successful, it could focus the efforts on root causes and on other foreign-policy issues that affect biodiversity. The foreign-policy community, in turn, should be willing to engage because biodiversity affects its more traditional political and economic concerns and because addressing biodiversity can advance foreign-policy interests.

National security is a term that was interpreted rather narrowly during the Cold War as related primarily to violent interstate conflict. Intrastate conflicts (wars

506

of national debilitation) were considered under this militaristic conception of security but were still viewed through the lens of proxy wars in the East-West conflict. National security in this context was related more to immediate bellicosity and the proximate causes of warfare than to underlying causes of a range of conflict not limited to orchestrated violence. Immediately after World War II, national security concerns were interpreted more broadly; economic stability was a centerpiece of the security strategy undergirding the Marshall Plan. With the coming of the Cold War, however, a broader view of national security was shunted aside in favor of the military standoff.

The end of the Cold War brought with it a lively debate over "new" security threats. A number of observers have called for "redefining security" to take into account an array of nonmilitary issues that pose fundamental threats to the health and well-being of populations or their national security. Jessica Mathews (1989), like Lester Brown (1977) and Richard Ullman (1983) before her, drew attention to links between environment and security and called for redefinition of *security* in her seminal *Foreign Affairs* article, "Redefining Security". They stressed the roles that environment and population variables could play, negatively or positively, in contributing to economic and political stability (see also Homer-Dixon 1994; 1999). Beyond the stability concern that still spoke to traditional security considerations, Mathews and others argued for a broadened conception of security that incorporated concern for individuals, society, and even ecosystems as a more meaningful response to post-Cold War threats (Myers 1993).

Some have warned against expanding the definition of national security to the point of meaninglessness (Deudney 1991; Gleditsch 1997; Levy 1995). If it means everything, it means nothing. Others have worried that linking environment and security merely amounts to a rhetorical ploy to grab budgetary resources and in fact presents a real threat that the environment might be militarized rather than security's being greened (Kakönen 1994; Wëver 1995). To avoid those pitfalls, sharp thinking is required to sort out the environmental issues that need to be considered with the highest priority and the ones that do not.

The relationship of biological diversity to the national interest and national security falls into four categories used in foreign-policy analysis: health and well-being of individuals; economic security; conflict, state capacity, and stability; and the role of security institutions. This analysis looks at the subject primarily from a generic viewpoint rather than solely from that of the United States. Another analysis (Westling 1999) appeared as this volume was in press.

HEALTH AND WELL-BEING OF INDIVIDUALS

Foreign-policy analysts are generally concerned about the protection of citizens at home and abroad from harmful effects of war, disease, and famine. Biodiversity loss can lead to disease, mortality, and food-supply problems, but, equally important, biodiversity can contribute to prevention of threats and enhance understanding of how to deal with them.

A well-elaborated contribution to health and well-being of individuals is that of the value of wild species to medicine, including pharmaceutical research (Grifo and Rosenthal 1997). This goes beyond the contribution to particular medicines (for example, the antibiotic cyclosporin or birth-control pills from a Mexican yam) to the contribution to research and development in the life sciences (particularly the health sciences). A most dramatic example is an enzyme from the bacterium *Thermus aquaticus*, originally discovered in a Yellowstone hot spring, which makes the polymerase chain reaction possible. This reaction, the development of which was honored by the Nobel Prize in chemistry in 1993,[1] is a rapid magnifying reaction that produces copious copies of genetic material in the space of hours. Among other things, it is fundamental to the Human Genome Project, with all its incalculable promise for human health, and is central in diagnostic medicine and a wide variety of biological research.

Second is the contribution of wild genes to agriculture and animal husbandry, which produce enormous benefits for people. This contribution is potentially greater today because genetic engineering essentially allows a gene to be transferred between any two species rather than only species that can be coaxed to interbreed. The importance of this contribution is evident in light of the need to feed the soaring human population by intensifying agriculture while reducing associated negative environmental affects. However, like any technology, it must be used carefully.

A third contribution of wild species to agriculture is that at the organism level, including pollination and integrated pest management that enhance agricultural production and health and save lives. An example of the latter was the identification, through the Consultative Group for International Agricultural Research, of a parasitic wasp from Paraguay that was the natural predator of the cassava mealy bug then on the verge of creating major famine in West Africa, where neither cassava, the mealy bug, nor the wasp was native (Herren and Neuenschwander 1991). Integrated pest management not only enhances agricultural yields (for example, it prevents billions of dollars of agricultural loss annually in the United States), but also reduces adverse environmental effects of pesticide use.

A fourth contribution of biological diversity to the health and well-being of individuals involves the physical threats stemming from the failure of ecosystem services (Myers 1996). A classic example is the flooding and loss of life in Bangladesh and India from deforestation further up the Ganges watershed in Nepal. The impact of Hurricane Mitch on Central America in 1998 was significantly aggravated by the loss of forest cover. A less well-known example, related to the ozone layer and the protection that it provides against UV radiation, is that a 1% increase in UV radiation causes a 10% increase in the incidence of cataracts; there is very little research on the implications of increased UV radiation for other forms of life and what they might mean for people.

Biological diversity contributes, sometimes directly and sometimes indirectly, to the growth of the life sciences. This goes way beyond medical research itself to involve important but serendipitous medical implications, such as how accidental

---

[1] The 1993 Nobel Prize for chemistry was awarded to Kary B. Mullis for his invention of the polymerase chain reaction (PCR) method.

contamination of a laboratory culture by Penicillium mold led to the discovery of antibiotics. The information that is contained in living organisms of value to the life sciences constitutes the library function of biological diversity and has implications as far afield as bioindustry and industrial ecology.

ECONOMIC SECURITY

Governments are naturally concerned about matters that affect the economic condition of their nations and people.

One concern about important national economic resources is the possible loss of monopoly because of international theft. A classic example—although not of actual theft—is the collapse of the Amazon rubber boom after rubber tree seeds were exported by Henry Wickham, an Englishman residing in the lower Amazon; the exported seeds provided the entire basis of the Malaysian rubber industry which supplanted Amazon rubber.

A second category, which might be far less obvious, is the effects of alien species that can cause serious problems when introduced into places where they are not native. Alien species are the second greatest cause of extinction after habitat destruction, but by and large they are not regarded as a security issue or as of great economic import. For example, the loss of the American elm through Dutch elm disease is probably viewed more as an aesthetic consequence than as an economic one. But, the collapse of the lake trout fishery in the Great Lakes because of the introduction of the lamprey and the clogging of the pipes of electric plants by the zebra mussel have clear economic consequences. The comb jelly *Mnemiopsis leidyi* (transported in ballast water from the Atlantic coastal waters of the New World) has short-circuited the food chain in the Black Sea and is now equal in biomass to the 250-million-dollar-a-year anchovy fishery that it has replaced (Carlton 1996); this clearly is an economic-security issue for the Black Sea nations. Sometimes, the combination of two alien species can create a problem that each alone does not. The zebra mussel accumulates polychlorinated biphenyls (PCBs) through filter feeding in the Great Lakes and could even have been a good bioremediator to clean up that pollutant. The introduction of a species of fish (a goby) that feeds on zebra mussels now opens the possibility that PCBs will work their way once again up the food chain with both human health consequences and economic consequences, including the need to close down the fishery (Jude 1996).

Ecosystem services provided by biological diversity provide a third connection with economic security. The Panama Canal, a strategic economic waterway, requires a freshwater supply if it is both to function as a canal and to provide a biological barrier between the Pacific Ocean and Caribbean Sea biotas. The freshwater supply, in turn, depends on the forests of the canal watershed. A Smithsonian scientist once calculated that total deforestation of that watershed would result in 3 million cubic meters of sediments entering the canal each year. Another example is the hydrological cycle of the Amazon basin, in which half the rainfall is generated internally largely because of the forest cover. The stability of the Amazon climate, and indeed that of central South America, depends on all the Amazon Pact nations' working together to maintain the integrity of that cycle.

A fourth connection between biological diversity and economic security involves physical damage to territory. Biodiversity loss can be both the consequence and the cause of such damage as in Hurricane Mitch in Central America in 1998. Nonetheless, if the more ambitious versions of the Hidrovia waterway project for the Parana Paraguay drainage went forward, the economic consequences could be similar to those engendered by modifications to the southern Florida ecosystem and the Mississippi drainage. The United States is now investing large sums to restore the South Florida ecosystem by reversing the effects of 50 years of independent decisions about water that have reduced sheet flow of water by 25%–50%. There was much greater Mississippi flood damage in 1993 than would otherwise have been the case, because of diking and other projects that altered the natural riverbed.

Another connection between biological diversity and economic security involves the relationships between genetic resources, science, and economic growth. For the United States and other advanced industrial nations, science and technology are essential to maintaining economic growth. Biological science and biotechnology, in particular, are sectors of research and development of major and growing importance. Access to genetic resources—the ability to use and study genetic material in or from other countries—is essential under appropriate rules and with due compensation, of course. Extinction, obviously, represents the ultimate loss of access, because living material no longer exists.

CONFLICT, STATE CAPACITY, AND STABILITY

Top foreign-policy and security concerns include avoidance of unnecessary conflict, coupled with preparedness in case of need, and efforts to maintain stability both outside and inside the state.

A traditional aspect involves the protection of strategic goods, usually thought of in terms of physical resources, such as oil and uranium. There might be instances in which these include genetic resources. From a historical perspective, Southeast Asian rubber is an example of a strategic target for the Japanese in World War II.

A second category would be conflicts arising over resources. There have been spats over fishery issues (involving, for example, Canada, Spain, and the outer continental shelf of North America), but there seem to be no examples of major interstate conflict arising over biological resources. In this context, it is ironic that nations will fight over a square meter of territory and ignore the loss of territory in cubic meters through soil erosion.

Biological resources can relate to defense preparedness. Certainly, access to rubber and quinine were essential to the Allied war effort in World War II, and antibiotics contributed in an important way as well. It is hard to see how biological resources will play as big a role in high-technology wars, except for the dark side of biological warfare. The threat of the latter is far greater than many recognize, and protocols and agreements are generally poorly developed and weak. A more interesting contribution of biological diversity might be as a source of intelligence information. The provenance of a Japanese submarine was once identified by algae scraped from its hull and analyzed by Ruth Patrick (personal communication). Many species have quite limited distributions and can therefore serve as useful sources of geographic information. Microbial species that can accumulate radionuclides can be used to assess compliance and noncompliance with nuclear nonproliferation.

A fourth and enormously important connection with biological resources is conflict prevention and confidence-building. Environmental cooperation between two states often leads to broader cooperation on seemingly more difficult issues. The common agenda on the environment between Brazil and the United States is an outstanding example. The water problems of Cyprus might present a first important subject of cooperation between North Cyprus and South Cyprus; certainly, the problem cannot be addressed without both parties. Binational peace parks can play an important role in reducing border tensions; in 1998, a peace park between Ecuador and Peru was a major element in resolving their territorial dispute.

The fifth link of biological resources, namely political tension between countries, is probably likely to be a contributing, rather than a causal, factor. An example would be US and Canadian tensions over the management of Pacific salmon stocks. An example of a causal factor and biodiversity loss as an associated consequence[2] was the El Salvador-Honduras soccer war, generally agreed to have been caused by problems with environmental refugees.

THE ROLE OF SECURITY INSTITUTIONS

Security institutions are generally not thought about from an environmental perspective, but biodiversity can have both positive and negative effects. The Department of Defense (DOD) now reviews any "significant military exercise" (a technical DOD term) for possible environmental effects. Although it is hard to see how this could weigh heavily during full war conditions (for example, the US military was not included in the greenhouse gas commitments negotiated under the climate convention at Kyoto, Japan, in 1997), there is substantial military activity during peacetime. DOD now works to conserve biological diversity on its extensive land holdings. Medea, a group of US scientists with security clearances, study data gathered by the US intelligence community[3] to see whether they

---

[2] For more on the linkages between environmental refugees and conflict, see Homer-Dixon and Percival 1996. In response to the grossly overpopulated and severely degraded land in their native land, Salvadorans had been gradually migrating into their less densely populated neighbor, Honduras. As the land continued to be more degraded and population continued to increase, more Salvadorans were crossing the Honduran border; this led to the Honduran government's expulsion of the migrants. War then broke out between the two countries in 1969.

[3] An environmental task force was established in 1992 by the Central Intelligence Agency to assess how crucial environmental issues could be solved through the use of the US national security apparatus. The task force brought together the group of 60 prominent US environmental and global-change scientists, know as MEDEA.

include useful environmental information. An example would be data on possible thinning of the Arctic ice cap as an early indicator of global warming.

CONCLUSIONS

Viewing through a traditional political-science lens, one is forced to conclude that environment (under the collective umbrella of biodiversity effects) is more often a contributing factor, with some other aspects of national interest and security, than a causal factor. The weak part of that conclusion is that although biodiversity loss is easy to ignore incrementally, for national interest and security the aggregate can be disastrous. For example, Haiti's major biodiversity loss is caused by almost complete deforestation, and loss and deforestation are clearly not in Haiti's national interest. Given present trends, the loss of biodiversity could also be disastrous on a global scale. The press of everyday problems makes it too easy, in Jessica Mathews's terms, for the urgent to override the important.

Scale and rate of change affect how we should view matters. As this forum met, there were gigantic smoke clouds from extensive fires in the Amazon, as well as the better-known vast fires in Indonesia. Together they mean that more of the world burned in 1997 than ever before in recorded history. That is hard to dismiss as not of high national interest and security concern. As Madeleine Albright has observed, threats to national security are no longer confined to armed threats.[4] They also come through the air, water, changing climate, and loss of biological diversity. The positive contributions of biodiversity and ecosystems—present and potential—and the negative effects of loss are so great that they merit much more serious attention. The "important"—the environment and biological diversity—has indeed become urgent.

ACKNOWLEDGMENTS

This paper required my learning about a field basically new to me, namely political science/foreign affairs, including its vocabulary. I am grateful to the Environmental Change and Security Project of the Woodrow Wilson Center for International Scholars for my tutorial, first and foremost to its founding director, P.J. Simmons. His successor, Geoff Dabelko, and Jessica Powers and Aaron Frank were all very helpful. J.P. Myers and Sarah Vogel helped with the information on zebra mussels, gobies, and PCBs. Kathleen Conforti was helpful in countless ways.

REFERENCES

Brown L. 1977. Redefining security. Worldwatch Pap No 14. Washington DC: Worldwatch Inst.

Carlton JT. 1996. Marine bioinvasions: the alteration of marine ecosystems by nonindigenous species. Oceanography 9:36–43.

---

[4] Remarks at Saint Michael's College, excerpted in the 1998 Environmental Change and Security Project Report.

Deudney D. 1991. Environment and security: muddled thinking. Bull Atom Sci April:23–8.
Gleditsch NP (ed). 1997. Conflict and the environment. Dordrecht: Kluwer.
Grifo F, Joshua R (eds). 1997. Biodiversity and human health. Washington DC: Island Pr.
Herren HR, Neuenschwander P. 1991. Biological control of cassava pests in Africa. Ann Rev Entonol 36:257–83.
Homer-Dixon T. 1999. Environment, scarcity, and violence. 2nd ed. Princeton, NJ: Princeton Univ Pr.
Homer-Dixon TF, Percival V. 1996. Environmental scarcity and violent conflict: briefing book. Toronto: Univ Toronto, The Project on Environment, Population, and Scarcity.
Homer-Dixon T. 1994. Environmental scarcities and violent conflict: evidence from cases. Int Security 19:5–40.
Jude DJ. 1996. Gobies: cyberfish of the 90's.
Kakönën J (ed). 1994. Green security or militarized environment. Brookfield: Darmouth.
Levy MA. 1995. Is the environment a national security issue? Int Security 20:35–62.
Mathews JT. 1989. Redefining security. For Affairs 68:162–77.
Myers N. 1996. Environmental services of biodiversity. Washington DC: National Acad Pr and Covelo CA: Island Pr.
Myers N. 1993. Ultimate security: the environmental basis of political stability. New York: WW Norton.
Patrick R. 1964. A discussion of natural and abnormal diatom communities. In: Jackson DF (ed). Algae and man. New York: Plenum Pr.
Patrick R. 1961. A study of the numbers and kinds of species found in rivers in eastern United States. Proc Acad Nat Sci Phila: 113:215–58.
Patrick R. 1962. Effects of river physical and chemical characteristics on aquatic life. J Amer Water Works Asso 54:544–50.
Ullman RH. 1983. Redefining security. Int Security 8:129–53.
UNEP [United Nations Environment Programme]. 1995. Global biodiversity assessment. Cambridge: Cambridge Univ Pr. 1140 p.
Westling AH. 1999. Biodiversity loss and its implications for security and armed conflict. In: Cracraft J, Grifo FT (eds). The living planet in crisis: biodiversity, science, and policy. New York: Columbia Univ Pr. p 209-16.
Wëver O. 1995. Securitization and desecuritization. In: Lipschutz RD (ed). On security. New York: Columbia Univ Pr. p 46-86.

BIODIVERSITY AND ORGANIZING FOR SUSTAINABILITY IN THE UNITED STATES GOVERNMENT

TIMOTHY E. WIRTH

United Nations Foundation, 1301 Connecticut Avenue NW, Suite 700, Washington, DC 20036

This paper is based on remarks made by Mr. Wirth as Under Secretary of State for Global Affairs at the Conference on Nature and Human Society at the National Academy of Sciences in Washington, DC, on 30 October 1997.

AFTER A DECADE of discussion on biodiversity through this Second National Forum on Biodiversity, *Nature and Human Society: The Quest for A Sustainable World*, it might be useful to look ahead. What do we want to have accomplished by the year 2007?

On October 28, 1997, the US stock market fell dramatically, caught in a tailspin that sent global markets reeling. The Hong Kong market stuttered and gasped, and morning television in the United States quoted overnight market changes. Economies all over Southeast Asia stumbled and fell, and the international financial institutions responded with billions of dollars. The news was on the front page everywhere in the world.

Meanwhile, the broadest fires in recent history were blazing in the Amazon, and the smoke from fires in Indonesia had spread over an area greater than that of the lower 48 states of the United States. El Niño was fingered, creating a convenient mask over the forces actually at the root of these crises. Negotiations for The Kyoto Protocol to the Framework Convention on Climate Change (1997) intensified, with greater stakes than any such international conference before. Yet, with few exceptions, those stories were back-page news, when they were covered at all, and certainly no one stepped in with billions of dollars. The contrast was sharp and significant.

Those two sets of events demonstrated the impact of globalization, which is intensifying the relationship between our economies and our environment. Consider the reaction generated when the markets crashed. But did anyone smell the forests burning? Did anyone hear the forests falling? We protect fragile economies and prop up failing currencies. But what about fragile ecosystems and failing species?

Certainly, if we are to have any hope of protecting the world's biological richness, we will have to do a much better job of getting people to listen and to understand—to listen to their home, Planet Earth, and to understand the connections between the health of the world's economies and the health of the resources on which those economies rely.

Economists, financiers, businessmen, and bankers will have to begin to recognize the costs hidden in exploiting the seas, the lands, and the air for short-term wealth. They will have to recognize that ecological systems are the very foundation of our society—in science, in agriculture, in social and economic planning. Five essential biological systems—croplands, forests, grasslands, oceans, and freshwaters—support the world economy. Except for fossil fuels and minerals, they supply all the raw materials for industry and provide all our food:

- Croplands supply food, feed, and an endless array of raw materials for industry, such as fiber and vegetable oils.
- Forests are the source of fuel, lumber, paper, and countless other products and house valuable watersheds that provide drinking water for growing urban areas.
- Grasslands provide meat, milk, leather, and wool.
- Oceans and freshwater produce food for people and resources for industry.

In the language of the business world, you could say that the economy is a wholly owned subsidiary of the environment. But when we pollute, degrade, and irretrievably compromise that ecological capital, we begin to do serious damage to the economy.

With that introduction, let me present a few ideas by focusing on the third Conference on Nature and Human Society, to be held in the year 2007.

GLOBALIZATION

By 2007, this forum should have a much better understanding of the impacts of globalization. Today, our economists know that we are profoundly remaking international trade and markets. "Globalization and international trade" has become a mantra, almost an ideology, promising a radiant future for us all.

But is there a dark side? Have we looked at other impacts? For example, are globalization and trade between the developed and developing worlds destroying subsistence agriculture? Are we co-opting Third World farmers into production for the international marketplace while their societies are made dependent on imported foods? The social and cultural consequences of this may be very serious.

Earlier this week, we heard that the number of languages spoken around the world has declined from 6,000 to 600 in this century alone. What else are we losing? What crops are gone? What about the knowledge of those crops? What of the indigenous people who carry this knowledge?

In 2007, we will be asking these questions more openly and aggressively, and the scientific community will have to be prepared to answer them.

POPULATION

If globalization is the first suggestion, certainly population is central as well. In 2007, we will know whether we have dealt with the urgency of the question. It is not a question of what to do, but of openly asking about population pressure. It is not always popular, but it must be done.

The growth of the world's population has slowed, but the base against which that rate applies is greater than ever before. Our planet is populated by the largest generation of youth in human history—and the next generation will be even larger. There are now roughly one billion teenagers in the world—900 million of who are in the developing world.[1] Even if average fertility were to fall rapidly to the replacement rate of 2.1, the sheer number of females giving birth over the next several decades will be so large that population will continue to grow rapidly for many years to come. This phenomenon—population momentum—will account for about half of anticipated population growth in the developing world through the year 2100.

At the International Conference on Population and Development, nations of the world agreed—and now must implement—an action plan that endorses a strategy to stabilize population growth by meeting the needs of individuals and addressing the range of factors that influence decisions about family size.

But acting around the world is not enough. We also must focus here at home, with special reference to our own consumption, disproportionate use of resources, and astonishing production of waste.

We must also understand better the concept of carrying capacity—how many of us can the earth sustain, in what lifestyle, and with what expectations? Obviously, population, like globalization, has a profound effect on biodiversity and on the purposes of the Conference on Nature and Human Society.

PERSISTENT ORGANIC POLLUTANTS

Third, I would raise the issue of persistent organic pollutants. At the Department of State, we have begun to explore this issue, and it has become one of our top priorities. We recently hosted an international meeting on land-based sources of marine pollution, and we are starting to focus on how we can affect this important issue.

Theo Colburn, of the World Wildlife Fund, and Diane Dumanowski and Pete Myers, of the W. Alton Jones Foundation, gave us a starting point in this discussion with *Our Stolen Future*. In 10 years, we will know whether this book is another *Silent Spring*. I believe that it is and that the research community will be deeply engaged at the next conference. How do toxicants travel? What are the

---

[1] World Population Data Sheet, Population Reference Bureau, 1998.

impacts? Are we poisoning ourselves? What are the implications for reproductive health?

RETHINKING BIODIVERSITY

Fourth, we will have gone a long way toward rethinking biodiversity, and perhaps we will be calling it something new. I'm not sure "ecosystem services" is much better. Maybe "nature's services"?

The point is that we have to tell the story better. Why do we preserve snail darters or kangaroo rats? Why do we study nematodes? How does the web of life fit together? And what does it do for the average citizen of the world?

On other issues, we have learned to tell the story:

- When the Cuyahoga River caught on fire, it became the poster event for the environmental movement.
- Asthma caused people to worry about their children and got us the Clean Air Act.
- Lead and learning were linked, and we removed lead from gasoline.
- Lakes were dying, and we understood acid rain and cleaned up our utilities.
- And maybe we will learn about global warming. Is El Niño the trailer for *Climate Change* the movie?

I predict that the link of nature's services to the science of biodiversity will become the way to tell the story. The links with economics will give us new tools to become loud messengers. And I can guarantee that until we all do a better job of telling the story, the Endangered Species Act will continue to be under attack and the Biodiversity Treaty will remain unratified for want of a two-thirds majority in the Senate.

One of the signal events of the third Conference on Nature and Human Society will be the awarding of a new prize, awarded for science in service to society. Perhaps we will call it the Ed Wilson Prize for Effective Individual Achievement, for the scientist who did the best job in translating his or her discipline to the public. Or the Peter Raven Award for Institutional Relevance, given to the scientific institution that best used its reach to advance public engagement in the preservation of the natural world.

No matter what the name, the point is this: For too long, those public-spirited scientists who sought to take their science outside the laboratory, to the public, to the television audience—or, Heaven forbid, to the political arena—have been punished. To tell the story, to popularize, to explain has somehow been unscientific; it sullied the profession, and those who did it were suspect and unpromotable. It is imperative that we as a society—and individual scientists—do a better job of rewarding those who translate their science, who bring it to the public's attention, and who foster broad public understanding.

My first tutor in thinking about science was Walter Roberts, a wonderful man and founder of the National Center for Atmospheric Science in Boulder. Walter taught me and others about the commitment of science in service to society, and he was right. Science is critical if our global society is going to develop sustainably.

REFERENCES

Carson R. 1962. Silent spring. Boston MA: Houghton Mifflin.
Colburn T, Dumanowski D, Myers P. 1996. Our stolen future. New York NY: Dutton.

EXAMPLES OF

SUSTAINABILITY

HOW TO GROW A WILDLAND:
THE GARDENIFICATION OF NATURE

DANIEL H. JANZEN
Joseph Leidy Laboratory, Department of Biology,
University of Pennsylvania, Philadelphia, PA 19104

THE BOTTOM-UP VIEW

THE FIRST fund-raising flyer, produced in a kitchen and nurtured by The Nature Conservancy and two academics, was titled *How to Grow a National Park*. Its cover depicted a cowpat with a newly germinated guanacaste tree seedling in the middle. In 1985, fund-raising efforts for tropical conservation centered on the argument that we must buy forest urgently because once it is cut down, it is gone forever. We argued the opposite for tropical dry forest, which once had covered at least half of the forested tropics. Human settlement had eliminated it so thoroughly that the only option was restoration through buying trashed remnants somewhere and restoring a portion that would be large enough to conserve an entire ecosystem. That "somewhere" focused on the 10,000-hectare Santa Rosa National Park in northwestern Costa Rica because we were familiar with it and its biology. The idea survived and grew because the Costa Rican community believed in it and worked for it and because the international community was willing to invest cash and labor to preserve the existence of important tropical nature.

In 1989, the idea became the Area de Conservacion Guanacaste (ACG) (http://www.acguanacaste.ac.cr). The operational word was "restore." The question was how to severely diminish four centuries of footprints of modern society and let the forest take back its land. We called the process restoration biology and biocultural restoration, but it was also secondary succession, regeneration, regrowth, reforestation, aforestation, farming, ranching, mitigation, recuperation, recovery, rehabilitation, and sustainability.

HOW COULD WE RESTORE THIS PARTICULAR TROPICAL DRY FOREST?

- Stop the anthropogenic fires.
- Restore the size.
- Integrate its socioeconomics with those of neighboring areas on all levels.
- Stop the ranching, farming, logging, and hunting.
- Pay the bills.

Stop the Anthropogenic Fires

Because this particular tropical dry forest does not have natural fires, we did not have the dilemma of deciding when to let it burn. The lands of the ACG have survived four centuries of clearing of forest and brush by repeated annual to semiannual anthropogenic fires during the 6-month dry season. In 1985, the 120,000 hectares of the ACG contained at least 50,000 hectares of highly inflammable old pastures and brushy fields. Every time a fire passed through it, more woody vegetation was eliminated. However, the general area had not been sufficiently successful to be a thoroughly cleared agroscape. Without fire, the remnants of forest within the open areas would be able to expand to restore the forest. Every farmer and rancher knew this, although biologists and conservationists were more skeptical.

Stopping the fires was not a technical issue or a biological question. The methods were straightforward: apply trucks, tractors, pumps, lots of brooms, radios and walkie-talkies, burned firebreaks, and fire lookouts. Rather, stopping the fires was a question of personnel management and motivation. It was a question of being there at 2 a.m. on Easter Sunday when your family and friends are at the beach; of working all night; of maintaining a lookout for 6 months, 24 hours a day. It was a question of working with the neighbors and of having them be the fire crew.

Elimination of the fire footprint was achieved by selecting about a dozen locally hired staff, giving them full responsibility, backing their budgetary needs, and giving them the opportunity to invent any schedule or administration—including going off site to combat fires on private neighboring land, strongly supporting a regionwide educational program about the value of eliminating fire, and calling on the regional police force and other volunteers when a particular fire got out of hand.

The ACG Fire Program and the ACG administration as a whole succeeded. Today, the brushy pastures and interdigitated remnants of dry forest in the ACG are virtually firefree and display at least 40,000 hectares of rapidly regenerating young forest. The seeds arrive by means of water, wind, birds, bats, rodents, ungulates, and carnivores.

Restore the Size

How big an area would be big enough? Parque Nacional Santa Rosa, part of one of Costa Rica's first ranches, was a 10,600-hectare island in the ghost of the dry forests that once extended from near Mazatlán, Mexico, to southern South America, with some rain forest intermingled here and there. What was that dry forest is now much of the neotropical agroscape and is clearly unrecoverable. All surviving neotropical dry forests are islands in that agroscape.

Santa Rosa was far too small for the survival of its ecological processes and dry-forest ecosystem—it contained only pieces of drainage basins, small portions of major habitats, and part of the contour, and it was virtually all edge. Also, it was far too small to absorb the many kinds of human footprints that would result from its becoming a local, national, and international garden. In particular, it needed to expand to the wetter east. Much of its more mobile biodiversity (insects and birds) migrates seasonally to the rain forests and cloud forests on and across the mountains to the east and return in the rainy season.

The ACG expanded until the dry forest was big enough. The border was not set by biological requirements, but by the reality of social resistance; it stopped where the very profitable portions of the agroscape began. This expansion incorporated other semiconserved islands of wildland (Sector Murcielago, Reserva Forestal Orosi, Parque Nacional Rincon de la Vieja, and Refugio de Vida Silvestre Isla Bolaños), and all the private lands in between—some 70 of them, ranging from small farms to large ranches—were purchased from squatters, absentee landlords, and land speculators.

On the one hand, this large-scale purchase of land was facilitated greatly by a rapid demise of the region's cattle industry, by the overall low quality of the regional agroscape, by Central American military turmoil, and by the socioeconomic reality that virtually all owners were willing to convert their land into more-profitable ventures elsewhere. Another major contribution was the moderate number of owners who believed that it was highly respectable to have their lands become national park, thus tolerating the minimal prices that the conservation community pays for existence value.

On the other hand, buying these private properties and displacing the employees intertwined the ACG inextricably with its neighbors. Houses on the ranches and farms became part of ACG's infrastructure, as did the dwellings of the former employees when they or their neighbors were hired as new ACG staff. The children of these former ranchers, farmers, and employees were among the pupils in the ACG Biological Education Program. ACG staff bought supplies in the local stores. The local decision-makers became members of the ACG's board of directors (Comite Local), a responsibility shared with the Ministry of the Environment and Energy (MINAE) and the staff of the ACG itself. From the start, the process of building the ACG was intrinsically an act of presence, quite different from an act of gazetting a large pristine wildland as a national park.

As the area of the ACG increased, so did the opportunities for its presence and socioeconomic integration. When a vandal sets a fire that burns 2,000 hectares of centuries-old African grass pasture, it is only a thin scar on the ACG landscape now, not the end of a project. If a deer is poached, it often can be shrugged off. When a soccer field or a picnic ground is needed, the land is there; after the schoolchildren in a biology course trample one 10-hectare section, they can trample another section while the first section recuperates. If 20 hectares of pasture is needed for the ACG's work horses, it is there. Does one need to become a biodegrader for 1,000 truckloads of orange peels a year? Build a new road for management? Put up a wind farm? Host an ecotourism program? Provide seeds for a mahogany-seed farm? Grow a carbon crop? Build a directory on the Internet for 235,000 species? Somewhere in these 120,000 hectares, such footprints may well be absorbable; in only 10,600 hectares, they rarely could be.

Today, the expansion of the ACG into the eastern rain forests and cloud forests has become part of the conservation solution to the effect of the drying and heating that the western dry forests of the ACG are suffering through global warming, an outcome that was unforeseen before 1992. During the 22 years of weather recorded in the ACG, 1997 was the driest and hottest year, and the trend continues. The rain forests and cooler cloud forests to the east have been a lifeboat for the dry forest on more than one level.

Integrate Its Socioeconomics with Those of Neighboring Areas on All Levels

The ACG is a 135,000-hectare terrestrial and marine garden that has 120 owner-employees, a US$1.6 million annual operating budget, and 3.3 million stockholders. It operates within the bylaws of incorporation of the state and, more specifically, within those of MINAE. The macroproduce of the ACG is the conservation of the biodiversity of its wildland and ecosystems into perpetuity. The process used to realize this goal is to be a major player in the national and local biodiversity industry, intertwined with the ecosystem industry: biodiversity development, ecosystem development, environmental-services development. All uses leave footprints, but this process calls for the unending quest for uses that are nondamaging. The ACG has come to peace with the reality that 5% of its biodiversity and ecosystems will be sacrificed to guarantee the existence of the remainder. This is the ACG wildland peace treaty that is being negotiated with the agroscape and the urban landscape.

Such a socioeconomic integration at the local, national, and international levels is sought through diverse activities. A few examples follow.

As the regional cattle industry has died over the last decade, the ACG's biodiversity and ecosystem industries have become part of the economic restoration in the region not only through cash flow, but also through offering relatively ceilingfree and diverse job opportunities that are far more in tune with modern society than were herding livestock and subsistence farming. The small neighboring town of Quebrada Grande is changing rapidly from a shopping center for cowboys to a suburb for the ACG that provides more urban activities. All ACG employees are Costa Rican, and 82% are from the immediate region; 42% are women. All are computerizing, all are networking, and all are exploring this new world of professional responsibility toward a goal—and the pain and opportunities these forces bring.

Since 1987, the ACG Biological Education Program has taught basic biology in the ACG's wildland habitats—expanding the responsibility beyond biocultural restoration into bioliteracy—to all 4th-, 5th-, and 6th-grade students, and now high-school students, in the vicinity of the ACG. Today, this means 42 schools and more than 2,000 students per year, 22% of the ACG's annual operating budget. It is widely rumored that the ACG has had an easy job because it is imbedded in a "tame populace," but this tameness was created deliberately.

As a result of the restoration of the original forest vegetation throughout the ACG, the watersheds are being restored for 11 major rivers that service all local towns and the irrigation systems for major agroscapes. This ACG water factory is becoming particularly crucial as global warming continues to heat up and dry out the region and as regional agriculture moves toward environmental control.

Also through restoration of the original forest vegetation throughout the ACG, atmospheric carbon is being farmed (see Costa Rica's P.A.P. in http//www.ji.org and http://www.unfccc.de). The ACG and its biodiversity and ecosystem industries thus become both the "green scrubber" and the insurance policy that the carbon will stay sequestered.

The ACG has been a major stimulus, supporter, training ground, and proving ground for many of the field activities of INBio (Instituto Nacional de Biodiversidad), the institution that has accepted major responsibility in the Costa Rican national biodiversity inventory, teaching of bioliteracy, and computerization of biodiversity management (http://www.inbio.ac.cr). Locally hired and trained parataxonomists and parabiodiversity prospectors working for ACG and INBio share the ACG facilities. These paraprofessionals are part of the intellectual and operational critical mass that carries forward the ACG's Research Program. The international taxonomic cleanup that swirls around INBio's national biodiversity inventory, in great part being carried forward by the nation's parataxonomists, is key to readying the taxonomic platform on which the ACG's biodiversity industry is based. At least 60% of Costa Rica's species occur within ACG's area, which comprises only 2% of the country. A directory of biodiversity on the ACG Web site is anticipated as the debut of this taxonomic platform.

The ACG grew out of Costa Rica's second-oldest national park and second-oldest hacienda. It has been a major stimulus and supporter for the rapidly evolving Sistema Nacional de Areas de Conservacion (SINAC) of MINAE, which is the administrative and technical integration of all of Costa Rica's conserved wildlands into 11 consolidated conservation areas. SINAC's wildlands constitute about 25% of the country and combine many traditional management categories into one: to save it without destroying it. Ecotourism is Costa Rica's largest crop. The ecotourist—whether a school child from Peoria or a researcher—is a better kind of cow, and the conservation areas are the pastures. SINAC was founded to forge a peaceful coexistence between the wildland garden and the agroscape and urban landscape. Nothing invites encroachment of neighbors more quickly than the impression of abandonment or disuse. Wildland biodiversity must have a national presence, a national farm.

The ACG is developing itself as a research-friendly platform for all ilks—local, national, and international. It is the place to find out a vast array of information. For example, how many times does a spider monkey scratch its left armpit (in the morning)? What species of plants do the caterpillars of rain-forest skipper butterflies eat? Can we clarify the species and genera of hundreds of species of water mites? What flowers do bats stick their heads into? Will a pharmaceutical company find its "gold" in a bottle of frozen baby ticks? How many eggs of the Ridley's sea turtle survive predation by vultures and coyotes? Where do species of plants live in a montane cloud forest? How can Cladocera be used to reduce the numbers of dengue-bearing mosquitoes? Do the parasitic wasps in the ACG reduce the density of leaf miners in the neighbor's orange orchard? How many children do current ACG staff have, and how many siblings did the parents have? How fast does an unburned pasture return to forest? How hard does the wind blow? Not only does this biodiversity and ecosystem research industry provide a type of high-yield ecotourism, but each of these research projects also carries the distant possibility of royalties—sometimes paid in fuel for the Biological Education Program, sometimes paid in votes by visitors, sometimes paid in cash from the pharmaceutical industry and other commercial users, and sometimes paid in sweat equity by the researchers themselves. Even my description of this pilot project in the survival of complex tropical wildland is yet another product of this farm.

My last example is that of a specific contract for biodiversity and ecosystem services between the ACG and Del Oro, a neighboring orange-juice company. The ACG is being paid for 20 years' worth of biological control agents, water, consulting, orange-peel degradation, and isolation from orange pests—US$480,000 in the coinage of 1,200 hectares of one of the biologically scarcest habitats in Costa Rica, the lowland transition forests between the Atlantic rain forest and the Pacific dry forest. This mutualism has other ramifications in the form of Del Oro's "green" orange juice that is now certified ECO-OK by Rainforest Alliance and has been made technically feasible through the environmental services provided by the ACG. This juice is penetrating the Costa Rican market, heading for the European market, and reinforcing the contemporary Costa Rican attitude of taking virtually its entire agroscape into sustainable development.

Stop the Ranching, Farming, Logging, and Hunting

The impact of everyday agroscape activities on the ACG was largely eliminated by stopping the fires and purchasing the land. The policies of a conserved wildland then regulate the tilling, weeding, and harvesting of this garden. A conservation-area garden has its public lands, its storage areas, its restricted sections, each with different rules, and each leaving different footprints, but no footprints are free. Early on, the ACG accepted that it would pay some small portion—say, 5%—of its biodiversity and ecosystem services to conserve the remaining 95% into perpetuity.

This viewpoint leads to paradoxical management decisions. In the late 1970s, when Santa Rosa was still very much a tiny semiconserved island in a great sea of agroscape, at least 2,000 semiferal cattle were living in its 10,600 hectares. Fires burned across virtually all of it every year, but it was a relatively stable mosaic pasture and remnant forest, as it had been for centuries. In a spate of classic national-park management, the cattle were removed, but no fire-control program was established. The introduced species of African pasture grasses then grew to 2 m high, and they provided fuel for the annual fires that began the steady, thorough process of forest removal.

The lesson was learned. The young ACG left the cattle on the pastures as the land was purchased in the middle 1980s and, at times, even leased browsing rights to as many as 7,000 additional cattle as biotic mowing machines. This kept the grass down as the nascent fire-control program came into its own. These newly firefree pastures filled even more rapidly with woody, shade-producing plants than did those without livestock. Could the cattle be left until full reforestation was accomplished? No, because their use as biotic mowing machines is not free. Their footprint is the trashing of the streams, rivers, and riparian vegetation unless they are fenced out of them at a greater cost than their market value. Ironically, however, a muddy dry-season waterhole with a horse standing in it dates back to the "natural" before the Pleistocene hunters and their carnivorous helpers took our megafauna. Eventually, some sector in the dry forest of the ACG will contain whatever Pleistocene megafauna can be recuperated.

I cannot overemphasize that a successfully conserved wildland is a garden. A topic in the news today is the restoration of forest for carbon farming. Carbon farming is not only forest restoration: Sale of the resulting carbon also can contribute to the operational costs of and provide investment capital for a conserved wildland. Just as tropical "debt-for-nature" swaps did not solve a nation's debt problems, but fueled some major conservation initiatives, carbon farming in conservation areas will not solve our greenhouse-gas problem, but it certainly can contribute to a holistic solution. This, in turn, brings up the many imaginative ways that the sequestered carbon can be harvested and "parked" elsewhere in buildings, furniture, and even underground deposits. Thus, a wildland tree becomes a long-term investment. Carbon harvesting and windthrows begin to merge in the nature of their footprints.

Pay the Bills

One can guard a large box of gold under the bed quite inexpensively, especially if no one else knows that it is there—it requires only some barbed wire, a gun, or a watchdog. The annual operating budget for Parque Nacional Santa Rosa in the middle 1980s was about US$120,000, including salaries, most of which was spent elsewhere, thus generating virtually no income for the region. Today, the ACG is 10 times as large, costs 10 times as much to operate, and generates diverse goods for barter and a large amount of cash for the region. It meets its costs through a combination of payment for services and interest income from its endowment. This endowment was established in the late 1980s through a combination of international donations for the existence value of the ACG and government subsidy as a "debt-for-nature" swap for both its existence value and its sustainable development. The future of the ACG depends heavily on its being able to seek reasonable compensation for the biodiversity and ecosystem services to the public and commercial sectors both independently and in consort with national-level and international-level projects. The new, landmark biodiversity-prospecting agreement between Yellowstone National Park and Diversa Corporation in California (http://www.wfed.org) is most welcome, as have been INBio's biodiversity-prospecting contracts with Merck and with the INBio–Cornell–Bristol-Myers

Squibb ICBG (International Cooperative Biodiversity Group) project (http://www.nih.gov/fic/res/lessons.htm; http://www.nih.gov/fic/res/icbg.htm).

Being 10 times as large as the original Parque Nacional Santa Rosa should, and does, bring massive economy of scale to the ACG. Why, then, is the annual budget 10 times as large? There are two reasons. First, the ACG is beginning to put its "box of gold" on the stock-and-bond market. This brings administrative costs: An Internet Web site is not free, a firefighter on call at 2 a.m. requires payment, and it costs to encourage a university-educated Costa Rican biologist to spend a lifetime as a 5th-grade teacher in a remote rain-forest town that is just constructing its first gas station. Second, the tropics long have been reputed to be a source of inexpensive local labor. Unfortunately, *local* is a geographic term that has come, unconsciously, to connote labor that can be compensated for in terms that would be appropriate for a mule. However, when one moves workers from the pastures and bean fields into computer work stations, the national inventory, and the halls of politics, the operating cost for personnel skyrockets. As Costa Rica becomes a sustainably developed country and realizes its human aspirations, its cost per citizen will be similar to that in the rest of the developed world.

Ironically, today we are quite concerned with internalizing environmental costs. The development of the ACG and many other Costa Rican institutions has made us all excruciatingly aware that internalizing the costs of biodiversity development and ecosystem-service development will require budgetary figures that were not anticipated by the societies that stand to gain in both the short and long terms. An enormous amount of labor and institutional subsidy has gone into the current projects of taxonomy, biodiversity prospecting, wildland administration, political decentralization, wildland-ecosystem engineering, and all the other things discussed here and in such international agreements as the Convention on Biodiversity.

THE TOP-DOWN VIEW

The exportable generalizations that we academicians and office-holders hold so dear are extracted easily from the details just discussed. In doing all of them, we were unconsciously creating a garden. The traits of the ACG have been and are being driven by the organic traits of the site itself, by the hard-wired genetic tendencies of humans to create more humans and their domesticated genomic extensions, by the specific culture in which the ACG is embedded, and by the global humanity in which *that* is imbedded.

A generalization of the top-down view is as follows:

• Restoring complex tropical wildlands is primarily a social endeavor; the technical issues are far less challenging.

• Survival of a complex wildland, whatever its origin, in the face of humanity's genes and domesticated genomic extensions, requires a major paradigm shift—we cannot afford to perceive the conserved area as "wild," which can be interpreted as "up for grabs."

• Sustainability of a wildland will be achieved only by bestowing garden status on it, with all the planning, caring, investing, and harvesting that implies.

• All use is effect, and all gardens are affected—restoration is "footprint absorption" by the garden, and it occurs on all levels.

• Planning, caring, investing, and harvesting in the wildland garden are achieved through a detailed understanding of biodiversity and its ecosystems and by simultaneous incorporation of a specific garden's social milieu at local, national, and international levels.

• The "achievable" is an ever-shifting and ever-negotiated n-dimensional hyperspace produced by the intrinsic traits of a specific wildland interwoven with the mosaic of social energies and agendas brought to bear on it.

To put it another way, we must use it or lose it; but when we use it, something must then restore it.

ACKNOWLEDGMENTS

The experiences and observations that have led to these reflections have been supported generously for 36 years by the US National Science Foundation, by the international scientific community, and by the government and people of Costa Rica. More specifically, the personnel of the Area de Conservacion Guanacaste (ACG), the Instituto Nacional de Biodiversidad (INBio), the Fundacion de Parques Nacionales (FPN), and the Ministerio del Ambiente y Energia (MINAE) have provoked and facilitated these thoughts. I particularly thank the Costa Rican team of Alvaro Umaña, Rodrigo Gámez, Alvaro Ugalde, Mario Boza, Alfio Piva, Pedro Leon, Luis Diego Gomez, Rene Castro, Randall Garcia, Johnny Rosales, Luis Daniel Gonzales, Karla Ceciliano, Jose Maria Figueres, Maria Marta Chavarria, Roger Blanco, Angel Solis, Isidro Chacon, Nelson Zamora, Jorge Corrales, Manuel Zumbado, Eugenia Phillips, Jesus Ugalde, Carlos Mario Rodriguez, Alonso Matamoros, Jorge Jimenez, Alejandro Masis, Ana Sittenfeld, Felipe Chavarria, Julio Quiros, Jorge Baltodano, Luz Maria Romero, and Sigifredo Marin, and all the parataxonomists of INBio, for their especially insightful and inspirational input over the last 12 years of development of these ideas. Although it is clear that the international cast of contributors to a concept of this nature is enormous, I particularly thank Winnie Hallwachs, Kenton Miller, Peter Raven, Tom Eisner, Jerry Meinwald, Ed Wilson, Don Stone, Paul Ehrlich, Hal Mooney, Kris Krishtalka, Jim Edwards, Gordon Orians, Monte Lloyd, Mike Robinson, Steve Young, Preston Scott, Leif Christoffersen, Odd Sandlund, Mats Segnestam, Eha Kern, Bernie Kern, Hiroshi Kidono, Frank Joyce, Ian Gauld, Jon Jensen, Murray Gell-Mann, Steve Viederman, Staffan Ulfstrand, Carlos Herrera, Steve Blackmore, Meridith Lane, Jim Beach, John Pickering, Amy Rossman, Bob Anderson, Terry Erwin, Don Wilson, Diana Freckman, Chris Thompson, Marilyn Roossnick, Luis Rodriguez, Dan Brooks, Charles Michener, Bob Sokal, John Vandermeer, Jack Longino, Rob Colwell, Chris Vaughan, and Tom Lovejoy for their investment in this process.

MEASURES TO CONSERVE BIODIVERSITY IN SUSTAINABLE FORESTRY: THE RÍO CÓNDOR PROJECT

MARY T. KALIN ARROYO

Departamento de Biología, Facultad de Ciencias, Universidad de Chile,
Casilla 653, Santiago, Chile

INTRODUCTION

FORESTS OCCUPY about 5,000 million hectares (Constanza and others 1997), the equivalent of one-third of all terrestrial ecosystems, and constitute a substantial fraction of Earth's vegetation. Some 60% of forest is at temperate (in a broad sense) latitudes (Constanza and others 1997), and that is where most forests are managed for timber and other commodities. Temperate forest is unequally distributed among the two hemispheres. Less than 10% occurs in the Southern Hemisphere (Arroyo and others 1996), this being concentrated mainly in the widely disparate areas of southern Chile and neighboring Argentina, New Zealand, and Tasmania.

Forests provide a wide range of ecosystem services and goods. The goods are wood, edible plants and fungi, medicinal plants, microorganisms with potential biological activity, ecotourism, and recreation. The services include maintenance of hydrological cycles and air and water quality, regulation of regional climate, nutrient cycling, soil conservation, carbon storage, provision of habitats for wildlife, and contributions to regional and local aesthetics. In a provocative paper attempting to calculate the monetary replacement value of the ecological services provided by Earth's ecosystems, forests were estimated to contribute 38% of total terrestrial ecosystem worth, the equivalent of $4.7 trillion per year, or $969/ha per year (Constanza and others 1997).

Forests, both temperate and tropical, house large amounts of biodiversity (figure 1). Apart from the more visible elements—such as birds, reptiles, mammals,

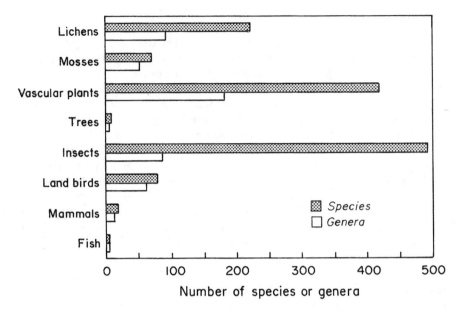

FIGURE 1   Biodiversity on the Río Cóndor property as now known, expressed in terms of species and generic richness for the main groups of organisms. Data on insects (mostly litter- and soil-dwelling) are for coastal forest only. Other figures based on sampling of entire property and nonforested and forested habitats. Amphibians do not occur in Tierra del Fuego. Data on fungi unavailable. Original data from Arroyo and others 1996.

and vascular plants—forests exhibit strong representation of the less conspicuous and often poorly known groups of organisms, such as bacteria, fungi, lichens, bryophytes, mollusks, and terrestrial arthropods. Some 70,000 species of fungi, well represented in forests, are recognized worldwide, but extrapolations suggest that there could be as many as 1,500,000 species (Hawksworth 1991). Some 2,000 species of ectomycorrhizal fungi are associated with Douglas fir alone in the Pacific Northwest (Marcot 1997). A high proportion of the estimated 16,000 bryophytes (Heywood and Watson 1995) also belong to forests. Some 1,200 species of beetles were collected on a single tree in Panama (Erwin 1982), and 492 species of insects, mostly from litter and surface soil layers, were found in coastal forest in Tierra del Fuego (Arroyo and others 1996). An estimated 7,000 species of arthropods (in comparison with 26 species of mammals, including bats) are found in late successional forests in the Pacific Northwest, containing a handful of trees (Marcot 1997). Tropical forests are evidently richer in tree species than their temperate counterparts, as indicated by a record of 473 tree species in a single hectare in Ecuador (Heywood and Watson 1995). However, the 1,200 tree species in temperate forests worldwide are not to be ignored (Heywood and Watson 1995). Temperate rain-forest trees, moreover, can show high diversity in vascular-plant epiphytes, as seen in the finding of 28 species on a single tree in New Zealand (Heywood and Watson 1995).

This essay addresses ecological sustainability and biodiversity conservation in a managed forest landscape. It refers to concrete actions for conserving biodiversity in a private sustainable forestry initiative, the Río Cóndor project in Tierra del Fuego, southern Chile. Less than 10% of the earth's terrestrial surface is protected, and conservation value in the past was more often influenced by scenic beauty and wilderness value than by biodiversity, such that existing protected areas are now often inadequately distributed for the protection of biodiversity (for example, Arroyo and Cavieres 1997). Long before objective assessments of existing protected areas can be realistically completed, very large areas of the world's remaining forests, especially in developing countries, will already have been submitted to some form of resource extraction as a result of economic pressures and social needs. This situation, added to the high biodiversity value of forests, suggests that it is time for scientists to pay greater attention to managed lands and to establish partnerships with sensitive users of the land, without whose collaboration the task of saving forest biodiversity will be very difficult.

ECOLOGICAL SUSTAINABILITY IN MANAGED FORESTS— AN EVOLVING PARADIGM

Newly recognized goods and ecosystem services of forests, high biodiversity content, and increasing consciousness of global climatic change have led society to question how forests are being used worldwide. In particular, recreation, scenic, and related amenity values have become central to the public's perception of the role and value of native forests in both developed and developing countries. These societal concerns, in turn, have been paralleled by substantial changes in scientific perception of sustainability as applied to managed forests.

Throughout much of this century, the management of forests focused closely on the extraction of a specific ecosystem good, wood, in keeping with the concept of *sustained yield*, defined as the management of a resource for *maximal continued production* consistent with the maintenance of a constantly renewable stock. Such strong emphasis on a particular resource of strictly utilitarian interest, while saving trees of commercial interest, has been less kind to other organisms, as shown, for example, by reduction of the diversity of tree species (Jarvinen and others 1977) and the risk of extinction of as many as 700 species of plants and animals because of past forestry practices (Wright 1995) in Finland. Some 1,487 plants and animals in Sweden associated with forest habitats are considered to have reached threatened status as a result of widespread application of forestry practices (Berg and others 1994). Specialist invertebrates in Fennoscandian boreal forests tend to disappear from local clear cuts while forest generalist species and numerous open-habitat species appear (Niemelä 1997). At the genetic level, selective logging increases inbreeding in tropical dipterocarp forests (Murawski and others 1994).

In the late 1980s and early 1990s, as greater knowledge of ecosystem processes and biodiversity function in forests became available, the idea of ecosystem-based forest management began to take hold (Arroyo and others 1996; Franklin 1995). An ecosystem approach to forest management, which will be referred to here as ecological sustainability (Arroyo and others 1996), calls for a shift away from the traditional focus of sustainable yield to one in which *all species and ecosystem processes* are given consideration and sustainable yield remains an important goal. The recognition that numerous elements of biodiversity in forests are essential for maintaining productive capacity is central to the concept of ecological sustainability. For example, ectomycorrhizal fungi are responsible for aiding nitrogen uptake and fixation by tree species; many lichens fix atmospheric nitrogen; some bryophytes act as sinks for nitrogen leachate; arthropods aid in nutrient cycling of down wood and are major decomposers, chewers, shredders, predators, and food sources in forest streams and rivers; fungi are important decomposers of woody debris; and birds and mammals can be dispersal agents of fruit and seeds (Marcot 1997). Under an ecosystem approach to forest management, moreover, natural forest variability is recognized in developing silvicultural prescriptions, disturbance regimes are mimicked as far as possible in selecting harvesting and regeneration methods, and maintenance of landscape integrity is sought through establishment of protection forest and other types of buffers and the protection of aquatic ecosystems.

The adoption of ecosystem management also brought home the fact that adjacent ecosystems can be interconnected through such processes as nutrient cycling and biotic links, so that effects in any one ecosystem can have eventual repercussions at higher levels in the biodiversity hierarchy (landscape and regional). For example, plant species are sedentary, but their pollen and seeds are often transported by animals that live for part of their annual cycle in adjacent vegetation types. Thus, reduction in the nectar-feeding birds in a managed forest, besides affecting the bird species, will have effects on plant species in an adjacent ecosystem. Reduction of soil microorganisms can slow natural decay and so affect nutrient cycling, but there will also be secondary effects on water quality and aquatic life downstream. Such linkages and feedbacks between different levels of the biological hierarchy oblige consideration at the species, ecosystem, landscape, and regional levels and mean that landowners must be equally concerned with aquatic and other ecosystems, in addition to those under management.

In the early 1990s, as a result of increasing CO_2 in the atmosphere because of the burning of fossil fuels and deforestation, the role of forests in maintenance of global carbon balance came into discussion, further modifying the expectations of ecological sustainability in managed forests. The conversion of forests to agricultural lands through burning releases carbon into the atmosphere; conversely, regenerating forests on managed or abandoned lands withdraw carbon. Although young and middle-aged forests accumulate more carbon than standing old-growth forests, the overall carbon balance in a harvested-forest landscape depends on the fate of wood harvested from old-growth forests (Houghton and others 1996). For example, mass-balance calculations for Pacific Northwest forests show that conversion of five million hectares of old-growth forest to younger plantations in Oregon and Washington in the last 100 years has produced a negative carbon balance because of the burning of slash and wood for fuel and the conversion of sawdust to paper with a short turnaround time for carbon release back into the atmosphere (Harmon and others 1990). Nevertheless, over broader areas of the Northern Hemisphere, the net effect of forest harvest and regrowth in temperate forests is considered to be zero to slightly positive (Houghton and others 1996). However, as projected from current land-use tendencies, the biomass and carbon stock in the equilibrium landscape that replaces Brazil's Amazonian forest after deforestation can be expected to have decreased by about 35% in relation to 1990 levels (Fearnside 1996). Under those conditions, collaboration in the regeneration and restoration of forests on managed lands in temperate areas and emphasis on wood products that permit carbon fixation for very long periods and conservation per se become important goals of sustainability in a managed-forest landscape.

As a result of rapid changes in the perception of the value of forests and equally rapid evolution of scientific ideas as to how forests should be managed, sustainable forestry has been steered in the direction of integrated resource management, or management that takes into account the multiple values of forests. Some aspects of those changes need to be kept squarely in perspective. Although many new measures are now being introduced into sustainable forestry for the protection of biodiversity and ecosystem processes, it has to be admitted that their effectiveness is largely unknown. The long-term studies required to test such effectiveness, which often must be on a spatial scale beyond the domain in which ecology is normally practiced, have not been undertaken to any great extent. It should be borne in mind that we have gone from an era of being concerned about a few tree species in the forest to one involving hundreds of species with different life-history properties, many different habitat requirements, and a demonstrated diversity of responses to harvesting at the population to regional levels (Arroyo and others 1996; Berg and others 1994; Jarvinen and others 1977; Murawski 1994; Niemelä 1997; Wright 1995). The task is not at all simple. One of the main problems is that describing the effects of forest harvesting on biodiversity has been the main focus until now, with far less emphasis on the more relevant manipulative research designed to find novel solutions to mitigate such effects. The most worthwhile studies clearly will be experiments conducted on managed lands themselves with untouched lands as controls. With those caveats in mind, an objective like biodiversity conservation with forest harvesting should be considered at the level of a *working hypothesis*. Until we know where the line is to be drawn—how much extraction of a commodity, such as wood, is possible while ecological and economic sustainability is maintained into perpetuity—the course of action must be to refine hypotheses by further observation and experiment, with the recognition that ecological sustainability is a long-range target.

One of the most urgent needs in the scientific domain is to develop predictive models for integrating various spatial scales to ascertain whether forests harvested on an ecosystem basis will recuperate to near their original ecological dynamics *and* biological content *and* aesthetic value and thereby be available for alternative uses—the ultimate aim of a *dynamic* interpretation of ecological sustainability. Knowledge of the limits that guarantee those last three conditions is important for developing countries, where the extraction of natural forest resources, even though undesired by large sectors of society, will often precede other, less resource-intensive uses, such as ecotourism and recreation (which, in requiring infrastructure and substantial new capital, are not always viable options at any given time). Determining these limits is particularly important in the Southern Hemisphere, where temperate forest is scarcer than in the Northern Hemisphere and thus not only is far more sensitive as a biome to inadequate management, but also, hectare for hectare, is under far greater demand for nonextractive uses. With respect to ecological sustainability, it is thus essential, first, to recognize that there are many scientific uncertainties and, second, to leave other options open, given that social perceptions of forest use will undoubtedly continue to change. A range of management strategies—from conservative levels of harvesting to the establishment of conservation safety networks, including permanent reserves, commitment to restoration and regeneration, long-term monitoring, and continued appraisal of results in the context of adaptive management—is essential to accommodate all the various situations. Application of the precautionary principle in combination with a multiple-use strategy has the advantage of allowing a landowner to switch to some alternative land use in the near future, if desired or if scientific findings suggest it to be the most appropriate pathway. The principles outlined here have been borne in mind by the group of Chilean scientists responsible for developing actions to conserve biodiversity in the Río Cóndor project, discussed below.

THE RÍO CONDÓR SUSTAINABLE-FORESTRY PROJECT

The Río Cóndor sustainable-forestry project entails land holdings comprising 273,000 hectares, at 54°S in Tierra del Fuego, Chile, of which 54% is forested (figure 2). It is the first forestry project in Chile in which the principles of modern ecological sustainability have been assumed. It is perhaps one of the more advanced anywhere, in terms of the diversity of strategies implemented to protect biodiversity well *before* commencement of harvesting and the commitment to long-term monitoring and research to test the effectiveness of such strategies (Arroyo and others 1995; Arroyo and others 1996; Pickett 1996). The Río Cóndor project, owned by the Trillium Corporation, Bellingham, USA, and registered through Forestal Trillium Ltd. in Chile, has committed, through voluntary stewardship principles, to a sustainable project based on production of quality wood and other forestry products of added value, protection of biodiversity and ecosystem processes, recognition of potential forest values, and creation of employment and other social benefits for people in the area. Wood chips, a primary forest product, will not be produced, and exotic tree species will not be planted for commercial purposes.

In the absence of scientific information, measures will be taken to generate data and postpone any action that would lead to environmental degradation until such data are available. A comprehensive monitoring program—embracing regeneration levels, soil conditions, water quality, rare and endangered species, indicator exotic species, and guanaco and beaver populations—will be carried out by specialists hired specifically for that purpose. In accordance with a commitment to incorporate scientific knowledge and ensure environmental compliance, the Río Cóndor project established an independent scientific commission (ISC) of

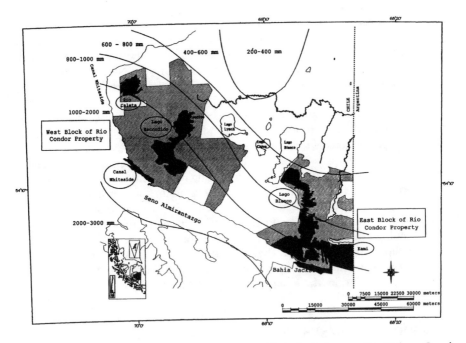

FIGURE 2   Location of four permanent reserves (dark shaded areas: Río Caleta, Canal Whiteside, Lago Escondido, Lago Blanco-Kami reserves) on the Río Cóndor property, Tierra del Fuego. Also shown (light shaded) are the two blocks of the Río Cóndor property (West and East blocks). Contours are isohyets (lines joining points that receive equal amounts of precipitation). Reserve boundaries were drawn on 25 March, 1999, by mutual agreement between Chilean scientists and the owners of the Rio Condor property.

botanists, zoologists, and forest ecologist through contact with the Chilean Academy of Sciences and retains a land steward (Arroyo and others 1996). The ISC functioned under a protocol signed by David Syre, owner of the Río Cóndor holdings, which guaranteed its rights and independence. The ISC effected extensive baseline studies and participated actively in designing the conservation strategy and monitoring program. Species and generic richness in several groups of organisms for the entire Río Cóndor property is shown in figure 1.

The Río Cóndor Forests and Landscape

Forests on the Río Cóndor property belong to the circumantarctic biome and include deciduous types (pure *Nothofagus pumilio*, pure *N. antarctica*, and *N. pumilio-N. antarctica*), a mixed evergreen-deciduous type (*N. betuloides-N. pumilio*), a pure evergreen type (pure *N. betuloides N. pumilio*), and a mixed evergreen type (*N. betuloides-Drimys winteri-Maytenus magellanica*). Although *Nothofagus* as a genus dates back to pre-Cretaceous times, recent molecular work (Manos 1997) shows that the three subantarctic species in Tierra del Fuego evolved recently. The Río Cóndor forests were consolidated since 5,000 years ago as climate became wetter in the Holocene and allowed replacement of drier steppe vegetation with forests (Heusser 1993). According to criteria given by Spies (1997), most of the inland forests on the Río Cóndor property would be classified as old-growth. However, coastal forests have been heavily affected in the past by selective logging, burning, and grazing; cattle grazing is still practiced in many inland valleys today. Many watersheds in the Río Cóndor forests have been heavily affected by the American beaver, *Castor canadensis*, liberated in Tierra del Fuego in 1946.

The forests, dominated primarily by one or two tree species, often lack a shrub stratum; biodiversity is moderate; and there is a conspicuous absence of sensitive groups, such as amphibians, salamanders, and very large mammals that are wholly dependent on the forest habitat. The Río Cóndor forests are thus appropriate for putting the principles of ecological sustainability into practice with a broad and precautionary management strategy and with an acceptable risk at baseline conditions. The relative simplicity of the forests, moreover, makes future monitoring realistic for landowners with respect to cost and effort. There are other positive aspects for organizing a sustainable landscape. Forests are interspersed with substantial extensions of subantarctic peat bogs, alpine, and lakes; the landscape is diverse and scenically beautiful. Those last elements have been used to maximal advantage, as will be seen below.

Building a conservation network to protect biodiversity. In accordance with ecological baseline studies carried out by 17 research teams, facilitative reserves in harvested forest, core reserves, and an extensive buffer system have been established on the Río Cóndor property (Arroyo and others 1995, 1996).

Measures to maintain biodiversity in situ in productive areas. One of the most challenging tasks in sustainable forestry is designing a set of measures to maintain viable populations of organisms *in the productive forest matrix itself.* No matter how simple the forest and how benign the harvesting method, organisms will be affected during silvicultural intervention, either temporarily because of physical elimination of populations or for very long periods because of elimination of specialized habitats in old trees that characterize old-growth forest or changes in microclimate.

To maintain ecosystem productivity, measures to conserve microorganisms, fungi, lichens, and soil arthropods involved in decomposition are particularly important. That objective has been sought in the Río Cóndor project by modifying the traditional shelterwood harvesting method used in *Nothofagus* forests in Chile. In harvested stands, aggregates (Franklin and others 1997) (facilitative reserves—see Arroyo and others 1996 for this concept) of mature trees will be retained permanently throughout the rotation cycle in addition to the 30–50% tree cover retained initially. Such aggregates, which maintain the original soil conditions and a more natural microclimate, are expected to be important for conserving of epiphytic lichens and mosses, such birds as the magellanic woodpecker that depends on old trees, and microorganisms that depend on woody debris in an advanced stage of decomposition. Many shade-loving herbaceous vascular plants, lichens, and mosses and the five small mammal species in the forest habitat are expected to survive temporarily in these aggregates and then be dispersed back into forest after harvesting. Extensive studies in the Río Cóndor forests have revealed that a high proportion of vascular plant species are abiotically dispersed and genetically self-compatible and thus are well adapted for rapid recolonization of the harvested forest matrix, as are the 68 species of mosses and over 200 species of lichens that disperse via spores or asexual propagules. Comparisons of virgin, 1-year harvested, and 8-year harvested forest showed that, although abundance differences arose, many native species either survived in or were able to return to the shelterwood matrix even without the aid of aggregates. Aggregates will be distributed across the entire harvested-forest landscape and are expected to greatly ease connectivity between harvested forest and other components of the conservation network, such as stream buffers and core reserves. That last point is very important in the Río Cóndor landscape, where spatial differentiation of habitats and species turnover along elevational gradients is low, differentiation of a true riparian zone is lacking, and species richness can be higher in the more open and warmer ecotonal habitats than in forest. Bearing in mind that core reserves established in forestry projects will never be large enough to account for the minimal viable population size of all organisms, the inclusion of aggregates in the harvested matrix should extend the effective safety net of reserves. Dispersing facilitative reserves across the productive landscape, of course, places limitations on their size and on the types of organisms that they will protect. Where the tradeoff lies between number (determining coverage) and size (determining structure and microclimate) is a matter for further research.

Woody debris and residual wood from harvesting will be left on the forest floor after harvesting, and the litter layer will be disturbed as little as possible. Debris and residual wood are important not only for their nutrient content, but also as habitat for small mammals, the endangered red fox, and several species of habitat-sensitive ground birds. These components also provide anchorage for incoming seeds and spores and the shaded conditions preferred by native herbaceous species. Opening of the Tierra del Fuego forests through harvesting was seen to be accompanied by an increase in exotic plants, including such aggressive species as *Taraxacum officinale*; dealing with exotic plants will probably constitute one of the more difficult problems. Hundreds of species of arthropods were found in the litter layer in the Río Cóndor forests. The success of these measures for conserving biodiversity in productive areas in the Río Cóndor landscape should be enhanced by the rotation cycle of around 90–110 years, the fact that there are no intermediate successional trees in these simple forests, and the shelterwood harvesting method itself, which is far more benign than clear-cutting. The fairly long rotation cycle in relation to maximal tree age in mature stands on good sites (about 150–250 years), made possible on the Río Cóndor property because it is very large, is expected to facilitate the return to near old-growth conditions and enable repeated dispersal events back into harvested forest. The use of the shelterwood harvesting method to retain 30–50% of tree cover until regeneration is fully established will further ease connectivity between individual aggregates.

Establishment of a system of permanent core reserves. Core reserves are a central part of the conservation strategy in the Río Cóndor project. These will perform multiple functions, including preservation of a representative sample of the main vegetation types on a regional scale; protection of specialist, rare, and endangered species; conservation of forest genetic material; protection of cultural values; provision of a resource for ecotourism and future research; and contribution to the aesthetic value of the Río Cóndor holdings. Core reserves will also act as facilitative reserves to replenish altered plant and animal populations in harvested forest in their own right, but this is expected to be more at the level of larger and more mobile organisms, such as mammals and birds, and a few bird-dispersed plant species.

Some 68,000 hectares—around 25% of the present holdings—has been assigned to preservation by the owners of the Río Cóndor property. The preserved land comprises four blocks (figure 2) that vary from an estimated 43,000–2,200 hectares. Reserves were selected through a process involving the participation of the ISC, an archaeologist, the present land steward of the Río Cóndor project, and Forestal Trillium Ltd. personnel (Arroyo and others 1996), after issue of a public statement on September 13, 1995, by the owners of the Río Cóndor property to create them. The reserves, established through a coarse-filter mode, include all five forest types on the Río Cóndor property and other nonforested vegetation types (matorral, subantarctic peat bogs, and high alpine) and span the east-west precipitation gradient across the Río Cóndor property. Together, the preserved areas contain 10,000 hectares of prime commercial-grade forest, and 17,000 hectares of unharvestable forest on steep slopes and of tree species not appropriate for harvesting. In establishing the Río Cóndor reserves, special attention was given to areas of high archaeological sensitivity along coastal areas and in the vicinity of the major lakes, in view of the 77 archaeological sites of Selk'nam affinity registered during baseline work. Additional considerations were continuity with other protected areas in the general region, such as Parque Nacional Tierra del Fuego, Argentina, boarding on the Lago Blanco-Kami Reserve; enhancement of areas of high aesthetic value, such as Fjord Almirantazgo (Canal Whiteside reserve); representation of altitudinal gradients; inclusion of parts of Atlantic- and Pacific-drained rivers (Lago Escondido reserve); and inclusion of watersheds ideal for long-term research on nutrient cycling (Lago Blanco-Kami reserve).

The reserves are expected to play an important role in protecting species in the face of regional conservation problems on the Río Cóndor property, such as *Pseudalopex culpaeus* (red fox, the largest mammal on the Río Cóndor property restricted to forest), *Maytenus disticha* and *M. magellanica* (two plant species with conservation problems), *Campephilus magellanicus* (magellanic woodpecker), several ground birds that require dark forest conditions, and a small number of vascular plants and small mammals endemic to Tierra del Fuego. The inclusion of important lakes, such as Lago Escondido, and part of Lago Blanco in the reserves places them well for the recreational activities and ecotourism contemplated in the Río Cóndor project.

Although 17% of 3.4 million hectares of *Nothofagus pumilio* forest in Chile is found in the National Protected Area System (CONAF 1997), the private Río

Cóndor reserves constitute the only preserved areas of this forest type in the far southern extreme of its distribution in Chile. Río Cóndor reserves also capture one of the richest alpine areas in Tierra del Fuego (Arroyo and others 1996) and a wide range of subantarctic peat bogs with a rich flora including rare and marginally distributed species. In equaling in size Parque Nacional Tierra del Fuego in adjacent Argentina, the Río Cóndor reserves constitute an important contribution to regional conservation in southern South America and the largest private conservation effort in a managed landscape in Chile. Apart from their in situ sustainability benefits, establishment of the Río Cóndor reserves constitutes a good example of collaboration by private landowners to complement inadequate spatial coverage of protected areas in a state-protected area system.

Ecological buffers. In addition to more conventional buffers (10-m strips around peat bogs, 50-m strips along the Río Cóndor and other streams, a 100-m coastal buffer, and restriction of harvesting on most slopes of over 45% and above 450 m in elevation), all *Nothofagus antarctica* forest and 60,000 hectares of subantarctic peat bogs on the Río Cóndor property have been considered in a buffer mode. *N. antarctica* forest is a natural buffer because of its occurrence in a wide range of ecologically marginal and ecotonal conditions, such as between the 450- to 700-m-elevation tree limit and forests of commercial interest; at the edges of peat bogs, streams, and lakes; and between *N. pumilio* forest and wet steppe. Sharing many species found in harvestible forests and being scattered widely throughout the Río Cóndor landscape, preserved *N. antarctica* forest will greatly increase coverage of the facilitative matrix. Such habitat similarity highlights a trend in the Tierra del Fuego forests for wide habitat tolerances. Indeed, many forest-dwelling species can also be found in wet steppe, in the alpine, and in disturbed secondary habitats, which, by agreement, will not be disturbed to any extent and thus will also play a facilitative role. Such low habitat specificity reflects a distinctive colonizing character in the postglacial biota of Tierra del Fuego. This feature is very favorable for the conservation of biodiversity in a managed-forest landscape in that other nonexploited vegetation types will contribute directly to the sustainability of the targeted forests.

The mostly *Sphagnum*-dominated, rain-fed peat bogs on the Río Cóndor property cover 22% of the landscape and are found in a wide variety of physiographic situations, from valley bottoms to slopes of over 30%. They contain some 107 vascular plant species, including rare species like *Tapeinia obscura* (Iridaceae); a high cover of fleshy fruited, bird-consumed species; many nitrogen-fixing lichens; 10 species of birds; and nesting sites for native geese—but no native mammals. The rationale for keeping peat bogs out of the productive universe is compelling. They play a key role in hydrology and nutrient cycling by providing continuous water supply to the forests in a landscape that has very few free-flowing streams. Peat bogs are recognized carbon sinks, containing (in the boreal forest zone) 108 times as much carbon per hectare as a forest (Gorham 1991). The subantarctic peat bogs of Tierra del Fuego have accumulated carbon over the same general period as their Northern Hemisphere counterparts and to similar depths (Heusser 1993) and thus can be assumed, in the absence of more detailed information, to be important carbon sinks. Because of the magnitude of stored carbon in peat bogs, their potential for contributing to global warming through CO_2 release by draining and harvesting is huge. It is probably not an exaggeration to state that the Río Cóndor property is centered on one of the largest carbon sinks in the Southern Hemisphere! The ISC submitted that the owners of the Río Cóndor property should seriously consider the possibility of placing these important sub-antarctic wetlands under RAMSAR in consideration of their regional hydrological significance and their role in maintaining global carbon balance.

OVERVIEW

In the Río Cóndor project, the scientific goal has been to combine protection with production in such a way as to ensure multiple sustainability benefits for a managed forested landscape at the stand, property, and regional levels and thus open the door to an integrated forestry project without foreclosing future options. The success of the series of actions that have been set into motion with the decided collaboration of the landowners at this remote location in the far southern temperate forests of South America depends heavily on maintaining the objectives of monitoring and future research. Such studies will have practical significance only if the knowledge generated is used to alter and implement management practices over time (Franklin 1995).

ACKNOWLEDGMENTS

Original baseline research was financed by Forestal Trillium Ltd., Chile, to which gratitude is expressed. The willingness of David Syre, President, Trillium Corporation, USA, to engage in ecologically sustainable forestry and forest preservation is acknowledged. This manuscript was written during the tenure of an Endowed Chilean Presidential Science Chair.

REFERENCES

Arroyo, MTK, Armesto J, Donoso C, Murúa R, Pisano E, Schlatter R, Serey I. 1995. Hacia un proyecto forestal ecológicamente sustentable: resumen ejecutivo. Rev Chilena Hist Nat 68:529–538.
Arroyo MTK, Cavieres L. 1997. The mediterranean-type climate flora of central Chile: what do we know and how can we assure its protection. Notic Biol 5(2):48–56.
Arroyo MTK, Donoso C, Murúa R, Pisano E, Schlatter R, Serey I. 1996 Toward an ecologically sustainable forestry project. Concepts, analysis and recommendations. Protecting biodiversity and ecosystem processes in the Río Cóndor Project, Tierra del Fuego. Santiago Chile: Dept Investigación y Desarrollo, Univ de Chile. 253 p.
Arroyo MTK, Jiménez H, Peñaloza A. 1996. Reservas Biológicas, Propiedad Río Cóndor, Tierra del Fuego. Criterios, reconocimiento en terreno, proposiciones, acuerdos. Unpublished report, Santiago de Chile, Available from University of Chile, Faculty of Sciences.
Berg A, Ehnström B, Gustafsson L, Hallingbäck T, Jonsell M, Weslien J. 1994. Threatened plant, animal and fungus species in Swedish forests: distribution and habitat associations. Cons Biol 8(3):718–31.
CONAF 1997 Chile. Catastro y evaluación de los recursos vegetacionales nativos de Chile. Santiago, Chile 12 p.

Constanza R, d'Arge R, de Groot R, Farber S, Grasso M, Hannon B, Limburg K, Naeem S, O'Neill RV, Paruelo J, Raskin RG, Sutton P, van den Belt M. 1997. The value of the world's ecosystem services and natural capital. Nature 387:253–60.

Erwin TL. 1982. Tropical forests: their richness in Coleoptera and other arthropod species. Coleopt Bull 36:74–5.

Fearnside PM. 1996. Amazonian deforestation and global warming: carbon stocks in vegetation replacing Brazil's Amazonian forest. Forest Ecol Manag 80:21–34.

Franklin JF. 1995. Sustainability of managed temperate forest ecosystems. In: Munasinghe M, Shearer W (eds.) Defining and measuring sustainability. The biogeophysical foundations. The United National University and the World Bank. p 355–85.

Franklin JF, Berg DR, Thornburg DA, Tappeiner JC. 1997. Alternative silvicultural approaches to timber harvesting: variable retention harvesting systems. In: Kohm A, Franklin JF (eds). Creating a forestry for the 21st century. The science of ecosystem management. Washington DC: Island Pr. p 111–39.

Gorham E. 1991. Northern peatlands: role in the carbon cycle and probable responses to climatic warming. Ecol Appl 1(2):182–95.

Harmon ME, Ferrell WK, Franklin JR. 1990. Effects on carbon storage of conversion of old-growth forest to young forests. Science 247:699–702.

Hawksworth DL. 1991. The fungal dimension of biodiversity: magnitude, significance and conservation. Mycolog Res 95:641–55.

Heusser C. 1993. Late quaternary forest-steppe contact zone, Isla Grande de Tierra del Fuego, subantarctic South America. Quat Sci Rev 12:169–77.

Heywood VH, Watson RT (eds.) 1995. Global biodiversity assessment. Cambridge UK: Cambridge Univ Pr. 1140 p.

Houghton JT, Meira Filho LG, Callander BA, Harris N, Kattenberg A, Maskell K (eds). 1996. Climate change 1995. The science of climate change. Cambridge UK: Cambridge Univ Pr. 572 p.

Jarvinen O, Kuusela K, Vaisanen R. 1977. Effects of modern forestry on the number of breeding birds in Finland in 1945–75. Silvia Fennici 11:284–94.

Manos PS. 1997. Systematics of Nothofagus (Nothofagaceae) based on rDNA spacer sequences (ITS): Taxonomic congruence with morphology and plastid sequences. Amer J Bot 84:1137–55.

Marcot BG. 1997. Biodiversity of old forests of the West: a lesson from our elders. In: Kohm A, Franklin JF (eds). Creating a forestry for the 21st century. The science of ecosystem management. Washington DC: Island Pr. p 87–109.

Murawski DA, Nimal Gunatilleke IAU, Bawa K. 1994. The effects of selective logging on inbreeding in Shorea megistophylla (Dipterocarpaceae) from Sri Lanka. Cons Biol 8(4):997–1002.

Niemelä J. 1997. Invertebrates and boreal forest management. Cons Biol 11(3):601–10.

Pickett S. 1996. Sustainable forestry in Chilean Tierra del Fuego. Trends Ecol Evol 11:450–1.

Spies T. 1997. Forest stand structure, composition and function. In: Kohm A, Franklin JF (eds). Creating a forestry for the 21st century. The science of ecosystem management. Washington DC: Island Pr. p 11–30.

Wright M. 1995. Death by a thousand. New Scientist 145:36–40.

CHEMICAL PROSPECTING:
THE NEW NATURAL HISTORY

THOMAS EISNER

Division of Biological Sciences, Section of Neurobiology and Behavior, Cornell University,
W347 Seeley F. Mudd Building, Ithaca, NY 14853

The smallest thing in nature is an entire world—Joan Miró

HARDLY ANYONE admits to being a naturalist any more. Natural history used to be the most respectable of professions, before fragmentation of the biological sciences created the multiplicity of subdisciplines that draw the allegiance of biologists today. Natural history, simply put, is the exploration of nature. It is the search for novelty in the biotic world—be it new species, new behaviors, new ecological interactions, new functions and structures, new materials, or anything else that remains hidden about organisms. Many contemporary biologists got their start as naturalists and are to this day naturalists at heart. They are driven as grownups to roam through nature to "have a look," just as they were as youngsters. But as professionals, they tend not to advertise the avocation because the professional establishment tends to belittle its importance. That is profoundly regrettable, for natural history has taken on new significance.

Given the state of our molecular understanding, virtually anything uncovered in nature can now be coupled to chemical knowledge. To discover new natural phenomena in today's world is to discover new chemicals and chemical processes and, by implication, new genetic capacities. Exploring nature is tantamount to "chemical prospecting", and such prospecting has great potential for reward (Eisner 1989–90, 1994a,b; Eisner and Beiring 1994). It can bring commercial benefit, as it has consistently in the search for medicines (Balick and others 1996; Grifo and Rosenthal 1997; Joyce 1994; Reid and others 1993); but most important, it can increase our appreciation of nature. Viewing nature as a source of applicable knowledge could have an enormous effect on conservation (Eisner 1989–90, 1994a; Reid and others 1993). It could lead to new goals for natural history and, by refocusing the process of discovery itself, could redefine the role of the naturalist explorer.

Exploration can be immensely enjoyable for the naturalist and can lead to unexpected findings. Moreover, discoveries that seem trivial can turn out to be valuable on reflection and further inquiry. Let me illustrate by example.

OIL-EATING BACTERIA

Hemisphaerota cyanea is a small blue chrysomelid, or leaf beetle, commonly found on palmetto plants (*Sabal* spp.) in the southeastern United States. As both larva and adult, it feeds on palmetto fronds. Anyone familiar with the beetle knows that the adult can offer considerable resistance to being picked off its plant. It clings with its feet, and does so with such tenacity that forces of upward of 200 times the beetle's weight might be needed to pull the insect off (figure 1a). The

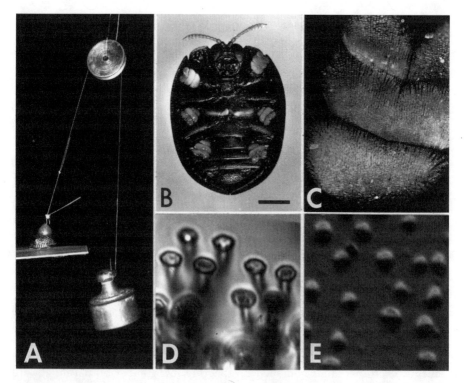

FIGURE 1   *Hemisphaerota cyanea.* (a) Beetle resisting load (2 g). (b) Ventral view of beetle, showing six broadly expanded foot soles. (c) Enlarged view of foot sole, showing bristles. (d) Enlarged view of bristle tips, showing adherent pads. (e) Droplets of oil relinquished by adherent pads on contact with substrate. (Bar = 1 mm.)

beetle ordinarily walks with a loose hold; it clamps down only when disturbed and by so doing can effectively counter the proddings of such enemies as ants.

H. cyanea secures its hold by adherence rather than anchorage. The "soles" of its feet bear a dense mat of bristles, whose terminal pads are wetted by oil (figure 1b-e). Collectively, the beetle's six legs have around 60,000 pads. The oil is produced by tiny glands that open at the bases of the bristles. During ordinary locomotion, the beetle treads lightly, touching the ground with only a small fraction of its pads. But when disturbed, it presses its six soles down flat, committing to contact with its entire complement of pads (Eisner 1972). Preliminary analyses showed the oil to consist of a mixture of long-chain unsaturated hydrocarbons.

We were interested in the adherence mechanism because relatively little was known about how insects secure their foothold during locomotion. *H. cyanea* was not unique in relying on wetted bristles for adherence. Many other insects, possibly including all chrysomelids, have bristle-bearing feet much like those of *H. cyanea*. But *H. cyanea* is exceptional in that it has bristles in formidable number, so it can use them for defense, as well as walking.

What turned out to be most interesting about *H. cyanea* is what we later discovered purely by chance: parasitic bacteria live in its feet and feed on the oil. We first noted them when we examined the oil droplets left in the beetle's wake when it walked on glass. We had some idea of the volatility of the oil and could estimate how long it would take for the droplets to evaporate, but they vanished more quickly than predicted. As the droplets disappeared, we noted the simultaneous appearance of distinct rod-shaped bacterial bodies at their margin (figure 2); they did not appear at the edges of droplets of various other oils that we placed on the glass as controls. We were intrigued by this finding but, lacking the appropriate expertise, failed to follow up by culturing and identifying the bacteria. It would be worth while to isolate these microorganisms. It is not unreasonable to presume that, as oil eaters, they could have an enzymatic trick or two hidden up their tiny sleeves.

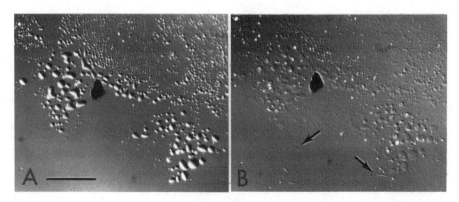

FIGURE 2 *Hemisphaerota cyanea.* Portion of beetle's oily footprint, photographed within hours of deposition (a) and days later (b). Note rod-shaped bacteria (arrows). (Bar = 50 μm.)

NEW BIOPOLYMERS

Many animals—including amphibians, earthworms, and mollusks—have a body coating of slime. The investiture protects against small predators, such as ants, which might be physically discouraged by sticky materials. Slugs use their slime to special advantage in that they are able to coagulate it locally at sites where they are attacked. This can be easily demonstrated: if a slug is gently poked with a toothpick, nothing much happens; but if the toothpick is simultaneously wiggled, the slug sets the coagulation mechanism in motion, and a rubbery blob forms around the tip of the probe. In my experience, this works with all slugs.

Needless to say, the coagulation mechanism, which indicates an underlying polymerization, provides effective protection against small mandibulate enemies. Ants and carabid beetles, for instance, are literally muzzled when they bite into a slug. They are thwarted the moment they bear down with their mandibles, and as they back away, their mouthparts are visibly encased in slime (figure 3a-b).

How the coagulation is effected remains a mystery. We have precise data (based on responses to electrical stimulation) indicating that the coagulation is triggered within a fraction of a second and that it is coincident (in at least some slugs) with the localized injection of crystalline material into the slime from specialized integumental cells. And we suspect that polymerization of proteins is involved. But we know essentially nothing about the chemical details.

Slug slime also has interesting physical properties that remain to be worked out. In coagulated form, for instance, slug slime sticks with remarkable tenacity to human skin, including wet skin. Sticky materials abound in nature and could constitute a fertile field for basic and applied research. Intriguing examples include the viscid spray of onychophorans (Alexander 1957), as shown in figure 3c-d; the gluey investiture of dalcerid caterpillars (Epstein and others 1994), in figure 3e; the slimy coating of some sawfly larvae (Eisner 1994c), figure 3f; and the sticky caudal secretion of sowbugs (Deslippe and others 1995–96) and some cockroaches (Plattner and others 1972).

SECRETS OF AN ENDANGERED SPECIES

Dicerandra frutescens (figure 4a) is a mint plant (family Lamiaceae) endemic to central Florida (Deyrup and Menges 1997). It is an inhabitant of the so-called Florida scrub, a highly interesting dry-land ecosystem characterized by sandy ridges, shrubby plants, a number of endemic vertebrates, and a wealth of insects (Deyrup and Eisner 1993). *D. frutescens* has a distinct, potent aroma—so potent that the plant, a small multibranched herb, can sometimes be spotted by its odor from meters downwind. Close examination of the plant revealed that it was virtually free of insect injury. Suspecting that the plant's odor repelled insects, we proceeded to locate its source, which turned out to be a terpenoid oil in tiny, hermetically sealed capsules on the leaves (figure 4c-e). The capsules, we thought, might function as chemical grenades. Insects would inevitably rupture them when biting into a leaf, causing the oil to spill out and become a deterrent.

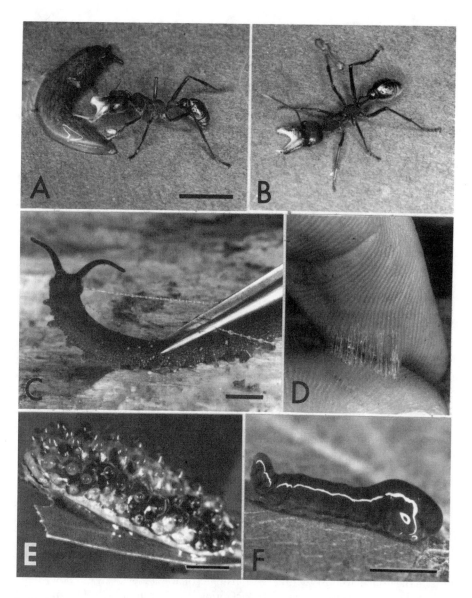

FIGURE 3   (a) Ant attack on slug. Ant (*Myrmecia* sp.) is withdrawing after biting slug and having its mouthparts gummed up with coagulated slime. (b) Minutes after encounter, ant is still attempting to rid itself of slime. (c) Unidentified onychophoran (from Panama) discharging strands of its sticky secretion in response to "attack" with forceps. (d) Onychophoran secretion quickly assumes rubbery consistency on exposure to air and becomes powerfully adhesive, even to human skin. (e) Dalcerid caterpillar (*Dalcerides ingenita*), showing dorsal coating of rubbery gelatinous warts. (f) Sawfly larva (*Calicoa cerasi*), showing glistening, sticky, integumental coating. [Bars = 0.5 cm (a), 0.5 cm (c), 2 mm (e), 5 mm (f).]

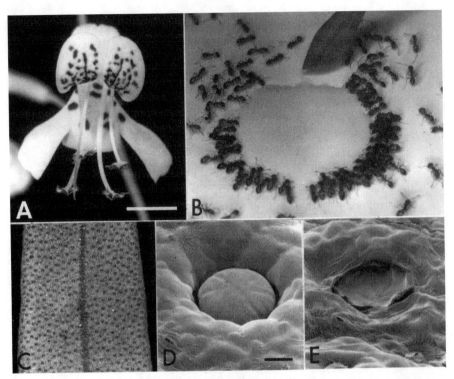

FIGURE 4 *Dicerandra frutescens.* (a) Flower. (b) Ants, feeding at sugary bait, dispersing in response to approach of freshly transected leaf of plant. (c) Portion of leaf, showing densely spaced "pearly" oil capsules. (d) Enlarged view of intact oil capsule. (e) Ruptured oil capsule. [Bars = 0.5 cm (a), 20 μm (d).]

A simple experiment showed that the grenades work. Ants attracted to a sugar source could be dispelled promptly if a *D. frutescens* leaf was suddenly brought into their vicinity (figure 4b). But the leaf had to be freshly transected. That is, the leaf had to be presented with some of its capsules already ruptured; intact leaves were tolerated by the ants.

Chemical work showed the oil to contain 12 volatile terpenoid components, of which the principal constituent, (+)-trans-pulegol, was a previously unknown natural product (Eisner and others 1990; McCormick and others 1993). Finding a new insect repellent was exciting, especially because *D. frutescens* was an unusual plant. The species had been discovered only in 1962, has a range of only a few hundred acres, and was already on the endangered species list (Middleton and Liittschwager 1994). Were most of its acreage not part of a protected site (the Archbold Biological Station), the plant would be in serious danger of being obliterated.

There was more to be found in this endangered species. Mycological work showed *D. frutescens* to contain upward of 20 endosymbiotic fungi. Some remain to be identified, and none has been exhaustively screened for bioactive materials;

but one has already been shown to be the source of an antifungal toxin (Chapela and Clardy, unpublished).

CONCLUSIONS

The information stored in the biotic world is boundless and has only begun to be accessed. Some 1.5 million species have been described, compared with the 10–20 million that are conservatively estimated to exist (Wilson and Peter 1988). Most of what we have to learn from nature remains to be discovered.

Who will make the discoveries, and for what purpose? Will it be the molecular biologist alone, the person who by virtue of a reductionist commitment wishes to access only the chemical and genetic "basics" of nature? I would argue that the naturalist has at least as much to contribute to this endeavor, in that field observation, beyond its descriptive aspects, also provides leads to molecular and genetic novelties. By using the above examples, I have tried to make this point. The dichotomy between molecular biology and natural history is artificial; the practitioners of these disciplines would do best to work in concert.

Natural history is in no fundamental way altered by its broadened molecular mission. What the naturalist has to offer will continue to have universal appeal (Greene 1986; Moffett 1993; Nuridsany and Pérennou 1996; Wilson 1984) and to contribute as it always has to such established disciplines as ecology, evolution, behavior, and systematics. But because discovery in nature has molecular implications these days, as well as vast potential for commercialization, the activity of the naturalist takes on new value. The naturalist, more than any other scientist, has the ability to list species by "chemical promise." By virtue of observational skills alone, naturalists have the capacity to sort out phenomena and point to those which might indicate the presence of chemicals (and genes) of potential interest to medicine, agriculture, or material sciences. This capacity makes naturalists extremely valuable members of the scientific enterprise, but they remain singularly unaware of their worth, just as the commercial establishment is ignorant of what they can provide.

Natural history is essentially deinstitutionalized. Hardly any academic courses teach how to discover in nature and how to assess, in conventional as well as molecular terms, the value of what simple observation can reveal. Naturalists will need to be trained worldwide and brought into the scientific mainstream. They are needed as explorers and as partners in applied science and industry. But most important, they are needed to speak for conservation. Ultimately, it is the explorer who is most aware of what we all stand to lose if the objects of exploration vanish. The example of the mint plant speaks for itself.

ACKNOWLEDGMENTS

My research has been supported largely by the National Institutes of Health. Collaborators who deserve special thanks are Jerrold Meinwald, Daniel J. Aneshansley, and Maria Eisner. I dedicate this paper affectionately to the memory of Rosalind Alsop (1940–1997), naturalist, research partner, and friend.

REFERENCES

Alexander AI 1957. Notes on onychophoran behavior. Ann Natal Museum 14:35–43.

Balick MJ, Elisabetsky E, Laird SA. 1996. Medicinal resources of the tropical forest. New York NY: Columbia Univ Pr.

Deslippe RJ, Jelinski L, Eisner T. 1995–96. Defense by use of a proteinaceous glue: woodlice vs. ants. Zoology 99:205–10.

Deyrup M, Eisner T. 1993. Last stand in the sand. Nat Hist 12/93:42–7.

Deyrup M, Menges ES. 1997. Pollination ecology of the rare scrub mint *Dicerandra frutescens* (Lamiaceae). Florida Sci 60:143–57.

Eisner T. 1972. Chemical ecology: on arthropods and how they live as chemists. Verh Deut Zool Gese 65:123–37.

Eisner T. 1989–90. Prospecting for nature's chemical riches. Iss Sci Tech 6:31–4.

Eisner T. 1994a. Chemical prospecting: a global imperative. Proc Amer Philos Soc 138:385–93.

Eisner T. 1994b. Bioprospecting. Issues Sci Tech 10:18.

Eisner T. 1994c. Integumental slime and wax secretion: defensive adaptations of sawfly larvae. J Chem Ecol 20:2743–9.

Eisner T, Beiring EA. 1994. Biotic exploration fund—protecting biodiversity through chemical prospecting. BioScience 44:95–8.

Eisner T, McCormick KD, Sakaino M, Eisner M, Smedley SR, Aneshansley DJ, Deyrup M, Myers RL, Meinwald J. 1990. Chemical defense of a rare mint plant. Chemoecology 1:30–7.

Epstein ME, Smedley SR, Eisner T. 1994. Sticky integumental coating of a dalcerid caterpillar: a deterrent to ants. J Lepidop Soc 48: 381–6.

Greene HW. 1986. Natural history and evolutionary biology. In: Feder ME, Lauder GV (eds). Predator-prey relationships. Chicago IL: Univ Chicago Pr. p 99–108.

Grifo F, Rosenthal J. 1997. Biodiversity and human health. Washington DC: Island Pr.

Joyce C. 1994. Earthly goods. New York NY: Little Brown.

McCormick KD, Deyrup MA, Menges ES, Wallace SR, Meinwald J, Eisner T. 1993. Relevance of chemistry to conservation of isolated populations: the case of volatile leaf components of *Dicerandra* mints. Proc Natl Acad Sci USA 90:7701–5.

Middleton S, Liittschwager D. 1994. Witness: endangered species of North America. San Francisco CA: Chronicle.

Moffett MW. 1993. The high frontier. Cambridge MA: Harvard Univ Pr.

Nuridsany C, Pérennou, M. 1996. Microcosmos: the invisible world of insects. New York NY: Stewart, Tabori and Chang.

Plattner H, Salpeter M, Carrel JE, Eisner T. 1972. Struktur und Funktion des *Drüsenepithels* der postabdominalen Tergite von Blatta orientalis. Z Zellforsch 125:45–87.

Reid WV, Laird SA, Meyer CA, Gámez R, Sittenfeld A, Janzen DH, Gollin MA, Calestous J (eds). 1993. Biodiversity prospecting: using genetic resources for sustainable development. Washington DC: World Resources Inst.

Wilson EO. 1984. Biophilia. Cambridge MA: Harvard Univ Pr.

Wilson EO, Peter FM(eds). 1988. Biodiversity. Washington DC: National Acad Pr.

CONSERVATION MEDICINE:
AN EMERGING FIELD

MARK POKRAS*
GARY TABOR*
MARY PEARL†
DAVID SHERMAN*
PAUL EPSTEIN‡

*Tufts University School of Veterinary Medicine,
200 Westboro Road, North Grafton, MA 01536
†Wildlife Preservation Trust International,
520 Locust Street, Suite 704, Philadelphia, PA 19102
‡Center for Health and the Global Environment, Harvard Medical School,
260 Longwood Avenue, Boston, MA 02115

THE LOSS OF biodiversity—the entire wealth of plant and animal species—is perhaps the most important problem that faces our fragile planet. Unwittingly and unremittingly, our species is in the process of bringing about an unprecedented biological disaster. In the wake of our growth and development lie hundreds of thousands of extinct species that are gone forever. The process of extinction continues, and today even larger numbers of species are threatened. Such losses undermine the ecological fabric that sustains the web of life, including human life. Ironically, this massive wave of species extinctions is foreclosing the discovery of new medicines and remedies from natural sources. Society has seemed ill-equipped to deal with these health crises, because we lack professionals who have the interdisciplinary skills to link the health issues of ecosystems, animals, and humans.

The health and well-being of people and other animals are threatened by the effects of humans on ecosystems, including the large-scale alteration and destruction of habitat, the decline and extinction of species, the alteration of ecological processes, the invasion of nonnative (alien) species, the continued economic emphasis on short-term bottom-line thinking, and the spread of contaminants and hazardous substances through all levels of the food chain. Animal health and human health are inextricably connected through the ecological realities that govern life on our planet.

The health of individuals, species, and populations and the more encompassing notion of environmental health represent a continuum of the way in which health concerns currently are defined. At all levels, the complexity of health issues is being revealed. The landscape of understanding includes a greater aware-ness of the synergism of cumulative effects and multiple stresses. As the scien-tific ability to study environmental perturbations and ecosystem dynamics im-proves, new patterns of disease transmission and alarming health effects are emerging.

WHAT IS CONSERVATION MEDICINE?

The term *conservation medicine* was first introduced by Koch (1996) to mean the study of the broad ecological contexts of health. It tries to relate concerns about the health of all living organisms to the integrity of ecosystems. The overlap of veterinary medicine, human medicine, and conservation biology forms the knowl-edge base for this field.

The field of veterinary medicine long has been recognized for its comparative approach. Until recently, it primarily addressed the health and productivity of animals owned by people. But veterinarians increasingly are concerned with turn-ing their skills to the health of wild animals and their habitats. Physicians also are recognizing that conservation of biodiversity is important, to protect species that provide a buffer against the emergence of pests and pathogens and that serve as potential medical models of and environmental sentinels for human health. Conservation biologists are working with veterinarians and physicians to expand beyond conventional paradigms of health and examine human and animal health through an ecological lens. By bringing the three disciplines together, new areas of research, education, policy, and training can be engaged. In short, solutions to future concerns about environmental health will be related to the development of effective interdisciplinary tools and modes of problem-solving.

Conserving the integrity of the biosphere is the applied goal of conservation medicine. It attempts to provide a cognitive framework for examining health functions within ecosystems. As health problems related to environmental deg-radation multiply and magnify in importance, health professionals increasingly will be relied on to comment on environmental strategies and to advise communities taking part in processes of environmental decision-making. In their publicly per-ceived roles as educators, all conservationists need to understand and articulate the linkage between human and animal health and intact ecosystems. Ultimately, as concerned citizens of the world, we must work together to define the appropri-ate balance between the needs of people, domestic animals, and wildlife in the face of finite amounts of energy, land, and other resources.

SOME CURRENT CHALLENGES: INTERFACES BETWEEN MEDICINE AND CONSERVATION

Emerging and Re-emerging Diseases

The influence of parasites and disease on human health and demographics rarely is questioned. Such recent books as *The Coming Plagues* (Garrett 1994) and such classics as *Rats, Lice, and History* (Zinsser 1963) have reflected our all-too-human interest in health threats that directly affect us and those close to us. Historically, conservationists and wildlife professionals have ignored or downplayed the effects of these same pressures on wildlife populations and natural systems until species became endangered.

However, as Garrett (1994) points out so eloquently, ecological perturbations are fast bringing down the barriers that once limited human-to-animal disease transmission. New variants of the cholera-causing organism, *Vibrio cholerae*, have been found moving in intercontinental patterns within marine algal blooms that are associated with red-tide phenomena and the periodic occurrence of El Niño–southern oscillation events (Epstein 1993). Strains of Hantavirus that have fatality rates of nearly 55% in humans have emerged in regions that exhibit disturbances of habitat and climate (Epstein 1995). Outbreaks of *Pfiesteria piscida*, a toxic dinoflagellate, in the Chesapeake Bay of Maryland recently have created headlines. Blooms of *Pfiesteria* associated with large-scale fish kills and disease in both people and animals have been linked to nutrient-rich agricultural runoff (Anonymous 1997; Barker 1997; Steidinger and others 1996).

In those instances, alarm has arisen because of concern for human health. The effects of the pathogens on populations of domestic or wild animals and natural ecosystems are poorly understood, but there are many examples in which health effects of human activities on animal populations are understood more clearly (Dobson and May 1982; Thorne and Williams 1988). For instance, the introduction of tuberculosis from humans to populations of orangutans and other endangered primates has serious implications for the long-term existence of these species in the wild (Jones 1982). Predators and diseases, plus disease vectors and reservoirs that people have introduced either purposefully or accidentally, have led to the extinction of many endemic Hawaiian bird species, and they threaten many more (van Riper and other 1986).

Chronic Toxic Pollutants

In parallel with the growing awareness of emerging infectious diseases, concerns about the effects of chronic exposure to toxic chemicals have surfaced as well (Colburn and others 1996). Bioaccumulation of selenium from agricultural runoff in the western United States has caused large-scale fish mortality, deformity and death in fish-eating birds and mammals, and the closure of some protected federal wildlife refuges (Botkin and Keller 1997). Endocrine-disrupting chemicals are suspected of producing widespread effects on the reproductive systems of fish, reptiles and amphibians, birds, and mammals, including humans (Colborn and Clement 1992). The bioaccumulation of persistent and widespread toxic substances may have effects that range from congenital defects to promotion of cancer, reproductive diseases, and increased susceptibility to disease (for example, immunological dysfunctions) (Botkin and Keller 1997; Colborn and Clement 1992; Colborn and others 1996). Although the precise mechanistic relationships between the biological activities of these toxic substances and disease are not understood completely, the trend is disturbing and underscores the need to examine the persistence of chemicals within the environment in an entirely new light.

Compromised Health of Ecosystems

At another level, the health of ecosystems is threatened by increased fragmentation of habitat, decreased ecological resilience, unbalanced proportions of predators and prey, introductions of alien species, changes in global climate, enhanced ultraviolet radiation, and the multitrophic-cascade effects related to disturbance and extinction (Carpenter and Kitchell 1993; Epstein 1993; Hollings 1996; Kreuss and Tscharntke 1994; Malcolm and Markham 1996). The integrity of ecosystems and the species they comprise is being undermined daily by incremental catastrophes.

Interdisciplinary Barriers

Each discipline approaches problem-solving from its own perspective and with its own set of inherent biases. Both medical and conservation professionals need to adopt new attitudes if truly creative interdisciplinary problem-solving is to occur. When asked what physicians and veterinarians needed to learn to play a more constructive role in conservation, one wildlife biologist recently remarked, "They should learn to leave their white coats and attitudes at home!" (A. Major, USFWS, pers. comm.). Clearly, we need to get to know each other better. A mutual respect for the knowledge and abilities of other professionals is an important prerequisite to progress.

THE GOALS OF CONSERVATION MEDICINE

The goal of conservation medicine is the integration of the diagnostic and problem-solving tools of medical professionals with the ecological and management knowledge of conservation professionals to preserve biodiversity and maintain the health of interdependent species (including humans).

One outcome of this synergy might be the creation of an integrated appraisal process for examining ecological health concerns. Such an appraisal process would try to incorporate the concepts of sustainability, life-cycle analysis, and systems thinking (Anderson and Johnson 1997; Clark 1993). By emphasizing the use of contextual knowledge in decision-making and diagnosis, an integrated health assessment could serve as such a tool. How this tool is defined and used deserves a separate, more extensive discussion. We mention it here as a possible example of how the talents and skills of multiple disciplines within the sciences and social sciences can be organized practically.

Conservation medicine is in its infancy. We are only beginning to define the tasks that can achieve its overall goal. Those tasks include the following:

• training environmentally literate health professionals;
• breaking down disciplinary barriers to communication and cooperation;
• establishing the scientific underpinnings of the interrelationships between human, animal, and environmental health;
• encouraging broad participation in public education; specific targets include policy-makers, voters, and children;

- being active in developing conservation and health policies that integrate human and animal concerns;
- encouraging broader definitions for concepts of health; and
- developing assistive technical applications, including
  - noninvasive diagnostic and therapeutic tools;
  - conservation-oriented reproductive biology, genetics, and medicine;
  - techniques to minimize the spread of exotic species and diseases;
  - development of epidemiological models that will integrate data on wildlife, human, and domestic animal health to improve understanding of the ecological dynamics of health and disease;
  - techniques for capture, restraint, anesthesia, and analgesia; and
  - techniques for determining age and marking and tracking individuals.

As natural communities shrink, wildlife populations decline and come under more stresses, and populations of humans and domestic animals grow, there are an increased number of health problems in all species and new opportunities for disease to cross taxonomic lines. Achievement of the tasks listed above will enable health professionals to develop the nontraditional skills and broad environmental concerns needed to work constructively as members of multidisciplinary conservation efforts.

CONCLUSION

> If it is granted that biodiversity is at high risk, what is to be done? The solution will require cooperation among professions long separated by academic and practical tradition.
>
> E.O. Wilson (1992)

Over time, the roles of veterinary and human medical practitioners have expanded with society's understanding of the relationships between species. The health community as a whole has a latent capacity to address environmental-health issues, but this will require new ways of thinking and new tools. We hope that through working with a diversity of environmental professionals, we can do for environmental health what medicine is trying to do for human and animal health: change the focus from the treatment of a pathological condition to the maintenance of health.

Conservation medicine can be characterized as a work in progress. It provides a framework for bringing the health-science professions into the realm of conserving biological diversity and ecosystems and for infusing conservation biology thinking into the health pedagogy. In the end, we hope to use this approach to help people understand that esoteric concepts like "conservation of biodiversity" are intimately connected to their own personal health and that of animals. We also hope that this paper will stimulate thinking and discussion and will lead to the further definition of how the medical perspective can bring added value to conservation efforts.

REFERENCES

Anderson V, Johnson L. 1997. Systems thinking basics: from concepts to causal loops. Cambridge MA: Pegasus Commun. 273 p.

Anonymous. 1997. Pfiesteria: federal and state agencies research toxic marine microorganism after human health problems arise. Chem Eng News 75(41):14.

Barker R. 1997. And the waters turned to blood. New York NY: Simon & Schuster.

Botkin D, Keller E. 1997. Environmental science—earth as a living planet. New York NY: J Wiley.

Carpenter SR, Kitchell JF. 1993. The trophic cascade in lakes. Cambridge UK: Cambridge Univ Pr.

Clark TW. 1993. Creating and using knowledge for species and ecosystem conservation: science, organizations, and policy. Persp Biol Med 36:497–525.

Colborn T, Clement C. 1992. Chemically-induced alterations in sexual and functional development. Princeton NJ: Sci Publ Co Inc.

Colborn T, Dumanoski D, Myers JP. 1996. Our stolen future. New York NY: Dutton. 306 p.

Dobson AP, May RM. 1982. Disease and conservation. In: Soulé M (ed). Conservation biology. Cambridge MA: Sinauer. p 345–65.

Epstein PR. 1993. Algal blooms in the spread and persistence of cholera. BioSystems 31:209–21.

Epstein PR. 1995. Emerging diseases and ecosystem instability: new threats to public health. Amer J Pub Health 85(2):113–15.

Garrett L. 1994. The coming plagues. New York NY: Farrar, Strauss and Giroux.

Hollings CS. 1996. Resilience of ecosystems: local surprise and global change. In: Clark WC, Munn RE (eds). Sustainable development of the biosphere. Cambridge UK: Cambridge Univ Pr. P 292–317

Jones DM. 1982. Conservation in relation to animal disease in Africa and Asia. In: Edwards MA, McDonnell U (eds). Animal disease in relation to animal conservation. Symp Zool Soc London 50:271–85.

Koch M. 1996. Wildlife, people, and development. Tropical Anim Health Prod 28:68–80.

Kruess A, Tscharntke T. 1994. Habitat fragmentation, species loss, and biological control. Science 264:196–9.

Malcolm JR, Markham A. 1996. Ecosystem resilience, biodiversity, and climate change: setting limits. Parks 6(2):38–49.

Steidinger KA, Burkholder JM, Smith SA. 1996. Pfiesteria piscida gen et sp nov a new toxic dinoflagellate with a complex life cycle and behavior. J Phycol 32(l): 157–61

Thorne ET, Williams ES. 1988. Disease and endangered species: the black-footed ferret as a recent example. Cons Biol 2(l):66–74.

van Riper III C, van Riper SG, Goff ML, Laird M. 1986. The epizootiology and ecological significance of malaria in Hawaiian forest land birds. Ecol Monogr 56:327–44.

Wilson EO. 1992. The diversity of life. Cambridge MA: Belknap Pr. 424 p.

Zinsser H. 1963. Rats, lice, and history. Boston MA: Little, Brown.

HOW COUNTRIES WITH LIMITED RESOURCES ARE DEALING WITH BIODIVERSITY PROBLEMS

JEFFREY A. MCNEELY

IUCN Biodiversity Policy Coordination Division,
rue Mauverney 28, 1196 Gland, Switzerland

IN THE DECADE since the groundbreaking publication of *Biodiversity* (Wilson and Peter 1988), we have made considerable progress in promoting the conservation of the world's diversity of genes, species, and ecosystems. That publication led to comprehensive new approaches to conservation, bringing information, knowledge, awareness, and ethics into a complex mixture of protected areas, agriculture, economics, intellectual-property rights, land tenure, trade, forestry, and so forth. It also led to the Convention on Biological Diversity (CBD), which now has been ratified by 172 countries (the United States is one of the handful of holdouts). It also has led to considerable scientific work in the field, as evidenced by this conference, numerous books and journals, and various other manifestations of interest and concern.

All this effort has led to greatly increased understanding about biodiversity and the threats to it. It is now well known that most of the world's species are found in the tropics, frequently in the countries that have the least financial, technical, and institutional means to conserve biodiversity (see table 1).

How, then, are the tropical countries coping with the challenge of conserving biodiversity? At least a partial answer is provided by table 2, which demonstrates that the tropical developing countries are making a substantial effort to establish protected areas, a major objective of which is to conserve biological diversity. In this effort, they are supported by the industrialized countries through various bilateral-aid agencies and through the Global Environment Facility, operated by the World Bank, United Nations Development Programme, and United Nations Environment Programme to provide several hundred US million dollars per year for biodiversity according to the priorities identified by the Conference of Parties of the CBD.

The CBD stresses the importance of international, regional, and global cooperation between states, intergovernment organizations, and the nongovernment sector in supporting action to conserve biological diversity and use biological resources sustainably. This is a clear recognition of the need of governments to collaborate with each other and with various kinds of multilateral and bilateral organizations if they are to be successful in their efforts to manage biological resources sustainably. Effects in one state—for example, consumption of such products as ivory, tiger bones, and medicinal plants—may affect biodiversity pro-

TABLE 1 The World's Most Species-Rich Countries by Rank

Mammals	Number of Species	Reptiles (continued)	Number of Species
1. Indonesia	515	6. Colombia	383
2. Mexico	449	7. Ecuador	345
3. Brazil	428	8. Peru	297
4. Democratic Republic of		9. Malaysia	294
Congo	409	10. Thailand/Papua New Guinea	282
5. China	394		
6. Peru	361	Amphibians	
7. Colombia	359		
8. India	350	1. Brazil	516
9. Uganda	311	2. Colombia	407
10. Tanzania	310	3. Ecuador	358
		4. Mexico	282
Birds		5. Indonesia	270
		6. China	265
1. Colombia	1721	7. Peru	251
2. Peru	1701	8. Democratic Republic of Congo	216
3. Brazil	1622	9. United States	205
4. Indonesia	1519	10. Venezuela/Australia	197
5. Ecuador	1447		
6. Venezuela	1275	Flowering Plants	
7. Bolivia	± 1250		
8. India	1200	1. Brazil	55,000
9. Malaysia	± 1200	2. Colombia	45,000
10. China	1195	3. China	27,000
		4. Mexico	25,000
Reptiles		5. Australia	23,000
		6. South Africa	21,000
1. Mexico	717	7. Indonesia	20,000
2. Australia	686	8. Venezuela	20,000
3. Indonesia	± 600	9. Peru	20,000
4. Brazil	467	10. Russian Federation	
5. India	453	(former USSR)	20,000

Source: McNeely and others 1990.

TABLE 2   Protected Areas of the World

Region	Area (km²)	Area Protected (km²)	% Protected
North America	23,433,902	2,654,814	11.3
Europe	5,105,551	506,602	9.9
North Africa and Middle East	13,118,661	476,812	3.6
Eastern Asia	11,789,524	447,773	3.8
Northern Eurasia	22,100,900	237,958	1.1
Sub-Saharan Africa	23,927,581	2,401,418	10.0
South and Southeast Asia	8,866,884	838,703	9.5
Pacific	573,690	21,661	3.8
Australia	7,682,487	837,929	10.9
Antarctica and New Zealand	13,625,961	52,256	0.4
Central America	542,750	104,084	19.2
Caribbean	238,620	31,995	13.4
South America	18,001,095	3,611,131	20.1
TOTAL	149,007,606	12,223,136	8.2

Source: McNeely and others 1994.

foundly in another. When species migrate between countries, wildlife populations are shared, making collaboration essential to their conservation. Furthermore, by definition, the obligations of the CBD for sharing technology and the benefits derived from the use of genetic material require cooperation between states.

The CBD specifically mentions the private sector and nongovernment organizations (NGOs), which include businesses, academe, citizen groups, and various kinds of private conservation organizations. The NGO community includes a large proportion of the world's leading scientists who are working on biodiversity issues and who have played a major role in advocating the need to conserve biodiversity. NGOs can bring commitment, innovation, clarity of purpose, and practical knowledge to environmental and developmental issues, and they often are especially effective at the local level.

This paper is a brief review of measures under the CBD for international cooperation to support national conservation efforts.

HOW THE CONVENTION ON BIOLOGICAL DIVERSITY PROMOTES INTERNATIONAL COOPERATION

While stressing national sovereignty over biodiversity, the CBD also strongly emphasizes international cooperation. It specifically recognizes that "the provision of new and additional financial resources and appropriate access to relevant technologies can be expected to make a substantial difference in the world's ability to address the loss of biological diversity." It also acknowledges that "special provision is required to meet the needs of developing countries, including the provision of new and additional financial resources and appropriate access to relevant technologies." Signatories acknowledge that "substantial investments are required to conserve biological diversity and that there is the expectation of a broad range of environmental, economic, and social benefits from those investments" (see Glowka and others 1994 for a guide to the CBD).

The CBD recognizes that the conservation of biodiversity and the sustainable use of biological resources are critically important for meeting the dietary, medicinal and other needs of the growing world population, for which purpose genetic resources and relevant technologies play an essential role.

Furthermore, the CBD expects that "the conservation and sustainable use of biological diversity will strengthen friendly relations among states and contribute to peace for humankind." This implicitly recognizes the principle of ecological security—that the peace and stability of a nation depend not only on its conventional military defenses, but also on its environmental stability. Environmental degradation within a country can result in social collapse and appalling human tragedies, leading to disputes within and between nations and even, ultimately, to war. In particular, overexploitation of resources shared between nations, such as water supplies and fish stocks, also can lead to conflict (see, for example, Homer-Dixon 1994). Therefore, stemming the loss of biodiversity contributes to peace and harmony between nations.

Elements of the CBD that are specifically relevant to international cooperation include the following.

- **Article 3. Principle.** "States have, in accordance with the Charter of the United Nations and the principles of international law, the sovereign right to exploit their own resources pursuant to their own environmental policies, and the responsibility to ensure that activities within their jurisdiction or control do not cause damage to the environment of other States or of areas beyond the limits of national jurisdiction."
- **Article 5. Cooperation.** "Each Contracting Party shall, as far as possible and as appropriate, cooperate with other Contracting Parties, directly or, where appropriate, through competent international organizations, in respect of areas beyond national jurisdiction and on other matters of mutual interest, for the conservation and sustainable use of biodiversity."
- **Article 8. In situ conservation.** "Each Contracting Party shall, as far as possible and as appropriate: (m) cooperate in providing financial and other support for in situ conservation. . . particularly to developing countries."
- **Article 9. Ex situ conservation.** "Each Contracting Party shall, as far as possible and as appropriate, and predominantly for the purpose of complementing in situ measures: (e) cooperate in providing financial and other support for ex situ conservation. . . . and in the establishment and maintenance of ex situ conservation facilities in developing countries."
- **Article 12. Research and training.** "The Contracting Parties, taking into account the special needs of developing countries, shall: (a) establish and maintain programmes for scientific and technical education and training in measures for the identification, conservation, and sustainable use of biological diversity and its components and provide support for such education and training for the specific needs of developing countries."

- **Article 13. Public education and awareness.** "The Contracting Parties shall: (b) cooperate, as appropriate, with other States and international organizations in developing educational and public awareness programmes, with respect to conservation and sustainable use of biological diversity."
- **Article 15. Access to genetic resources.** "2. Each Contracting Party shall endeavour to create conditions to facilitate access to genetic resources for environmentally-sound uses by Contracting Parties and not to impose restrictions that run counter to the objectives of this Convention. 4. Access, where granted, shall be on mutually agreed terms and subject to the provisions of this Article. 5. Access to genetic resources shall be subject to prior informed consent of the Contracting Party providing such resources, unless otherwise determined by that Party. 6. Each Contracting Party shall endeavour to develop and carry out scientific research based on genetic resources provided by other Contracting Parties with the full participation of, and where possible in, such Contracting Parties."
- **Article 16. Access to and transfer of technology.** "Each Contracting Party. . . undertakes to provide and/or facilitate access for and transfer to other Contracting Parties of technologies that are relevant to the conservation and sustainable use of biological diversity or make use of genetic resources and do not cause significant damage to the environment. Access to and transfer of technology to developing countries shall be provided and/or facilitated under fair and most favourable terms, including on concessional and preferential terms where mutually agreed."
- **Article 17. Exchange of information.** "The Contracting Parties shall facilitate the exchange of information, from all publicly available sources, relevant to the conservation and sustainable use of biological diversity, taking into account the special needs of developing countries."
- **Article 18. Technical and scientific cooperation.** "The Contracting Parties shall promote international technical and scientific cooperation in the field of conservation and sustainable use of biological diversity, where necessary through the appropriate international and national institutions."
- **Article 20. Financial resources.** "The developed country Parties shall provide new and additional financial resources to enable developing country Parties to meet the agreed full incremental costs to them of implementing measures which fulfill the obligations of this Convention and to benefit from its provisions."

WHAT DEVELOPING COUNTRIES ARE DOING FOR THEMSELVES

This list of internationally agreed principles might imply that the developing countries are dependent on the largesse of the developed countries to take care of their own biodiversity. Nothing could be farther from the truth. Virtually all countries in the tropics have implemented a wide range of measures to conserve their own biodiversity and to use their biological resources sustainably. Of course, they can do even better if they receive additional support, but many of them are turning difficult circumstances to their advantage by using innovative and cost-effective approaches to conservation and sustainable use.

One issue of particular interest, because it affects both cultural diversity and biological diversity, is the role of indigenous groups and local communities in managing protected areas.

Indigenous peoples often have cultural values and institutions that differ from those of the dominant culture within which they are found. As Alcorn (1997) has pointed out, most indigenous peoples are politically marginal groups that are known variously as tribals, hill tribes, or other such terms. They often claim property rights to ancestral lands and waters and the right to retain their own customary laws, traditions, languages, and institutions, as well as the right to represent themselves through their own institutions. Furthermore, indigenous peoples that live in areas that are important for conservation are linked closely to their local resource base and frequently have developed resource-management systems and social institutions that are responsive to environmental feedback. Thus, their local knowledge has a particular contribution to make to protected-area management.

But the primary reason why managers of protected areas in the tropics are recognizing the decision-making authority of indigenous peoples is that they have prior rights over the lands and waters in which protected areas are being established, and many would assert that such peoples have rights to make decisions about how to manage their ancestral lands. Indeed, article 8(j) of the CBD says, "Subject to its national legislation, [each Contracting Party shall] respect, preserve, and maintain knowledge, innovations, and practices of indigenous and local communities embodying traditional lifestyles relevant for the conservation and sustainable use of biological diversity and promote their wider application with the approval and involvement of the holders of such knowledge, innovations, and practices and encourage the equitable sharing of the benefits arising from the utilization of such knowledge, innovations, and practices."

In most parts of the tropics, rural villagers believe that they have historical rights to the land and resources that governments have declared "protected" in the national interest (for example, Vandergeest 1996). No areas of "empty" land exist that could be managed free of human influence, although most governments have followed the European model of claiming all forests to be the property of government. This conflict has led to the wide recognition that conservation of biodiversity cannot succeed unless it is linked to economic opportunities and investments aimed at those who otherwise might threaten the viability of protected areas through their activities in pursuit of their livelihood.

The increasing attention given to local communities does not imply necessarily that local communities are the major threat to protected areas and the biodiversity they support. In fact, in most tropical countries, the major threats to protected areas come from outside influences, such as government-supported timber concessions, road-building activities, agricultural subsidies, mining concessions, dam construction, expanding populations, air and water pollution, and (in the longer term) climate change. Most such problems need to be addressed as part of regional planning and central government policy rather than protected-area management.

People can be expected reasonably to institute their own conservation measures when they are the primary decision-makers and beneficiaries. Numerous examples can be cited from various parts of the world (for example, Birckhead and others

1992; Kemf 1993; Kothari and others 1996; Stone 1991; UNEP 1988; Wells and Brandon 1992; West and Brechin 1991; Western and others 1994). These examples support the general point that earning the support of local communities means giving them a stake in the success of a well-managed protected area.

When areas within the traditional territories of indigenous peoples are managed as limited-access extractive reserves, they may be considered legitimate protected areas worthy of international recognition. In Australian "indigenous protected areas", for example, land tenure is vested with the aboriginal people, but the land usually is managed by the National Conservation Agency under a leasing arrangement.

In Nicaragua, the Miskito people have formed their own NGO, "Mikupia", to manage the Miskito Coast Protected Areas, overseen by a commission that includes four representatives from the national government, one from the regional government, one from the Mikupia, and two from the Miskito communities (Barzetti 1993).

In Peru, the 322,500-hectare Tamshiyacu-Tahuayo Communal Reserve contains no permanent settlements (Bodmer and others 1991). It is divided into a fully protected core area and an area of subsistence use. Actions voluntarily implemented by the local people to control exploitation include prohibition of the use of nets and lances in the oxbow lakes of the reserve during low-water seasons, limitations on fishing technology, prohibition of commercial fisheries, and prohibition of the use of fish poisons. Fish populations in the area appear to be rebuilding, and the local communities are benefiting directly from their self-imposed management programs.

In the Philippines, the Kalahan Education Foundation, a local NGO established by the Ikalahan Tribe, is implementing an integrated program of community forest management and the extraction of nontimber forest products, leading to the production of jams and jellies from forest fruits, the extraction of essential oils, the collection and cultivation of flowers and mushrooms, and the manufacture of furniture. The foundation is based on the Kalahan reserve, which supports about 550 Ikalahan families that live within the 14,730-hectare reserve of ancestral land.

In eastern Indonesia, many fishing villages have established a form of marine protected area called *petuanang* as part of a body of traditional resource-management practices known as *sasi*. The pentuanang has certain closed seasons and is carefully managed in terms of permitted fishing techniques, and only certain types of fishing gear are permitted (Spiller 1997). However, more recently, the demand for increased production of fish for trade and export has weakened the control of village leaders in managing traditional resource-management systems, although modern approaches to participatory planning could help resurrect the traditional management systems that worked well for many generations.

In Papua New Guinea, where about 97% of land is in community ownership, the government has established wildlife-management areas where local communities voluntarily agree to certain controls on exploitation (Eaton 1985). Each wildlife-management area has a wildlife management committee with representatives from local communities and from resource-management agencies of the government. These committees have instituted such measures as royalties for the taking of deer, ducks, and fish by outsiders; hunting restrictions, such as forbidding all nontraditional hunting methods, the use of shotguns, and the use of dogs; prohibition of the collection of crocodile eggs; fishing restrictions, such as forbidding the use of commercially manufactured nets, hurricane lamps, and fish poisons; and restrictions on logging. In all areas, the rules enacted tend to promote traditional practices and authority.

Interestingly enough, many developing-country governments are finding that conservation actually pays, especially through tourism. For example, Galapagos National Park generated direct revenues of US$3.7 million in 1995. The Galapagos National Park kept about a third of the receipts, and the rest was used to support protected areas on the mainland of Ecuador (Southgate 1996). Some protected-areas systems in the Caribbean do even better, largely because of dive tourism. Divers spend about US$30 million per year at the Bonaire Marine Park in the Netherlands Antilles, US$14 million in protected areas in the British Virgin Islands, more than US$53 million per year in marine protected areas in the Cayman Islands, and US$23 million in Virgin Islands National Park on St. John (OAS/NPS 1988).

Not surprisingly, some governments are turning to the private sector to help earn greater benefits from tourism. For example, through the Zambia Privatisation Agency, the Zambian National Parks and Wildlife Service offered some 25 prime locations in the parks on competitive-tender lease. These locations include sites in the Mosi-Oa-Tunya National Park, at Victoria Falls, and in the South Luangwa National Park, Kafue National Park, and Blue Lagoon National Park. Sites include government-owned lodges, camps, and other tourist attractions.

Some private tourism companies also are seeking ways to contribute to protected areas. One illustration of corporate approaches to funding conservation through tourism is Operation Eye of the Tiger which has been established with funding from Outdoor India Tours Pvt. Ltd., New Delhi, and has links with Kentucky Fried Chicken in the United States. This operation has pledged to create disturbance-free habitats for tigers, to carry out ecodevelopment and conservation education, and to promote research on the tiger, its habitat, and its allied species.

The National Parks Trust of South Africa has negotiated an agreement with the Conservation Corporation, a private group, for the management of the Ngala Game Reserve. This led to the establishment in 1992 of the first "contract reserve" between Kruger National Park and a private enterprise, giving the Conservation Corporation exclusive rights for operating tourist activities in 14,000 hectares of the park. The fees paid to the park are used for wildlife management, research, educational programs, and community-based projects adjacent to Kruger National Park (Borrini-Feyerabend 1996).

HOW NGOS ARE SUPPORTING PROTECTED AREAS IN THE TROPICS

Recognizing that governments are unable to take full responsibility for all protected areas, NGOs have stepped in in many countries to provide their flexible and creative approaches to overall plans for national protected-area systems. NGOs are playing a particularly important role in Latin America (Redford and Ostria 1995), including the following:

- The Programme for Belize (PFB) has been given responsibility for management of the 92,614-hectare Rio Bravo Conservation and Management Area and for holding the land in trust for the people of Belize. Originally supported by private donations, PFB hopes to earn sufficient revenue from forest products and tourism to become self-sustaining.
- In Guatemala, the Fundacion Defensores de la Naturaleza was given authority in 1990 by the Guatemalan Congress to manage the operations and administration of the Sierra de las Minas Biosphere Reserve (236,300 hectares), including the work of the park guards; it is in charge of management decisions, including training, infrastructure, and communications, under the supervision of the National Council of Protected Areas.
- In Panama, the Asociacion Nacional Para la Conservacion de la Naturaleza has an agreement with Panama's Institute for Natural Renewable Resources (INRENARE) to demarcate the boundaries of the Darien Biosphere Reserve (597,000 hectares), to train and equip park personnel, to install infrastructure, and to conduct biological inventories.
- In Bolivia, the Fundacion Amigos de la Naturaleza (FAN) has been granted a 10-year management contract by the National Department for the Conservation of Biodiversity for the Noel Kempff Mercado National Park (927,000 hectares). FAN is responsible for hiring rangers, building infrastructure, and helping to reduce poaching.
- In Colombia, the Fundacion Pro-Sierra Nevada de Santa Marta is responsible for managing three areas within the Sierra Nevada de Santa Marta National Park (300,000 hectares), including land-protection and community-outreach activities.
- In Ecuador, the Fundacion Natura has a formal agreement with the Ministry of Agriculture to participate and collaborate in the management of protected areas, working on training staff and raising funds, including facilitating a debt-for-nature swap valued at US$10 million.
- In Paraguay, the Fundacion Moises Bertoni is legally responsible for managing the Mbaracayu Forest Nature Reserve (63,000 hectares).

Governments are beginning to give greater legal recognition to the role of NGOs in protected areas. In 1993, the congress of Colombia passed a law that recognized the role of civil society in conservation and named private reserves as legal conservation units. Colombia now has some 120 private protected areas that are mobilized into the Network of Private Nature Reserves, an NGO that comprises private farmers and landowners, community organizations, agricultural cooperatives, and other NGOs.

In the Philippines, partnerships have been formed between the public and private sectors by integrating the assistance of NGOs into the management of protected areas at national and local levels. A new NGO, known as the NGO for Integrated Protected Areas, Inc. (NIPA), has been established to recruit and coordinate local support activities, to provide technical assistance, to monitor implementation, and to assist in the establishment and implementation of a livelihood fund that will be used to support village socioeconomic-development projects and employment activities designed to reduce pressures on the protected areas. NIPA now is supporting work at 10 high-priority protected areas in the Philippines, establishing protected-area management boards consisting of local governments, NGOs, and representatives of indigenous peoples. NIPA has recruited local NGOs to assist with field activities, community organizing, and strengthening of the protected-area management boards. Progress is promising, and communities are now aware of the need to integrate conservation and development activities.

NGOs also are involved in supporting the effective management of Indonesia's Kerinci-Seblat and Lore Lindu National Parks (Elliott and others 1993). In Kerinci-Seblat, four provincial NGO alliances are working on soil- and water-conservation projects in five park-boundary villages; in Lore Lindu, four small NGO alliances are implementing a range of community-development activities in the Lake Lindu enclave. These NGOs are providing an effective channel for reaching local communities, a critical element in the success of integrated conservation and development projects. However, the NGOs are not self-supporting. They require access to technical expertise, training in technical and managerial skills, and funding for overhead and field activities. Java-based national NGOs and international NGOs, such as The Nature Conservancy, are serving as intermediaries between donors and the local NGOs, thereby overcoming some of the operational constraints that grassroots NGOs face when working with donors.

One example of a grassroots NGO working in support of a protected area is the Foundation for Community Development in Indonesia's Wasur National Park. This new local NGO has a field staff of community organizers; a management board of community representatives, teachers, and other informal leaders; and a steering committee that is composed of government officials, a representative of the World Wildlife Fund (WWF), the head of the local government, and community representatives. Although the legal status of the foundation is just being established, it is already working in the park to help local communities meet their immediate economic needs (Barber and others 1995).

At the opposite end of the spectrum is a remarkable new quasi-NGO, the Leuser International Foundation (LIF). In 1995, this private nonprofit organization was granted a 7-year, renewable, exclusive "conservation concession" for a contiguous area that includes the existing Gunung Leuser National Park (905,000 hectares), 505,000 hectares of protection forest, and 380,000 hectares of production forest in Sumatra, Indonesia. This concession grants LIF the right to manage and coordinate activities for conservation and sustainable development within the ecosystem, on the basis of objectives and work plans that are reviewed and approved by the minister of forestry. The long-term objective of the project is to transform the area within the boundaries of the ecosystem into an expanded national park that has multiple use zoning, as mandated in Indonesia's Conservation Act of 1990 (Rijksen and Griffiths 1995). The government concluded a financing agreement with the European Union in May 1995, under which the European Union has provided a grant of US$40.6 million—to be matched by US$22.5 mil-

---

lion from the Indonesian government—to LIF for its conservation activities. Under this concession, LIF essentially is assuming the government's role of making and implementing conservation and development policy for a particular site, albeit within a framework of government supervision. Discussions also have been held within the ministry of forestry concerning a possible expansion of LIF's concession to include a monopoly on selling the value of Leuser's carbon-sequestration function on the international market that is beginning to develop under the impetus of the Framework Convention on Climate Change (Barber and Nababan 1997).

In Nepal, the King Mahendra Trust for Nature Conservation (KMTNC), a semiautonomous, nongovernment, nonprofit organization, has been established for the purpose of conserving, preserving, and managing nature and its resources in an effort to improve the quality of life of the Nepali people. KMTNC is designed to raise funds for the development and management of protected areas and to execute projects; it has established associated national trusts in the UK, Japan, the Netherlands, France, Germany, and Canada. It has worked in Sagarmatha National Park, Chitwan National Park, the Annapurna Conservation Area Project, and elsewhere on various aspects of protected-area management. KMTNC is managed by a board of directors that comprises various senior government and nongovernment officials and several representatives of the international community.

A major innovation for Nepal is enabling the Annapurna Conservation Area (762,900 hectares) to be managed by KMTNC, which is able to raise money directly from overseas (especially from WWF) and has considerable autonomy, enabling it to bypass many of the procedures associated with government agencies and to execute projects with a relatively slim and flexible bureaucracy. In the Annapurna Conservation Area, KMTNC has an autonomous and substantial role in managing an innovative multiple-use conservation area that is probably a unique arrangement for an NGO in Asia. Its main management objectives include forestry and wildlife conservation, alternative-energy development, community education, and tourism, and it fully involves local residents in the planning and management of the natural resources of the area. Management costs are supported by entrance fees charged to tourists in the conservation area (US$13/day).

These examples show that NGOs, supported by both domestic and international funding, have played important roles in conserving biodiversity in various parts of the tropics. They often provide an extremely useful supplement to government-organized initiatives.

INTERNATIONAL COOPERATION TO PROVIDE FINANCIAL SUPPORT TO BIODIVERSITY IN THE TROPICS

It is widely appreciated that insufficient funds are being invested to conserve biodiversity and that innovative approaches are required for generating the additional financial support required for implementing the CBD (Li 1995; Newcombe 1995; World Resources Institute 1989). The need for additional resources arises from the imbalance between a country's need for capacity-building and provision of basic infrastructure for conserving biodiversity and the country's ability to mobilize resources. Resources can be augmented through existing mechanisms, such as the fiscal system, user charges, resource rental fees, and privatization, as well as through such new mechanisms as environmental taxes and betterment charges. Even so, it appears that domestic resources in most developing countries will continue to be inadequate for financing the conservation of biodiversity, because of the limited tax and capital base of many of these countries, their underdeveloped taxation systems and weak capital markets, and the need to divert resources to servicing foreign debt. The reasons external financial resources are needed to conserve biodiversity are listed in figure 1.

In this section of this paper, I briefly surveyed promising innovations in financing the conservation of biodiversity and described each financial tool and the policies, technologies, and entrepreneurial initiatives that make the tool successful. I estimated the importance of each tool, described limits to its wider use, and identified actions that could enhance that tool's leverage. My emphasis was on innovative tools that are relatively poorly known.

Why external financial resources are needed to conserve biodiversity:

- *Equity.* Many of the benefits of biodiversity flow to all citizens of the world, but the costs tend to fall to the countries that have only limited financial resources.
- *Capital constraints.* Because at least some developing countries have insufficient resources, external financing is needed to bridge the gap between the demand (both private and public) for the conservation of biodiversity and the domestic supply of funding to support that conservation.
- *Cash flow.* Although many investments in conserving biodiversity will provide substantial benefits, the full benefits may not be realized for many years, however the costs need to be paid today, necessitating long-term bridge financing, which is difficult to obtain in developing countries.
- *Supporting policy reform.* Financing often is required to cushion the short-term effects of policy reforms required to move toward sustainable use of biological resources, to compensate those adversely affected by the new policies, or to build consensus for the reforms.
- *Covering foreign-exchange components.* Many investments in biodiversity may involve foreign-exchange components to build the confidence of investors and to leverage domestic sources of financing. Generating foreign exchange by exploiting biological resources may be contrary to the objectives of the Convention on Biological Diversity; external investment may reduce the need for such exploitation.
- *Benefits.* Conservation services are provided to the global community by developing countries; and financial support can help poor countries or avoid irreversible losses of biodiversity that may be highly valued after those countries become more wealthy.

FIGURE 1 Reasons external financial resources are needed to conserve biodiversity.

This discussion seeks to help the widest range of investors who could and should have a hand in crafting and using these financial tools. They include the full spectrum of those both active and potentially active in the conservation of biodiversity: the international governing system; national governments; the private sector, both national and multinational; and NGOs, both local and international. Table 3 gives an overview of the characteristics of various funding mechanisms (McNeely and Weatherly 1996; Panayotou 1995).

CONCLUSIONS

With the global economy now dependent on the reliable flow of biological resources from all parts of the world, international cooperation is essential for ensuring that biological resources are used in a sustainable way that leads to the conservation of biological diversity. Such cooperation can produce many benefits, but these depend, above all, on adequate investments in the field of biodiversity. The wealthy industrialized countries have recognized that they can benefit from biological resources that are found in developing countries, whose economic conditions do not enable them to invest adequately in conserving biodiversity. The developing countries are showing a remarkable capacity for innovation, as the examples from local communities and NGOs have shown, but they need funding. The Convention on Biological Diversity is one means of determining the kinds of activities that are most suitable in which to invest. Clearly, international cooperation in implementing the Convention on Biological Diversity will lead to increased support of the developing countries whose own efforts at conservation are helping to make the world a better place for all people to live in.

REFERENCES

Alcorn JB. 1997. Indigenous peoples and protected areas. In: Borrini-Feyerabend G (ed). Beyond fences: seeking social sustainability in conservation. Gland Switzerland: IUCN. p 44–9.

Barber CV, Afiff S, Purnomo A. 1995. Tiger by the tail? reorienting biodiversity conservation and development in Indonesia. Washington DC: World Resources Inst; Jakarta Indonesia: WALHI and Pelangi Inst.

Barber CV, Nababan A. 1997. Eye of the tiger: conservation policy and politics on Sumatra's last rainforest frontier. Jakarta Indonesia: World Resources Inst and WWF-Indonesia Prog.

Barzetti V (ed). 1993. Parks and progress: protected areas and economic development in Latin America and the Caribbean. Washington DC: IUCN and Inter-American Development Bank.

Birckhead J, de Lacy T, Smith L (eds). 1992. Aboriginal involvement in parks and protected areas. Canberra Australia: Austral Inst Aboriginal and Torres Strait Islander Stud.

Bodmer R, Penn J, Fang TG, Moya I. 1991. Management programmes and protected areas: the case of the Reserva Comunal Tamshiyacu-Tahuayo, Peru. Parks 1(l):21–5.

Borrini-Feyerabend G. 1996. Collaborative management of protected areas: tailoring the approach to the context. Gland Switzerland: IUCN.

Borrini-Feyerabend G, Brown M. 1997. Social actors and stakeholders. In: Borrini-Feyerabend G (ed). Beyond fences: seeking social sustainability in conservation. Gland Switzerland: IUCN. p 3–7.

Eaton P. 1985. Tenure and taboo: customary rights and conservation in the South Pacific. In: Third South Pacific national parks and reserves conference: report, vol II. South Pacific Regional Environment Programme. New Caledonia: Noumea. p 164–75.

Elliott J, Khan A, Saad Z. 1993. Developing partnerships: a study on NGO-donor linkages in Kerinci-Seblat and Lore Lindu National Parks. Jakarta Indonesia: PACT.

TABLE 3   Advantages and Disadvantages of Various Funding Mechanisms for Biodiversity

Funding Mechanism	Advantages	Disadvantages
I. The International Governing System		
Charging for use of the global commons	• Potential for vast amounts of funds • User pays	• Requires international agreement; difficult to attain • Needs new institutions to manage funds
Joint implementation	• Large amounts of funds primarily for forest biodiversity • Links biodiversity with climate change	• Requires unprecedented levels of coordination • Tacitly accepts continued high consumption of fossil fuels in North • Funds available only for direct forest management
International taxation	• Potential for vast amounts of funds • Can influence policies to be more supportive of biodiversity	• May not be World Trade Organization-compatible; requires political will • Funds may be diverted to purposes unrelated to biodiversity
Funds from trade in tropical timber	• Could raise US$1.5 billion per year with no effect on final product prices • Provides incentives for improved forest management	• Consumer countries forgo important tax revenues • Needs internationally agreed monitoring and enforcement
II. Governments		
Taxes and charges	• Can generate substantial funds with existing structures • Can build on "polluter-pays" and "beneficiary-pays" principles • "Green" taxes can change consumer behavior in favor of biodiversity without increasing total tax burden	• Many governments resist hypothecated taxation • Taxpayer resistance • Biodiversity-rich areas are often distant from sources of funding
Tradable permits	• Can generate billions of dollars of funding • Can change behavior affecting biodiversity • Specifies opportunity costs and provides mechanism for beneficiaries to pay them	• Administratively demanding • Behavioral changes might last only as long as payments continue • Difficult to translate to international level
Privatization and property rights	• Property rights give responsibility to people living closest to the resources • Assigning shares of privatized state corporations to conservation endowments helps retain public accountability	• Government monitoring of resource management in remote areas is difficult • Why use for biodiversity instead of for other needs? • Privatizing can destroy effective community-based management systems
Debt-related measures	• Can generate funds in national currencies and slightly reduce debt burdens	• Some resentment of "conditionality"

continues

TABLE 3 Continued

Funding Mechanism	Advantages	Disadvantages
III. The Private Sector		
Transfer of development rights and credits	• Involves private sector in joint-implementation measures that may benefit biodiversity	• Biodiversity benefits are a side issue
Prospecting rights and biological royalties	• Significant funds could be generated by discoveries of new drugs or other substances from nature • Utility of biological resources can be increased, thereby providing incentives for conservation	• Needs effective international agreements on intellectual-property rights and royalties • Long lead time • Difficult for royalty income to reach field level • Bureaucratic complications may lead to overregulation, which stifles innovation and exploration
"Green" investments	• Private sector invests in biodiversity as result of enlightened self-interest • Funds generated regularly from sales	• Weak capacity in some countries to regulate private sector • Requires appropriate incentives from government
IV. NGOs		
"Debt-for-nature" swaps	• Generates significant funds in national currency • Can be used to endow trust funds for long-term investment	• Discounted debts now less available • Can be inflationary
Targeted fund-raising	• Allows public willingness to pay to be tapped in support of biodiversity • Can build strong alliance between NGOs, public sector, and private sector	• Requires significant investment in fund-raising • Needs sympathetic government regulations, such as tax deductions

Source: McNeely and Weatherly 1996; Panayotou 1995.

Glowka L, Burhenne-Guilmin F, Synge H. 1994. A guide to the convention on biological diversity. Env Pol Law Paper No 30. Gland Switzerland: IUCN.

Homer-Dixon TF. 1994. Environmental scarcities and violent conflict: evidence from cases. Intl Security 19(1):5–40.

Kemf E (ed). 1993. Protecting indigenous people in protected areas. San Francisco CA: Sierra Club.

Kothari A, Singh N, Suri S (eds). 1996. People and protected areas: towards participatory conservation in India. New Delhi India: Sage.

Li S. 1995. Sources of funding for the Convention on Biological Diversity. In: McNeely JA (ed.) Biodiversity conservation in the Asia and Pacific region. Asian Development Bank, Manila, Philippines. p. 304–19.

McNeely JA, Miller KR, Reid WV, Mittermeier RA, Werner TB. 1990. Conserving the world's biological diversity. Gland Switzerland: IUCN and Washington DC: WRI, CI, WWF, the World Bank.

McNeely JA, Harrison J, Dingwall P (eds). 1994. Protecting nature: regional reviews of protected areas. Gland Switzerland: IUCN.

McNeely JA, Weatherly WP. 1996. Innovative funding to support biodiversity conservation. Intl J Soc Econ 23(4–6):98–124.

Newcombe K. 1995. Financing innovation and instruments: contribution of the investment portfolio of the pilot phase of the Global Environment Facility. In: McNeely JA (ed). Biodiversity conservation in the Asia and Pacific region. Manila Philippines: Asian Dev Bank. p 320–41.

OAS/NPS [Organization of American States/USDOI National Park Service]. 1988. Inventory of Caribbean marine and coastal protected areas. Washington DC: OAS/NPS.

Panayotou T. 1995. Matrix of financial instruments and policy options: a new approach to financing sustainable development. Paper presented to Second Expert Group Meeting on Financial Issues of Agenda 21, Glen Cove NY, 15–17 February 1995.

Redford KH, Ostria M. 1995. Parks in peril sourcebook. Arlington VA: The Nature Conservancy.

Rijksen HD, Griffiths M. 1995. Leuser development programme master plan. Amsterdam Netherlands: AIDEnvironment and IBN-DLO.

Southgate D. 1996. What roles can ecotourism, non-timber extraction, genetic prospecting, and sustainable timber production play in an integrated strategy for habitat conservation and local development? Washington DC: Inter-Amer Dev Bank.

Spiller G. 1997. Community-based coastal resources management in Indonesia. Sea Wind 11(2):13–9.

Stone RD. 1991. Wildlands and human needs: reports from the field. Washington DC: World Wildlife Fund.

UNEP [United Nations Environment Programme]. 1988. People, parks, and wildlife: guidelines for public participation in wildlife conservation. Nairobi Kenya: UNEP.

Vandergeest P. 1996. Property rights in protected areas: obstacles to community involvement as a solution in Thailand. Env Conserv 23(3):259–68.

Wells M, Brandon K. 1992. People and parks: linking protected areas management with local communities. Washinton DC: World Bank, WWF, and USAID.

Wilson EO, Peter FM (eds). 1988. Biodiversity. Washington DC: National Acad Pr.

West PC, Brechin SR (eds). 1991. Resident peoples and national parks: social dilemmas and strategies in international conservation. Tucson AZ: Univ Arizona Pr.

Western D, Wright RM, Strum S (eds). 1994. Natural connections: perspectives in community-based conservation. Washington DC: Island Pr.

World Resources Institute. 1989. Natural endowments: financing resource conservation for development. Washington DC: World Resources Inst.

BIODIVERSITY AND SUSTAINABLE HUMAN DEVELOPMENT: THE COSTA RICAN AGENDA

RODRIGO GÁMEZ
SANDRA RODRÍGUEZ
ANA ELENA VALDÉS
Instituto Nacional de Biodiversidad,
Apartado 22, Santo Domingo de Heredia, Costa Rica 3100

BIODIVERSITY, particularly tropical biodiversity, has been the focus of public attention in recent years, and much has been written about the compelling arguments that support ensuring its conservation into perpetuity. It is clear that biodiversity must be conserved for ethical, aesthetic, spiritual, and economic reasons. Numerous national and international agreements address these issues. The specific need to harmonize conservation with the socioeconomic development of populations is addressed by the Convention on Biological Diversity.

It is then necessary to face the challenge of implementing biodiversity conservation at a country level and in its social, political, cultural, and economic contexts. Regrettably, there are very few examples of how countries can attain the desired balance between conservation and development.

This paper describes the historical and continuing efforts of Costa Rica in its quest for a model of development that simultaneously allows the conservation of its biological patrimony and satisfies the basic requirements of its population. These efforts are now an integral part of the emerging "sustainable human development" initiative and paradigm, which are expected to guide the country into the next century as it faces the challenges of today's changing world.

Costa Rica is a biologically rich but economically poor, small, developing tropical country that has consolidated as a democracy in the absence of an army and has given high priority to investment in health, education, and welfare. In spite of numerous financial and organizational limitations, which are typical of developing countries, Costa Rica is making substantial advances in integrating biodiversity values into the mainstream of its development.

THE ROOTS OF THE QUEST FOR A SUSTAINABLE
MODEL OF DEVELOPMENT IN COSTA RICA

The relationship between humanity and nature constitutes an issue of growing concern in Costa Rica. The concern stems from the features that have characterized the country's course of development and from the prevailing values and paradigms that are expected to guide Costa Rican society's development in the future.

The aspirations of Costa Rican society are well interpreted in the following paragraph (Arias 1989):

> When we work for development, we are seeking an austere and fair life style. We want a society where everybody can satisfy at least his/her basic needs. We do not aspire to a model of development above our possibilities, nor to a society of welfare for a few and of suffering for many. We are neither a part of the armament race, nor a part of an uncontrolled race of economic growth at any cost, that threatens the environment or subdues our people to pressures that weaken our social convenience. We are looking for peace based on the absence of misery, for a democracy more and more participatory and for access to the welfare education provides.

Costa Rica's quest for sustainable biodiversity development is part of a broader initiative of national sustainable human development. In the last 50 years, Costa Rica has followed a development path unique in its region, characterized by a stable political system based on a disarmed democratic government, high economic growth rates (table 1), and substantial advances in social indicators. The product of a sustainable social policy, this process has resulted in high life expectancy and low levels of illiteracy. The proportion of low-income homes was reduced by more than half in 36 years (from 1960 to 1996), and infant mortality to less than one-fourth of what it was in 1960; the human population more than doubled in the same period (MIDEPLAN 1997; Proyecto Estado de la Nación 1994).

The country has naturally made errors. One is that Costa Rica's development model has been based to a great extent on nonsustainable use of natural resources, which has caused the rapid depletion of a substantial portion of the country's

TABLE 1   Costa Rica's Evolution Indicators, 1960–1996

Indicator	Unit	1960	1996
Human development indicator	Coefficient	0.55	0.88
Population	1,000	1,199.00	3,202.44
Low-income homes	%	50.00	21.60
Life expectancy at birth	years	62.50	75.60
Infant mortality	1,000	68.00	12.90
Gross national product, per capita	1990 US$	1,080.00	2,222.00
Primary forest cover	%	56.30	22.30

Source: 1960 data modified from Proyecto Estado de la Nación 1994; 1996 data modified from Ministerio de Planificación Nacional y Política Económica 1997.

forest cover. Between 1950 and 1970, Costa Rica lost one-third of its primary forest cover (Hartshorn and others 1982).

The agriculture sector is and has been in the last 50 years one of the main engines of economic development and a generator of the gross national product. However, agricultural development has been based on subsidies and incentives to increase production without agroecologic limitations and considerations and has resulted to a great extent in degradation of land and loss of forest (Fournier 1991; Gámez 1989; Hartshorn and others 1982).

The urgent need to address the environmental problems became increasingly evident, particularly in the first half of this century. Between 1960 and 1980, the country witnessed the strong emergence of a conservation movement; public, academic, and private sectors gradually became involved in different types of efforts and initiatives to address specific aspects of this crisis (Fournier 1991; Gámez and Ugalde 1988; Hartshorn and others 1982). This coincided with a growing national and international interest in the country's natural history and in preserving its biodiversity (Gómez and Savage 1982). It is notable that it was in the old National School of Agriculture (later, the Agriculture College of the University of Costa Rica) that the roots of Costa Rica's environmental thinking emerged between 1940 and 1950 (Fournier 1991).

The National Forest Service, the National Park Service, and the Wildlife Service were formally established between 1970 and 1980 under the Ministry of Agriculture and provided an adequate legal framework that enabled the creation and management of national parks and other categories of protected areas. The successful development of the National Park Service and the rapid consolidation of the protected areas in the country's institutional framework are historic landmarks in Costa Rica's quest for a harmonious relationship with nature (Gámez and Ugalde 1988; Gómez and Savage 1982).

Although the following years witnessed a substantial increase in the size and number of protected areas and deforestation rates began to decline, the numerous environmental problems steadily intensified, as well as the organizational and financial problems of the government's environmental agencies, and with them public awareness of their implications increased. Costa Rica needed a comprehensive and integrated natural-resource management program.

The first formal effort to address the need for a congruent national policy for natural-resource management appeared in 1974 in the First National Congress of Renewable Natural Resources (Fournier 1991; Universidad de Costa Rica 1974). However, it was not until 1977 that the need for a new development model that would "achieve a greater level of well being for a greater number of Costa Ricans" was formally addressed at the highest political levels in Costa Rica, during the symposium "The Costa Rica of the Year 2000", convened by the Ministerio de Planificación Nacional y Política Económica (the Ministry of Planning), or MIDEPLAN. Environmental concerns were included as an inherent component of the country's well-being (Ministerio de Cultura Juventud y Deportes y Oficina de Planificación Nacional y Política Económica 1977).

In 1986, the Ministerio de Recursos Naturales, Energía y Minas (Ministry of Natural Resources, Energy, and Mines) or MIRENEM, was established, and the national parks, forest, and wildlife services were transferred to the new ministry, signaling the top political priority assigned to these activities. MIRENEM rapidly consolidated, meeting the urgent need for a political environmental authority in the country that would integrate conservation efforts and define environmental policy.

Between 1987 and 1989, MIRENEM initiated the first formal national process to formulate a conservation strategy for Costa Rica's sustainable development. The process provided an opportunity to analyze the environmental issues in the broader context of the country's social and economic development. For the first time, biodiversity emerged as a key issue in the new view of sustainable development (MIRENEM 1989).

In 1987, a Biodiversity Office was created in MIRENEM to define "a new strategy and conservation program for Costa Rica's wildlands" (Gámez and others 1993). The new strategy and program began to emerge from a highly participatory analysis that capitalized on the country's environmental concerns and accomplishments and on the nearly two decades of conservation experience.

The historical process summarized here led to important changes in biodiversity-conservation thinking, policy, and accomplishments. To reinforce its role in environmental issues, the ministry assumed new responsibilities; its name was changed to Ministerio del Ambiente y Energía (Ministry of Environment and Energy), or MINAE, and the Sistema Nacional de Areas de Conservación (the National System of Conservation Areas), or SINAC, was legally consolidated (Ley Orgánica de Ambiente 1995).

Although this paper focuses on the roles of SINAC and the Instituto Nacional de Biodiversidad (the National Institute of Biodiversity), or INBio, in biodiversity conservation, the contribution of private organizations has been decisive in strengthening Costa Rica's environmental movement. Historically, nongovernment organizations like Fundación de Parques Nacionales, Fundación Neotrópica, the Tropical Science Center, and the Organization for Tropical Studies have played leading roles in supporting the ministry's biodiversity-conservation efforts. Fundación de Parques Nacionales has become a financing entity that has in diverse ways enabled the resources required to buy and administer lands for many conservation areas.

THE NEW STRATEGY AND ORGANIZATION FOR SUSTAINABLE BIODIVERSITY DEVELOPMENT

The new Costa Rican strategy is based on the premise that the best way to conserve biodiversity is to use it sustainably to promote the country's intellectual, spiritual, social, and economic development. The implementation of this strategy requires three overlapping tasks: save large wildlands, determine what biodiversity is found in these wildlands, and use the biodiversity in a sustainable manner (Gámez 1991; Gámez and others 1993; Janzen 1992).

The fulfillment of the first task demanded the reorganization of existing institutions and the emergence of new ones. The national park, forest, and wildlife services evolved into SINAC (MIRENEM 1992, 1994). The country was divided into 11 geographic sectors that were called conservation areas (SINAC 1997). The definition of the conservation areas represents a first approach to managing entire bioregions or ecosystems at the national level comprising three categories of land use described below. A positive result of the new policies of rational use of the natural resources is the decrease in the average deforestation rate from 40,000 ha/year in 1986 to 8,000 ha/year in 1994 and an increase in the dense forest cover, which until 1984 had been decreasing continuously (MIDEPLAN 1997).

The country is viewed as divided into three major categories of land use: wildlands conserved for their biodiversity, the agro-pastoral-forestry landscape, and the urban landscape. The three categories of land use are expected to provide different types of equally valuable goods and services and to coexist in harmony so that the activities of one do not harm the others (Presidencia de la República 1994).

Costa Rica is saving representative samples of the species and ecosystems present in the country through a system of protected wildlands within the conservation areas. Nearly 24% of the country is protected under different categories of management, and 11.8% is national parks and reserves (579,412 ha). The remaining areas are forest reserves and wildlife refuges under private ownership (García 1996). There are nearly 170 private reserves, representing a notable contribution by the private sector to biodiversity conservation and a clear indication of the increasing understanding of wild biodiversity's social and economic values (MIDEPLAN 1997).

The choice of the particular wildlands to protect has, with notable recent exceptions, been based not on scientific ecological considerations, but mostly on a complex combination of economic and political opportunities (García 1996; Janzen 1992). What has been saved probably includes all that could be protected at the time. Most of but not all the species and ecosystems in the country are thus protected. A strategy and action plan that aims to resolve the problem that some species and ecosystems are unprotected (technical proposals for territorial ordering aimed at the conservation of biodiversity known as Proyecto GRUAS) was recently formulated (García 1996).

According to the existing legislation (SINAC 1997 Asamblea Legislativa), SINAC is a decentralized administrative system in which each conservation area groups and manages state-owned protected wildlands and is responsible for the management of forests and wildlife in private wildlands.

The governance of the conservation areas is under reorganization. Community participation is expected to be incorporated in different ways at the local, regional, and national levels. External national and international conservation authorities are being named to serve as advisers to conservation areas (Asamblea Legislativa).

The fulfillment of the second and third tasks, knowing the biodiversity in the wildlands and using the knowledge to promote its nondestructive use, required the creation of a new organization, Instituto Nacional de Biodiversidad, or INBio. A major factor in INBio's creation was the urgent need for an organization solely responsible for conducting a national biodiversity inventory of the protected wildlands, centralizing the resulting information, and promoting sustainable use (Gámez 1991; Gámez and others 1993). The value and importance of inventory activities conducted in Costa Rica for over a century by national and international scientific organizations and individuals were recognized. Nevertheless, this approach presented the practical problem of scattered biological specimens and information, as well as diverse and discontinuous inventory approaches, which together made the integration and management of the information difficult. For strategic reasons, INBio was established as a private, public-interest, nonprofit organization. Conceptually, INBio offers an innovative form of direct participation of the civil society in biodiversity conservation and management in direct collaboration and coordination with the government. SINAC and INBio work in close partnership in a strategic alliance supported by a periodically updated legal collaborative agreement that stipulates the rules and regulations that guide the partners' activities (Sittenfeld and Gámez 1993). Both organizations have assumed the leadership and responsibility for implementing the sustainable-biodiversity-development initiative described in this paper. Developing the biodiversity resource base is an ambitious and complex task that no institution can possibly attain by itself. It demands the establishment of strategic alliances and partnerships with widely different sectors of society, nationally and internationally, as an inherent component of any socially sustainable scheme. That means interactive work among, for example, the scientific-academic, economic, industrial, political, agricultural, educational, tourist, conservationist, mass-media communication, urban, and rural sectors. Partnerships and alliances for the ulterior purpose of biodiversity conservation into perpetuity demand, in many cases, drastically changed views, attitudes, and traditions that are ingrained in the core activities of many sectors of society, including the scientific and academic sectors.

The direct involvement of all sectors of society in the implementation of the sustainable-biodiversity-development initiative is of paramount strategic importance. All sectors must perceive and play a direct role and be actors rather than spectators. For example, entities that traditionally have concentrated the decision-making power must delegate authority and responsibility, and nongovernment organizations must share roles historically played by government agencies.

INBio's parataxonomists program is a good and successful example of the rural sector's direct involvement in a scientific activity previously considered almost exclusively pertinent to the scientific-academic sector. It has succeeded, among other reasons, because of the acceptance by the scientific sector of this mutually beneficial partnership (Janzen 1992; Reid and others 1993).

The sustainability paradigm must also be applied to the institutions responsible for conducting the sustainable-biodiversity-development process, which demands strong and viable organizations that are capable of dealing with complex problems. That means implementing organizational development schemes that define and follow the institutional mission, guide strategic planning and reengineering processes, and seek financial security.

Both SINAC and INBio are addressing those issues, SINAC in the context of a government entity when the government is down-sizing and budgets are being drastically reduced, and INBio as a nongovernment organization dependent entirely on its own capabilities to conceive and implement initiatives and raise and generate necessary funds.

Since its inception in 1987, SINAC's institutional organization and wildland management have been undergoing a dynamic process of change. SINAC is still far from consolidation and is required to face the challenge of responding to the new conceptual premise of biodiversity conservation through its sustainable use. As a government organization, SINAC has been affected by the problems of inefficiency and inefficacy in public administration common in developing countries. In spite of these internal difficulties and the country's growing environmental problems, SINAC has made substantial advances toward the implementation of the new philosophy and organization. These include a more congruent perspective on and criteria for natural-resource management and conservation, decentralization, and the staff's growing perception of SINAC's role as a public-service organization (SINAC 1997).

SINAC seeks to achieve the final delimitation of the national territory protected for its biodiversity and the integration of the Costa Rican system of wildlands as part of the initiative of a Mesoamerican Biological Corridor (García 1996; Proyecto Corredor Biológico Mesoamericano Informe Técnico Regional, unpublished).

In the quest for a sustainable model of development, the government of Costa Rica established the Sistema Nacional de Desarrollo Sostenible (National System for Sustainable Development), or SINADES, which is headed by a Sustainable Development Council. SINADES integrates different sectors of society—including the government, industry, universities, and civil society—and serves as a forum for analysis and discussion. It also responds to the issues of Program 21, Convention on Climate Change, and the convention Biological Diversity that the country has signed and ratified. The Ministry of Planning functions as the executive secretariat of SINADES.

To address specific topics of sustainability, specialized consultative commissions were created. One of these, the Comisión Asesora de Biodiversidad (the Biodiversity Advisory Commission), or COABIO, was responsible for policy and planning issues related to the implementation of the Convention on Biological Diversity at the national level (Gámez and Obando 1995). COABIO was composed of 13 experts on the different topics of the convention. It also serves as the technical-scientific advisory body to the government in all international activities of the Conference of the Parties of the Convention on Biological Diversity. With the approval of the new biodiversity law in 1998, COABIO was replaced with the Comisión Nacional para la Gestión de la Biodiversidad (National Commission for Biodiversity Management) CONAGEBIO.

COABIO, in coordination with SINAC and INBio, assessed the state of knowledge of Costa Rican biodiversity and analyzed the accomplishments and failures of the initiatives historically conducted in conservation by the different national public and private organizations. This study will lead to a reformulation of the country's biodiversity strategy and action plan, highlighting gaps where actions are required or need to be corrected. COABIO also collaborated directly in formulating biodiversity legislation.

SUSTAINABLE BIODIVERSITY DEVELOPMENT IN PRACTICE: BRINGING THE POTENTIAL BENEFITS OF BIODIVERSITY IN THE WILDLANDS TO COSTA RICAN SOCIETY

The environmental services supplied by protected areas that were taken for granted for many decades, such as water production and fixation of carbon dioxide, are now starting to be valued and included in national accounts. These services have an important potential to increase the income generated as payments from industry and other sectors of society as the real costs are internalized.

Biodiversity as a source of information also has great potential. INBio was created in 1989 solely for the purpose of conserving Costa Rica's biodiversity into perpetuity. This pilot project responded to the needs to accelerate generation of knowledge of wildland biodiversity and to promote the use of this knowledge as a tool for the country's economic, social, and intellectual development (Gámez 1991, 1996; Gámez and Gauld 1993; Gámez and others 1993; INBio 1994, 1995, 1996, 1997; Janzen 1992; Sittenfeld and Lovejoy 1995).

One of the major steps toward making biodiversity accessible to society has been INBio's design and implementation of an innovative biodiversity-inventory method in collaboration with SINAC. This has produced a high-quality taxonomic reference collection of over 2 million specimens of Costa Rica's arthropods, plants, mollusks, and fungi. The effort has resulted in substantial increases in the knowledge of the country's biodiversity. INBio's inventory through collaborative efforts has described new species, new records of species described elsewhere in the tropics, and new distribution records for known species. The inventory is conducted by teams composed of parataxonomists in the field, technicians, curators, and national and international expert taxonomists. Specimens are collected by parataxonomists in biodiversity stations in the conservation areas, giving the inventory a wide geographic spatial coverage and continuity through time as field collection occurs continuously throughout the year.

The parataxonomists' participation has important social implications in that the inventory is in itself an educational experience and a vehicle for intellectual promotion for an important rural sector of the population (Gámez 1996; Janzen 1992). Rather than playing marginal roles, rural residents have become main actors in the scientific effort to know the biodiversity of the country. It also means building up local scientific capability and direct sources of knowledge in the conservation areas, which are then available to the schools of neighboring rural communities. Being a parataxonomist also brings in economic benefits to rural families. The economic benefits have multiplying effects in the rural communities, which rapidly perceive the benefits of the activities conducted in the protected wildlands.

The on-the-job training received by national expert curators with basic degrees in biology has enabled the institution to develop and consolidate an increased taxonomic capacity while conducting the inventory process. This has occurred under circumstances in which the country did not have the time or the financial resources available to build up a core group of taxonomic specialists with higher academic degrees, as would normally be expected in a rich industrial country (Janzen 1992).

The success of INBio's inventory method has been made possible by the active and permanent collaboration of an increased number of expert taxonomists from North America and Europe and their institutions. For the international collaborator, this mutually beneficial initiative translates into training parataxonomists and technicians, tutoring local curators, and identifying properly curated specimens, which are often sorted to the morpho-species level. It also represents an extremely efficient use of visiting taxonomists' time.

Conceptually, the inventory represents the first step in making biodiversity in the wildlands available for social and economic uses. It is a user-oriented inventory guided by objectives not always compatible with interests in scientific academic sectors (Gámez 1996; Janzen 1992), as illustrated in recent publications (Gámez and others 1997; Kaiser 1997).

Biodiversity-information management is the core of INBio. In the inventory process, field samples are accompanied by basic data indicating where, when, by whom, and how the specimens were collected. The raw data are enhanced via a process involving the use of scientific and technological know-how in chemistry, taxonomy, geography, information management, and other fields. The information generated is provided in an appropriate format to economic and intellectual users (GIS, multimedia field guides, books, lectures, tours, and so on); at the same time, it constitutes feedback for the process of generating more information (INBio 1997).

Biodiversity prospecting appears in profile as one of the industrial goals for the 21st century, and biodiversity-rich tropical developing countries, such as Costa Rica, have a unique opportunity to lead the process (Mateo 1996; Sittenfeld and Lovejoy 1995). Even before the emergence of the Convention on Biological Diversity, INBio's policies recognized the need to establish collaborative research agreements and mutually beneficial partnerships with industry in the developed world, as stated by Eisner (1989). Those policies set guidelines for working with commercial partners under mutually agreed-on terms that recognized the country's ownership of the materials, the need for technology transfer and scientific capacity-building, the equitable sharing of benefits derived from the commercialization of products, and the strategic need to contribute from the beginning to SINAC's conservation activities (Janzen and others 1993; Sittenfeld and Gámez 1993).

The above considerations have emerged from internal analysis and discussions with the government, political, and private sectors and in accordance with prevailing advanced thinking and existing legislation. For those reasons, INBio's initiatives have been supported by four consecutive government administrations since 1989. However, INBio underestimated the difficulties of communicating the complex nature of collaborative research agreements in bioprospecting to the general public. That and ideological factors account to a large extent for the concerns that emerged among some national and international groups after the pioneer agreement with Merck and Company in 1991.

In addition to Merck's agreement (Reid and others 1993; Sittenfeld 1995), INBio has entered several collaborative research agreements with corporations in the pharmaceutical, cosmetic, biotechnological, and agricultural sectors (Mateo 1996; Sittenfeld and Artuso 1995). As a result, the organization has developed substantial knowledge and expertise in the complex array of subjects involved in bioprospecting, including legislation, terms of agreement, business negotiation, science and technology, and information required from inventories.

On the basis of the experience and know-how gained by the organization, INBio will need to increase its scientific and technological capacities substantially to change from a reliable partner capable of providing a wide variety of extracts from diverse organisms to a partner capable of providing chemically defined molecules with known biological activities determined through bioassays. This possibility appears more likely as the institution enters innovative types of partnerships with national universities and other organizations that have stronger scientific and technological capacities.

The future need to focus on problems of national relevance in agriculture, health, and industry is also clear to INBio. These initiatives might not be perceived as having high priority from a financial point of view, but they are certainly important from a local social perspective. INBio's experiences in bioprospecting have served to formulate national policy and legislation, as exemplified by Costa Rica's new biodiversity legislation. This knowledge has also been shared with others in African and Latin American countries through technical workshops.

The experiences of the conservation areas, INBio, and other organizations, point to four major categories of social and intellectual users and uses: ecotourism, management of wildlands, political decision-making, and education.

The tourist boom experienced by Costa Rica is closely linked to the attraction to its natural beauty and protected areas. Tourism is the country's main source of foreign income—greater than coffee, cattle, and banana production (MIDEPLAN 1997). Current and future trends highlight the need to increase the competitiveness of the country in ecotourism by adding substantially more information value to the activity. Such value should be reflected in the information made available to visitors through guided tours, field guides, CD ROMS, and other forms of interactive presentations and learning experiences. Parks and reserves should logically be equally prepared to deal with increased national and international visits. However, with the benefits of nature-oriented visits come environmental threats. Costa Rica needs to improve its policies and regulations, particularly those related to wildland visitation. To deal properly with those issues, area managers need suitable information and the institutional capacity to introduce it in their management plans. INBio's inventory information emerges as the logical source of information for the conservation areas.

Information for political decision-makers, for both national and local governments, has high priority in Costa Rica. If biodiversity needs to rank high in political agendas, politicians need not only be more educated on the subject, but also have adequate information readily available for sound policy-making. This information should also be available to public constituencies as a whole.

The SINAC-INBio partnership is addressing and beginning to implement initiatives congruent with the preceding notions. As stated in the introductory section of this paper, education in Costa Rica has historically had top national priority and been a major factor in the country's particular course of development. The solution to the complex problems associated with the conservation of biodiversity into perpetuity and its sound use in the context of the sustainable-human-development initiative depends heavily on a bioliterate population. "Bioliteracy" is evolving as the leading idea in INBio's emerging educational activities, now enthusiastically endorsed by Costa Rica's Ministerio de Educación Publica (Ministry of Public Education, or MEP).

Bioliteracy is defined as an experiential process that guides a person to understand biodiversity and to adopt a principle of respect for life in all forms. This basic understanding fosters changes in behavior that enable harmonious relations with nature to achieve sustainable human development (INBio 1996). Bioliteracy is equated to *literacy* in its conventional meaning and so must be part of the basic educational process that allows a person to read and write, add and subtract, and, in this case, learn the basics of nature's language. The bioliteracy initiative seeks the consolidation of moral values and the development of new attitudes toward nature, in the sense formulated by Wilson (1992). The development of the proper method for inculcating bioliteracy is part of a pilot project conducted jointly by INBio and two rural public schools under MEP's supervision.

The experimental activities include workshops, field trips, and interactive learning with computer aids. INBio is building on several national experiences, such as the National Computers in Education Program, a joint initiative of the Omar Dengo Foundation and the Ministry of Education; the Biological Education Program of the Guanacaste Conservation Area; and the methodological know-how of the parataxonomists program.

This new environmental ethic is a fundamental pillar of a sustainable society and a sustainable world. Bioliteracy addresses the complex problems of valuation of biodiversity and its key role in sustainability.

What Costa Rica has done in its quest for a sustainable-human-development scheme is due largely to the prolonged investment in peace and development of human capabilities. The integration of the environment variable, represented by its rich biodiversity, is congruent with the country's values and expectations. Intellectual and economic international collaboration has been fundamental in complementing the country's efforts in its quest. Costa Rica's experience constitutes a pilot project for many other tropical developing countries. Furthermore, the Central American region can benefit from the information and knowledge generated by the Costa Rican experience. It might also be a viable example of compliance with the terms of the Convention on Biological Diversity.

REFERENCES

Arias O. 1989. In: Memoria primer Congreso Estrategia de Conservación para el desarrollo sostenible de Costa Rica. San José Costa Rica: Ministerio de Recursos Naturales, Energía y Minas. p 22.

Asamblea Legislativa 1998, Ley de biodiversidad 7788.

Eisner T. 1989. Prospecting for nature's chemical riches. Iss Sci Tech 6(2):31–4

Fournier LA. 1991. Desarrollo y perspectiva del movimiento conservacionista Costarricense, San José, Costa Rica: Editorial de la Universidad de Costa Rica. 113 p.

Gámez R. 1989. Threatened habitats and germplasm preservation: a Central American perspective. In: Knutson L, Stoner AK (eds). Beltsville Symposia in Agricultural Research 13 Biotic Diversity and Germplasm Preservation, Global Imperatives. The Netherlands: Kluwer. p 477–92.

Gámez R. 1991. Biodiversity conservation through facilitation of its sustainable use: Costa Rica's National Biodiversity Institute. Trends Ecol Evol 6: 377–8.

Gámez R. 1993. Wild biodiversity as a resource for intellectual and economic development: INBio's pilot project in Costa Rica. In: Sandlund OT, Schei PJ (eds). Proceedings of the Norway/UNEP Expert Conference on Biodiversity. Oslo Norway: Directorate for Nature Management/Norwegian Institute for Nature Research. p 141–51.

Gámez R. 1996. Inventories: preparing biodiversity for non-damaging use. In: di Castri F, Younés T (eds). Biodiversity, science, and development: towards a new partnership. Wallingford UK: CAB International. p 180–3.

Gámez R, Ugalde A. 1988. Costa Rica's national park system and the preservation of biological diversity: linking conservation with socioeconomic development. In: Almeda F, Pringle CM (eds). Tropical rainforests: diversity and conservation. San Francisco CA: Academy of Sciences and AAAS Pacific Division. p 131–42.

Gámez R, Piva A, Sittenfeld A, León E, Jiménez J, Mirabelli G. 1993. Costa Rica's conservation program and National Biodiversity Institute (INBio). In: Reid W, Sittenfeld A, Laird , Janzen DH, Meyer CA, Gollin MA, Gámez R, Juma C (eds). Biodiversity prospecting: using genetic resources for sustainable development. Washington DC: World Resources Inst. p 53–67.

Gámez R, Gauld ID. 1993. Costa Rica: an innovative approach to the study of tropical biodiversity. In: LaSalle J, Gauld I (eds). Hymenoptera and biodiversity. Wallingford UK: CAB International. p 329–36.

Gámez R, Obando V. 1995. Proyecto para la formulación de planes y estrategias de biodiversidad — formación de la Comisión Asesora en Biodiversidad (COABIO) informe final primera etapa. Unpublished

Gámez R, Janzen DH, Lovejoy T, Solórzano R. 1997. Costa Rican all-taxa survey. Science 277: 1–148.

García R. 1996 Propuesta técnica de ordenamiento territorial con fines de conservación de biodiversidad. Informe de país: Costa Rica proyecto corredor biológico Mesoamericano. Santo Domingo de Heredia Costa Rica: Ministerio del Ambiente y Energía. 114 p

Gómez, LD, Savage JM. 1982. Searchers on that rich coast: Costa Rican field biology. In: Janzen DH (ed). Costa Rican natural history. Chicago: Univ Chicago Pr. p 1–11.

Hartshorn GS, and others. 1982. Costa Rica country environmental profile: a field study. San José Costa Rica: Tropical Science Center.

INBio [Instituto Nacional de Biodiversidad]. 1994. Memoria anual 1993. Santo Domingo de Heredia, Costa Rica: Instituto Nacional de Biodiversidad.

INBio [Instituto Nacional de Biodiversidad]. 1995. Memoria anual 1994. Santo Domingo de Heredia, Costa Rica: Instituto Nacional de Biodiversidad.

INBio [Instituto Nacional de Biodiversidad]. 1996. Memoria anual 1995. Santo Domingo de Heredia, Costa Rica: Instituto Nacional de Biodiversidad.

INBio [Instituto Nacional de Biodiversidad]. 1997. Memoria anual 1996. Santo Domingo de Heredia, Costa Rica: Instituto Nacional de Biodiversidad.

Janzen DH 1992. A south-north perspective on science in the management, use, and economic development of biodiversity. In: Sandlund OT, Hindar K, Brown AHD (eds). Conservation of biodiversity for sustainable development. Oslo Norway: Scandinavian Univ Pr. p 27–54.

Janzen DH, Hallawachs W, Gámez R, Sittenfeld A, Jiménez J. 1993. Research management policies: permits for collecting and research in the tropics. In: Reid W, Sittenfeld A, Laird SA, Janzen DH, Meyer CA, Gollin MA, Gámez R, Juma C (eds). Biodiversity prospecting: using genetic resources for sustainable development. Washington DC: World Resources Inst. p 131–57.

Kaiser J. 1997. Unique all taxa survey in Costa Rica "self-destructs." Science 276: 865–996.

Ley Orgánica del Ambiente 7554. 1995. San José Costa Rica: La Gaceta 215.

Mateo N. 1996. Wild biodiversity: the last frontier? the case of Costa Rica in the globalization of science. In: Bonte-Freidheim C, Sheridan K (eds). The place of agricultural research. The Hague: International Service for National Agricultural Research. p 73–82.

Ministerio de Cultura, Juventud y Deportes y Oficina de Planificación Nacional y Política Económica. 1977. La Costa Rica del año 2000. San José Costa Rica: Impr Nacional. 711 p.

MIDEPLAN [Ministerio de Planificación Nacional y Política Económica]. 1997. Costa Rica panorama nacional 1996: balance anual, social, económico y ambiental. San José Costa Rica: MIDEPLAN. 284 p.

MIRENEM [Ministerio de Recursos Naturales, Energía y Minas]. 1989. Memoria 1er congreso estrategia de conservación para el desarrollo sostenible de Costa Rica. San José Costa Rica: Ministerio de Recursos Naturales Energía y Minas.

MIRENEM [Ministerio de Recursos Naturales, Energía y Minas]. 1992. Sistema nacional de áreas de conservación: un nuevo enfoque. San José Costa Rica: Ministerio de Recursos Naturales Energía y Minas.

MIRENEM [Ministerio de Recursos Naturales, Energía y Minas]. Servicio de Parques Nacionales. 1994. Estrategia global para el sistema nacional de áreas de conservación. San José Costa Rica: Ministerio de Recursos Naturales Energía y Minas.

Presidencia de la República. 1994. Del bosque a la sociedad: un nuevo modelo costarricense de desarrollo en alianza con la naturaleza. Janzen DH, Sancho E, Lovejoy A (tr). San José Costa Rica: EUNED. 236 p.

Proyecto Estado de la Nación. 1994. Reflexiones generales en torno al desarrollo humano sostenible. In: San José CR. Estado de la nación en desarrollo humano sostenible: un análisis amplio y objetivo sobre la Costa Rica que tenemos a partir de los indicadores más actuales. Impr Lara Segura. p 1–12.

Reid W, Sittenfeld A, Laird SA, Janzen DH, Meyer CA, Gollin MA, Gámez R, Juma C (eds). 1993. Biodiversity prospecting: using genetic resources for sustainable development. Washington DC: World Resources Inst. 341 p.

Sittenfeld A. 1995. INBio-Merck collaborative biodiversity research agreement. Costa Rica: partnerships in practice. London UK: Dept of the Environment.

Sittenfeld A, Gámez R. 1993. Biodiversity prospecting by INBio. In: Reid W, Sittenfeld A, Laird SA, Janzen DH, Meyer CA, Gollin MA, Gámez R, Juma C (eds). Biodiversity prospecting: using genetic resources for sustainable development. Washington DC: World Resources Inst. p 69–97.

Sittenfeld A, Artuso A. 1995. A framework for biodiversity prospecting: the INBio experience. Aridlands Newsletter, the University of Arizona Spring/Summer 37:8–11.

Sittenfeld A, Lovejoy A. 1995. INBio's biodiversity prospecting program: generating economic returns for biodiversity conservation. Final compendium for a practical workshop on biodiversity prospecting for Cameroon, Madagascar, and Ghana, INBio, 24 April–2 May. 15 p.

SINAC [Sistema Nacional de Areas de Conservación]. Unidad Técnica del Sistema Nacional de Areas de Conservación. 1997. El sistema nacional de áreas de conservación de Costa Rica concepto, funciones y avances en su implementación. San José Costa Rica. Ministerio de Ambiente y Energía.

Universidad de Costa Rica. 1974. Primer congreso nacional sobre conservación de recursos naturales renovables. San José Costa Rica: Facultad de Agronomía, Univ Costa Rica.

Wilson EO. 1992. The diversity of life. Cambridge MA: Belknap. 424 p.

THE NATIONAL BIODIVERSITY
INFORMATION SYSTEM OF MEXICO

JORGE SOBERÓN
PATRICIA KOLEFF
Comision Nacional Para el Conocimiento y Uso de la Biodiversidad,
Fernández Leal #43, Conabio, Barrio de la conchita, Coyoacan, Mexico CP 04020

INTRODUCTION

DESPITE ITS novelty and sheer scope, the concept of biodiversity has already been used as a basis of multilateral treaties, global funds, national strategies, and many other political and scientific initiatives. Foremost among the actions that countries are expected to undertake to preserve biodiversity is the creation of inventories (Reid and others 1992; SA2000 1994) and information systems (Olivieri and others 1995; United Nations Environmental Programme, UNEP; WCMC 1996; http://www.unchs.unon.org) to organize the huge body of biodiversity data already in existence. Without powerful informational tools, the task of protecting, managing, and using biodiversity at national levels is impossible. Many peasant and indigenous cultures manage their biological resources successfully with nonmodern information tools (Berlin and others 1973; Haverkort and Millar 1994), but the increase in the temporal, spatial, and taxonomic scales implied by the full concept of biodiversity leads us to the use of scientific, modern tools for the knowledge and use of biodiversity.

In March 1992, the Mexican government created a national commission, (Comision Nacional para el Conocimiento y Uso de la Biodiversidad [CONABIO] 1992, http://www.conabio.gob.mx), with the task of coordinating the national biodiversity inventory and the associated databases and information systems. In this presentation, we outline its conceptual framework and current status, focusing on the role of the users and providers of the data. Mexico's National Biodiversity Information System is not yet finished in a strict sense, but many of its components are already in operation and providing many services, which we describe here.

A BIODIVERSITY INFORMATION SYSTEM

An information system can be defined as a structured set of processes, personnel, hardware, and software to turn data into usable information (WCMC 1997). That definition forces us to focus on a number of issues. First, data are provided by channels usually under the control of people. In the biodiversity field, the people are the geographers, taxonomists, ecologists, geneticists, foresters, traditional physicians, wholesalers in natural products, and so on, that generate raw or aggregated data about any of the levels of biodiversity. A biodiversity information system (BIS) must have clearly defined and operative relations with the providers of the data.

Second, what "information" means is determined by a set of potential or actual users. For example, the list of the Latin binomials of medicinal plants in a given municipality (or region) for which a national market exists might be useless to inhabitants who lack scientific training, whereas the same data presented in the form of a guide with common regional names and illustrations might be highly informative to them. Similarly, to a national-level decision-maker, a map of endemic species richness might be much more useful than an equation fitted to the raw data, which might be packed with information for a macroecologist; and the raw data themselves could be useful to a taxonomist. The output of a BIS should be defined in close contact with the main users of the system.

Third, the technical specifications of a BIS are likely to be exacting and difficult. Because of its multiscale features, data will appear in a number of formats, from those of geographical information systems (GISs) to text files, images, taxonomic datasets, genomic information files, and so on. Also, some categories of data are large. For example, the Digital Elevation Model of Mexico at 1:250,000 is almost 500 megabytes (Mb). Some curatorial databases are also of significant size. Two examples of CONABIO illustrate this: a database with 403,507 records of specimens of vascular plants with 26 fields from 85 projects is 111 Mb and the bird database with 250,283 records of specimens and 164 fields is 88 Mb. The structure of a BIS should respond to the complexity and magnitude of the problem at hand.

Many BISs are already in operation (WCMC 1996). Among the better known are the Nature Conservancy Heritage Program (Jenkins 1988; http://www.tnc.org/), the World Conservation Monitoring Centre (WCMC, http://www.wcmc.org.uk/), the Australian government's Environmental Resources Information Network (ERIN, http://kaos.erin.gov.au/erin.html), and the Costa Rican Instituto Nacional de Biodiversidad (InBio) system (http://www.inbio.ac.cr/). One of the main criteria for characterizing a BIS is whether it is based on raw, "atomic data" or on interpreted information. For example, the set of locations where a species has been recorded is less interpreted than a researcher's rendering of the area of distribution of the species. BISs that are explicitly based on atomic data are the

Australian ERIN (Chapman and Busby 1994) and Costa Rica's InBio. The system being developed in CONABIO belongs to this class.

THE NATIONAL BIODIVERSITY INFORMATION SYSTEM OF MEXICO

The main task given to CONABIO by the presidential act that created it was to coordinate the inventory of Mexican biodiversity and to develop and maintain the information system for it. CONABIO started by holding workshops with potential users and providers of the information. Such consultations have been maintained, formally or informally, to the present. We also reviewed a number of existing BISs.

Most of the needs of users were related to variations on the symmetrical themes of "What entities are present here?" and "What exists and where?" In those two questions, "entities" and "what" can mean a species, a higher-level taxon, or an attribute of them, such as "mammals," "federally listed butterflies," "medicinal plants," or "migratory birds." "Where" and "here" refer to arbitrary polygons and regions naturally or politically defined, such as a given ecoregion, a state, a municipality, or a protected area. Those questions call for simple "distributional" information. For example,

- What endangered species exist in a state, municipality, protected area, or ecoregion?
- What are the holdings of particular taxa in particular museums or collections (generally speaking, curatorial data for taxonomic groups present in Mexico)?
- What tree species in a given ecoregion are present in one of the major army nurseries?
- What butterfly and bird species are present in a municipal natural park?
- In what region(s) can a given species produced in the army nurseries be planted?
- In what region(s) is there a high likelihood of the presence of some endangered (or otherwise defined) species?
- What municipalities cover a given ecoregion?

A second class of information is not distributional but is associated with particular taxa. Examples of this more complex, "verbal" information are

- information on production technology for particular species (mainly useful tree species);
- chemical or clinical data on medicinal plants;
- indigenous knowledge about species;
- toxicological and first-help data on poisonous species;
- markets (buyers, certifiers, and exporters) for useful species, such as nontimber forest products;
- demographic or genetic data on particular species or groups, such as whales, cacti, or other "charismatic" taxa;
- traffic data on species regulated under the Convention on International Trade in Endangered Species of Wild Fauna and Flora (CITES);

- images, pictures, illustrations, recordings, and multimedia data related to con-spicuous species; and
  - general information about protected areas, including images and tourist data.

A third class of information has a temporal component. It is related to trends in the sizes of regions or populations. Obvious examples are

- rates of change of particular types of vegetation into other types (such as in deforestation) and
  - changes in the sizes of populations of particular species.

The provenances, updating regimes, quality, and structure of the data required to fulfill the needs described above are highly heterogeneous. A BIS capable of responding to such a set of demands probably does not exist, although many ex-isting systems are quite capable of answering questions on some levels in the biodiversity scale. Some systems are based entirely on bibliographic information, such as the Indian Indira Gandhi Conservation Monitoring Centre (http://www.wcmc.org.uk/igcmc/) and Napralert at the University of Illinois (Farnsworth 1988); metadata systems, such as the National Biodiversity Information Infrastruc-ture of the Department of the Interior in the United States (http://www.nbs.gov/); mixed systems, such as The Nature Conservancy in the United States and several Latin American countries (Jenkins 1988; http://www.tnc.org/); systems of state scope, such as the Gap Analysis Program (GAP, http://www.gap.uidaho.edu/gap/); systems oriented wholly to taxonomic information, such as the Expert Center for Taxonomic Identification (ETI, http://turboguide.com/data2/cdprod1/doc/cdrom.frame/002/607.pub.Expert.Centre.for.Taxonomic.Identification.ETI.html) and the PLANTS National Database (http://plants.usda.gov/plants/); and at least one system with a world scope, WCMC (http://www.wcmc.org.uk/). A scheme of a hypothetical, ideal BIS capable of answering all types of biodiversity questions is depicted in figure 1.

The hypothetical scheme that appears in figure 1 has a realistic interpretation. The core element is the data associated with the *specimen*—the atomic data re-ferred to before. Particularly important from the perspective of a BIS are the *georeference* and *taxoreference* associated with a specimen (Colwell 1996). The georeference is the data that specify the locality where the specimen was collected. It is subject to a variety of errors and imprecision. Without resorting to modern geographic positioning system (GPS) technology, the georeference might be ob-tained to a precision of a few kilometers. GPS increases the resolution to a few hundred meters or better. The georeference expressed in coordinates provides the most flexible link to information that is spatially structured.

The taxoreference is the expression of a hypothesis on the current position of a specimen within the system of biological taxonomy (Bisby 1995). Besides being hypothetical, the taxoreference is also subject to errors and imprecision. There-fore, this taxonomic information must be updated and modified periodically by specialists. The scientific names are the essential language to communicate about biodiversity (May 1995; Patrick 1996; Thompson 1996), and the taxoreference is

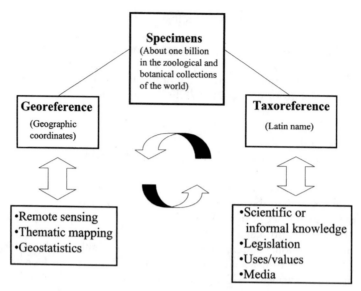

FIGURE 1 Scheme of ideal biodiversity information system. Information should be available on different scales, temporal and spatial, and from different perspectives. Georeferenced and taxoreferenced specimen databases provide links required to move among scales and points of view.

the link to the world of bibliographic data, legislation, markets, population information, and so on.

Taken together, the georeference and the taxoreference provide the links between sets of data that have a geographical structure and sets that are normally associated with a Latin binomial. From the perspective of a national BIS, the information on the millions of specimens in herbariums, museums, and scientific collections constitutes the backbone that allows movement along the many levels in the structure of biodiversity (Soberón and others 1996). That is why the atomic data on a specimen are so powerful and why CONABIO has since its creation been working on assembling the specimen-data backbone.

Specimen Information on Mexican Species

The biological specimens that have been collected in Mexico are deposited in 190 Mexican institutional collections and in other countries. According to an inventory of the taxonomic activities that CONABIO organized in 1996, there are about 10 million holdings in Mexican collections, in different stages of curation and data capture; the sizes and geographical distributions of the collections in Mexico are uneven. Depending on the taxonomic group, the proportions of specimens held in Mexican and foreign institutions vary. Some collections of Mexican specimens in foreign institutions are very important. For example, the Birds Database (Peterson and others 1997), which has information on about 300,000 specimens, is 80% foreign in origin, with perhaps a further 10% of Mexican collections still to be included. Foremost among the countries with holdings of Mexican specimens are the United States, Canada, and the United Kingdom; several European countries also have important collections in some taxa.

CONABIO's botanical databases, in contrast, are being compiled mainly in two Mexican collections, principally in the National Herbarium of the Instituto de Biologia, UNAM (MEXU, IBUNAM), and the Herbarium of the Escuela Nacional de Ciencias Biológicas of the National Polytechnic Institute (ENCB, IPN). Those two collections contain 1.5–2.0 million specimens.

The task of computerizing and georeferencing the data in the collections is significant. Since its creation, CONABIO has obtained about 230 databases, which contain data on more than 4,250,000 specimens of plants and animals.

The cost of obtaining specimen information is roughly constant per specimen, so computerizing large holdings is much more cost-efficient than small collections.

The information on vertebrates is extensive and probably covers most of the world collections of specimens that have been collected in Mexico. Despite this, large gaps and aggregation of collection sites along roads are still present (figure 2). This is important in countries with a large spatial turn component of diversity that requires extensive sampling over most of the country (Sarukhán and others 1996).

250   0   250   500

Km

FIGURE 2   Collecting localities of terrestrial vertebrates. Data came from 43 databases with 514,178 records (26 projects on mammals, 142,923 records; eight projects on amphibian and reptiles, 80,741 records; and nine projects on birds, 290,514 records). Gaps and roads become apparent by aggregation of dots.

Major botanical collections have been more difficult to computerize, and it was only in 1996 that the ENCB herbarium began the task. IBUNAM has partial computerization of some taxonomic groups or regions. Medium-size herbariums like the Instituto de Ecologia of Xalapa (XAL, around 250,000 specimens) and the small herbariums of the Asociacion Mexicana de Orquideología (AMO, around 110,000 specimens) and Centro de Investigaciones Científicas de Yucatán (CICY, almost 45,000 specimens) are almost totally computerized. XAL and AMO were pioneers in computerizing collections in Mexico. Other collections have been partially computerized, but the addition of data from computerized collections to CONABIO's databases is slowing down because a large proportion of Mexican specimens are still not curated or determined and because a few significant collections both in Mexico and abroad have not started comprehensive computerization of their specimens. When this task is finished, new increases in specimen data will have to come mostly from new explorations.

Quality Control

Information for the specimen databases is obtained from projects undertaken by universities or research groups and is externally peer-reviewed. But an independent process of quality control is required because data can be subject to faulty determination, unstable taxonomy (McNeill 1993), equivocal georeferencing (Chapman and Busby 1994), and other problems (WCMC 1996). It is possible to spot a large number of those by "inconsistency analysis" (Murguía 1996), whereby records are checked for intrarecord consistency, proper spelling and synonymy, taxonomic nestedness, or geographic consistency (Margules and Austin 1995; Margules and Redhead 1995; Stockewell, in Hart 1997).

The coordinator role of CONABIO is important although all the information that has been integrated came from specialists. CONABIO has been responsible for maintaining and updating the data, obtaining authority files, developing inconsistency-spotting routines, and organizing a network for sharing updated data.

The databases that have gone through the full process of data-quality control are included in the large "container" called BIOTICA (http://www.conabio. gob.mx), which has links to the GIS and to bibliographic information. Soon it will be linked to information on markets and legislation.

The data model for BIOTICA was developed in CONABIO with the assistance of a number of users (mainly taxonomists). The increasing number of providers using the same model, already more than 105, is greatly reducing the number of inconsistencies. Some data providers still use their own systems or commercial data managers; their information can be included in BIOTICA, but the number of inconsistencies tends to be larger than when BIOTICA is used.

Uses of the Mexican National Biodiversity Information System

Although the Mexican National Biodiversity Information System is still unfinished as an integrated system, many of its components are fully operational, and it is already providing services. Most services answer requests for information about the distributional questions noted earlier. Every month, around 30 requests are made by telephone, e-mail, or fax or personally. The providers of the data are informed about who used their information and what for, and the users can ask the providers for more details. (Some information is protected for various reasons.) A single request might require information from up to 20 databases, each the product of years of expert work and sometimes the result of centuries of accumulated institutional efforts.

CONABIO's home page (http://www.conabio.gob.mx) receives an average of 2,000 hits per day. The page is linked to some databases already released, many GIS thematic maps of Mexico, and the results of a number of analyses like a biodiversity priority-setting of Mexico and a guide to species illegally traded through Mexico.

The Future of the Mexican National Biodiversity Information System

User demands require that CONABIO's BIS include information about protected species (including trends in populations) and useful or marketable species and provide a higher level of resolution of cartography, including time series for vegetation cover in some areas. Therefore, the next steps should probably focus on

• higher resolution for priority areas (the priority-regions workshop [http://www.conabio.gob.mx/textos/prior.htm] yielded 155 regions covering about 20% of Mexico as those still promising for conservation efforts; the workshop used 1:4,000,000 cartography and pinpointed areas on which there was a serious lack of information);
• monitoring (this may be done on different scales, the simplest using satellite images to monitor changes in vegetation);
• updating of databases and catalogs;
• completing the computerization of the national collections and promotion of the extensive use of the system as a powerful tool for scientific purposes, management, and communication;
• repatriation of information and strengthening relations with foreign museums to ensure collaboration;
• data on useful species (trees, medicinals, and ornamental, food and nontimber forest products), which are especially important for peasant communities, small ranchers and farmers, and national and international biotechnology industries (CONABIO has started a project to create an information system for 600 such species. It will be based on data already obtained for the reforestation program, but it will also include data on uses, production techniques, ecological requirements, images, and so on); and
• a similar effort for 300 protected species to add to the current database on the CITES species.

Perspectives on a Regional Biodiversity Information System

An example of cooperation and information-sharing among neighbor countries was recently initiated with the support of the Commission for Environmental Cooperation of the NAFTA countries. Following the lead of previous efforts developed by Julian Humphrys, formerly of Cornell University, a pilot study was finished in 1997 that was based on data from the Mexican Bird Atlas (300,000 records), the Breeding Bird Survey (160,000 and 100,000 more in collections), and the US Breeding Bird Survey (15,000 records). This pilot example, the North American Biodiversity Information Network (NABIN), demonstrates the feasibility of accessing data distributed in several independent institutions. It is an example of the benefits of sharing information (Peterson and others 1997). One of the goals achieved was the development of a common catalog based on previous efforts (American Ornithological Union, http://www.itis.usda.gov/itis/). Another interesting development was the capacity to do bioclimatic modeling by sending the results of queries to the machine at the San Diego Super Computing Center. NABIN demonstrated the feasibility of creating a large-scale, multicountry distributed BIS. It will open the doors to larger efforts, such as the Inter-American Biodiversity Information Network being discussed by several countries.

CONCLUSION

We have described an information system based on specimen data and the uses that nonbiologists might have for the information. In Mexico, assembling the data required the participation of hundreds of Mexican taxonomists, ecologists, agronomists, and geneticists. The Mexican Government, through CONABIO, had to support not only the creation of databases, but also a large part of the basic activities of the researchers, such as purchase of cabinets and equipment, visits to foreign institutions, and field trips. Maintaining the information system will require continued support for the creators and maintainers of the information. The cost, although great for the country, has remained moderate relative to the large increase in the value of the information in the collections, which is now available and being used by an unprecedented number of people.

The specimen data are the core of a multiscale BIS. Despite the enormous holdings of systematic institutions all over the world, large gaps in our knowledge about the biota of the planet remain. Therefore, we will need more funds and concentrated efforts to computerize existing collections and increase the pace of exploration (SA2000 1994). The example of CONABIO shows that the task is feasible and should be tackled on a global basis.

ACKNOWLEDGMENTS

We are grateful to the many people at CONABIO who worked to make this presentation possible. We thank especially Raul Jimenez, CONABIO's director of systems, for the GAP analysis; Hesiquio Benitez, subdirector of external services; Carlos Alvarez and all the personnel in the Biotic Inventories Area who worked on the GIS. We are also very grateful to the providers of the data we have presented here. Their collective effort is what makes biodiversity information systems possible.

REFERENCES

Berlin BD, Breedlove E, Raven P. 1973. General principles of classification and nomenclature in folk biology. Am Anthro 75:214–42.

Bibby CJ, Collar NJ, Crosby MJ, Heath MF, Imboden C, Johnson TH, Long AJ, Sttatersfield AJ, Thirgood SJ. 1992. Putting biodiversity on the map: priority areas for global conservation. Cambridge UK: International Coun for Bird Preservation.

Bisby FA (coord). 1995. Characterization of biodiversity. In: Heywood VH (ed). Global biodiversity assessment. Cambridge UK: Cambridge Univ Pr. p 21–106.

Butterfield BR, Csuti B, Scott JM. 1994. Modeling vertebrate distributions for gap analysis. In: Miller RI (ed). Mapping the diversity of nature. Oxford UK: Chapman & Hall. p 53–68.

Chapman AD, Busby JR. 1994. Linking plant species information to continental biodiversity inventory, climate modeling and environmental monitoring. In: Miller R (ed). Mapping the diversity of nature. London UK: Chapman & Hall. p 179–94.

Colwell RK. 1996. Biota. The biodiversity database manager. Sunderland MA: Sinauer.

Farnsworth NH. 1988. Screening plants for new medicines. In: Wilson EO, Peter FM (eds). Biodiversity. Washington DC: National Acad Pr. p 83–97.

Hart D. 1997. New communities will benefit from HPC technology. Gather/Scatter 13(3):14–5.

Haverkort B, Millar D. 1994. Constructing biodiversity: the active role of rural people in maintaining and enhancing biodiversity. Etnoecologica 2(3):51–64.

Jenkins Jr RE. 1988. Information management for the conservation of biodiversity. In: Wilson EO, Peter FM (eds). Biodiversity. Washington DC: National Acad Pr. p 231–8.

Margules CR, Austin MP. 1995. Biological models for monitoring species decline: the construction and use of databases. In Lawton J, May RM (eds). Extinction rates. Oxford UK: Oxford Univ Pr. p 183–96.

Margules CR, Redhead TD. 1995. BioRap. Guidelines for using the BioRap methodology and tools. Dickson Australia: CSIRO.

May RM. 1995. Conceptual aspects of the quantification of the extent of biological diversity. In: Hawksworth DL (ed). Biodiversity measurement and estimation. London UK: Chapman & Hall. p 13–20.

McNeill. 1993. Instability in biological nomenclature: problems and solutions. In: Bisby FA, Russell GF, Pankhurst RJ (eds). Designs for a global plant species information system. Oxford UK: Clarendon Pr. p 94–108.

Murguía M. 1996. Jerarquización de las metodologías de validación de datos de georreferencia. CONABIO. Mexico.

Peterson T, Navarro A, Warner R, Pisanty Y, Kennedy J. 1997. Pilot project North American bird information network. North American Biodiversity Information Network. CCA.

Olivieri ST, Harrison J, Busby JR (Coord.) 1995. Data and information management and communication. In: Heywood VH (ed). Global biodiversity assessment. Cambridge UK: Cambridge Univ Pr. p 21–106.

Patrick R. 1996. Systematic: a keystone to understanding biodiversity. In: Reaka-Kudla ML, Wilson DE, Wilson EO (eds). Biodiversity II: understanding and protecting our biological resources. Washington DC: Joseph Henry Pr. p 199–211.

Reid W, Barber C, Miller K. 1992 Global biodiversity strategy. Washington DC: World Resources Inst, The World Conservation Union, UNEP.

SA2000. 1994. Systematics agenda 2000: charting the biosphere. Technical report. produced by systematics agenda 2000. A consortium of the American Society of Plant Taxonomists, the Society of Systematic Biologists, and the Willi Hennig Society, in cooperation with the Association of Systematics Collections.

Sarukhán J, Soberón J, Larson J. 1996. Biological conservation in a high beta-diversity country. In: di Castri F, Younès T (eds). Biodiversity, science, and development: towards a new partnership. Cambridge UK: CAB International.

Soberón J, Llorente J, Benítez H. 1996. An international view of national biological surveys. Ann Missouri Bot Gard 83: 562–73.

Thompson FC. 1996. Names: the keys to biodiversity. In: Reaka-Kudla ML, Wilson DE, Wilson EO (eds). Biodiversity II: understanding and protecting our biological resources. Washington DC: Joseph Henry Pr. p 199–211.

Wilson EO. 1996. Introduction. In: Reaka-Kudla ML, Wilson DE, Wilson EO (eds). Biodiversity II: understanding and protecting our biological resources. Washington DC: Joseph Henry Pr. p 1–3.

WCMC [World Conservation Monitoring Centre]. 1996. Guide to information management in the context of the convention on biological diversity. Nairobi Kenya: UNEP.

WCMC [World Conservation Monitoring Centre]. 1997. Darwin initiative handbooks. Cambridge UK: WCMC.

COMMUNITY INVOLVEMENT AND SUSTAINABILITY:
THE MALPAI BORDERLANDS EFFORT

WILLIAM MCDONALD
1555 12th Street, Douglas, AZ 85607
RONALD J. BEMIS
Natural Resources Conservation Service, USDA,
RRT 1 Box 226, Douglas, AZ 85607

THE MALPAI BORDERLANDS GROUP is a grassroots, landowner-driven organization that is attempting to implement ecosystem management on nearly one million acres of unfragmented landscape in southeastern Arizona and southwestern New Mexico along 70 miles of the Mexican border. The elevation of the area ranges from about 4,500 to 8,500 feet. The San Bernardino and Animas valleys, along with the Peloncillo and Animas mountain ranges, lie within the boundaries of the Malpai Borderlands. Annual precipitation here averages 12–20 in. Put succinctly, it is high and dry. Nonetheless, this area of remarkable biological diversity is home to numerous wildlife species. Perhaps most remarkable is that fewer than 100 people live in a region that is half the size of Rhode Island. One observer called it a "working wilderness".

The Malpai Borderlands is cattle ranching country, and ranching has kept this country open for the last century. In the Southwest, ranching depends on the existence of large amounts of open-space landscape. Many of the resident families are descended from the homesteaders who established ranches here around the turn of the 20th century. The diversity of land ownership is nearly as great as that of the country itself: 53% is privately owned, and 47% is owned by the federal or state government. The 320,000-acre Gray Ranch (which·is predominantly private land) skews the percentage of total private land in the Malpai Borderlands. The other, smaller ranches that make up the remainder of the area range between 15,000 and 40,000 acres; most contain more than 50% government-owned land, which is leased for grazing, and, when combined with the private lands, make economic units. The intermingled character of the ownership guarantees that government policies regulating the use of state and federally owned land will determine the fate of the private land. In turn, the fate of the private land, which generally contains the most reliable sources of water and other advantages (this was, after all, the land picked by the homesteaders as the best available), will determine the open-space future of the surrounding and intermingled government land.

In the fall of 1991, a small group of ranchers in the Malpai Borderlands met with a group of individuals from the environmental community at the headquarters of a ranch owned by the Glenn family, known as the Malpai Ranch. These ranchers were concerned about the future of the big open landscape that is their homeland and wanted to get together with some of the critics of livestock-grazing in the West to see whether they shared any concerns and, perhaps, could find some common ground. This group, calling itself the "Malpai Group," continued to meet at different ranch homes over a 2-year period. They were joined by scientist Raymond Turner, who has spent his life researching and recording changes in the landscape of the Southwest through this century.

Two types of common ground were identified. One was a mutual concern about the possibility of fragmentation of the region. On the fringes of the area, some ranches already had sold to subdivision. Neither ranching nor many wildlife species prosper in an area that is fragmented by development. Second was a concern about the seemingly inexorable encroachment of woody species on the grasslands. The group believed that some human activities contributed to this occurrence; fire suppression was identified as perhaps both the most damaging and the most easily changed. It was generally acknowledged that truly sustainable ranching might be the only hope for holding this landscape together in the future. In the arid West, ranching is the only livelihood based on human adaptation to wild biotic communities.

The group was unsure of its next step but believed that, whatever it would be, it should be driven by good science, contain a strong conservation ethic, be economically feasible, and be initiated and led by the private sector, with the agencies coming in as partners rather than with the private sector as their clients.

The suppression of a small brushfire by a federal agency, over the objection of the ranch manager whose intermingled private land was involved, proved to be the catalyst that took the group to the next step. Another meeting was held at the Malpai Ranch, this time with 30 area landowners in attendance. From that meeting came a request to the agencies to join with the landowners in creating a comprehensive fire-management plan for the region. The landowners even took the first step, presenting the agencies with a map of all the different individual ranches. Each ranch map showed the owner's preference for a response to a fire—let it burn, decide at the time, or suppress immediately. The agencies reacted positively to the request. A meeting with representatives of all the land-management agencies followed, and the parties committed to embark on an ecosystem approach to all resource management in the area, including fire. This enthusiasm by the agencies for a privately led initiative surprised many, but with thought, it made sense. It is truly ludicrous to expect government land-management agencies to take the lead, with shrinking budgets, conflicting internal agendas, outside litigation, and partisan politics pulling them first in one direction, then in another, not to mention the consistently high turnover of personnel in key positions. One highly placed agency official remarked, "We just don't want to get in your way." In a supporting role, however, the resources and expertise that dedicated agency employees can contribute is invaluable.

While this effort was beginning, the largest ranch in the area was changing hands. The Nature Conservancy (TNC) had bought the Gray Ranch a few years before to keep it from being broken up and possibly subdivided. Now, TNC was preparing to sell the Gray Ranch to a local ranching family. The Hadley family, which had spent 20 years on the Guadalupe Canyon Ranch and had considerable resources beyond its cattle operation, was to be the buyer. The Hadleys created a private organization, the Animas Foundation, with which to purchase and manage the Gray Ranch. Maintaining the vast open-space character of the ranch was important to both the Hadleys and TNC, so part of the purchase agreement included conservation easements, to be held by TNC on the private lands of the Gray Ranch. These easements stipulate that the ranch can never be subdivided.

John Cook, a TNC senior vice president, negotiated the sale. The Hadleys introduced Cook to some of their ranching neighbors, and he became intrigued and inspired by the fledgling Malpai Group. TNC generally was looked on with disfavor by most of the ranchers in the area, primarily because of their displeasure with TNC's practice of buying private land and then reselling it to the federal government. But TNC had done something different with the Gray Ranch, and the ranchers were impressed with John Cook's sincerity and his obvious love of the land. TNC was potentially a formidable partner, bringing to the table good science, a history of good working relations with the agencies, organizational skills and energy, a link to foundations and other donors, and even top-notch legal advice. At the ranchers' request, John Cook and some of his colleagues began working with the group. Some of the ranchers, fearful that TNC would take over the Malpai Group, dropped out at this point. The remainder believed, however, that this was the team to move ahead with. Thus, at the very time TNC was giving up its land holdings in the area, it was asked to remain.

In the spring of 1994, the Malpai Borderlands Group came into being as a nonprofit organization. The group has a nine-member board of directors and counts as its cooperators all landowners in the area who wish to work with the group, all government agencies engaged in any way with the borderlands area, three universities, TNC, and various scientists. The Natural Resources Conservation Service (NRCS) assigned Ron Bemis, a senior range conservationist, to work with the Malpai Borderlands Group in both states as the only NRCS field-level employee in the nation to have two-state responsibility. Likewise, the US Department of Agriculture Forest Service assigned its senior range conservationist for the Coronado Forest, Larry Allen, to work with the group. In addition, two districts of the Bureau of Land Management work closely with the Malpai Borderlands Group. In fact, the voluntary commitment of all agencies to work together with this landowner-driven group toward mutual goals has been one of the hallmarks of our effort.

Early on, the Malpai Borderlands Group formulated the following goal statement: "Our goal is to restore and maintain the natural processes that create and protect a healthy, unfragmented landscape to support a diverse, flourishing community of human, plant, and animal life in our Borderlands Region. Together, we will accomplish this by working to encourage profitable ranching and other traditional livelihoods which will sustain the open-space nature of our lands for generations to come." Everything done by the group must be consistent with these stated goals. Actions taken so far have resulted in better communication between landowners, between landowners and the agencies, and even between agencies.

Part of the group's success has come from its insistence on involving the best available science in whatever it does. The link with science had its start with Ray Turner and his colleagues at the Desert Laboratory in Tucson. The science link expanded when TNC became involved with the group and then was boosted again when the Forest Service's Rocky Mountain Research Station obtained a large grant to work in the area.

The Malpai Borderlands Group has formed a science advisory committee that reviews and oversees the various research projects going on in the region and includes such researchers as James Brown, of the University of New Mexico. The immediate past president of the Ecological Society of America, Dr. Brown has maintained a long-term project in an adjacent valley, studying, among other things, the individual effects of various birds and mammals on the area's landscape. The committee recently helped establish a standardized range-monitoring protocol for use by the group's cooperators.

Among the various research projects is a rehabilitative effort set up by the Forest Research Station, the NRCS, and a private landowner on 150 acres of very-eroded land adjacent to a creek. The presence of significant archaeological artifacts on the site has prevented the use of mechanical means to address the erosion problem. The research station funded a survey by the University of Oklahoma, which found evidence on the site of a history of fairly intensive human activity dating back to AD 1000. Without the use of mechanized equipment, and not wishing to introduce exotic grass species, the landowner was stymied about how to rehabilitate and protect the site. Native grass seed is nearly 20 times more expensive than seed of adapted exotics, and it often does not pioneer well. The decision was made to use the landowner's cattle herd to affect the erosion site intensively by feeding native grass hay, raised at the NRCS Plant Materials Center, to the cattle at the site for 3 days, after which the site would be fenced off and rested for an as-yet-undetermined period to monitor the results. This project is just one example of how cooperation has allowed for an action that the landowners, researchers, agency, or university would have been unable to do alone.

Another example has occurred on a neighboring ranch, where the Magoffin family became concerned for the welfare of an amphibian, the Chiricahuan leopard frog, which is listed as threatened. During a recent drought, the water source that was habitat for the frogs on the ranch began drying up. The Magoffins began hauling water to the frogs as a stopgap and also began consulting with herpetologist Cecil Schwalbe, of the University of Arizona, about how best to protect the frog in the future. According to Dr. Schwalbe, the biggest threat to the leopard frog is predation by introduced species, such as bullfrogs, that live in aquifers and waters on public lands. He believes that the best chance in the future for the leopard frogs is in isolated sources of water on private land, such as the Magoffins' ranch. Working together, the Magoffins, Dr. Schwalbe, the Arizona Game and Fish Department, the NRCS, and the Malpai Borderlands Group designed, funded, and created a permanent water source at the site of the frogs' jeopardized habitat and at one other site on the ranch where they are known to exist on the ranch. These waters were designed so that they also could be used in the ranch operation, making this a win for all concerned. A high-school biology class in nearby Douglas, Arizona, has collected tadpoles from the Magoffin sites and is raising them with the idea of distributing them to other isolated waters on private land in the region; the hope is that this program will obviate the eventual listing of this species as endangered.

In March 1996, Warner Glenn, owner of the Malpai Ranch, encountered a jaguar in the Peloncillo Mountains. Armed with a pistol, he shot several times with a camera instead. As the big cat was leaving his sight, he realized that he faced a dilemma. The jaguar was proposed for listing as endangered in the United States. If he went public with his story and his photographs of the jaguar, the resulting attention might lead to the designation of the area in the future as critical habitat, a designation that could affect the two activities on which his livelihood depends, hunting with dogs and grazing cattle. After a lifetime of hunting mountain lions, he felt a kinship with the big cats and a fascination with the jaguar as well as a concern for its future. The deciding factor was Warner Glenn's faith in the ability of the Malpai Borderlands Group to make it turn out right.

After a meeting with the appropriate agencies, the Malpai Borderlands Group became active with a coalition of other organizations and individuals in drawing up a conservation agreement for the jaguar in Arizona and New Mexico. Officially sponsored by the game and fish departments of both states, the agreement was attacked by activists as simply a ruse designed to subvert listing the jaguar as endangered. Despite this, although the jaguar is now listed, the conservation agreement and the working group that drew it up live on.

At the invitation of the Malpai Borderlands Group, world-renowned big-cat researcher Alan Rabinowitz visited to survey the site of the jaguar encounter, as well as the corridor that runs from the Peloncillos to the Sierra Madres in Mexico. Rabinowitz's opinion is that the Peloncillos and the neighboring Sierra San Luis are not true habitat for the jaguar. The true habitat lies to the south, which is where resources and efforts should be directed (Rabinowitz 1997). He did, however, help Warner Glenn set up some trip cameras in an effort to record any further visits by jaguars.

The Malpai Borderlands Group also met with representatives of an activist organization well known for suing the government for species listings and critical-habitat designations. While professing to have no current interest in pursuing critical-habitat designation in the United States for the jaguar, the activists vowed that they would pursue endangered-species listing for the leopard frog, regardless of the success of the group's efforts to restore the population, which has dampened that effort somewhat.

The most successful, yet most frustrating, type of work for the Malpai Borderlands Group has concerned the use of fire. Tree-ring studies by Tom Swetnam and subsoil studies by Owen Davis, both with the University of Arizona, yielded evidence that fire historically affected nearly all sites in the Malpai Borderlands at least once a decade (Swetnam and Baisin 1995). Today, this area may be one of very few in this country where a large-scale attempt could be made to replicate that frequency of fire. In fact, during the last 4 years, because of the relationships developed by the Malpai Borderlands Group between the neighbors, the agencies, and the rural fire departments, more naturally ignited fires have been allowed to burn. About 120,000 acres have been affected, including two prescribed burns. One advantage of prescribed burning is that it permits studies to be done before and after. For both burns, various studies are looking at the effects on different plant and animal species. These fires were ignited during the normal pre-monsoon fire season, when lightning strikes often occur, mimicking natural fire as nearly as possible. All the fires, natural and prescribed, have tended to leave behind a burned and unburned mosaic pattern, allowing for side-by-side comparison.

The first prescribed burn presented several political challenges. The targeted area lay in two states and involved coordination with six agencies in both states. In addition, a wilderness-study area was involved, and because of the international boundary, Mexico needed to be consulted. With a Herculean effort by everyone involved, the planning was actually completed in 8 months and the burn itself was quite successful.

Although the second burn did not involve anywhere near the jurisdictional difficulties of the first, the attempt nearly ran aground when ecosystem management came into conflict with single-species management. The two ranchers involved voluntarily withheld grazing from their forest allotments to build the fine-fuel load high enough to affect the woody species, but the consultation under section 7 of the Endangered Species Act between the Forest Service and the Fish and Wildlife Service over the possible effect of fire on three species listed as endangered dragged on for 2 years. Eventually, the disagreements between biologists in the two agencies over the possible effects on one species, the desert-blooming agave, became so heated that the Malpai Borderlands Group requested that Jamie Clark, national head of ecological services for the Fish and Wildlife Service (now its director), visit the site to help resolve the debate. Negotiation led to the establishment of plots for before-and-after studies to be funded by the Forest Research Station. This avoided any further stalemate and the fire was ignited in the pre-monsoon period.

This experience taught us several things, principally that the site-by-site approach is just too costly. The planning for this burn cost about $20 per acre for the Forest Service alone, but it cost only $3 per acre to actually perform the burn. Because the cost of consultations was the primary factor in driving up the cost of planning, it became clear that an alternative approach to prescribed burning was desirable. What has emerged is a comprehensive programmatic approach that will identify and attempt to resolve as many concerns as possible for the entire area before planning begins for a specific burn. We hope that this can be accomplished in a 2-year period, after which the process of burning on specific sites will be expedited and require a minimal number of consultations. At this point, while awaiting the data from current and future studies, the Malpai Borderlands Group feels positive about the results of the burns, both natural and prescribed. Early results show a considerable immediate effect on the woody species and the rejuvenation of the grasses, resulting in more ground cover.

Unfortunately, not everyone has waited for the data. The herpetologist who led a before-burn study on the endangered ridgenose rattlesnake has issued a report recommending critical-habitat designation for the snake and recommending against future prescribed burning in the Peloncillo Mountains at elevations above 5,000 ft. The report also recommends that livestock-grazing on all Forest Service allotments that contain ridgenose rattlesnake habitat be restricted to midwinter. Even before this report had been reviewed by those for whom it was intended, and well before the Malpai Borderlands Group became aware of it, these recommendations were incorporated by a US Fish and Wildlife Service herpetologist into a court-ordered biological opinion on grazing for two Bureau of Land Management districts that cover nearly one-third of the land area of Arizona. Even though the report itself (which is final but not published yet) states that the effects of grazing on the habitat of the ridgenose rattlesnake are unknown, it recommends midwinter grazing only. The snake-survey team shut down its study within a week after the burn, well before the monsoon rains and the resulting revegetation of the site began, permitting no opportunity to study even the short-term effects of the fire on the habitat, but the report still recommends no burning above 5,000 feet. Only one of 13 collared snakes died in the fire, and it was not a ridgenose rattlesnake. The survey team itself was responsible for the loss of two snakes during the course of its research.

Given the facts, what is the basis for the no-burn recommendation? Why is this recommendation part of a biological opinion on grazing? We believe that with the force of the Endangered Species Act behind them, some individuals within the Fish and Wildlife Service have been abusing the power of the act increasingly in recent years to force their will, with little regard for science. For instance, peer review is not required for opinions expressed in section 7 consultations. Under pressure from the courts, biological opinions are being thrown together with the flimsiest of scientific underpinnings. We believe that these opinions are destructive and counterproductive to collaborative efforts like ours. The opinion on the ridgenose rattlesnake effectively prevents any prescribed burning in the Peloncillo mountains, and its grazing recommendations potentially could affect some ranching operations to the point of jeopardizing their continuation as ranches, possibly putting thousands of acres at risk for development. This "shoot-from-the-hip science" hardly encourages private landowners to want researchers to come onto their ranches. The trust and openness that have characterized the efforts of the Malpai Borderlands Group to this point are threatened, encouraging nonparticipating landowners to remain nonparticipating. The few who believe that the safest policy toward endangered species is to "shoot, shovel, and shut up" will stay convinced of the certitude of their position, and they may even gain some converts. In such an atmosphere of mistrust, for instance, it will be difficult for landowners to have the confidence to place leopard frogs willingly on their private land. Landowners must know that the Endangered Species Act will not be used retroactively to restrict the activities on which their livelihood depends.

We believe that rigid single-species management in our biologically diverse world is wrong. Whether the species is a ridgenose rattlesnake, a willow flycatcher, or a beef cow, management for one species alone is narrow-minded, shortsighted, ineffective, and, in fact, harmful.

Will this unfortunate action ultimately blow apart the efforts in the Malpai Borderlands? We hope not. The Malpai Borderlands Group is positioned uniquely to bring to bear the scientific rigor and influence necessary to address this abuse. If the principals are willing to come together to talk and to work for as long as it takes for all concerns to be addressed fairly, the confidence and trust that must exist for a collaborative effort to work can return.

From Montana to Hawaii to Brazil, the "radical-center" approach of the Malpai Borderlands Group is regarded by many as the best—and maybe the only—hope for our remaining wildlands. However, reasonable people in both the public and private sectors must be allowed to work together in pursuit of creative solutions to issues about the land as they occur. If they are not allowed this flexibility, all the government policies and global treaties that can be dreamed up will amount to only so much hot air and wasted paper and ink.

Writing in support of the approach of the Malpai Borderlands Group, James Brown stated, "Ranchers, conservationists, government-agency employees, research scientists, and the American public all have much to lose if the present climate of distrust, disagreement, and interference is perpetuated. All have much to gain through interaction, cooperation, and collaboration" (Brown and McDonald 1995). Which will be our legacy? The generations to come will be the biggest losers or winners.

REFERENCES

Brown JH, McDonald W. 1995. Livestock grazing and conservation on southwestern rangelands. Cons Biol 9:1646.

Rabinowitz A. 1997. The status of jaguars in the United States: trip report. New York NY: Wildlife Conservation Society, Bronx Zoo. p 3–5.

Swetnam TW, Baisin CH. 1995. Historical fire occurrence in remote mountains of southwestern NM and northern Mexico. Gen Tech Rept INT-320. Ogden UT: USDA Forest Service Intermountain Research Station. p 153–6.

INDEX